Handbook of Research on Aspects and Applications of Incompressible and Compressible Aerodynamics

Sathish K. Kumar
Nehru Institute of Engineering and Technology, India

Naren Shankar Radhakrishnan
Vel Tech Rangarajan Dr. Sagunthala R&D Institute of Science and Technology, India

A volume in the Advances in Mechatronics and Mechanical Engineering (AMME) Book Series

IGI Global
PUBLISHER of TIMELY KNOWLEDGE

Published in the United States of America by
IGI Global
Engineering Science Reference (an imprint of IGI Global)
701 E. Chocolate Avenue
Hershey PA, USA 17033
Tel: 717-533-8845
Fax: 717-533-8661
E-mail: cust@igi-global.com
Web site: http://www.igi-global.com

Library of Congress Cataloging-in-Publication Data

Names: Kumar, Sathish, 1984- editor. | Radhakrishnan, Naren Shankar, 1984- editor.
Title: Handbook of research on aspects and applications of incompressible and compressible aerodynamics / Sathish K. Kumar and Naren Shankar Radhakrishnan, editors.
Description: Hershey, PA : Engineering Science Reference, an imprint of IGI Global, [2022] | Includes bibliographical references and index. | Summary: "This book is for readers interested in aerodynamics, starting with information on standard atmosphere and basic concepts, then covering the equations and mathematical models used to describe and characterize flow fields, as well as their thermodynamic aspects and applications to highlight the relation between concepts and their use in aircraft design"-- Provided by publisher.
Identifiers: LCCN 2022008788 (print) | LCCN 2022008789 (ebook) | ISBN 9781668442302 (h/c) | ISBN 9781668442326 (ebook)
Subjects: LCSH: Compressibility. | Aerodynamics.
Classification: LCC TL574.C4 A87 2022 (print) | LCC TL574.C4 (ebook) | DDC 629.132/32--dc23/eng/20220503
LC record available at https://lccn.loc.gov/2022008788
LC ebook record available at https://lccn.loc.gov/2022008789

This book is published in the IGI Global book series Advances in Mechatronics and Mechanical Engineering (AMME) (ISSN: 2328-8205; eISSN: 2328-823X)

British Cataloguing in Publication Data
A Cataloguing in Publication record for this book is available from the British Library.

All work contributed to this book is new, previously-unpublished material. The views expressed in this book are those of the authors, but not necessarily of the publisher.

For electronic access to this publication, please contact: eresources@igi-global.com.

Advances in Mechatronics and Mechanical Engineering (AMME) Book Series

J. Paulo Davim
University of Aveiro, Portugal

ISSN:2328-8205
EISSN:2328-823X

Mission

With its aid in the creation of smartphones, cars, medical imaging devices, and manufacturing tools, the mechatronics engineering field is in high demand. Mechatronics aims to combine the principles of mechanical, computer, and electrical engineering together to bridge the gap of communication between the different disciplines.

The **Advances in Mechatronics and Mechanical Engineering (AMME) Book Series** provides innovative research and practical developments in the field of mechatronics and mechanical engineering. This series covers a wide variety of application areas in electrical engineering, mechanical engineering, computer and software engineering; essential for academics, practitioners, researchers, and industry leaders.

Coverage

- Tribology and surface engineering
- Biologically Inspired Robotics
- Control Methodologies
- Intelligent Sensing
- Computational Mechanics
- Control Systems Modelling and Analysis
- Computer-Based Manufacturing
- Autonomous Systems
- Manufacturing Methodologies
- Medical Robotics

IGI Global is currently accepting manuscripts for publication within this series. To submit a proposal for a volume in this series, please contact our Acquisition Editors at Acquisitions@igi-global.com or visit: http://www.igi-global.com/publish/.

The Advances in Mechatronics and Mechanical Engineering (AMME) Book Series (ISSN 2328-8205) is published by IGI Global, 701 E. Chocolate Avenue, Hershey, PA 17033-1240, USA, www.igi-global.com. This series is composed of titles available for purchase individually; each title is edited to be contextually exclusive from any other title within the series. For pricing and ordering information please visit http://www.igi-global.com/book-series/advances-mechatronics-mechanical-engineering/73808. Postmaster: Send all address changes to above address.

Titles in this Series

For a list of additional titles in this series, please visit: www.igi-global.com/book-series

Advances in Computational Approaches in Biomechanics
Pritam Pain (Haldia Institute of Technology, India) Sreerup Banerjee (Haldia Institute of Technology, India) and Goutam Kumar Bose (Haldia Institute of Technology, India)
Engineering Science Reference • © 2022 • 306pp • H/C (ISBN: 9781799890782) • US $260.00

Developing Charging Infrastructure and Technologies for Electric Vehicles
Mohammad Saad Alam (Aligarh Muslim University, Aligarh, India) Reji Kumar Pillai (India Smart Grid Forum, India) and N. Murugesan (India Smart Grid Forum, India)
Engineering Science Reference • © 2022 • 343pp • H/C (ISBN: 9781799868583) • US $240.00

Multidisciplinary Perspectives on Green Electromobility and Charging Stations
Yu Miao (East China University of Sciencer and Technology, China)
Engineering Science Reference • © 2022 • 305pp • H/C (ISBN: 9781799839972) • US $225.00

Electric Vehicles and the Future of Energy Efficient Transportation
Umashankar Subramaniam (Prince Sultan University, Saudi Arabia) Sheldon S. Williamson (Ontario Tech University, Canada) Mohan Krishna S. (Alliance University, India) and Febin Daya J. L. (Vellore Institute of Technology, India)
Engineering Science Reference • © 2021 • 293pp • H/C (ISBN: 9781799876267) • US $240.00

***Modeling** and Optimization of Solar Thermal Systems Emerging Research and Opportunities*
Jagadish (National Institute of Technology, Raipur, India) and Agnimitra Biswas (National Institute of Technology, Silchar, India)
Engineering Science Reference • © 2021 • 215pp • H/C (ISBN: 9781799835233) • US $205.00

Recent Trends on Electromagnetic Environmental Effects for Aeronautics and Space Applications
Christos D. Nikolopoulos (Department of Electronic Engineering, School of Engineering, Hellenic Mediterranean University, Greece)
Engineering Science Reference • © 2021 • 285pp • H/C (ISBN: 9781799848790) • US $295.00

Modern Machining Processes and Techniques for Aerospace Materials
Anand Pandey (Manipal University Jaipur, India)
Engineering Science Reference • © 2021 • 300pp • H/C (ISBN: 9781799845102) • US $215.00

IGI Global
PUBLISHER of TIMELY KNOWLEDGE

701 East Chocolate Avenue, Hershey, PA 17033, USA
Tel: 717-533-8845 x100 • Fax: 717-533-8661
E-Mail: cust@igi-global.com • www.igi-global.com

List of Contributors

Table of Contents

Detailed Table of Contents

Section 1
International Standard Atmosphere

Suthagar S., Vel Tech Rangarajan Dr. Sagunthala R&D Institute of Science and Technology, India
Gopalakrishnan K., Anna University, Tirunelveli, India
Kumaran T., Vel Tech Rangarajan Dr. Sagunthala R&D Institute of Science and Technology, India
Praveen Kumar, Department of Energy Studies and Construction Management, Wollege University, Ethiopia

The atmosphere is the layer of gases surrounding the Earth and retained under its gravity. The atmosphere of our planet is made up of blankets of air that contain numerous gases such as oxygen, nitrogen, carbon dioxide, and others. Earth's atmosphere can be differentiated into five regions based on their physical properties such as temperature and pressure. Troposphere, stratosphere, mesosphere, ionosphere or thermosphere, and exosphere are these zones. The atmosphere provides the platform for these aerospace vehicles, which have taken off from Earth to operate at different altitudes. To investigate and understand the performance of flight tests, wind tunnel tests, and the design of aerospace vehicles or any other flying objects, there is a need for standard values to evaluate the parameters of the airplane or any other objects influenced by aerodynamic forces. Finally, the purpose of this chapter is to help the reader understand what the standard atmosphere is and how it can be utilized to analyze aeronautical vehicles.

Surekha Rathi Samundi D., Bharath Institute of Higher Education and Research, India

The atmosphere is the layer of gases that surrounds the planet. The atmosphere is retained by the gravity of the planet. Hence, it is also called planetary atmosphere. The performance of the aircraft and rockets depends on the physical properties of the atmosphere in which they fly. It is therefore advisable to study the variation of pressure, temperature, and density with the altitude. The real atmosphere is composed of dust, water vapor, and moisture, and it never remains constant. Hence, a hypothetical model called a standard atmosphere was employed. This chapter elaborates the international standard atmosphere, atmospheric boundary layer, and the stability of the atmosphere.

Section 2
Incompressible Flow Aerodynamics

Here, the authors explain the Magnus effect. The ball is deflected in the same direction as the rotation. The most common exposure and welcome statement of the Magnus effect is that a spinning object creates a vortex of fluid swirling around it. On the side where the movement of the vortex is in the same direction as the direction of the flow to which the object is exposed, the speed will increase. On the opposite side, where the directions are opposite, the speed will decrease. It is explained here, according to Bernoulli's principle, that the pressure is lower on the side with the greatest velocity, and therefore, there is an unbalanced force orthogonal to the flow of the fluid.

One of the most important design impacts in aircraft is an airfoil. This airfoil is also considered the cross-section of the wing, so the flow characteristics of air depend on the airfoil's shape. There are some airfoil theories we can consider for the design consideration while designing the wing. The airfoil theories describe the nature of fluid flow over the wing based on the angle of attack for variable speed conditions. This chapter deals with airfoil terminology and the theories of airfoil like the methodology of conformal transformation, Cauchy-Riemann relations, complex potential, Kutta-Joukowski, thin airfoil theory, and their applications. The theories mentioned above explain transformation and its applications, Kutta condition, Kelvin's circulation theorem, starting vortex creation of airfoil for variable speed considered for aircraft, and the advantages of the optimization.

Wing design is a very complicated and intricate issue. It is not feasible to cover everything in this chapter; however, it is possible to discuss some of the essential ideas that underpin design for high lift and low drag. Lift may be increased in four ways for fixed air characteristics and free-stream speed: increased wing area, increased angle of attack, increased camber, increased circulation through the use of high-momentum fluid. One of the most important applications of potential flow theory was the study of lifting surfaces such as aircraft wings, since the boundary conditions on a complex geometry can significantly complicate any attempt to tackle the problem via analytical techniques, which involves some simplification assumptions in order to arrive at a solution. These assumptions will be related to the concept of three-dimensional thin wing issues in this chapter.

Viscosity is a property that expresses the internal drag of a fluid to motion; impact of viscosity states the statics and flows. Statics means whenever fluids at zero velocity have no relative movements between layers of fluid and thus du/dy = 0. At the time there is no shear stress and viscosity of the fluid is free. Fluid viscosity plays a major role on the fluid floating in it. The authors focused on solids and fluids and the no slip condition, momentum transfer through molecular motion, shear stress and viscosity, Couette flow, and Poiseuille flow. Here the authors made a discussion the Newtonian viscous flow, and the statement of Newton's law of viscosity was examined. The discussion has been extended up to viscosity and the effect of their temperature and impact of increasing in temperature has been explained along with surface tension.

This chapter describes the basic concepts of aerodynamics, evolution of lift and drag, types of drag, reduction of wing tip vortices, non-planar wing concepts for increased aerodynamic efficiency, various methods for determination of aerodynamic forces of an airplane, classification of wind tunnels, blower balance tunnels, and a case study report on aerodynamic force measurement of the non-planar wing systems. To increase the aerodynamic efficiency of the monoplane configuration, the 'C-wing' configuration is presented in this chapter. The aim is to prove, at all angles of attack, C-wing produces a higher (L/D) ratio than straight wing for the same wetted surface area. The aerodynamic characteristics of three different wing models with NACA-64215 aerofoil such as straight wing, C-wing, and inverted C-wing at different angles of attack and low Reynolds number are shown. The inverted C-wing created more lift but produced more vibration, which may lead to lesser structural integrity.

Section 3
Compressible Flow Aerodynamics

Flow separation and wind gusts affect the aerodynamic performance of low-Reynolds number flyers with a chord-based Reynolds number of 105 or below. Active flow control provides information into fluid dynamics as well as potential vehicle performance enhancements. Aircraft engines, as well as stationary flow devices, exhibit undesirable flow states that have a significant impact on noise output and aerodynamic loss. Vibrations and aerodynamic performance are likely to suffer as a result of flow and boundary layer separations. Flow management can be an effective approach for lowering noise levels, increasing efficiency and thereby lowering fuel usage. Various forms of flow can be controlled using both passive and active flow control strategies. The active alteration of aerodynamic flows utilizing small time-dependent actuators in well-chosen places is defined as adaptive flow control. Micro-adaptive flow control has a significant overall system benefit in a variety of large-scale applications, which are presented.

The chapter addresses an extensive numerical investigation for the optimization of static pressure recovery in the dump diffuser of annular combustor for modern aero-gas turbine engines. The modelling and simulation processes are accomplished through the commercial ANSYS fluent module with SST k-epsilon (k-ε) turbulence model with the atmospheric conditions prevailing at 9 km altitude. The effect of velocity changes on the flow recirculation and vortex mixing phenomena is also addressed with novel dome-shaped optimization at the velocity of 25 m/s. The dome geometry can also be considered as a flame tube head, and the influence of vortex pattern and pressure distributions caused by the different dome shapes are analyzed using CFD. The dome-shaped optimization strategy addressed herein could initiate more potential studies on dump diffusers with hot and cold flow conditions to enhance the overall combustion efficiency in the near future.

Preface

The branch of dynamics that deals with the movement of air and the forces acting around the object is called as Aerodynamics. Aerodynamics mainly concern the forces that the gaseous fluids; particularly air, exerted on the object moving through it. Fluids are classified as either compressible or incompressible. Incompressible fluids do not experience significant changes in density during flow. It is generally accepted that liquids are incompressible; water is a perfect example. On the other hand, compressible fluids experience noticeable changes in density. In general, gases are compressible; air is the most commonly recognized compressible liquid. Compressible gases exhibit many interesting properties such as shock waves in steady state, which are not present in incompressible fluids.

Temperature changes are another factor that causes incompressible and compressible flows to differ. In a compressible flow, the temperature tends to be constant. A compressible flow, however, can cause substantial changes in temperature that can lead to an energy exchange between modes.

Mach number is a dimensionless parameter, expressed as the ratio of flow velocity to local velocity of sound. It is possible to obtain the effect of compressibility in fluid flow when the flow Mach number increases beyond 0.3. Air is compressible and sound travels at 343 meters per second for 20°C and 1 atm. Water, on the other hand, has a small density change and its sound velocity is higher than that of air. When water is kept at 20°C and 1 atm, sound travels at approximately 1,480 meters per second. Thus, liquids are considered to be incompressible fluids.

The basic concepts, various aspects and applications of incompressible flow aerodynamics, compressible flow aerodynamics are the subject of study in this book.

A compressible flow is usually characterized by a Mach number greater than 0.3. Incompressible fluids, such as air, have a speed slower than Mach 0.3, despite being gases. Compressing a gas below Mach 0.3 is not considered compressible (in the thermodynamic sense). Flows that are compressible possess more complexities when compared to incompressible flows.

Mach number can be differentiated into various types. They are

Subsonic ($0.3 < M < 0.8$)
Transonic ($0.8 < M < 1.2$)
Supersonic ($1.2 < M < 5$)
Hypersonic ($M > 5$)

Under subsonic flow condition, the Mach number is between 0.3 and 0.8. Pressure and density are weakly related, so no shocks will occur within this regime of flow. During supersonic conditions, the flows are characterized by the downstream pressure effects. The upstream flows are unaffected by any conditions encountered downstream.

Modern flights are impossible without aerodynamics. To study the movement of any matter related to any gases or air, aerodynamics is the source which will explain all the flow behavior and the flow characteristics. It's not easy to notice the way air interacts with the cars when we are travelling at slow speeds or on a calm day. But when we are traveling at high speeds during extremely windy conditions, air resistance (also called as drag) has a profound impact on a car's performance, acceleration and fuel efficiency. At this point aerodynamics comes to play. If the car is designed by considering the airflow understanding its behavior, it will improve the engine and can also reduce accelerating difficulties. In real life, there are number of illustrations of aerodynamics starting from, flying a kite to race cars, golf balls, riding on a bike, gliders, aircrafts, gas turbines, rockets, seed spraying, drones flight, soccer balls, etc. are some other real life examples. Principles of aerodynamics are also applied by the engineers in buildings, bridges. Nevertheless, the main interest of aerodynamics is in designing an aircraft and automobiles.

This book explains the fundamentals of physics behind the flow aerodynamics, which is vital for developing the knowledge and understanding of compressible and incompressible flows. With this background, the book is directed towards the students, research scholars, lectures of aeronautical/ aerospace engineering as a target audience who are interested to study about the flow behavior and characteristics at various aspects.

Based on the theme of the book, the chapters have been divided into three different sections. Section 1 deals with International Standard Atmosphere which includes chapter 1 and 2, Section 2 is grouped as Incompressible Flow Aerodynamics in which chapters 3 to 7 are included and Section 3 Compressible Flow Aerodynamics which incorporates from chapter 8 to 18.

Chapter 1 titled as "ISA" explains about the various properties and behavior of atmosphere. As a vehicle is operated in the air the properties of air that have an effect on the performance of a aircraft need to be understood. An aircraft doesn't fly at zero altitude rather it flies above 40000ft. Moreover each vehicle like gliders, passenger aircrafts, military aircrafts, missiles, rockets, reentry vehicles fly at different altitude. Hence understanding the atmosphere properties like pressure, density, temperature from sea level conditions to the top layer of the atmosphere has become the foremost factor in order to clearly determine the flow physics of different vehicles at different speed regimes. This chapter will give the readers an in-depth knowledge about all the properties of the atmosphere

Chapter 2 gives a review on the various aspects of earth's atmosphere. The properties of atmosphere vary with altitude. Pressure, temperature, density vary with the altitude. The pressure decreases as the altitude increases thereby decreasing the temperature with respect to the lapse rate. The instrumentations used in any aircrafts are calibrated for standard atmosphere. The atmospheric boundary layers, stability of atmosphere are explained in this chapter. The variation of the pressure, temperature, density with respect to the altitude is listed in this chapter for all the layers of the atmosphere.

Chapter 3 deals with a two-dimensional inviscid incompressible flow concept. Inviscid incompressible flow provides a thorough introduction to elementary flows. This chapter explains the fundamental equations, basic theorems for understanding the inviscid incompressible flow. The limitations are also discussed in this chapter. Further Magnus effect and its related forces are also explained in this chapter.

Chapter 4 explains about airfoil, its theory and its applications. An airfoil is the cross section of a wing. Aircraft is usually designed in the shape of an airfoil which is used to increase lift by altering the speed at which air passes over the wings. Designing an airfoil has a fundamental theory and conditions to satisfy the aerodynamic performance characteristics based on the nature of the aircraft. The nomenclatures of an airfoil are explained in detail along with some essential features like angle of attack, incidence angle.

The various methods used for designing an airfoil are also explained in this chapter. The chapter also gives details about various mathematical expressions to understand a complex geometry.

Chapter 5 explains about finite wing theory. There are two types of wing: infinite wing and finite wing. A two-dimensional wing is an infinite wing, whereas a three dimensional wing is a finite wing. A finite wing is called three dimensional because air can travel around the wingtip, creating trailing vortices. A three-dimensional flow cannot be obtained in a two dimensional wing. On a real aircraft this condition is indeed not possible because an infinite wing cannot be built in reality. In aerodynamic point of view, trailing vortices results in reducing the slope taken between angle of attack and lift coefficient curve. Hence, this chapter explains the forces, theories and the mathematical expression required for designing a finite wing along with some real problem and solutions.

Chapter 6 explains about the viscous flow and its effects. Viscosity plays a major role on fluids when it is flowing. This chapter explains about the Reynolds number, boundary layer thickness, boundary layer separation effect of pressure gradient which are considered as a factors affecting viscosity. The chapter also explains the theory based on which the effects of viscous flows are studied.

Chapter 7 describes the measuring techniques of aerodynamic forces using blower balance tunnel at low Reynolds number. Basics concepts of aerodynamics are explained in order to understand the flow properties considered for measuring. Evaluation of lift, drag, types of drag and various concepts behind determining the aerodynamic forces are also emphasized in this chapter.

Chapter 8 deals with the thermodynamics of fluids in motion. This chapter provides a fundamental knowledge to understand the concept behind the compressible flow aerodynamics. Work and energy transfer, various laws of thermodynamics, principles of energy conversion and fluid compressibility are explained in this chapter.

Chapter 9 explains about the flow through supersonic nozzle and glimpses of supersonic jet. The flow physics inside a supersonic nozzle is explained in detail which gives a thorough knowledge for the readers regarding the nozzle flow. The chapter starts with explaining the basic theories and laws to be considered while designing a supersonic nozzle. Further the chapter also provides a numerical analysis of different nozzles in-order to brief the flow characteristics.

Chapter 10 explains about the shock reflection and interactions in practical cases like space shuttle, hypersonic air breathing vehicles like ramjet and scramjet. When a vehicle or objects travels at a speed more than the speed of sound, shock waves is produced. Shock waves are a compressed pressure wave or simply a compression wave. In reality, shock waves are characterized by a sudden or an abrupt change in the flow properties and behaviour. If a shock wave is formed there is always a rapid raise in pressure, temperature and density. The pressure deflection is one of the main phenomena which will help us to understand the shock deflection better has also been focused in this chapter. The shock running, polar shocks are briefed in this chapter. The shock wave which is formed when impinges on the wall, the shock tends to reflect. This is called shock reflection. This shock reflection has been explained in this chapter which will give the readers an in-depth understanding about the various concept related to shock reflection. One must have knowledge about shock interaction when focusing on shock reflection. Hence this chapter also covers the concept of shock interactions in detail.

Chapter 11 explicate about the high-speed flow over airfoils, wings, and airplane configurations. Different types of supersonic wing, area rule relations for designing a proper supersonic aircraft is explained in this chapter. Unlike a low-speed flow, the high-speed flow physics is a bit complex to understand. This chapter helps the readers to understand the basic concepts and flow behavior of a high-speed aircraft with a help of numerical analysis over a supersonic aerofoil.

Chapter 12 gives a detailed introduction about hypersonic aerodynamics and its flow characteristics. On the sea level, sound travels at about 340m/s. If it goes faster than that then it's called as supersonic. In the natural world, asteroids and meteors coming to earth travel at hypersonic speed. Space shuttles, re-entry vehicles also travel at hypersonic speed. In recent year's aircrafts, missiles travelling at hypersonic speed being built. Therefore, understanding the flow characteristics and physics behind hypersonic flow is very important. This chapter explains about the hypersonic flow physics, vehicle design considerations and its constraints. The chapter also provides a brief explanation about the difference in the supersonic aerodynamics and hypersonic aerodynamics. In addition, it also explains about thin shock layer, entropy layer, temperature, low density. Flying any vehicle at hypersonic speed becomes difficult as the air molecules tend to break down. In hypersonic speed the heating effects become really important as the temperature increases. Hence this chapter explains about the heating effects.

Chapter 13 is a study of drag reduction on a hypersonic vehicle using aerospike. If a vehicle travels more than five times faster, it's called hypersonic. Flying any vehicle at hypersonic speed becomes difficult as the air molecules tend to break down. This leads to a generation of high temperature. Generally hypersonic vehicles are designed with a blunt shaped forebody, in order to reduce the heat at the body surface. This gave rise to increased wave drag at the forebody. Many researches are carried out for the past years and developments are done which will reduce the drag over the hypersonic vehicle. Aerospike is one technique which is attached at the vehicles forebody in order to reduce the drag. This chapter aims to review the research work focusing on aerospike technique carried out in the past decades.

Chapter 14 takes an initiative to explain about various shapes used in aero fields which evolved from the nature. Those shapes are grouped as bio-mimic shapes. Shapes of fishes, birds and terrestrial animals are used to explain about the optimization of aerodynamic forces in this chapter. This chapter provides knowledge for the reader's concerning the development of bio mimic flight configurations.

Chapter 15 explains the adaptive flow controls techniques in aerodynamics. This chapter enlightens the knowledge of the readers in optimizing a design in large scale applications which is done by an interdisciplinary approach which includes fluid physics, structural mechanics, and material science.

Chapter 16 is a computational investigation on the influence of dump gap and novel flame tube geometries in the hybrid diffuser performance. This chapter carries out a numerical analysis for the optimization of static pressure recovery in the dump diffuser of annular combustor for a modern aero-gas turbine engines. The geometrical configurations, meshing, solution setup and the results are broadly examined in this chapter.

Chapter 17 explores to investigate about the effect of roughness over an airfoil. Numerical investigations are carried out over an airfoil at three different roughnesses and at different angle of attacks. The basis of airfoil, nomenclatures, different airfoil profiles and the performance is also explained before going into the numerical investigation. This chapter will help the readers to optimize the effect of roughness at an optimized angle of attack.

Vehicle aerodynamics is the study of the aerodynamics of vehicles such as cars, trucks, race cars, bullet trains and so on. Vehicle aerodynamics and aircraft aerodynamics are completely different due to the reasons such as shape difference, operating altitude, operating velocity, degrees of freedom, and many more. The study on vehicle aerodynamics will enable us to understand how to reduce drag and how to prevent undesired lift forces. Chapter 18 touches upon all these aspects.

This book is specially designed for the audience who are interested in compressible and incompressible flows. This book is intended to offer the necessary information and knowledge for the readers in the field of aerodynamics. The chapters are explained in detail starting from low speed to high speed in all aspects along with the application related to the real time scenarios.

Sathish K. Kumar
Nehru Institute of Engineering and Technology, India

Naren Shankar Radhakrishnan
Vel Tech Rangarajan Dr. Sagunthala R&D Institute of Science and Technology, India

Acknowledgment

We would like to thank IGI Global for giving us the opportunity to edit a book. We also thank Ms. Katie McLoughlin, Assistant Development Editor-Book Development, IGI Global for supporting us throughout the development of this book. We thank the authors who have put tremendous efforts in this book project. We would also thank the reviewers who gave unbiased review on time for successful completion. The generous support from Nehru Institute of Engineering and Technology, Coimbatore, India and Vel Tech Rangarajan Dr. Sagunthala R&D Institute of Science and Technology, Chennai, India, allowed contributing an exciting interesting material to new researchers and students. We hope this book would be a necessary material for engineers. Special thanks to Ms. S. Irish Angelin for her support and motivation which has made this book possible.

Section 1
International Standard Atmosphere

Chapter 1
Role of Altitude in the Design of Aerospace Vehicles

Suthagar S.
Vel Tech Rangarajan Dr. Sagunthala R&D Institute of Science and Technology, India

Gopalakrishnan K.
Anna University, Tirunelveli, India

Kumaran T.
Vel Tech Rangarajan Dr. Sagunthala R&D Institute of Science and Technology, India

Praveen Kumar
Department of Energy Studies and Construction Management, Wollege University, Ethiopia

ABSTRACT

The atmosphere is the layer of gases surrounding the Earth and retained under its gravity. The atmosphere of our planet is made up of blankets of air that contain numerous gases such as oxygen, nitrogen, carbon dioxide, and others. Earth's atmosphere can be differentiated into five regions based on their physical properties such as temperature and pressure. Troposphere, stratosphere, mesosphere, ionosphere or thermosphere, and exosphere are these zones. The atmosphere provides the platform for these aerospace vehicles, which have taken off from Earth to operate at different altitudes. To investigate and understand the performance of flight tests, wind tunnel tests, and the design of aerospace vehicles or any other flying objects, there is a need for standard values to evaluate the parameters of the airplane or any other objects influenced by aerodynamic forces. Finally, the purpose of this chapter is to help the reader understand what the standard atmosphere is and how it can be utilized to analyze aeronautical vehicles.

DOI: 10.4018/978-1-6684-4230-2.ch001

INTRODUCTION

Just can't see anything aerospace vehicles such as Airplanes, rockets, and Missiles through our naked eye unless there is no atmosphere in our mother earth. These gases exist in nature in the atmosphere and vary at different altitudes as well as the different locations on the earth. Different gas molecules present in the atmosphere influence the variation of the atmosphere's features. As a result, the physical properties of the earth's atmosphere vary depending on altitude, it will vary with the variation of altitudes as well as the location of the earth. Earth's atmosphere is not just only a platform for aerospace vehicles, but it also provides free energy resources such as oxygen to fuel the aerospace vehicles to operate in the earth's atmosphere. Furthermore, there is no clear dividing line between these regions and each region eventually combines with its neighbors. These regions are differentiated and demarcated at different altitude levels based on the physical properties calculated by using the hydrostatic equation (Anderson & Bowden, 2005).

In this chapter, we'll go over everything in detail, how these physical properties are calculated, how these regions are defined and they can be used to understand the properties with a variation of altitude through graphical image representation. Aerospace vehicles such as aircraft, rockets, and missiles cannot fly unless there is no atmosphere on the earth or any planet. The vehicles operated using the atmosphere may be categorized into two: atmospheric vehicles, such as planes and helicopters, are one example, parachutes, etc., which always operate within a sensible atmosphere i.e., within the stratosphere region and it will not move beyond a particular altitude because beyond this region there will be the absence of required gas properties, which is necessary for atmospheric vehicles and another one is space vehicles such as satellites, interplanetary probe, exo-planet space probe, re-entry vehicles tend to operate outside the sensible atmosphere and it can be obey operated any altitude level. As these aerospace vehicles are encountering the atmosphere during their mission, they might be influenced by the atmospheric properties. As a result, the features of the atmosphere must be considered (Jaganraj et al., 2015).

The atmosphere of the Earth is not a static system; it is constantly changing. Physical qualities of the atmosphere, such as pressure and temperature, are affected by altitude. Location (longitude and latitude), time of day, weather, and perhaps even solar sunspot activity are all factors to consider. To design and investigate the performance of these aerospace vehicles operated through our earth's atmosphere. Variations in these physical features of the environment must be taken into consideration as a precondition. Without knowledge of these physical properties, it is impossible to design and flight aerospace vehicles. The determinants of these properties are temperature, pressure, and density. These physical properties are used to calculate the various parameters of aerospace vehicles such as aerodynamic forces, gravitational forces, airspeed, etc. The gravitational forces and aerodynamic forces encountered and experienced by aerospace vehicles vary with their altitudes. This is the fact that because of atmospheric properties viz; pressure, density, and temperature also change with altitudes. It implies that these forces are strong functions of atmospheric properties. For this purpose, these standard values are agreed upon through international agreement and named as International Standard Atmosphere (ISA) (Anderson & Bowden, 2005).

International Standard Atmosphere (ISA) refers to an imaginary set of values assigned to the physical properties of the atmosphere at different altitude levels through an international agreement under the International Civil Aviation Organization (ICAO). Though these agreement numbers don't represent actual conditions at any particular time, they can act as a guideline. It is vital to study the standard atmosphere, which helps us to plan and design perfect aerospace vehicles such as aircraft, Rocket, Missiles, etc., by investigating and understanding the performance of aerospace vehicles and various flying objects of the atmosphere at different altitudes. Without any standard values, it is very difficult to design and investigate the performance of airplanes or any other flying objects in the earth's atmosphere. Therefore, values prescribed under the International standard of the atmosphere (ISA) are needed to investigate and understand the airplane's performance is influenced by the physical features of the earth's atmosphere. As a result, a set of values for atmospheric characteristics has been agreed upon, which represent the measurement of average year-round conditions (Jaganraj et al., 2015). These figures are used to compare the performance of different planes as well as the performance of the same plane throughout different days of flight testing. The standard atmosphere will provide us mean values for pressures, temperatures, densities, and other attributes as a function of height; these values have previously been determined using an experimental balloon and sounding-rocket measurements, as well as a mathematical model of the atmosphere.

The standard atmosphere's principal role is not to reflect regular atmospheric conditions. Therefore, its primary goal is to compile tables of common reference conditions that aeronautical professionals all over the world can use in an organized manner. The above chapter will also go over, some definitions of different terms such as absolute altitude, geometric altitude, Geo-potential altitude and get the knowledge about the same. Some fundamental physics definitions are also used and an equation (the hydrostatic equation) that are needed tools for creating the numbers in the standard atmosphere tables. Through the use of standard formulas derived from basic equations, we will solve some problems to understand how to calculate the standard values (Chanthavane, 2021). Then we move to discuss how these values in the tables are obtained. These values provide a base reference to calculate the value of various parameters of the aerospace study.

We can't see aerospace vehicles such as Airplanes, rockets, and Missiles through our naked eye unless there is no atmosphere in our mother earth (Daidzic, 2015). Generally, the atmosphere is the layer of gases surrounded by the earth and retained under its gravity. The atmosphere of our planet is made up of blankets of air that contain various gas such as oxygen, nitrogen, carbon dioxide, and others. Several gases present in the atmosphere vary at different altitudes as well as the different locations of the earth. The features of the atmosphere are influenced by the presence of different gas molecules in the atmosphere. As a result, the physical qualities of the earth's atmosphere are not uniform at all altitude levels; but will vary depending on altitude and the earth's location. Because aerospace vehicles operate in the earth's atmosphere, it is vital to have a deep understanding of the characteristics of this environment to correctly operate the vehicles.

Earth's atmosphere is not just only a platform for operating aerospace vehicles, but it also provides free energy resources such as oxygen to fuel the aerospace vehicles to operate in the earth's atmosphere (Kurzke & Halliwell, 2018).

Earth's atmosphere has contained various regions and it can be differentiated based on their physical properties such as temperature and pressure. This chapter will go over the properties of the earth's atmosphere in distinct places in detail, how these regions are defined and how they can be used to understand the properties with a variation of altitude through graphical image representation. Further, we will learn in this chapter about various aerospace vehicles operated in different regions of the atmosphere.

This chapter further will discuss International Standard Atmosphere (ISA), some definitions of different terms such as absolute altitude, geometric altitude, Geo-potential altitude and get the knowledge about the same.

We also use certain essential physics concepts and an equation (the hydrostatic equation) to generate the data in the standard atmospheric tables. Through the use of standard formulas derived from basic equations, we will solve some problems to understand how to calculate the standard values. Then we move to discuss how these values in the tables are obtained. These values provide a base reference to calculate the value of various parameters of the aerospace study. Finally, the purpose of this chapter is to help the reader understand what the standard atmosphere is and how it can be utilized to analyze aeronautical vehicles (Cavcar, 2000).

THE ATMOSPHERE OF THE EARTH

The atmosphere of the earth is a blanket of different gases that can be classified into five zones depending on specific intrinsic qualities (please note) (Figure 1). These are the five locations of the atmosphere of the earth: Troposphere, Stratosphere, Mesosphere, Ionosphere or Thermosphere, and Exosphere (Ananthasayanam & Narasimha, 1987). There is no clear dividing line between such regions, and each ultimately combines with just beside it. These regions are differentiated and demarcated at different altitude levels based on the physical properties calculated by using the hydrostatic equation.

Figure 1. Regions of Earth atmosphere
Source: URL(https://www.google.com/search?q=Regions+of+Earth+Atmosphere&source=lnms&tbm=isch&sa=X&ved=2ahUKEwjLw9GTqbb3AhXflbcAHXb_CNYQ_AUoAXoECAEQAw&biw=1366&bih=657&dpr=1#imgrc=9NLSqfBMthxhKM)

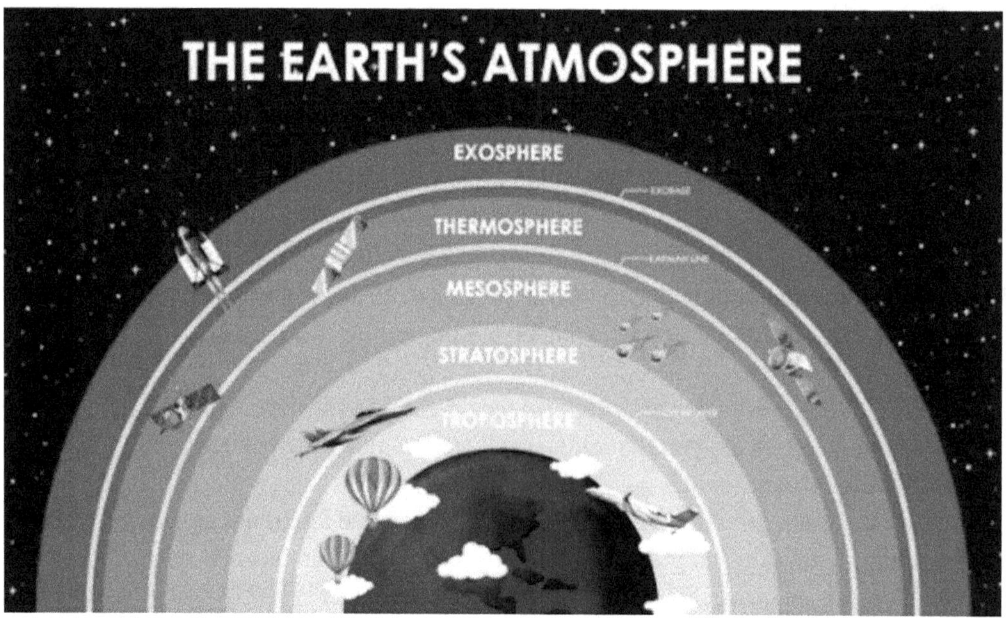

The Troposphere

This is the planet's first and closest zone near the surface, and it is the smallest of the five regions, yet it contains the majority of the atmosphere's mass, approximately 75%. This region provides a habitat for various living organisms including human beings. It is characterized by turbulent air conditions, as well as a reduction in pressure and temperature as altitude rises (Chanthavane, 2021). The temperature drops linearly at a rate of about 6.5 K / kilometer. The tropopause is the highest point in the troposphere. At the poles, the tropopause is about 9 kilometers high, whereas, at the equator, it is about 16 kilometers high.

The Stratosphere

The second zone extends again from tropopause to a distance of around 50 kilometers. Although there are high-velocity winds in this area, they are not extremely gusty. Until about 25 kilometers, the temperature remains constant, at which point it begins to rise. This region plays a vital role and provides a platform for operating aerospace vehicles such as aircraft, helicopters, parachutes, etc. This region also contains the ozone layer, which protects all living species, including humans, on our planet from dangerous ultraviolet sun radiation via Ozone molecules. The stratopause is the highest point in the stratosphere.

The Mesosphere

The mesosphere is the third and middle layer of the atmosphere, extending from the stratopause to nearly 80 kilometers. In this area, the temperature drops to around -900C. The pressure and density of air in the mesosphere are extremely low, but thc air nevertheless has the same composition as at sea level. The mesopause is the mesosphere's highest point (Tsuchiya et al., 2019).

The Ionosphere or Thermosphere

This zone stretches from the mesopause to around 1000 kilometers. The presence of ions and free electrons distinguishes it. At 110 km, the temperature rises to around 0°C, then to around 1000°C at 150 km, and finally to around 1780°C at 700 km (see in Figure 2). In this area, electrical phenomena such as the aurora borealis can be seen.

The Exosphere

This is the earth's final and highest layer of atmosphere. This region contains a very small number of molecules. Gradually, the region merges with interplanetary space.

IMPORTANCE OF ATMOSPHERE FOR AEROSPACE VEHICLES

Aerospace vehicles such as aircraft, rockets, and Missiles cannot fly unless there is no atmosphere on the earth or any planet. The atmosphere provides the platform for these aerospace vehicles, which have taken off from earth to operate at different altitudes (Daidzic, 2019). The vehicles operated using atmosphere may be two types of classification:

Figure 2. Temperature & pressure changes in the atmosphere of the earth are typical
Source: URL(https://www.google.com/search?q=Temperature+%26+pressure+changes+in+the+atmosphere+of+the+earth+are+typical&tbm=isch&ved=2ahUKEwj_28mUqbb3AhUbjtgFHfQnC_QQ2cCegQIABAA&oq=Temperature+%26+pressure+changes+in+the+atmosphere+of+the+earth+are+typical&gs_lcp)

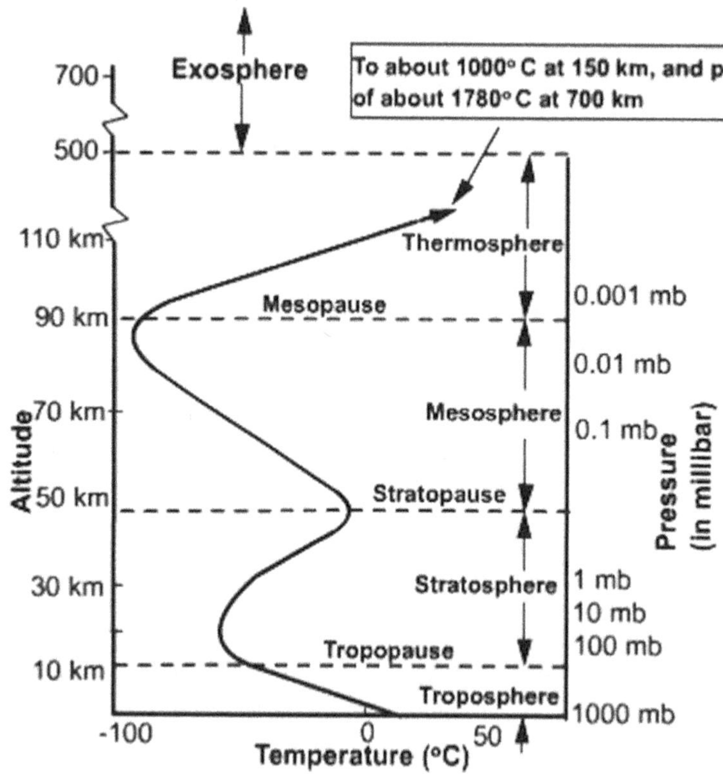

One is atmospheric vehicles such as planes, helicopters, and parachutes, which always operate in a safe environment, i.e., within a reasonable atmosphere. within stratosphere region and it will not move beyond a particular altitude because beyond this region, there will be the absence of required gas properties, which is necessary for atmospheric vehicles and another one is space vehicles such as satellites, interplant probe, exo-planet space probe, re-entry vehicles tend to operate outside the sensible atmosphere and it can be operated at any altitude level (Ananthasayanam & Narasimha, 1987). As these aerospace vehicles are encountering the atmosphere during their mission, it might be influenced by the atmospheric properties, as a result, it is vital to consider the qualities of the atmosphere.

Need of Standard Atmosphere

The atmosphere of the Earth is not a static system; it is constantly changing. Altitude, longitude and latitude, time of day, period, and even solar global temperature are all factors to consider that affect physical properties of the atmosphere such as pressure, temperature, and density. To research and design the performance of these aerospace vehicles operated through our earth atmosphere, Variations in these physical properties of the atmosphere should be taken into consideration as a prerequisite. Without adequate knowledge of these physical properties, it is impossible to design and flight aerospace vehicles.

The determinants of these properties are temperature, pressure, density (Ananthasayanam & Narasimha, 1987). These physical properties are used to calculate the various performance parameters of aerospace vehicles such as aerodynamic forces, gravitational forces, airspeed, etc. Therefore, these properties are used to develop a standard atmosphere, which is used for determining the performance of aerospace vehicles.

The gravitational forces and aerodynamic forces always encountered and experienced by aerospace vehicles vary with their altitudes. This is because air variables such as pressure, density, and temperature vary with altitude. It implies that these forces are strong functions of atmospheric properties (Kaushik, 2019).

To investigate and understand the performance of flight tests, wind tunnel tests, and the design of aerospace vehicles or any other flying objects, there is a need for standard values to evaluate the parameters of the airplane or any other objects influenced by aerodynamic forces.

For this purpose, these standard values are agreed through international agreement and named as the environment by the international standard atmosphere (ISA).

International Standard Atmosphere (ISA)

As a result of an international agreement controlled by the International Civil Aviation Organization, the International Standard Atmosphere (ISA) is an imagined set of values given to the physical properties of the atmosphere at various altitude levels (ICAO). These accepted characteristics, while not representing actual conditions at any given time, might be used as a reference. It is vital to study the standard atmosphere, which helps us to plan and design perfect aerospace vehicles such as aircraft, Rocket, Missiles, etc., by investigating and understanding the performance of aerospace vehicles and various flying objects of the atmosphere at different altitudes (Daidzic, 2019).

Without any standard values, it is very difficult to design and investigate the performance of airplanes or any other flying objects in the earth's atmosphere. As a response, values established upon by the International Standard Atmosphere (ISA) are required to investigate and understand an airplane's performance, which is influenced by the physical qualities of the earth's atmosphere. As a result, a set of air quality values has been agreed upon, which represent measures of average year-round conditions. These figures are used to look at the various planes and the performance with the same plane during flight testing on different days (Kurzke & Halliwell, 2018).

The standard atmosphere will provide us with the overall mean of pressures, temperatures, densities, and several other characteristics such as a function of height; these values were previously calculated using an experimental balloon and sounding-rocket measurements combined with a mathematical model of the atmosphere.

The standard atmosphere's principal role is not to reflect normal environmental conditions. Its primary objective is to produce tables of common reference circumstances that aeronautical professionals all around the world can use in a structured manner.

It should be noted, however, that several standard atmosphere values exist, those that were produced by various organizations at various dates using different test data in the models. However, the variations are insignificant for all practical purposes below 30 km altitude, which is the range of modern airplanes. The 1959 ARDC model atmosphere is a widely used standard atmosphere. (ARDC refers to the former Air Research and Development Command of the United States Air Force, which is now the Air Force Research Laboratory.) This book's atmospheric tables are based on the ARDC model environment from 1959.

DEFINITION OF ALTITUDE

We all know that altitude refers to the height above the ground. However, even though it is available in broad terms, we must define the term properly for engineering purposes. As a result, we define and use three separate altitudes in the following sections: absolute, geometric, and geopotential altitudes, as well as terminology like pressure, temperature, and density altitudes (Kaushik, 2019).

Absolute Altitude (h_a)

The height is the distance between the earth's center and the altitude. Example: Its measurement on the tape would be maybe we could drill a hole in the ground to the planet's core & Our tape was expanded till it reached the middle, by definition known as, the absolute altitude (ha)

Geometric Altitude (h_G)

The altitude as a ratio to sea level or in other words, the height above sea level. Example: Imagine we're at Marina Beach in Chennai, India, with the ground level at the same level as the sea. The geometric altitude (h_G) is the measurement made on the tape, if we came flying in a helicopter and dropped a tape down to the earth at sea level at the point—The geometric height above sea level, specifically.

Figure 3, Schematic diagram represents geometric altitude and an absolute altitude
Source: Book (Anderson, J.D., "Introduction to Flight")

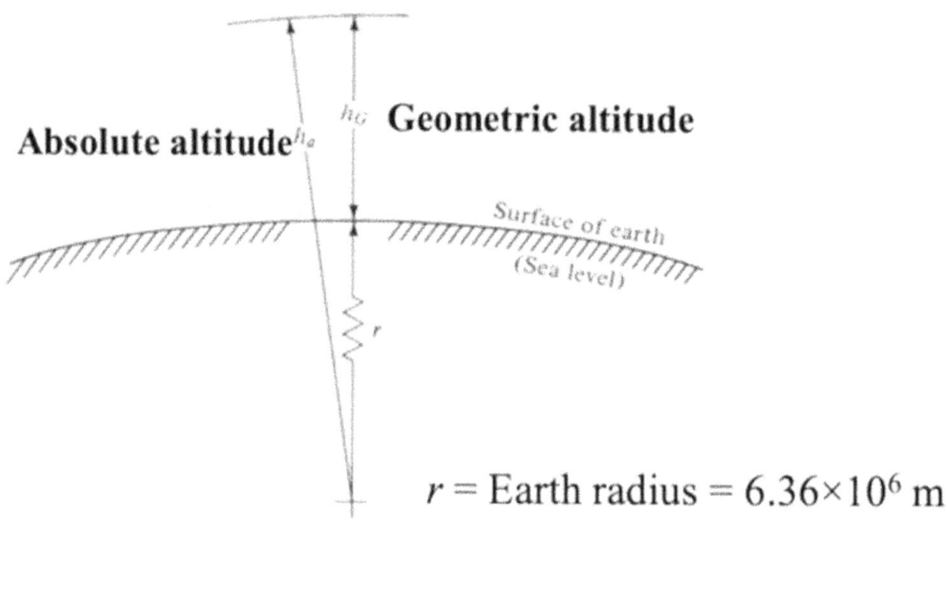

Figure 3 shows that the absolute altitude is equal to the sum of the geometric altitude and the earth's mean radius. This relationship can be expressed mathematically as

$$p = p(h) \tag{1}$$

Where,

r_e or r is the mean radius of the earth

Geopotential Altitude (h)

The geometric altitude has been adjusted to reflect gravity variations.

GRAVITY AND ALTITUDE RELATIONSHIP

As we now got familiarized with various altitude concepts, it is very important to understand the variation of acceleration due to gravity with altitudes, because, for space flight, the absolute altitude varies with gravity. Newton's universal gravitational laws will be used an evaluate (Anderson & Bowden, 2005).

Gravity varies inversely with the proportion to the distance from the earth's center, according to Newton's law of gravitation. For a certain absolute height ha, at sea level gravitational acceleration is go, and the gravitational acceleration at a local level is; the relationship between g_a and g_o is as follows

$$g_a = g_o \left[\frac{r_e}{h_a} \right]^2 = g_o \left[\frac{r_e}{r_e + hg} \right]^2 \tag{2}$$

Why are these attitudes and their dependent parameters necessary to discuss?

These concepts are required to express the thermodynamic properties of the atmosphere (P, ρ, T) as a function of altitudes.

HYDROSTATIC EQUATION

We'll now assume a model for the hydrostatic equation that includes a fluid element of air at rest, allowing us to determine the differences in p, T, and ρ as a function of level.

Figure 4 shows a small stationary fluid element of air. We'll use an element with rectangular faces for this example, with sides of unit length on the top and bottom faces and infinitesimally small height on the side faces dh_G. The pressure p is sensed on the bottom face, resulting in an upward force of $P \times 1 \times 1$ a force applied to a fluid element. The top face has a slightly higher altitude (measured as the distance, dh_G); Due to the reason that pressure fluctuates with altitude, the pressure on the top and bottom faces will differ by an infinitesimally small amount dp. As a result, the pressure is felt at the top p+dp is ft. It causes the fluid element to be forced downward. Furthermore, the fluid element's volume is $(1)(1)dh_G = dh_G$ and because ρ is the mass of the fluid element is simply its mass per unit volume $\rho(1)(1)dh_G = \rho dh_G$. If the local gravity acceleration is ga when the fluid element's weight is $g_a \rho dh_G$ as represented in Figure 4. Because the fluid element is stationary, the three forces depicted in Figure 4 top and bottom pressure forces, and mass equalize.

Figure 4. Force diagram for the hydrostatic equation
Source: Book (Anderson, J.D., "Introduction to Flight")

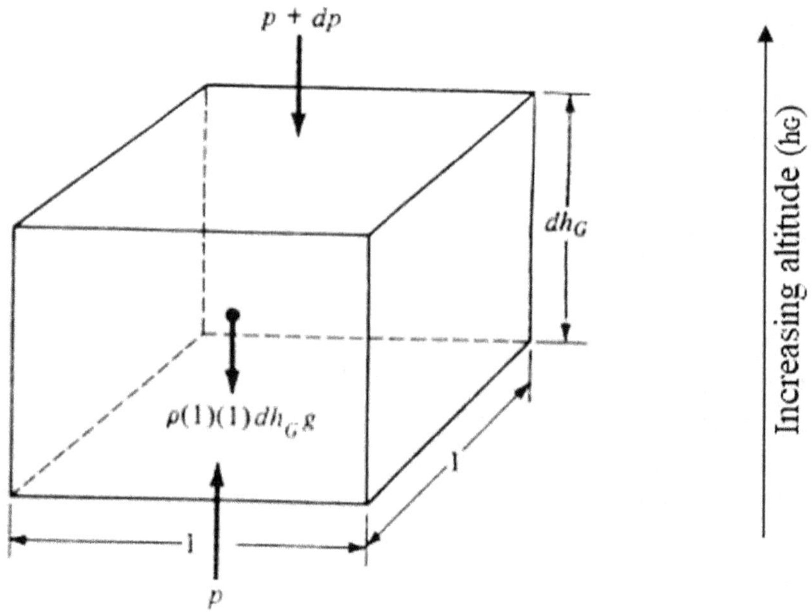

Hence

$$p = p + dp + \rho g_a dh_G$$

Thus

$$dp = -\rho g_a dh_G \quad (3)$$

The hydrostatic equation, which applies to any density fluid, such as seawater or air in the atmosphere, is Equation (3).

Within the simplest form, equation (3) is a differential equation; it attaches an infinitesimally small change in pressure. A change in altitude of an infinitesimally small magnitude correlates to a change in altitude of a corresponding magnitude, wherein differential calculus's language, dp and are differentials. It is important to note that in Equation (3), g is a variable, which g depends as determined by Equation (3).

Equation (3) should be integrated to give us what we want for it to be useful: the change in pressure as a function of altitude $p = p\ (h_G)$. We assume that g is constant throughout the atmosphere and that it is the same as it is at sea level g_o, to make the integration easier. In aeronautics, this is somewhat of a historical convention.

As a result, we may write Equation (3) as

$$dp = -\rho g_o dh \quad (4)$$

To make Equations (3) and (4) numerically equivalent, however, the altitude h in Equation (4) must differ slightly from Equation (3) is used to compensate for the fact that g_a differs significantly from go. Suddenly, a new altitude h has been identified, known as the geopotential altitude, which differs from the geometric altitude. Consider a particular geometric altitude, h_G to better understand the concepts of geopotential altitude, where the pressure value. Let's now raise an infinitesimal amount to the geometric altitude dh_G, as a result of which the new geometric altitude is $h_G + dh_G$. The pressure is lower at this new altitude p+dp, where is the value dp is given by Equation (3). Let us now put this same value in Eq. (4).

If we divide Equation (4) by (3), we get

$$1 = \frac{\left(\frac{g_o}{g_a}\right)}{\left(\frac{dh}{dh_G}\right)}$$

Obviously, since g_o and are different, dh and dh_G must be separate; that is, the absolute values of dh and dh_G at corresponds to the same pressure change dp, are dissimilar. As a result, the numerical values have changed dh and h_G. Different values relate to the same actual physical position in the atmosphere.

Equation (4) establishes a "fictitious" altitude to create a geopotential altitude. Many standard atmosphere tables, on the other hand, give their data in terms of geopotential altitude, and it's important to know the difference. The imaginary altitude that is physically consistent with the assumption is known as geopotential altitude of g_a= constant = g_o.

GEOPOTENTIAL ALTITUDES (h) AND GEOMETRIC ALTITUDES (h_G)

We should get the relationship between geometric and geo-potential altitude by dividing Equation (4) by Equation (3) using the hydrostatic equations from Equation (3) and (4).

$$1 = \frac{\left(\frac{g_o}{g_a}\right)}{\left(\frac{dh}{dh_G}\right)}$$

$$dh = \left(\frac{g_a}{g_o}\right) dh_G \tag{5}$$

$$dh = \left(\frac{r_e^2}{\left(r_e + hg\right)^2}\right) dhg_a \tag{6}$$

Integration dh in Equation (6) up to any given value of from sea level h (different from h_G) geometric altitude h_G at a given point in the atmosphere, we've seen

$$\int_0^h dh = \int_0^{h_g} \left(\frac{r_e^2}{(r_e + hg)^2} \right) dhg_a$$

$$\int_0^h dh = \int_0^{h_g} \frac{dh_g}{(r_e + h_g)^2}$$

$$h = r_e^2 \left(\frac{-1}{r_e + h_g} \right)_0^{h_g}$$

$$= r_e^2 \left(\frac{-1}{r_e + h_g} + \frac{1}{r_e} \right)$$

Thus,

$$h = \left(\frac{r_e}{r_e + h_G} \right) h_G \tag{7}$$

Geopotential altitude is h and geometric altitude is h_G. The required relationship between the two altitudes is as follows. Before we get relations such as p=p(h), We may make utilize of Equation (7) therefore integrate with p to h_G.

Modeling of the Standard Atmosphere

We have been studying various altitudes up to this point. We'll now go on to discuss the typical environment. As previously stated in this chapter, researchers conducted experiments with sounding rockets and hot air balloons to explore the fluctuation of temperature with altitude earlier.

Figure 5 shows a typical pattern of a temperature change as a function of altitude. There are certain vertical lines (known as constant temperature or isothermal areas) and inclined lines that are consistent (known as gradient regions).

The plot of temperature variation with altitude (Figure 5) was established using experimental data. However, there is no analogous experiment for the thermodynamic parameters that we are interested in (pressure and density). As a result, we use the temperature V_s altitude plot in Figure 5 to estimate the pressure and density variation with altitude.

Figure 5. Temperature variation with altitude in the standard atmosphere
Source: Book (Anderson, J.D., "Introduction to Flight")

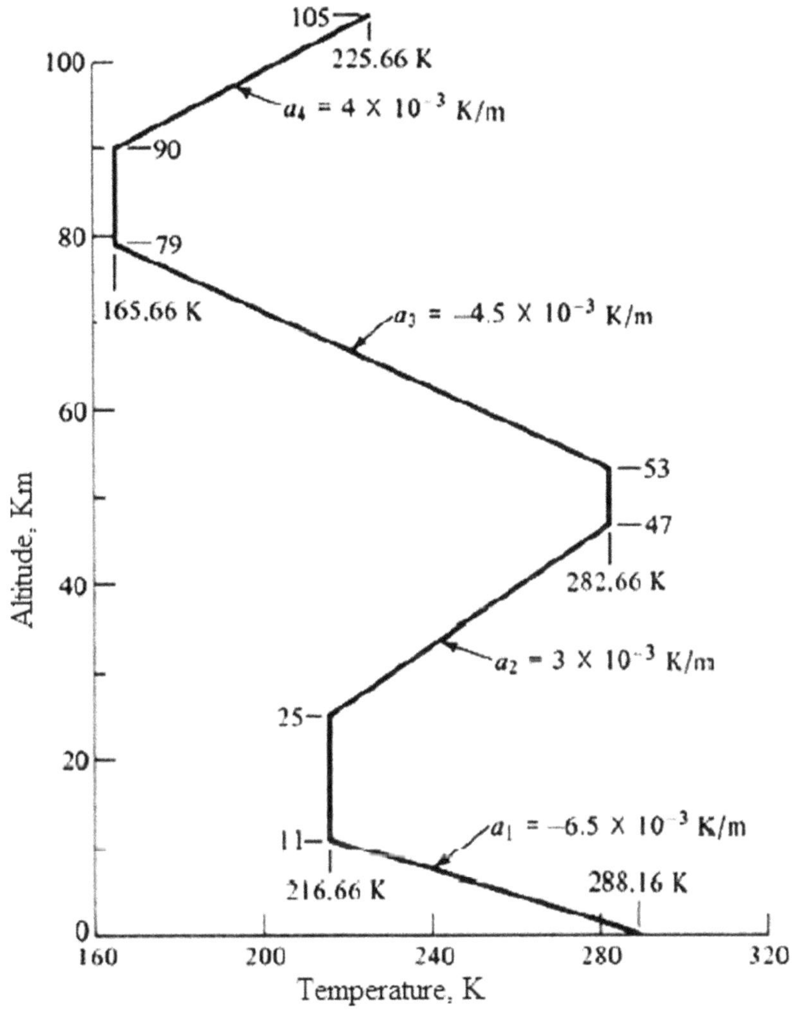

The hydrostatic equation has been updated.

$$dp = -\rho g_o dh \tag{8}$$

The hydrostatic equation has been divided, $P=\rho RT$ we have

$$\frac{dP}{P} = -\left(\frac{\rho g_o dh}{\rho RT}\right) = -\left(\frac{g_o}{RT}\right)dh \tag{9}$$

First, we'll look at the isothermal zone of the standard atmosphere, which is defined by the vertical lines in Figure 5 and drawn in Figure 6, and where the temperature remains relatively constant. At the base of the isothermal area, temperature, pressure, and density are shown. Figure 6 are T_1, P_1, and ρ_1

respectively. In terms of geopotential h_1, the base is located at a given altitude. Consider a location in the isothermal layer above the base with a variable altitude of h. By integrating Equation (9) between h_1 and h, we may get the Pressure h When the parameters in Equation 8 are substituted, we get:

$$\int_{P_1}^{P} \frac{dP}{P} = -\frac{g_o}{RT}\int_{h_1}^{h} dh \tag{10}$$

Figure 6. Isothermal layer
Source: Book (Anderson, J.D., "Introduction to Flight")

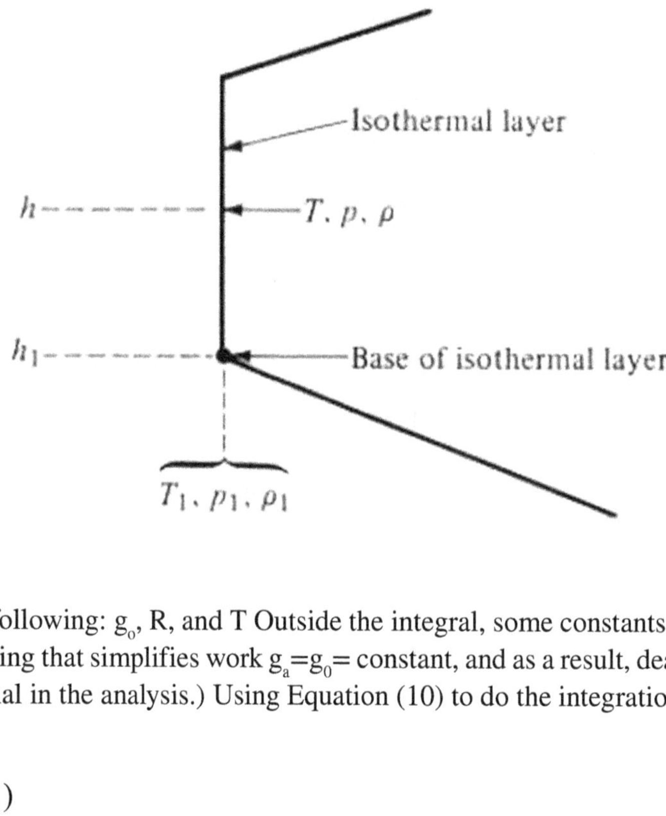

Take note of the following: g_0, R, and T Outside the integral, some constants can be used. (This exemplifies how assuming that simplifies work $g_a = g_0 =$ constant, and as a result, dealing with geopotential altitude h it is essential in the analysis.) Using Equation (10) to do the integration, we get

$$\ln\frac{P}{P_1} = -\frac{g_o}{RT}(h - h_1)$$

Alternatively, we may write

$$\frac{P}{P_1} = e^{-[g_o/(RT)](h-h_1)} \tag{11}$$

Using the same approach as before, the equation of state,

$$\frac{P}{P_1} = \frac{\rho T}{\rho_1 T_1} = \frac{\rho}{\rho_1}$$

Thus,

$$\frac{\rho}{\rho_1} = e^{-[g_o/(RT)](h-h_1)} \tag{12}$$

Equations (11) and (12) show the changes of p and ρ versus geopotential altitude for the standard atmosphere's isothermal layers.

Figure 7 Gradient layer
Source: Book (Anderson, J.D., "Introduction to Flight")

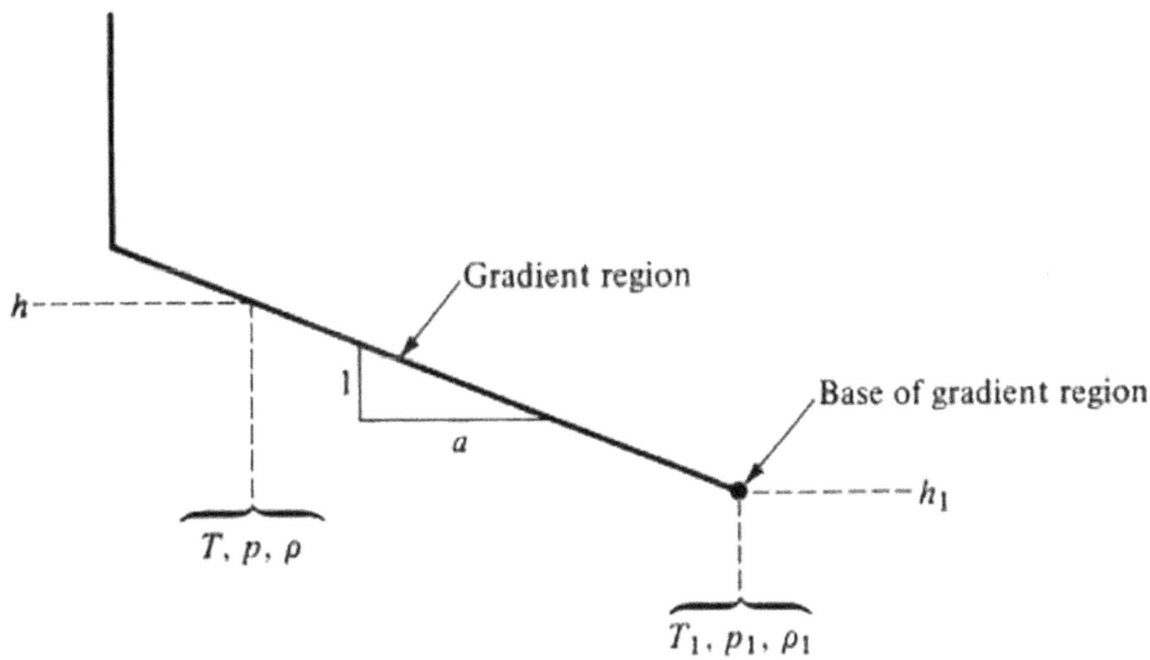

Since we look at the gradient area, as shown in Figure 7, we can see that the temperature variation is linear and can be expressed geometrically as,

$$\frac{T-T_1}{h-h_1} = \frac{dT}{dh} \equiv a \tag{13}$$

Where an is a constant calculated from the defined temperature variation in Figure 5 for each region. The lapse rate for the gradient zone is commonly referred to as the value of a.

$$dh = \frac{1}{a} dT \tag{14}$$

Equation 9 is obtained by substituting the value of dh from Equation 14 into Equation 9,

$$\frac{dP}{P} = -\frac{g_o}{aR}\frac{dT}{T} \tag{15}$$

Equation 15 yields when the gradient layer's base (shown in Figure 7) is integrated with a point at a certain height h in the gradient layer, h_G

$$\ln\frac{P}{P_1} = -\frac{g_o}{aR}\ln\frac{T}{T_1}$$

$$\frac{P}{P_1} = \left(\frac{T}{T_1}\right)^{-\frac{g_o}{aR}} \tag{16}$$

Using the equation of state, we can write P = ρRT,

$$\frac{P}{P_1} = \frac{\rho T}{\rho_1 T_1} \tag{17}$$

Equation 16 can also be rewritten as Equation 18 to,

$$\frac{\rho T}{\rho_1 T_1} = \left(\frac{T}{T_1}\right)^{-\frac{g_o}{aR}}$$

$$\frac{\rho}{\rho_1} = \left(\frac{T}{T_1}\right)^{-\frac{g_o}{aR}-1}$$

$$\frac{\rho}{\rho_1} = \left(\frac{T}{T_1}\right)^{-\left(\frac{g_o}{aR}+1\right)} \tag{18}$$

Table 1. Caption should be sentence case with no ending punctuation if only one sentence

Parameters	Gradient layer	Isothermal layer
Pressure	$\frac{P}{P_1}=\left(\frac{T}{T_1}\right)^{-\frac{g_o}{\alpha R}}$	$\frac{P}{P_1}=e^{-[g_0/(RT)](h-h_1)}$
Density	$\frac{\rho}{\rho_1}=\left(\frac{T}{T_1}\right)^{-\left(\frac{g_o}{\alpha R}+1\right)}$	$\frac{\rho}{\rho_1}=e^{-[g_0/(RT)](h-h_1)}$

The standard atmosphere will always be dependable on a regular day since the assumptions that were used to build the mathematical models are not ignored. On a non-typical day, however, this model cannot be trusted. As a result, two additional altitudes must be defined using the standard atmospheric model, which will be described in the following component of this course (Daidzic, 2015).

Remember that T varies linearly with altitude and follows the provided relationship.

$$T = T_1 + a(h - h_1) \tag{19}$$

Equation (19) results in $T=T(h)$ when it's plugged into the gradient layers

Equation (16), we've got $p=p(h)$; Alternatively, using Equation (18) we've got $\rho=\rho(\mathrm{h})$.

Let will see just how the standard atmosphere is created. Starting at sea level (see Figure 5) (h=0), where standard pressure, density, and temperature values at sea level — P_s, ρ_s, and T_s, respectively—are

$$P_S = 1.01325 \times 10^5 N/m^2 = 2116.lb/ft^2$$

$$\rho_S = 1.2250 \mathrm{K}g/m^3 = 0.002377 slug/ft^3$$

$$T_S = 288.16K = 518.69^{o}R$$

For the first gradient zone, these are the starting values. To obtain T values as a function of h, use Eq. (19) until T = 216.66 K, which appears at h = 11.0 km. Equations (16) and (18) will be apply to these T values in the first gradient layer to get the equivalent P and ρ. Then, using Equations (11) And (12), start at h = 11.0 km as the base of the first isothermal section (see Figure 4) until h = 25 km, which is the base of the next gradient zone, to compute values of p and ρ against h. We will establish a table of values for the standard atmosphere using Figure 5 and Equations (11), (12), and (16) to (19).

Appendix A has a table for SI units & Appendix B contains a table for English engineering unit. Look more closely at these table and get to know them. They set the tone for the rest of the day. In the first column, the geometric altitude is supplied, while the matching geopotential altitude is given in the second column equation (7). Based on the previous discussion, the relevant standard values of temperature, pressure, and density for each altitude are given in the third through fifth columns, respectively (Anderson & Bowden, 2005).

It's important to remember that the standard atmosphere is just that: a reference environment. It doesn't accurately represent the true atmospheric qualities that may exist at any given location and time.

Appendix A (Anderson & Bowden, 2005), for example, states that at a 3 km altitude (geometric), p = 0.70121 × 105 N/m^2, T = 268.67 K, and ρ = 0.90926 kg/m^2. In practice, if you could levitate yourself to 3 km above sea level right now from where you are, you would most probably experience a p, T & ρ different from the results calculated from Appendix A (Anderson & Bowden, 2005). As will be demonstrated in the book's later sections, the standard atmosphere only permits us to simplify test results and computations to a handy, agreed-upon reference.

We shall now can better understand the meaning and significance of the geometric altitude h_G, and the geopotential altitude, h. Equations (11) to (19) are used to compute the changes in characteristics in the standard environment. These equations are calculated with the simplifying assumption of a constant value of gravity acceleration equal to its value at sea level, g constant go. As a result, the altitude in these equations is the geopotential altitude by definition, h. Examine these equations once more—you'll see that go and h, not g and h_G occur in these equations. The primary reason for establishing the geopotential altitude is the simplicity provided by assuming a constant value of g. This is the only period we'll use geopotential altitude in this chapter, and it'll be to calculate the data in Appendices A and B. Moreover, because h and h_G equation relates them (7) we can always obtain the geometric altitude h_G, that corresponds to a certain geopotential altitude value, h. The geometric altitude h_G, on either side, is the actual height above sea level, making it more convenient. As a result, the first column in Appendices A and B is h_G and the entries are example h_G (Kaushik, 2019). The corresponding values of h are given in the second column, and these are the values utilized in Equations (11) to (19) to obtain the corresponding numbers for P, ρ, and T.

PRESSURE, TEMPERATURE, AND DENSITY ALTITUDES

We may now establish three new "altitudes"—pressure, temperature, and density altitudes—with the tables of Appendices A and B in hand. The simplest way to obtain this is to lead by example. Assume we're on a plane travelling at a certain, geometric altitude (Daidzic, 2015).

For the purposes of this discussion, the value of our actual altitude is irrelevant. However, we measure the exact outside air pressure at this altitude to be 6.16104 N/m^2. The standard height that corresponds to a pressure of 6.16104 N/m^2 is 4 km, according to Appendix A. As a result, we may state that we're flying at a pressure height of 4 km. At the same time, we record the real outdoor air temperature as 265.4 K. The standard height that corresponds to a temperature of 265.4 K is 3.5 km, according to Appendix A (Anderson & Bowden, 2005). As a result, we estimate we're flying at a temperature altitude of 3.5 kilometers. As a result, we're flying at a pressure altitude of 4 km and a temperature altitude of 3.5 km at the same time, despite the fact that our real geometric altitude is a different figure. In the same line, the definition of density altitude is given (Anderson & Bowden, 2005). These values—pressure, temperature, and density altitudes—are only handy figures that are connected to the actual P, T, and for the actual altitude at which we are flying using Appendix A or B (Anderson & Bowden, 2005).

Illustration 1

At a geopotential altitude of 14 km, calculate the standard atmospheric values of T, P, and ρ.

Solution

Remember that T is the standard atmosphere's specified variation. As a result, we may go straight to Figure 5 and see that at h = 14 km,

T = 216.66K

Equations (11) to (19) must be used to piece together the various levels from sea level to the exact altitude with which we are engaged in order to derive p and ρ. The first section (from Figure 5) is a gradient zone from h = 0 to h = 11.0 km, begin at sea level. The rate of lapse is

$$a = \frac{dT}{dh} = \frac{216.66 - 288.16}{11.0 - 0} = -6.5K / km$$

Or a= -0.0065K/m

As a result, using equations (16) and (18), which are for a gradient zone with sea level as the basis (hence P_1 = 1.01 × 105 N/m² and ρ_1 = 1.23 kg/m³),

We notice this at h = 11.0 km

$$P = P_1\left(\frac{T}{T1}\right)^{-\frac{g_o}{(aR)}} = (1.01\times10^5)\left(\frac{216.66}{288.16}\right)^{-\frac{9.8}{[-0.0065(287)]}}$$

Where, g_0 = 9.8 m/s² in SI units. Hence p (at h = 11.0 km) = 2.26×104 N/m².

$$\rho = \rho_1\left(\frac{T}{T1}\right)^{-\left[\frac{g_o}{(aR)}+1\right]} = (1.23)\left(\frac{216.66}{288.16}\right)^{-\left\{\frac{9.8}{[-0.0065(287)]}+1\right\}}$$

= 0.367kg/m^3, at h =11.0 km.

This p and ρ values then form a basis for the very first isothermal zone (see Figure 4). Equations (11) and (12) are the equations for the isothermal zone, where now P_1 = 2.26 ×104 N/m² and ρ_1 = 0.367 kg/m³. For h = 14 km, h – h_1 = 14 – 11 = 3 km = 3000 m.

From Equation (11),

$$P = P_1 e^{-\left[\frac{g_o}{(RT)}\right](h-h_1)} = (2.26\times10^4)e^{-\left[\frac{9.8}{[287(216.66)]}\right](3000)}$$

P = 1.41 × 10⁴N/m^2.

From Eq. (12),

$$\frac{\rho}{\rho_1} = \frac{p}{p_1}$$

Hence,

$$\rho = \rho_1 \frac{p}{p_1} = 0.367 \frac{1.41 \times 10^4}{2.26 \times 10^4} = 0.23 kg / m^3$$

Within a round-off tolerance, these values match the values in App. A. Note: This example proves how the figures in Appendices A & B (Anderson & Bowden, 2005) were calculated.

Illustration 2

Does pressure decrease faster in the gradient regions or in the isothermal regions? The pressure drops with increasing altitude in both the gradient and isothermal zones of the standard atmosphere.

Solution

Consider an infinitesimally small increase in altitude, dh. The corresponding infinitesimally small change in pressure is dp, and is given by Equation (9), repeated here:

$$\frac{dP}{P} = -\left(\frac{g_o}{RT}\right) dh$$

To interpret the physical meaning of the differential relationship given by Equation (9), consider a given altitude h where the pressure is p. If we increase altitude by an infinitesimally small amount, dh, the corresponding infinitesimally small change in pressure is dp. The ratio dp/p is the fractional change in pressure. (You can also interpret this as a "percentage change" in pressure, which in reality is given by 100 (dp/p)). The rate of change of this proportion in relation to altitude change, dh, is represented as

$$\frac{\frac{dp}{p}}{dh} = -\frac{g_o}{RT} \tag{20}$$

Obtained from Equation (9). To properly answer the question posed in this example, we need to evaluate the value of $\frac{\frac{dp}{p}}{dh}$ in the isothermal regions and the gradient regions. Clearly, from Equation (20), this value depends only on the local temperature at the given altitude h.

From this, we make the following observations:

In the first gradient region, where T decreases with altitude (see Figure 5), the absolute value becomes larger as h increases (i.e., the pressure decreases at a faster rate). For example, at the base of the first gradient region, where h = 0 and T = 288.1 6 K, we have, from Equation (20)

$$\frac{\frac{dp}{p}}{dh} = -\frac{g_o}{RT} = -\frac{9.8}{(287)(288.16)} = -1.185\times10^{-4} \text{ Per meter}$$

At the top of the first gradient region, where h = 11 km and T = 216.66 K, we have

$$\frac{\frac{dp}{p}}{dh} = -\frac{g_o}{RT} = -\frac{9.8}{(287)(216.66)} = -1.576\times10^{-4} \text{ Per meter}$$

Clearly, in the first gradient region, the pressure decreases at a faster rate as h increases.

In contrast, in the isothermal region, because T is constant in this region, the pressure decreases at the same rate with altitude; that is, from h = 11 km to h = 25 km, the value per meter; does not change with altitude.

However, examining the second gradient region in Figure 4, Where T increases with an increase in h, the pressure decreases at a slower rate as h increases.

Conclusion: The value of T determines the fractional rate of pressure change concerning the altitude at every altitude.

Illustration 3

What are the altitudes of pressure, temperature, and density of an airplane flying at an altitude where the absolute pressure and temperature are 4.72×104 N/m^{2}and 255.7 K, respectively?

Solution

Look in App. A for the standard altitude value that corresponds to the pressure altitude P = 4.72×104 N/m^2. This is 6000 m. Hence

Pressure altitude = 6000 m = 6 km

Look in App. A for the standard altitude value corresponding to the temperature altitude T = 255.7 K. This is 5000 m. Hence

Temperature altitude = 5000 m = 5 km

We must first calculate ρ the density altitude using the equation of state:

$$\rho = \frac{p}{RT} = \frac{4.72 \times 10^4}{287(255.7)} = 0.643 \text{ Kg / m}^3$$

Considering Appendix, A and interpolating from there 6.2 and 6.3 km, the standard altitude values that correspond to ρ = 0.643 Kg / m^3 is about 6240 m. Hence,

Density altitude = 6240 m = 6.2 km

It's important to note that temperature altitude isn't a found this very interesting. Because there are numerous values for the altitude-versus-temperature function, the result for temperature altitude might well be 5.0, 38.2, or 59.5 kilometers. Only the lowest temperature altitude value is utilized in this section.

Illustration 4

At an altitude that equated to a pressure height of, a level-flight maximum-velocity run was performed 30,000 feet and a density altitude of 28,500 feet at the same time, according to the flight test data for a specific airplane. Calculate the air temperature at the altitude at which the plane was flying during the test.

Solution

From Appendix B (Anderson & Bowden, 2005):
For pressure altitude = 30,000 ft

p = 629.66 lb/ft^2

For density altitude = 28,500 ft

ρ = 0.9408x10-3slug/ft^3

These are the p and ρ values that occurred at the same time at the altitude at which the plane was traveling. As a result of the equation of state,

$$T = \frac{p}{\rho R} = \frac{629.66}{(0.94082 \times 10^3)(1716)} = 390^o R$$

Illustration 5

The ambient temperature in the air ahead of an airplane in flight is 240 K. At what temperature altitude is the airplane flying?

Solution

The purpose of this example is to show the ambiguity of the use of temperature altitude. First, just examine Figure 4. Go to the abscissa and find T = 240 K. Then, simply cast your eyes upward. Within the scale of this figure, three different altitudes have a temperature of 240 K. Using App. These altitudes are (to the nearest entry) 7.4 km, 33 km, and (returning to Figure 4) about 63 km. Of course, the airplane cannot be at all three altitudes simultaneously. We conclude that the definition of temperature altitude has limited usefulness.

CONCLUSION

A standard atmosphere table, such as the ones included in Appendices A and B of this book (Anderson & Bowden, 2005)., will prove to be among the most useful references we have throughout our careers in aerospace engineering. It is necessary for calculating the performance of an airplane. It is necessary for a rational comparison of flight test information from different sources. It helps to put data from various wind tunnel facilities on a common basis.

Also, the equations used to compile the standard atmosphere can be programmed into our hand calculator, freeing us from having to read the tables. The tables, however, are particularly useful for carrying out "back-of-the-envelope" engineering calculations.

No table of the standard atmosphere existed at the time of the Wright brothers. They did not need one because all their work was done essentially at sea level. For their calculations of lift and drag, they did, however, need a value of the ambient air density.

This they had indirectly through a now-anachronistic empirical factor called "Smeaton's coefficient, "which was based in part on the value of sea-level density, as well as a fairly exact figure of Smeaton's coefficient, as measured by Samuel Langley at the Smithsonian Institution. (For more details, see John Anderson, Aerodynamics: A History and Its Impact on Flying Machines, New York: Cambridge University Press, 1997. By the time of World War, I, however, airplanes were regularly flying at altitudes of 10,000 ft and higher, and the lack of a standard table of the variation of atmospheric properties with altitude was becoming a real stumbling block for airplane designers. This prompted the big push for the compilation of standard atmospheric data.

The equations used for the compilation of the standard altitude tables for air, as developed in this chapter, are the same as used for the calculation of the properties throughout foreign planetary atmospheres. This should come as no surprise, as the physics underlying the calculation of atmospheric properties on earth is the same as on Venus, Jupiter, and so forth. Therefore, this chapter is relevant to space flight and the design of space vehicles.

Finally, we emphasize that the tables of the standard atmosphere in Appendices A and B did not simply come out of thin air. The values tabulated there were obtained from the application of physics, as embodied in the equation of state and the hydrostatic equation some of the important themes presented in this chapter are included below to help and simply reinforce this concept:

1. The standard atmosphere is established to provide a consistent reference point for flight testing, wind tunnel data, and general aerospace vehicle design and performance.

2. A temperature difference with altitude based on a mean of experimental data determines the standard atmospheric parameters. Using physics laws, the pressure and density changes with altitude are calculated from this empirical temperature fluctuation. The hydrostatic equation is one of several laws:

$$dp = -\rho g dh_G$$

3. The pressure and density changes in the isothermal zones of the standard atmosphere are given by

$$\frac{P}{P_1} = \frac{\rho}{\rho_1} = e^{-[g_o/(RT)](h-h_1)}$$

4. The pressure and density differences in the standard atmosphere's gradient zones are given by and, respectively.

$$\frac{P}{P_1} = \left(\frac{T}{T_1}\right)^{-\frac{g_o}{aR}}$$

$$\frac{\rho}{\rho_1} = \left(\frac{T}{T_1}\right)^{-\left(\frac{g_o}{aR}+1\right)}$$

Such that, $T=T_1+a(h-h_1)$ and the provided lapse rate is denoted by the symbol a.

5. The pressure altitude is equivalent to real in flight or the laboratory, ambient pressure is a factor to consider in the standard atmosphere.

For example, if a flow's ambient pressure is 393.12 lb/ft^2, independent of where it is or what it is doing, according to reports, the flow corresponds to a pressure height of 40,000 feet (see Appendix B). Density and temperature altitudes can be done in the same way.

Real Issues

1. The pressure, density, and temperature at 13 km in the standard atmosphere are 1.9399104 N/m^2, 3.119410-1 kg/m^3, and 216.66 K, respectively. Using these values, calculate the standard atmospheric pressure, density, and temperature at a height of 20 km, then verify them with the standard altitude tables.
2. Consider a plane that is flying at a real altitude. 2.65104 N/m^2 and 220 K, are the surface external pressure and temperature, correspondingly. What are the elevations of pressure and density?
3. A modern plane's pilot radios the ground to verify she's in level flight at a standard altitude of 30,000 feet during a flying test. What is the air pressure in the area where the plane is flying?

4. Consider a plane traveling at a higher altitude of 35,500 feet & 33,000 feet above density. Calculate the temperature of the outside air.

REFERENCES

Ananthasayanam, M. R., & Narasimha, R. (1987). A proposed international tropical reference atmosphere up to 1000 km. *Advances in Space Research*, *7*(10), 117–131. doi:10.1016/0273-1177(87)90084-6

Anderson, J. D., & Bowden, M. L. (2005). Introduction to Flight (6th ed.). McGraw-Hill Higher Education.

Cavcar, M. (2000). The international standard atmosphere (ISA). Anadolu University.

Chanthavane, S. L. (2021). *Space Radiation Interactions in Earth's Atmosphere, Low Earth Orbit, and Deep Space* (Doctoral dissertation). Rice University.

Daidzic, N. E. (2015). Efficient general computational method for estimation of standard atmosphere parameters. *International Journal of Aviation, Aeronautics, and Aerospace*, *2*(1), 3.

Daidzic, N. E. (2019). On atmospheric lapse rates. *International Journal of Aviation, Aeronautics, and Aerospace*, *6*(4), 2.

Jaganraj, R., Boopathy, G., & Varun, V. (2015). Experimental Investigation of Passive Flow Control on Bluff Bodies. *International Journal of Applied Engineering Research: IJAER*, *10*(8), 19793–19798.

Kaushik, M. (2019). The Standard Atmosphere. In *Theoretical and Experimental Aerodynamics* (pp. 3–25). Springer. doi:10.1007/978-981-13-1678-4_1

Kurzke, J., & Halliwell, I. (2018). Gas Properties and Standard Atmosphere. In *Propulsion and Power* (pp. 613–618). Springer. doi:10.1007/978-3-319-75979-1_14

Tsuchiya, T., Kanamori, M., & Takahashi, T. (2019). Numerical analysis of linear wave propagation in the atmosphere with temperature gradient for Mach cutoff reproduction. *Japanese Journal of Applied Physics*, *58*(SG), SGGB01.

Chapter 2
A Review on Various Aspects of Earth's Atmosphere

Surekha Rathi Samundi D.
https://orcid.org/0000-0001-9458-2125
Bharath Institute of Higher Education and Research, India

ABSTRACT

The atmosphere is the layer of gases that surrounds the planet. The atmosphere is retained by the gravity of the planet. Hence, it is also called planetary atmosphere. The performance of the aircraft and rockets depends on the physical properties of the atmosphere in which they fly. It is therefore advisable to study the variation of pressure, temperature, and density with the altitude. The real atmosphere is composed of dust, water vapor, and moisture, and it never remains constant. Hence, a hypothetical model called a standard atmosphere was employed. This chapter elaborates the international standard atmosphere, atmospheric boundary layer, and the stability of the atmosphere.

INTRODUCTION

Earth's atmosphere (Figure 1) contains water in all phases, which is considered as one of the essential factors for the development of life. The air in this model is assumed to be dry, clean and at rest (i.e., no wind or turbulence) (Talay, 1975) with respect to the earth. An internationally accepted model was introduced by the International Civil Aviation Organization (ICAO) in 1952 and is defined as ICAO standard atmosphere or International standard atmosphere (ISA). International standard atmosphere (ISA) is a reference atmospheric model that describes the changes of the ideal gas properties of earth's atmosphere (gaseous envelope that surrounds the planet) with the altitude and elevation. It has been designed to give a common reference for atmospheric properties at various altitudes. Indicators in Aircraft Instruments and the specifications of the performance of the aircraft are derived using ISA as a reference. Hence all the aircraft performance instrumentation is calibrated for the standard atmosphere. In the standard atmosphere, the sea level pressure is 101325 N/m^2, density is 1.225 kg/m^3, temperature is 15°C and the acceleration due to gravity is 9.81 m/s^2 (Calvert, 1990).

DOI: 10.4018/978-1-6684-4230-2.ch002

Figure 1. Atmosphere

PHYSICS OF ATMOSPHERE

Atmosphere and Aerodynamics

What is the atmosphere? Earth's atmosphere is commonly termed as air. Air is the mixture of several gases (Figure 2) that is primarily composed of 78.084% of Nitrogen (N_2) and 20.947% of Oxygen (O_2) and 0.977% of other gases by volume and 75.46% of Nitrogen and 23.19% of Oxygen and 1.355 of other gases by mass (Tomasi et. al, 1998). The other gases comprise small amounts of Argon (Ar), Carbon dioxide (CO_2), Neon (Ne), Helium (He), Krypton (Kr), Xenon (Xe), Hydrogen (H_2), Nitrogen monoxide (N_2O), Methane (CH_4), Ozone (O_3), Sulphur dioxide (SO_2), Nitrogen dioxide (NO_2) and Iodine (I_2) (Auld and Srinivas, 2008). Over a range of altitude, the proportion varies little, nevertheless the atmosphere may be regarded as a homogeneous gas of uniform composition. So, the gaseous envelope that surrounds the planet is the atmosphere.

Figure 2. Composition of atmosphere

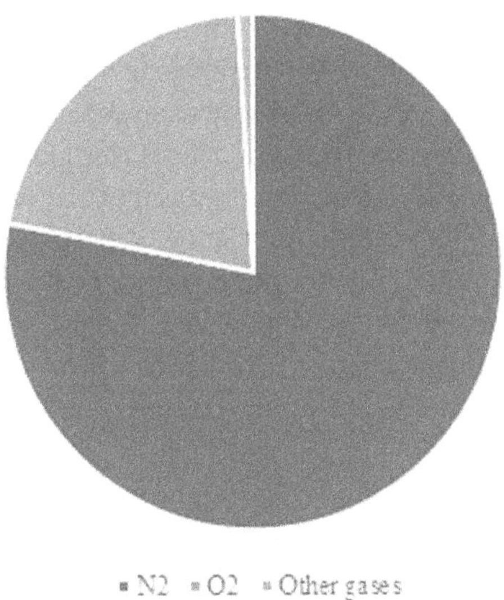

A substance that moves and deforms continuously as shear stress is applied is called fluid. Since air follows this definition, air is considered as fluid. As the aircraft operates in the air, the physical properties of the atmosphere that affects the performance of the aircraft must be understood. The study of the interaction between the moving object and atmosphere is Aerodynamics. Air flows past an object (Figure 3), changes the pressure exerted on the object. Also, the viscosity of the air leads to the existence of the shear force that resists the flow. As a result of these pressure and the shear force, the object experiences the aerodynamic forces (Lift and drag) and moment (Anderson and Bowden, 2005). Hence the understanding of interaction of the air with the moving object is important.

Figure 3. Flow past the airfoil and wedge

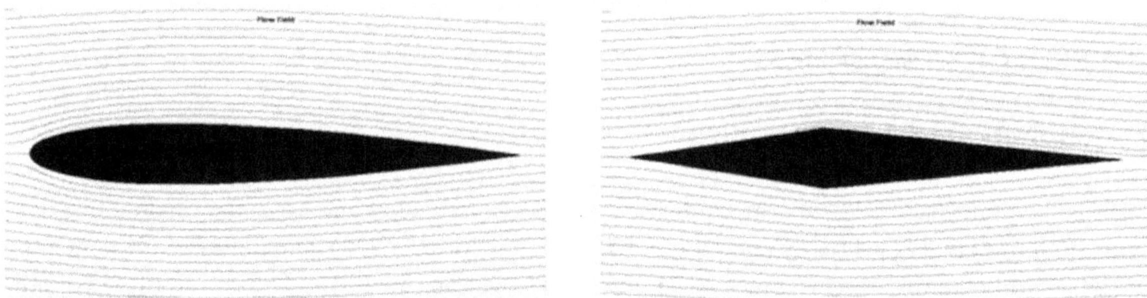

Structure of Atmosphere

As per the International Standard Atmosphere, the average atmospheric pressure at the sea level and the absolute atmospheric mass is 1.01325 X 10^5 Pa or 1 atm and 5.1480 X 10^{18} kg respectively. The value of atmospheric mass is about 2.5% less than would be inferred from the mean sea level and the earth's area of 51.0072 X 10^9 hectares. Atmospheric pressure is the total weight of the air above the unit area at the point where the pressure is measured. Thus, the atmospheric pressure depends on the weather and the location. The atmospheric mass of about 50% is distributed below 5.6 km, 90% of mass below 16 km and 99.9997% below 100 km (Calvert, 1990). An imaginary line at 100 km is termed as the Karman line. It marks the beginning of space i.e.; it separates the atmosphere and the space. (Note -The human travellers are called as Astronauts beyond the Karman line) North American X15, a X-Plane series of Experimental aircraft set an altitude record in 1963 by reaching the edge of outer space i.e., 108 km.

Based on the temperature's variation with altitude, the atmosphere can be divided into five major layers. By understanding the variation of temperature with the altitude, we can learn about how the atmosphere works. The layers are – from lowest to highest (Figure 4): Troposphere, Stratosphere, Mesosphere, Thermosphere and Exosphere.

Troposphere

The troposphere extends from the earth's surface to, on average, about 11 km in altitude, with its height higher at the equator and lower at the poles. Why does the temperature go down, as the altitude increases from the earth's surface within this layer? Because, most of the heat in the troposphere is generated by the transfer of the energy from the earth's surface. Most aviation takes place in this layer.

Stratosphere

The stratosphere is located between 11 km and 47 km above the earth's surface. It is the highest part of the atmosphere where the jet planes fly. The pilots prefer to fly in the lower layer of the stratosphere, because of the less air turbulence (Labitzke, 1972, Von Zahn et. al, 1998, Meriwether and Gerrard, 2004, Thayer and Livingston, 2008). In the stratosphere, the temperature increases with altitude. What is the source of the heat? The direct heat source is the sun. Ozone layer – converts UV energy into heat energy. Because of the ultra violet radiation, the temperature increases with the increase in altitude.

Figure 4. Layers of the atmosphere

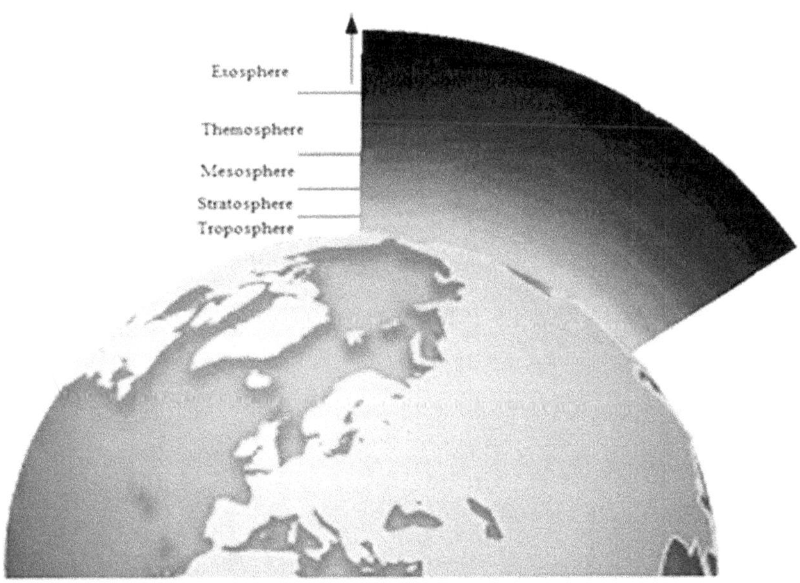

Mesosphere

This is located above the stratosphere and is extended up to 84.9 km above the earth's surface. The top layer of the mesosphere is Mesopause, the coldest part of the atmosphere with the temperature of -90° C. The absorption of the electromagnetic radiation from the sun is very low, since there are very few air molecules in this layer. Hence the temperature decreases with increase in altitude The air in the mesosphere is much thinner i.e., less dense than the stratosphere below. This thin air prevents the air from warming much.

Thermosphere

This is located between 84.9 km and 700 km above the earth's surface. The lowest part of the thermosphere is the ionosphere. The high energy X-rays and UV rays from the sun are absorbed in this layer, increasing the temperature of this layer. Akin to this, the low density of the molecules in the thermosphere also increases the temperature.

Exosphere

This is located between 700 km and 10,000 km above the earth's surface and is considered to be the final frontier of earth's gaseous envelope. This layer doesn't behave like a gas because the air in this layer is very very thin i.e., the density of the molecules is extremely low. Hence the particles from this layer escape into space.

Variation of Properties With Altitude in ISA

Atmosphere is a dynamically changing system. The standard atmosphere gives a mean value of the properties as a function of altitude. These values are obtained from the combination of experimental balloon and the sounding rocket measurements with the mathematical model of the atmosphere.

Figure 5. Variation of temperature with altitude (Source: Manual of ICAO Standard Atmosphere, 1994)

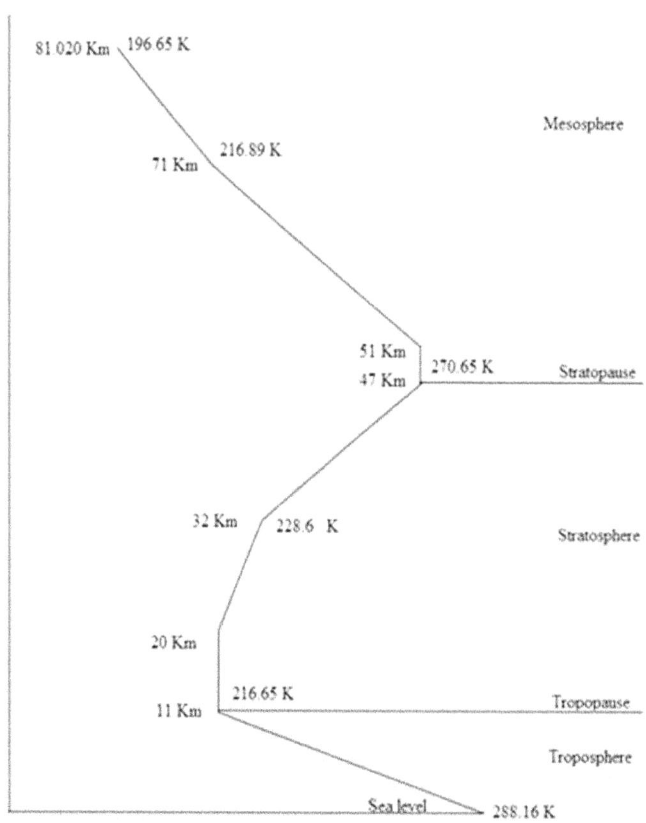

Among various ISA tables (Marvin (1925), Willis Ray Gregg (1922), Diehl (1925), ARDC (1959)), ICAO Standard Atmosphere, 1994 ISA table is followed here. The properties of the ICAO standard atmosphere have been calculated assuming the air to be a perfect gas and free from the dust and moisture. It is based on the values of temperature (Figure 5), pressure and density of the air at mean sea level. This standard atmosphere reflects the mean atmospheric condition to a reasonable degree.

The aerodynamic forces experienced by the aircraft vary with the atmospheric properties viz., pressure, temperature and density. The atmospheric properties vary with the altitude. Similarly, the gravitational force also varies with the altitude. Hence, it is necessary to specify the altitude explicitly.

Altitude is the vertical distance above the reference, but did you know the classification of the altitude? Altitude can be classified into (ICAO, 1993):

1. **Geometric altitude:** It is measured from the sea level. It is denoted by h.
2. **Absolute altitude:** It is measured from the centre of the earth. It is denoted by h_a.

$$h_a = h + r \quad (1)$$

3. **Geopotential altitude:** It is measured from the sea level that accounts for the variation of gravity with latitude and altitude. The local acceleration due to gravity depends on the altitude and the relation between the acceleration due to gravity (g_o) at the sea level and the acceleration due to gravity (g) at an absolute altitude is given by,

$$g = g_o \left[\frac{r}{h_a} \right]^2 \quad (2)$$

where r is the nominal radius of the earth, 6.356766 X 10^{16} m.

Relationship Between Geometric and Geopotential Altitude

The hydrostatic equation is given by,

$$dP = -\rho g dh \quad (3)$$

where, dP is the hydrostatic pressure and dh is the geometric altitude.

The above equation has to be integrated in order to obtain the hydrostatic pressure at a particular geometric altitude. Since, the density and the acceleration due to gravity are the function of altitude, the integration becomes more complex. In order to simplify this, the geometric altitude (h) has been replaced by the geopotential altitude (H). Thus, the modified hydrostatic equation is given by,

$$dP = -\rho g_o dH \quad (4)$$

By solving equations (1) and (5), we get

$$dH = \frac{g}{g_o} dh$$

$$\int_0^H dH \int_0^h \frac{r^2}{(r+h)^2} dh \text{ (since } h_a = h + r\text{)}$$

$$H = r^2 \int_0^h \frac{dh}{(r+h)^2}$$

$$= r^2 \int_0^h (r+h)^{-2} dh$$

$$= r^2 \left[\frac{-1}{r+h} \right]_0^h$$

$$\boldsymbol{H} = \frac{rh}{\boldsymbol{r}+\boldsymbol{h}}$$

This equation gives the relationship between geopotential and geometric altitude.

Modelling the Atmosphere

The plot of variation of temperature with the altitude is shown in Figure 5. As mentioned above, these values are obtained from the combination of experimental balloon and the sounding rocket measurements with the mathematical model of the atmosphere. A similar plot of variation of pressure and density with the altitude is not available. Hence, we need to obtain the pressure and density from the temperature-altitude plot.

For calculating the variations of properties with the altitude, the modified hydrostatic equation and the perfect gas laws have been used. Being static with the earth, the atmosphere is subject to acceleration due to gravity. The conditions of air static equilibrium are determined by the hydrostatic equation. Since the air is the combination of gases, it follows the laws of gases. The perfect gas law is given by,

$$P = \rho RT \tag{5}$$

where,

P is the pressure
R is the Specific gas constant (287.052 J/kg K)
T is the absolute temperature.

Considering the temperature variation into account, we categorize the layers of the atmosphere as isothermal layer and gradient layer. If the temperature remains constant with the geopotential altitude, it is termed as an isothermal layer and if the temperature varies linearly with the altitude, then it is termed as temperature gradient layer.

Gradient Layer

In this layer, the temperature varies with the geopotential altitude and the temperature variation is given by,

$$\frac{dT}{dH} = \frac{T - T_1}{H - H_1} \equiv \alpha$$

α is the lapse rate or vertical temperature gradient, T_1 and H_1 are the temperature and the geopotential altitude of the lower limit of the concerned layer.

$$T = T_1 + \alpha(H - H_1) \tag{6}$$

$$dH = \frac{1}{\alpha} dT$$

By substituting dH in the modified hydrostatic equation (4),

$$dP = -\frac{\rho g_o}{\alpha} dT$$

Figure 6. Gradient layer

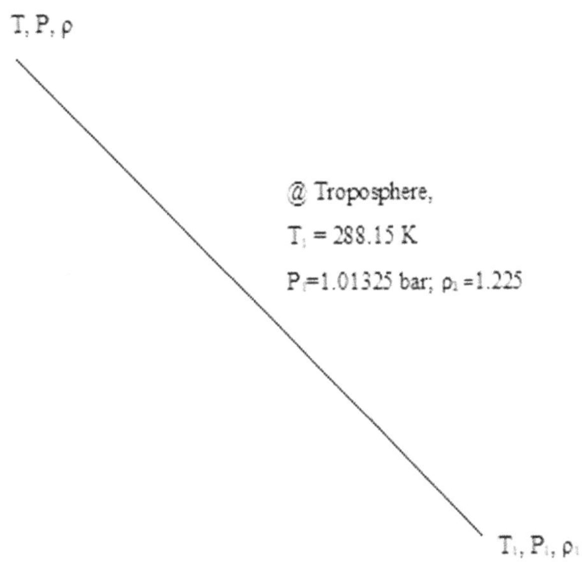

Now divide the above equation by equation (5),

$$\frac{dP}{P} = -\frac{g_o}{\alpha RT} dT$$

$$\int_{P_1}^{P} \frac{dP}{P} = -\int_{T_1}^{T} \frac{g_o}{\alpha RT} dT$$

$$\int_{P_1}^{P} \frac{dP}{P} = -\frac{g_o}{\alpha R} \int_{T_1}^{T} \frac{dT}{T}$$

$$ln\left(\frac{P}{P_1}\right) = -\frac{g_o}{\alpha R} ln\left(\frac{T}{T_1}\right)$$

$$\frac{P}{P_1} = \left(\frac{T}{T_1}\right)^{-\frac{g_o}{\alpha R}} \quad (7)$$

In order to obtain the value of density,

$$P = \rho RT \quad (8)$$

$$P_1 = \rho_1 RT_1 \quad (9)$$

$$\frac{P}{P_1} = \frac{\rho}{\rho_1}\frac{T}{T_1}$$

$$\frac{\rho}{\rho_1} = \frac{P}{P_1} \times \frac{T_1}{T}$$

Therefore, $\frac{\rho}{\rho_1} = \left(\frac{T}{T_1}\right)^{-\frac{g_o}{\alpha R}-1}$

$$\frac{\rho}{\rho_1} = \left(\frac{T}{T_1}\right)^{-\left(\frac{g_o}{\alpha R}+1\right)} \quad (10)$$

Equations (7) and (10) are used to calculate the pressure and density at the required altitude in the gradient layer.

Isothermal Layer

By dividing the modified hydrostatic equation by the perfect gas equation, we get

$$\frac{dP}{P} = -\frac{g_o}{RT}dH$$

$$\int_{P_1}^{P}\frac{dP}{P} = -\int_{H_1}^{H}\frac{g_o}{RT}dH$$

$$\int_{P_1}^{P}\frac{dP}{P} = -\frac{g_o}{RT}\int_{H_1}^{H}dH$$

$$ln\left(\frac{p}{P_1}\right) = -\frac{g_o}{RT}\left(H - H_1\right)$$

$$\frac{P}{P_1} = e^{-\frac{g_o}{RT}(H-H_1)}$$

Akin to the gradient layer, we obtain the density from the perfect gas equations,

$$\frac{\rho}{\rho_1} = \frac{P}{P_1}\times\frac{T_1}{T}$$

Since temperature is constant, $T=T_1$

$$\frac{\rho}{\rho_1} = \frac{P}{P_1} = e^{-\frac{g_o}{RT}(H-H_1)} \quad (11)$$

Equation (11) is used to calculate the pressure and density at the required altitude in the isothermal layer.

The standard atmosphere will always be reliable on a normal day, because the assumptions are not violated. But on a non-standard day, this model cannot be completely relied upon. During a non-standard day, the pressure, temperature, density and the other properties may vary. Hence, it is necessary to define new types of altitudes based on the standard atmosphere model.

Imagine that an airplane is flying at a particular geopotential altitude on a non-standard day. If you measure the air pressure at that altitude, it may be 61640 Pa (assume). From the ISA table, the corresponding altitude to the pressure of 61640 Pa is found to be 4 km. We may say this as, that the airplane is flying at the pressure altitude of 4 km. Thus, we can say that, the **pressure altitude** is the altitude an airplane would be at, if it was a standard day. Similarly, the **temperature altitude** is the altitude in the standard atmosphere which corresponds to the actual temperature encountered in flight. For example, if the temperature of the air, no matter where it is, is 265.4 K, then the altitude corresponding to this temperature is 3.5 km. Similarly, the **density altitude** is the altitude in the standard atmosphere which corresponds to the actual temperature encountered in flight.

ATMOSPHERIC BOUNDARY LAYER

ABL is the lowest part of the atmosphere. The contact of the atmosphere with the earth's surface has the direct impact on the atmospheric boundary layer. This portion of atmosphere is strongly influenced by the air which has been in direct contact with the surface. This height is termed as the thickness of ABL. Closest to the ground, within the troposphere, about 1-2 km from the surface lies the ABL. Hence all the momentum exchanges between the earth's surface and the atmosphere takes place through the boundary layer. This exchange may occur in mechanical and thermal forms. (Stull, 2015)

The mechanical contact between the air and surface arises from the friction exerted by the wind against the surface. This shearing between the wind and the surface creates turbulence. In the absence of thermal processes, the thickness of ABL depends on the level of turbulence in the lowest atmosphere.

The thermal contact between the air and surface has its source in the solar radiation. Since the air is largely transparent, we see long distances through the air. Thus, the solar radiation traverses the atmosphere and reaches the surface. In contrast, the land surface is opaque and hence it heats up and radiates the heat away. Since the earth's surface is much cooler than the sun, the frequency of the emitted heat from the earth is much lower, i.e., in the infrared range of frequency. Hence, this radiation is invisible and some portion of it escapes into space but the maximum portion is retained in the atmosphere because of the absorption of heat by water vapour, CO_2 and other gases (Avelar et.al 2012)

Convective Behaviour of ABL

During night, the sun is absent and it generally ends with a nocturnal boundary layer (NBL). The thickness of this layer depends on the prevailing wind velocity and the surface roughness. Above this layer, the air is stratified, because the maximum portion of the heat escapes to space during night.

The sun shines during the day and the atmosphere begins to be heated from the earth's surface. Convection occurs when the warm air rises and cool air sinks. The convective motion of the thermal process overtakes the wind shear turbulence of the mechanical process. Thus, the stratified layer created during the night is replaced by the convective boundary layer (CBL) by the process of penetrative convection.

As time passes, the CBL grows eventually and reaches the maximum height. At sunset, the convection stops because of the suppression of heat radiation from the earth's surface. The CBL becomes the undisturbed fossil mixed layer called the residual layer. The wind causes friction near the earth's surface and a new NBL is created.

ATMOSPHERIC STABILITY

Consider an air parcel (pocket of air). It is held in equilibrium by two forces: Buoyancy force and gravitational force. The buoyancy force lifts the air parcel upwards and the gravitational force pulls the air parcel downwards. It is the tendency of the atmosphere to either inhibit or promote vertical motion. (Stull, 2015)

Determination of Atmospheric Stability

Atmospheric stability measures the gravitational resistance of an atmosphere to vertical displacements. The stability of the atmosphere is determined appropriate to the vertical temperature and moisture in the troposphere. Consider an air parcel at the earth's surface. The air parcel in the sampled atmosphere does not exchange heat across the boundary, hence the process is adiabatic. As the air rises up, it will adiabatically expand and cool. The rate of cooling with the increase in altitude depends on whether the temperature reaches the dew point and consequent saturation.

Dry Adiabatic Lapse Rate

The temperature of the air parcel drops as it is lifted up. If the air parcel is not saturated i.e., if the relative humidity is less than 100%, then unsaturated air parcel cools at the constant rate of 9.75°C/km. The rate at which the air parcel cools is the lapse rate and since the air parcel is unsaturated, it is termed as dry adiabatic lapse rate. The environmental lapse rate as per ISA is 6.5° C/km. If the dry adiabatic lapse rate is greater than the environmental lapse rate, then the air parcel sinks, that is the vertical displacement is damped. It is considered stable. The environmental lapse rate greater than the dry adiabatic lapse rate favours the vertical displacement and is considered unstable. Atmospheric layer is neutrally stable, if the environmental lapse rate and the dry adiabatic lapse rate are same.

Moist Adiabatic Lapse Rate

As the parcel is lifted, it will cool at the dry adiabatic lapse rate until the saturation occurs. If the air parcel is saturated, i.e., if the temperature of the air parcel reaches the dew point temperature, then the water vapor will condense to form a cloud. During the process of condensation, the higher energy state of the water vapor is changed to the lower energy state of liquid water. During this phase change, the excess energy gets released in the form of latent heat. The latent heat of condensation is $2.5 \text{ X } 10^6 J/Kg$, which means $2.5 \text{ X } 10^6$ Joule of energy is released during the condensation of one kg of water vapor. As the latent heat released during the process of condensation is added to the air parcel, the air parcel will cool at a slower rate than the dry adiabatic lapse rate. It is known as the moist adiabatic lapse rate. The saturated air parcel cools at a constant rate of 4.5° C/km. This means that the air parcel cools at a slower rate (4.5° C/km) when it is saturated, than at the higher rate (9.75° C/km) when it is unsaturated.

If the moist adiabatic lapse rate is greater than the environmental lapse rate, then the temperature of the air parcel is cooler than the temperature of the environment i.e., the air parcel is dense than the surrounding air and hence it will not be able to rise. Hence the atmosphere is said to be absolutely stable.

If the moist adiabatic lapse rate is less than the environmental lapse rate, then the temperature of the air parcel is warmer (less dense) than the environment and the air parcel will lift off. Hence the atmosphere is said to be absolutely unstable. The level or the altitude at which the air parcel becomes warmer than the environment is called the Level of Free Convection (LFC). Above LFC, the air parcel will continue to cool at a moist adiabatic rate, and it accelerates upwards, no longer requiring an external lifting force.

If the environmental lapse rate lies between the dry and moist adiabatic lapse rate, then it is said to be conditionally unstable (Figure 7). During the saturated condition, the atmosphere would be unstable because the moist adiabatic lapse rate is less than the environmental lapse rate. In case of unsaturated condition, the atmosphere would be stable because the environmental lapse rate is greater than the dry adiabatic lapse rate.

Figure 7. Stability of the atmosphere

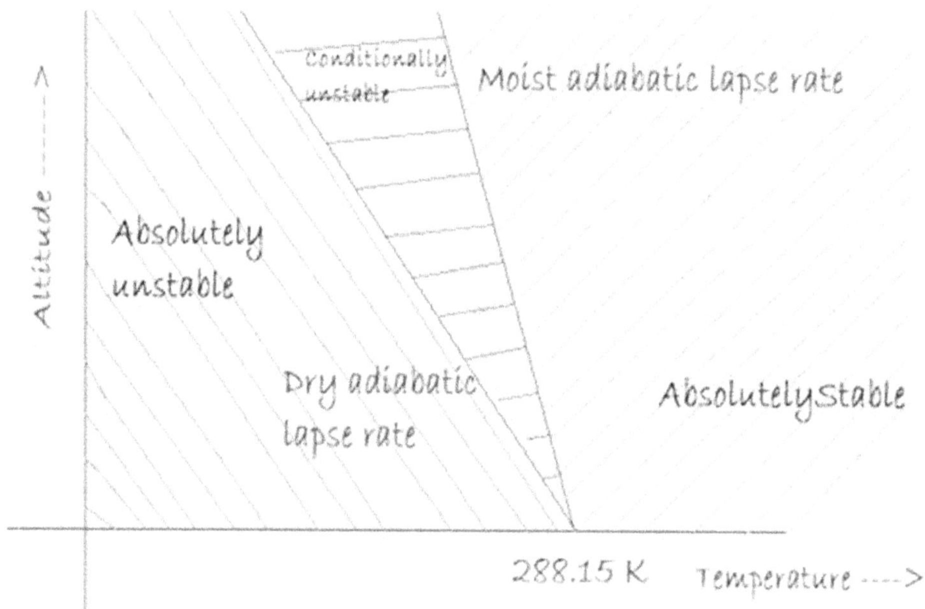

Atmospheric Instability and Cloud Development

Tendency of the atmosphere to either advance or obstruct vertical motion of the air, state the stability of the atmosphere. The air parcel has been forced to rise initially by certain methods, The air that rises by these methods, keeps rising further on its own account is instability. The air rises up from the earth's surface by various methods that include convection, convergence, orographic lifting and frontal lifting (Zhang, 2015) (Stull, 2015).

1. **Convection:** As the earth's surface is heated by the solar radiation, the warm air rises upwards (Figure 8). It refers to the motion of the air in the vertical direction.

Figure 8. Lifting by convection

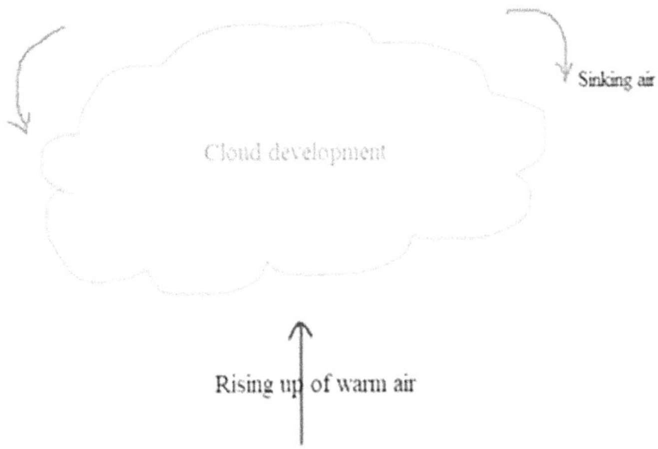

2. **Convergence:** Lifting of air by convergence is an atmospheric condition occurs when the air near the surface flows together and is pushed upwards i.e., convergence of air along the earth's surface (Figure 9) . When the air converges, it is forced to lift the entire layer of air. Large scale convergence may lift the air, across hundreds of kilometres. Lifting of convergence is typically much weaker than the lifting by convective process.

Figure 9. Lifting by convergence

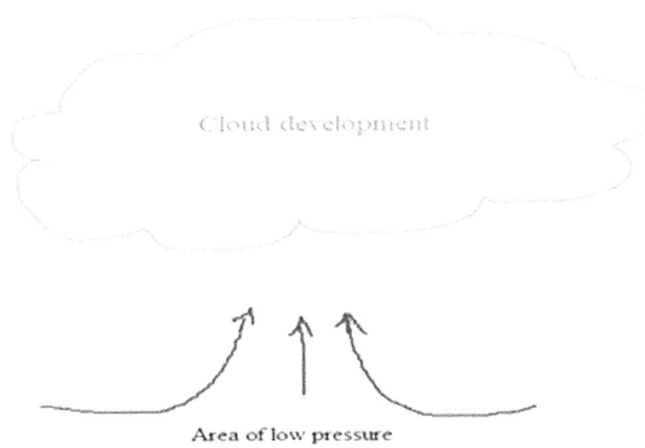

3. **Orographic lifting:** It is the process of lifting due to topography. When the air is resisted by a mountain, it is lifted up and flows along the slope of the mountain (Figure 10). The air cools as it rises and when the temperature of the air reaches the dew point temperature, the water vapor condenses and forms clouds.

Figure 10. Lifting due to topography and frontal lifting

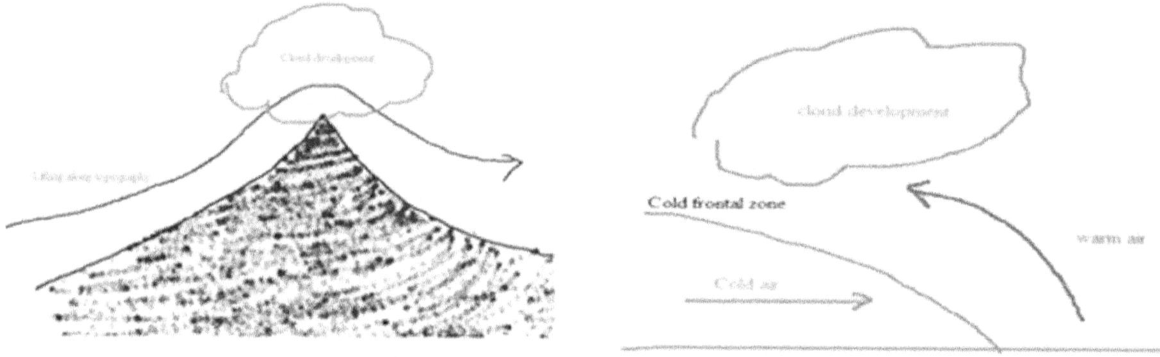

4. **Frontal lifting:** It refers to the lifting of air along the frontal boundaries. A front is the transition zone between the warm and cold air masses. If the cold air advances towards the warm air, then it is termed as cold front. The cold and dense air masses lift the warm air ahead of it. As it rises, the warm air cools and condenses to form clouds.

If the warm air moves towards the cold air, then it is called a warm front. The warm and less dense air rises up and flows over the cold air ahead of it. As this warm air rises, it cools and condenses and forms clouds.

Table 1. Atmospheric propertied in the international standard atmosphere

Geopotential altitude (m)	Temperature (K)	Pressure (Pa)	Density (kg/m³)
	Troposphere		
0	288.16	101325.0	1.2250
100	287.51	100129.4	1.2133
200	286.86	98945.2	1.2017
300	286.21	97772.4	1.1901
400	285.56	96610.9	1.1786
500	284.91	95460.6	1.1673
600	284.26	94321.4	1.1560
700	283.61	93193.2	1.1448
800	282.96	92076.0	1.1336
900	282.31	90969.7	1.1226
1000	281.66	89874.1	1.1116
1100	281.01	88789.3	1.1008
1200	280.36	87715.0	1.0900
1300	279.71	86651.4	1.0793
1400	279.06	85598.2	1.0686

continues on following page

Table 1. Continued

Geopotential altitude (m)	Temperature (K)	Pressure (Pa)	Density (kg/m³)
1500	278.41	84555.3	1.0581
1600	277.76	83522.8	1.0476
1700	277.11	82500.6	1.0372
1800	276.46	81488.5	1.0269
1900	275.81	80486.4	1.0166
2000	275.16	79494.4	1.0065
2100	274.51	78512.3	0.9964
2200	273.86	77540.0	0.9864
2300	273.21	76577.5	0.9765
2400	272.56	75624.7	0.9666
2500	271.91	74681.5	0.9569
2600	271.26	73747.9	0.9472
2700	270.61	72823.7	0.9375
2800	269.96	71909.0	0.9280
2900	269.31	71003.6	0.9185
3000	268.66	70107.4	0.9091
3100	268.01	69220.4	0.8998
3200	267.36	68342.5	0.8905
3300	266.71	67473.7	0.8814
3400	266.06	66613.8	0.8723
3500	265.41	65762.8	0.8632
3600	264.76	64920.6	0.8543
3700	264.11	64087.2	0.8454
3800	263.46	63262.5	0.8365
3900	262.81	62446.4	0.8278
4000	262.16	61638.8	0.8191
4100	261.51	60839.7	0.8105
4200	260.86	60049.1	0.8020
4300	260.21	59266.7	0.7935
4400	259.56	58492.7	0.7851
4500	258.91	57726.8	0.7768
4600	258.26	56969.1	0.7685
4700	257.61	56219.4	0.7603
4800	256.96	55477.8	0.7522
4900	256.31	54744.1	0.7441
5000	255.66	54018.3	0.7361
5100	255.01	53300.3	0.7282
5200	254.36	52590.0	0.7203

continues on following page

Table 1. Continued

Geopotential altitude (m)	Temperature (K)	Pressure (Pa)	Density (kg/m³)
5300	253.71	51887.5	0.7125
5400	253.06	51192.5	0.7048
5500	252.41	50505.1	0.6971
5600	251.75	49825.2	0.6895
5700	251.10	49152.7	0.6819
5800	250.45	48487.6	0.6745
5900	249.80	47829.8	0.6670
6000	249.15	47179.3	0.6597
6100	248.50	46535.9	0.6524
6200	247.85	45899.6	0.6452
6300	247.20	45270.4	0.6380
6400	246.55	44648.3	0.6309
6500	245.90	44033.0	0.6238
6600	245.25	43424.7	0.6168
6700	244.60	42823.1	0.6099
6800	243.95	42228.3	0.6030
6900	243.30	41640.3	0.5962
7000	242.65	41058.9	0.5895
7100	242.00	40484.0	0.5828
7200	241.35	39915.7	0.5762
7300	240.70	39353.9	0.5696
7400	240.05	38798.5	0.5631
7500	239.40	38249.5	0.5566
7600	238.75	37706.8	0.5502
7700	238.10	37170.3	0.5439
7800	237.45	36640.0	0.5376
7900	236.80	36115.9	0.5313
8000	236.15	35597.9	0.5252
8100	235.50	35085.9	0.5190
8200	234.85	34579.8	0.5130
8300	234.20	34079.7	0.5069
8400	233.55	33585.5	0.5010
8500	232.90	33097.1	0.4951
8600	232.25	32614.4	0.4892
8700	231.60	32137.5	0.4834
8800	230.95	31666.2	0.4777
8900	230.30	31200.6	0.4720
9000	229.65	30740.5	0.4663

continues on following page

Table 1. Continued

Geopotential altitude (m)	Temperature (K)	Pressure (Pa)	Density (kg/m^3)
9100	229.00	30285.9	0.4607
9200	228.35	29836.8	0.4552
9300	227.70	29393.0	0.4497
9400	227.05	28954.7	0.4443
9500	226.40	28521.6	0.4389
9600	225.75	28093.8	0.4335
9700	225.10	27671.3	0.4283
9800	224.45	27253.8	0.4230
9900	223.80	26841.5	0.4178
10000	223.15	26434.3	0.4127
10100	222.50	26032.1	0.4076
10200	221.85	25634.8	0.4026
10300	221.20	25242.5	0.3976
10400	220.55	24855.0	0.3926
10500	219.90	24472.4	0.3877
10600	219.25	24094.6	0.3829
10700	218.60	23721.5	0.3780
10800	217.95	23353.0	0.3733
10900	217.30	22989.3	0.3686
Tropopause			
11000	216.65	22630.1	0.3639
Lower Stratosphere			
11100	216.65	22276.1	0.3582
11200	216.65	21927.6	0.3526
11300	216.65	21584.5	0.3471
11400	216.65	21246.9	0.3417
11500	216.65	20914.5	0.3363
11600	216.65	20587.3	0.3311
11700	216.65	20265.2	0.3259
11800	216.65	19948.2	0.3208
11900	216.65	19636.1	0.3158
12000	216.65	19328.9	0.3108
12100	216.65	19026.5	0.3060
12200	216.65	18728.9	0.3012
12300	216.65	18435.9	0.2965
12400	216.65	18147.5	0.2918
12500	216.65	17863.6	0.2873
12600	216.65	17584.1	0.2828

continues on following page

Table 1. Continued

Geopotential altitude (m)	Temperature (K)	Pressure (Pa)	Density (kg/m³)
12700	216.65	17309.0	0.2783
12800	216.65	17038.2	0.2740
12900	216.65	16771.7	0.2697
13000	216.65	16509.3	0.2655
13100	216.65	16251.0	0.2613
13200	216.65	15996.8	0.2572
13300	216.65	15746.5	0.2532
13400	216.65	15500.2	0.2492
13500	216.65	15257.7	0.2453
13600	216.65	15019.0	0.2415
13700	216.65	14784.0	0.2377
13800	216.65	14552.8	0.2340
13900	216.65	14325.1	0.2304
14000	216.65	14101.0	0.2267
14100	216.65	13880.4	0.2232
14200	216.65	13663.2	0.2197
14300	216.65	13449.5	0.2163
14400	216.65	13239.1	0.2129
14500	216.65	13032.0	0.2096
14600	216.65	12828.1	0.2063
14700	216.65	12627.4	0.2031
14800	216.65	12429.9	0.1999
14900	216.65	12235.4	0.1967
15000	216.65	12044.0	0.1937
15100	216.65	11855.6	0.1906
15200	216.65	11670.1	0.1877
15300	216.65	11487.5	0.1847
15400	216.65	11307.8	0.1818
15500	216.65	11130.9	0.1790
15600	216.65	10956.8	0.1762
15700	216.65	10785.4	0.1734
15800	216.65	10616.7	0.1707
15900	216.65	10450.6	0.1680
16000	216.65	10287.1	0.1654
16100	216.65	10126.1	0.1628
16200	216.65	9967.7	0.1603
16300	216.65	9811.8	0.1578
16400	216.65	9658.3	0.1553

continues on following page

Table 1. Continued

Geopotential altitude (m)	Temperature (K)	Pressure (Pa)	Density (kg/m³)
16500	216.65	9507.2	0.1529
16600	216.65	9358.5	0.1505
16700	216.65	9212.1	0.1481
16800	216.65	9067.9	0.1458
16900	216.65	8926.1	0.1435
17000	216.65	8786.4	0.1413
17100	216.65	8649.0	0.1391
17200	216.65	8513.7	0.1369
17300	216.65	8380.5	0.1348
17400	216.65	8249.4	0.1327
17500	216.65	8120.3	0.1306
17600	216.65	7993.3	0.1285
17700	216.65	7868.2	0.1265
17800	216.65	7745.1	0.1245
17900	216.65	7624.0	0.1226
18000	216.65	7504.7	0.1207
18100	216.65	7387.3	0.1188
18200	216.65	7271.7	0.1169
18300	216.65	7158.0	0.1151
18400	216.65	7046.0	0.1133
18500	216.65	6935.8	0.1115
18600	216.65	6827.3	0.1098
18700	216.65	6720.5	0.1081
18800	216.65	6615.3	0.1064
18900	216.65	6511.8	0.1047
19000	216.65	6410.0	0.1031
19100	216.65	6309.7	0.1015
19200	216.65	6211.0	0.0999
19300	216.65	6113.8	0.0983
19400	216.65	6018.2	0.0968
19500	216.65	5924.0	0.0953
19600	216.65	5831.3	0.0938
19700	216.65	5740.1	0.0923
19800	216.65	5650.3	0.0909
19900	216.65	5561.9	0.0894
20000	216.65	5474.9	0.0880
Mid Stratosphere			
20100	216.75	5389.3	0.0866

continues on following page

Table 1. Continued

Geopotential altitude (m)	Temperature (K)	Pressure (Pa)	Density (kg/m³)
20200	216.85	5305.0	0.0852
20300	216.95	5222.1	0.0838
20400	217.05	5140.6	0.0825
20500	217.15	5060.3	0.0811
20600	217.25	4981.3	0.0798
20700	217.35	4903.6	0.0786
20800	217.45	4827.2	0.0773
20900	217.55	4752.0	0.0761
21000	217.65	4677.9	0.0748
21100	217.75	4605.1	0.0736
21200	217.84	4533.4	0.0725
21300	217.94	4462.9	0.0713
21400	218.04	4393.5	0.0702
21500	218.14	4325.2	0.0690
21600	218.24	4258.1	0.0679
21700	218.34	4191.9	0.0669
21800	218.44	4126.9	0.0658
21900	218.54	4062.9	0.0647
22000	218.64	3999.9	0.0637
22100	218.74	3937.9	0.0627
22200	218.84	3876.9	0.0617
22300	218.94	3816.8	0.0607
22400	219.04	3757.7	0.0597
22500	219.14	3699.6	0.0588
22600	219.24	3642.4	0.0579
22700	219.34	3586.1	0.0569
22800	219.44	3530.7	0.0560
22900	219.54	3476.2	0.0551
23000	219.64	3422.5	0.0543
23100	219.74	3369.7	0.0534
23200	219.84	3317.7	0.0526
23300	219.94	3266.6	0.0517
23400	220.04	3216.2	0.0509
23500	220.14	3166.7	0.0501
23600	220.23	3118.0	0.0493
23700	220.33	3070.0	0.0485
23800	220.43	3022.8	0.0478
23900	220.53	2976.3	0.0470

continues on following page

Table 1. Continued

Geopotential altitude (m)	Temperature (K)	Pressure (Pa)	Density (kg/m³)
24000	220.63	2930.5	0.0463
24100	220.73	2885.5	0.0455
24200	220.83	2841.2	0.0448
24300	220.93	2797.6	0.0441
24400	221.03	2754.7	0.0434
24500	221.13	2712.5	0.0427
24600	221.23	2670.9	0.0420
24700	221.33	2630.0	0.0414
24800	221.43	2589.7	0.0407
24900	221.53	2550.1	0.0401
25000	221.63	2511.1	0.0395
25100	221.73	2472.7	0.0388
25200	221.83	2434.9	0.0382
25300	221.93	2397.7	0.0376
25400	222.03	2361.0	0.0370
25500	222.13	2325.0	0.0364
25600	222.23	2289.5	0.0359
25700	222.33	2254.6	0.0353
25800	222.43	2220.2	0.0348
25900	222.53	2186.4	0.0342
26000	222.62	2153.1	0.0337
26100	222.72	2120.3	0.0332
26200	222.82	2088.1	0.0326
26300	222.92	2056.3	0.0321
26400	223.02	2025.0	0.0316
26500	223.12	1994.3	0.0311
26600	223.22	1964.0	0.0306
26700	223.32	1934.2	0.0302
26800	223.42	1904.8	0.0297
26900	223.52	1875.9	0.0292
27000	223.62	1847.5	0.0288
27100	223.72	1819.5	0.0283
27200	223.82	1791.9	0.0279
27300	223.92	1764.8	0.0274
27400	224.02	1738.0	0.0270
27500	224.12	1711.7	0.0266
27600	224.22	1685.9	0.0262
27700	224.32	1660.4	0.0258

continues on following page

Table 1. Continued

Geopotential altitude (m)	Temperature (K)	Pressure (Pa)	Density (kg/m³)
27800	224.42	1635.3	0.0254
27900	224.52	1610.6	0.0250
28000	224.62	1586.3	0.0246
28100	224.72	1562.3	0.0242
28200	224.82	1538.8	0.0238
28300	224.92	1515.6	0.0235
28400	225.01	1492.7	0.0231
28500	225.11	1470.2	0.0227
28600	225.21	1448.1	0.0224
28700	225.31	1426.3	0.0220
28800	225.41	1404.9	0.0217
28900	225.51	1383.7	0.0214
29000	225.61	1362.9	0.0210
29100	225.71	1342.5	0.0207
29200	225.81	1322.3	0.0204
29300	225.91	1302.4	0.0201
29400	226.01	1282.9	0.0198
29500	226.11	1263.7	0.0195
29600	226.21	1244.7	0.0192
29700	226.31	1226.1	0.0189
29800	226.41	1207.7	0.0186
29900	226.51	1189.6	0.0183
30000	226.61	1171.8	0.0180
30100	226.71	1154.3	0.0177
30200	226.81	1137.0	0.0175
30300	226.91	1120.0	0.0172
30400	227.01	1103.3	0.0169
30500	227.11	1086.8	0.0167
30600	227.21	1070.6	0.0164
30700	227.31	1054.6	0.0162
30800	227.40	1038.9	0.0159
30900	227.50	1023.4	0.0157
31000	227.60	1008.2	0.0154
31100	227.70	993.2	0.0152
31200	227.80	978.4	0.0150
31300	227.90	963.8	0.0147
31400	228.00	949.5	0.0145
31500	228.10	935.4	0.0143

continues on following page

Table 1. Continued

Geopotential altitude (m)	Temperature (K)	Pressure (Pa)	Density (kg/m³)
31600	228.20	921.5	0.0141
31700	228.30	907.8	0.0138
31800	228.40	894.3	0.0136
31900	228.50	881.0	0.0134
32000	228.60	867.9	0.0132
Upper Stratosphere			
32100	228.88	855.0	0.0130
32200	229.16	842.4	0.0128
32300	229.44	829.9	0.0126
32400	229.72	817.7	0.0124
32500	230.00	805.6	0.0122
32600	230.28	793.7	0.0120
32700	230.56	782.0	0.0118
32800	230.84	770.6	0.0116
32900	231.12	759.2	0.0114
33000	231.40	748.1	0.0112
33100	231.68	737.2	0.0111
33200	231.96	726.4	0.0109
33300	232.24	715.8	0.0107
33400	232.52	705.3	0.0105
33500	232.80	695.0	0.0104
33600	233.09	684.9	0.0102
33700	233.37	675.0	0.0101
33800	233.65	665.2	0.0099
33900	233.93	655.5	0.0097
34000	234.21	646.0	0.0096
34100	234.49	636.7	0.0094
34200	234.77	627.5	0.0093
34300	235.05	618.4	0.0091
34400	235.33	609.5	0.0090
34500	235.61	600.7	0.0089
34600	235.89	592.1	0.0087
34700	236.17	583.6	0.0086
34800	236.45	575.2	0.0085
34900	236.73	566.9	0.0083
35000	237.01	558.8	0.0082
35100	237.29	550.8	0.0081
35200	237.57	543.0	0.0079

continues on following page

Table 1. Continued

Geopotential altitude (m)	Temperature (K)	Pressure (Pa)	Density (kg/m³)
35300	237.85	535.2	0.0078
35400	238.13	527.6	0.0077
35500	238.41	520.1	0.0076
35600	238.69	512.7	0.0075
35700	238.97	505.4	0.0074
35800	239.25	498.2	0.0072
35900	239.53	491.2	0.0071
36000	239.81	484.2	0.0070
36100	240.09	477.4	0.0069
36200	240.37	470.6	0.0068
36300	240.65	464.0	0.0067
36400	240.93	457.5	0.0066
36500	241.21	451.0	0.0065
36600	241.50	444.7	0.0064
36700	241.78	438.4	0.0063
36800	242.06	432.3	0.0062
36900	242.34	426.2	0.0061
37000	242.62	420.3	0.0060
37100	242.90	414.4	0.0059
37200	243.18	408.6	0.0058
37300	243.46	402.9	0.0058
37400	243.74	397.3	0.0057
37500	244.02	391.8	0.0056
37600	244.30	386.3	0.0055
37700	244.58	381.0	0.0054
37800	244.86	375.7	0.0053
37900	245.14	370.5	0.0053
38000	245.42	365.4	0.0052
38100	245.70	360.3	0.0051
38200	245.98	355.4	0.0050
38300	246.26	350.5	0.0049
38400	246.54	345.6	0.0049
38500	246.82	340.9	0.0048
38600	247.10	336.2	0.0047
38700	247.38	331.6	0.0047
38800	247.66	327.0	0.0046
38900	247.94	322.6	0.0045
39000	248.22	318.1	0.0045

continues on following page

Table 1. Continued

Geopotential altitude (m)	Temperature (K)	Pressure (Pa)	Density (kg/m³)
39100	248.50	313.8	0.0044
39200	248.78	309.5	0.0043
39300	249.06	305.3	0.0043
39400	249.34	301.1	0.0042
39500	249.62	297.0	0.0041
39600	249.91	293.0	0.0041
39700	250.19	289.0	0.0040
39800	250.47	285.1	0.0040
39900	250.75	281.3	0.0039
40000	251.03	277.5	0.0038
40100	251.31	273.7	0.0038
40200	251.59	270.0	0.0037
40300	251.87	266.4	0.0037
40400	252.15	262.8	0.0036
40500	252.43	259.3	0.0036
40600	252.71	255.8	0.0035
40700	252.99	252.3	0.0035
40800	253.27	249.0	0.0034
40900	253.55	245.6	0.0034
41000	253.83	242.3	0.0033
41100	254.11	239.1	0.0033
41200	254.39	235.9	0.0032
41300	254.67	232.8	0.0032
41400	254.95	229.7	0.0031
41500	255.23	226.6	0.0031
41600	255.51	223.6	0.0030
41700	255.79	220.6	0.0030
41800	256.07	217.7	0.0030
41900	256.35	214.8	0.0029
42000	256.63	212.0	0.0029
42100	256.91	209.2	0.0028
42200	257.19	206.4	0.0028
42300	257.47	203.7	0.0028
42400	257.75	201.0	0.0027
42500	258.03	198.4	0.0027
42600	258.32	195.8	0.0026
42700	258.60	193.2	0.0026
42800	258.88	190.7	0.0026

continues on following page

Table 1. Continued

Geopotential altitude (m)	Temperature (K)	Pressure (Pa)	Density (kg/m³)
42900	259.16	188.2	0.0025
43000	259.44	185.7	0.0025
43100	259.72	183.3	0.0025
43200	260.00	180.9	0.0024
43300	260.28	178.5	0.0024
43400	260.56	176.2	0.0024
43500	260.84	173.9	0.0023
43600	261.12	171.6	0.0023
43700	261.40	169.4	0.0023
43800	261.68	167.2	0.0022
43900	261.96	165.0	0.0022
44000	262.24	162.9	0.0022
44100	262.52	160.8	0.0021
44200	262.80	158.7	0.0021
44300	263.08	156.7	0.0021
44400	263.36	154.6	0.0020
44500	263.64	152.7	0.0020
44600	263.92	150.7	0.0020
44700	264.20	148.8	0.0020
44800	264.48	146.8	0.0019
44900	264.76	145.0	0.0019
45000	265.04	143.1	0.0019
45100	265.32	141.3	0.0019
45200	265.60	139.5	0.0018
45300	265.88	137.7	0.0018
45400	266.16	135.9	0.0018
45500	266.44	134.2	0.0018
45600	266.73	132.5	0.0017
45700	267.01	130.8	0.0017
45800	267.29	129.1	0.0017
45900	267.57	127.5	0.0017
46000	267.85	125.9	0.0016
46100	268.13	124.3	0.0016
46200	268.41	122.7	0.0016
46300	268.69	121.2	0.0016
46400	268.97	119.6	0.0015
46500	269.25	118.1	0.0015
46600	269.53	116.6	0.0015

continues on following page

Table 1. Continued

Geopotential altitude (m)	Temperature (K)	Pressure (Pa)	Density (kg/m³)
46700	269.81	115.2	0.0015
46800	270.09	113.7	0.0015
46900	270.37	112.3	0.0014
Stratopause			
47000	270.65	110.9	0.0014
Temperature at sea level - 288.16 K			
Pressure at sea level - 101325 Pa			
Density at sea level - 1.225 kg/m³			
Specific gas constant, R = 287.052 J/kg K			
Standard acceleration due to gravity - 9.806 m/s²			

(Source: Atmospheric and Space Environments CoS, 2010).

REFERENCES

Anderson, J. D., & Bowden, M. L. (2005). *Introduction to flight.* McGraw Hill Book Company.

Avelar, A. C., Fisch, G., & Marciotto, E. (2012). Atmospheric Boundary Layer Simulation in a Wind Tunnel for Analysis of the Wind Flow at the Centro de LanÃ § amento de AlcÃĒntara. In *4th AIAA Atmospheric and Space Environments Conference* (p. 2930). Academic Press.

Calvert, J. G. (1990). Glossary of atmospheric chemistry terms (Recommendations 1990). *Pure and Applied Chemistry, 62*(11), 2167–2219. doi:10.1351/pac199062112167

CoS. (2010). *Guide: Guide to Reference and Standard Atmosphere Models* (AIAA G-003C-2010 (2016)).

International Civil Aviation Organization (ICAO). (1993). *Manual of the ICAO Standard Atmosphere: extended to 80 kilometres (262 500 feet)* (Vol. 7488). International Civil Aviation Organization. Atmospheric and Space Environments.

Labitzke, K. (1982). On the interannual variability of the middle stratosphere during the northern winters. *Kisho Shushi. Dai2shu, 60*(1), 124–139.

Meriwether, J. W., & Gerrard, A. J. (2004). Mesosphere inversion layers and stratosphere temperature enhancements. *Reviews of Geophysics, 42*(3).

Stull, R. B. (2015). *Practical meteorology: An algebra-based survey of atmospheric science*. Academic Press.

Talay, T. A. (1975). Introduction to the Aerodynamics of Flight (No. NASA-SP-367). Academic Press.

Thayer, J. P., & Livingston, J. M. (2008). Observations of wintertime arctic mesosphere cooling associated with stratosphere baroclinic zones. *Geophysical Research Letters, 35*(18).

Tomasi, C., Vitake, V., & De Santis, L. V. (1998). Relative optical mass functions for air, water vapour, ozone and nitrogen dioxide in atmospheric models presenting different latitudinal and seasonal conditions. *Meteorology and Atmospheric Physics*, *65*(1), 11–30.

Von Zahn, U., Fiedler, J., Naujokat, B., Langematz, U., & Krüger, K. (1998). A note on record-high temperatures at the northern polar stratopause in winter 1997/98. *Geophysical Research Letters*, *25*(22), 4169–4172.

Zhang, M. H. (2015). *Atmospheric Circulation and Wind Systems*. Wiley. doi:10.1002/9781118900116.ch9

Section 2

Incompressible Flow Aerodynamics

Chapter 3
Introduction to Two-Dimensional Inviscid Incompressible Flow

Kaliappan S.
https://orcid.org/0000-0002-5021-8759
Velammal Institute of Technology, Chennai, India

Raj Kamal M. D.
Velammal Institute of Technology, Chennai, India

Balaji V.
Loyola Institute of Technology, India

Socrates S.
Velammal Institute of Technology, Chennai, India

Andrii Kondratiev
O. M. Beketov National University of Urban Economy, Ukraine

ABSTRACT

Here, the authors explain the Magnus effect. The ball is deflected in the same direction as the rotation. The most common exposure and welcome statement of the Magnus effect is that a spinning object creates a vortex of fluid swirling around it. On the side where the movement of the vortex is in the same direction as the direction of the flow to which the object is exposed, the speed will increase. On the opposite side, where the directions are opposite, the speed will decrease. It is explained here, according to Bernoulli's principle, that the pressure is lower on the side with the greatest velocity, and therefore, there is an unbalanced force orthogonal to the flow of the fluid.

DOI: 10.4018/978-1-6684-4230-2.ch003

INTRODUCTION

The Magnus effect is an ancient phenomenon in fluid mechanics. Isaac Newton seems to have explained it first, (Cayzac et al., 2011) an overview of the Magnus effect of projectiles and missiles is given. The first part of the paper is dedicated to explaining the physical mechanisms that determine the Magnus effect. (Johnson, 1986) The Magnus effect has been shown to be well known and somewhat well understood with respect to the flight of spherical projectiles. (Kenyon, 2016) The equations for the magnus force on a rotating and translating solid cylinder in a liquid are created for two different liquid models. In both cases, the flow is stable, frictionless and no vortices are formed or emitted behind the cylinder. However, while Model 1 assumes freedom of rotation, Model 2 does not, but explicitly uses the centrifugal force acting on the curved streamline above the cylinder. The Magnus force in Model 2 is 15% greater, which is probably greater than can be explained by the approximations made in the model. Observation is needed to determine which model is closest to the truth. (de Sá, 2021) The stable, non-rotating flow of a uniform flow of a non-compressible, non-viscous liquid on a solid exerts no force on the solid. This work presents a useful derivation of this result, known as the D'Alembert's paradox. (Mohebbi and Sellier, 2014) This paper proposes a new method for implementing Kutta conditions in a non-viscous, incompressible flow (potential flow) with no rotation on the wing. In contrast to the common method, this method is not based on the panel method. It is based on a finite difference scheme formulated on a bounding grid using Type-Elliptic-Grid-Generation-Technique. The proposed algorithm is a novel and rapid way to implement the Kutta condition by computing the stream function on the wing surface with derived equations for the wing with both a finite trailing edge angle and a tapered trailing edge. Use the method. The results obtained show excellent agreement with the results of the analytical and panel methods, thus confirming the accuracy and accuracy of the proposed method (Tong et al., 2021) Fluctuations in wall shear stress (WSS) during the interaction of oblique shock waves with a flat turbulent boundary layer are investigated by direct numerical simulation (DNS) at Mach 2.25. The numerical results are in very good agreement with previous experiments and DNS data in terms of turbulence statistics, wall pressure, and skin friction. Fluctuating WSS properties such as probability density function (PDF), frequency spectrum, spatiotemporal correlation, and convection velocity are systematically analysed.

MAGNUS EFFECT

Magnus effect, creation of a sidewise force on a spinning cylindrical or spherical solid submergesin a fluid, when there is a relative movement between the spinning bodyand the fluid.

A spinning frame flowing via a fluid deviates from it's directlydirectiondue tostrainvariations that generate withinside the fluid because ofpaceadjustmentsprecipitatedwith the aid of using the spinning frame. The Magnus impact is a selected manifestation of Bernoulli's theorem: fluid strain decreases at factorswherein fluid pace increases. In the case of a rotating ball withinside the air, the rotating ball includesa fewair with it. Viewed from the ball's position, air rushes in all directions. Drag at theaspect of the ball that spins withinside the air (with inside thecourse the ball travels) slows down the float of air, at the same time as drag on the alternativeaspecthastens the float of air. Higher strainat theaspectwherein airflow decreases forces the ball in the direction of the low-strainarea on the otheraspect, whereinthere's a relative boom in airflow.

Figure 1.

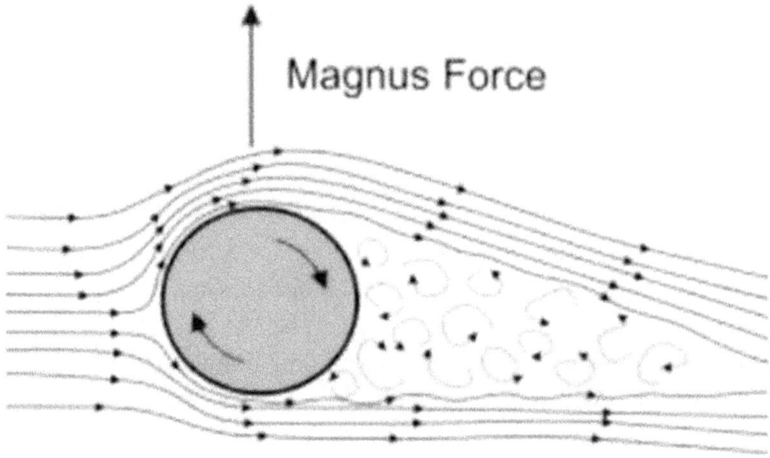

For two alternative fluid models, a formula for the Magnus force on a rotating and translating solid cylinder in a fluid is constructed. The flow is constant and frictionless in both circumstances, with no eddies forming or shedding behind the cylinder. Model one, on the other hand, is based on the assumption of irrotationality, whereas model two does not, instead relying on the centrifugal force acting on the curving streamlines above the cylinder. Model two's Magnus force is 15% greater in magnitude, which is likely more than can be explained by approximations within the models (Kenyon, 2016) A weak asymmetry between the left and right sides is caused by spin. The projectile is pushed to the right by the spinning wall, which deviates the transversal free stream to the left (Cayzac et al., 2011) With spinning objects, the Magnus effect can be plainly visible as air is dragged rapidly around one side. Magnus effect is exemplified by the air swings seen in sports such as soccer and baseball. On the boundary layer of yawing and spinning projectiles, a weak asymmetry is found, but only at low incidences. The high spin creates isolated vortex effects at high incidences. The lateral force produced by these vortex asymmetries adds up to the total Magnus effect. The insertion of fins may be a way to alleviate these lateral forces. The Magnus effect is used commercially in rotor ships, where propulsive force is generated by airflow around enormous revolving cylinders (Virk and Virk, 2018)

Application of Magnus Effect

The carry of magnets is a very vital case, On the whole study of the physics of many ball-related sports. In addition, it is essential in the field of defense, where this effect helps to examine the effects of rotating guided missiles. On the other hand, it has some engineering applications, especially in the design of aircraft and motor boats.

The Magnus impact is in particular utilized in video games which include football, golf, cricket, tennis, baseball and plenty of more. This idea is crucial in know-how the physics in the back of many ball sports.

Some plane are constructed the use of the Magnus impact to raise the use of a rotating cylinder with insidethe front of the wing, permitting them to fly at decrease horizontal speeds

Used in outside ballistics. The aggregate of the lateral wind element reasons a magnetic pressure to be implemented to the bullet.

Rotor ships use flatter rotors that are mast-like cylinders installed vertically at the deck of a ship. These assist the propulsion. With the wind beginning from the side, anahead thrust is created because of the magnus impact.

The Magnus effect has been known for a long time, although it has just a few applications. The first commercial application was in 1924, when a ship with large cylinders instead of sails was built in Germany (Virk and Virk, 2018).

Figure 2.

Application of Magnus Effect in Sports

Here we explain the Application of Magnus Effect in Sports

Baseball

All principal league pitchers concentrate on a Magnus impact event, even thoughthey willnow no longer-recognise it. Throwing Bent Balls, Thumb, and Slippery Balls - Magnus does it all. If a glassdesires to fold a ball of dough, he sends it out very fast, spinning it to the edges that spin in the direction of the dough. The ensuing magnetic pressuremovements the ball out of its linear route and far from the bar of mass.

Golf

The excellent golfers with inside theinternationalcan't hit extra than 250 yards without the Magnus effect. Upon impact, it creates a rotating wedge-fashioned impeller in a golfing ball. The motive force makes the ball spin (regularly backwards) at over 4000rpm at top notch speed! The form of the golfing ball additionally encourages an extended stroke, developingmuch less turbulence across the ball while it flies with inside the air. These factors, the roll of the again and the form of the ball, significantlyhave an effect on a golfer's capacityto persuade the ball.

Tennis

Tennis gamers like baseball and golfers use the Magnus impact to convey out the greatof their opponent. A tennis ball with aexcessive coefficient of floor friction and comparatively low mass can bend critical-lywhile struck correctly. In fact, Isaac Newton changed Heinrich Magnus' concept after looking tennis healthy in 1672. Hitting a tennis ball at an attitude with the racket reasons the ball to spin, inflicting it to alternatepathbecause it travels.

Modern Application of Magnus Effect

The Magnus Spherical Airship

This space-age flying automobilechanged intoconstructed as a prototype in 1982 and changed into re-named the Magnus round aircraft. This lighter-than-air craft created a big payload with the aid of using spinning its large ball at the relationshipfactors on the perimeters of the ship. The elevator created with the aid of using the rotating ball may be used to moveitems and people.

Magenn Air Rotor System (MARS)

This flying electric powered generator is referred to as the Magenn Air Rotor System (MARS). The drum frame of the system rotates with the wind, even as the mills rotate at the relationshipfactorspro-ducing electricity. "Helium (inert and inert lighter than fuel line in air) maintains the air rotor excessive for the fine winds, and its rotation additionallyreasons the Magnus effect. This offersextraliftmaintains the toolstrong and holds it in position. In a managed and limited place, and reasons it to upward thrust above your head rather than drifting withinside the blowing wind.

Ballistics of Paintball

Advanced long-range paintball barrels eject the paintball while spinning backwards. Like throwing the cursor at a baseball, spinning the paintball backwards creates a magnus effect, causing the ball to fly long distances. The magnus effect is proportional to the speed of the bullet, so eventually when the paintball slows down it loses its effect and falls to the ground. These modern barrels can double the range of a paintball bullet

Alembert's Paradox

A theorem of fluid mechanics, which states that no pressure acts on an item at regularspeed in aim-mediately line via a big mass of incompressible and faulty fluid this isto start withshifting at rest (or in uniform motion). This reputedly paradoxical assertionmay be understood via way of means ofinformation that non-viscous beverages do now no longer exist. If such fluids existed, there could be no innerbodily mechanism to use upelectricity into heat. Therefore, no pressuremay becarried out to the body, due to the factpaintingswith inside the fluid is completedwithoutgrowing the internetelectricity of the fluid. This paradox stems from the capabilitygo with the driftprinciple of an incompressible and invisible vortex loosego with the drift. According to the principle of capability flux, numerousbasic currents may be superimposed on everydifferent to generate complicated currents.

For example, the aggregate of easygo with the drift and double go with the driftoffersgo with the driftwithoutcarry in a round cylinder, as shownbelow

Figure 3.

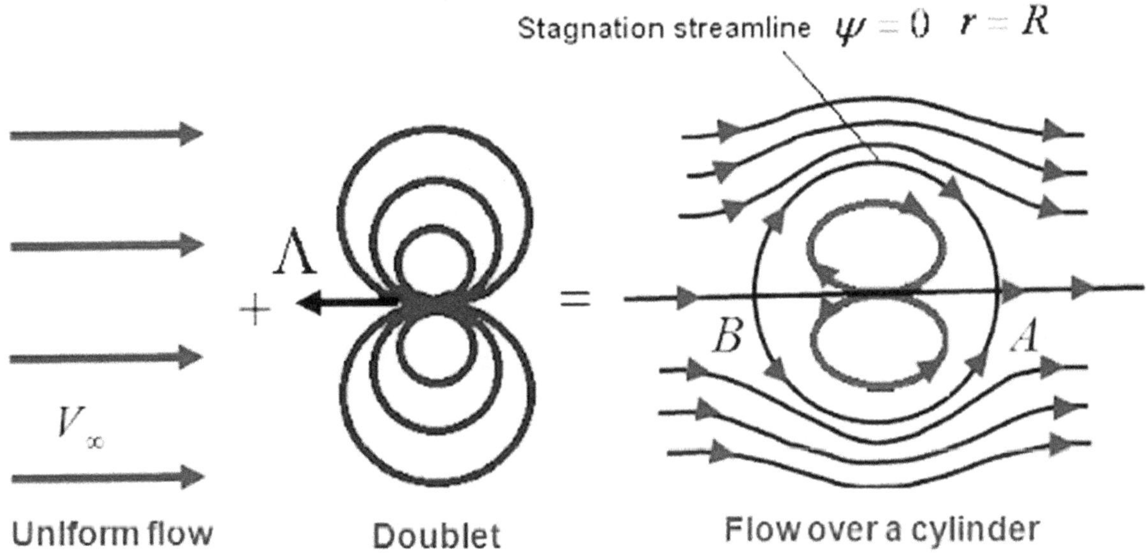

From the figure above, it is clear that the flow in the upper and lower halves of the cylinder is regular and symmetrical, with no lift. Similarly, the flow is symmetrical in the left and right halves of the cylinder, resulting in zero resistance. Does this imply that the flow of a viscous incompressible fluid through the cylinder creates no resistance? If so, the potential theory must be flawed. Consider another instance of upward flow in a cylinder. Perhaps there will be some resistance if there is a finite lift.

This flow of time is not symmetrical in the upper and lower halves of the cylinder and assumes that the lifting force will be finite. What about resistance? In the left and right halves of the cylinder, the flow remains symmetric. The pressure distribution at the front of the cylinder is precisely balanced by the pressure distribution at the rear of the cylinder. Of course not, the resistance is still zero.

Prandtl argued that the adhesion boundary condition produces a considerable shift in flow velocities in a narrow layer near the body wall at high velocities and high Reynolds numbers. This causes vorticity to occur in the boundary layer, as well as viscous kinetic energy dissipation. The separation of solid body streams is caused by the dissipation of energy, which is absent from all viscous theories. Due to the viscous shear stress in the wall, the low pressure in the wake creates form resistance, which can be larger than frictional resistance.

Figure 4.

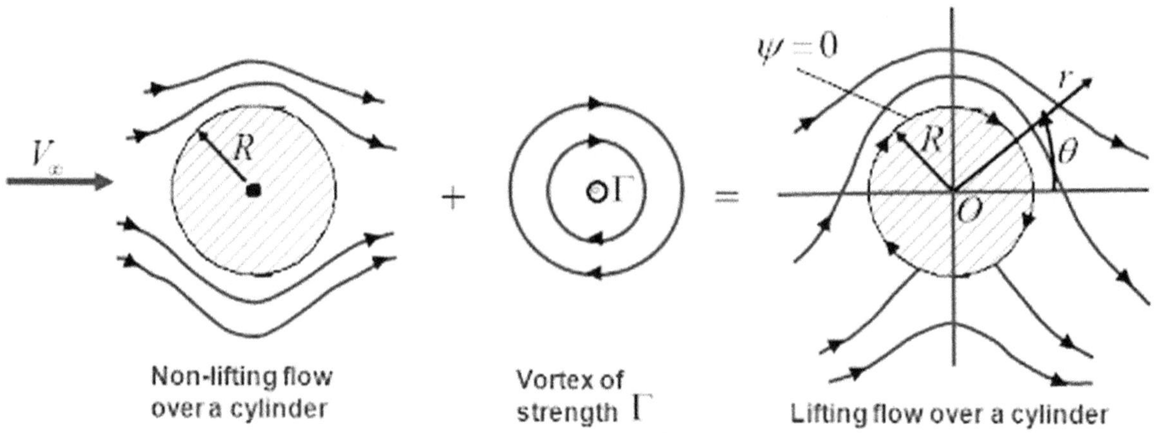

The actual flow in the cylinder is about the same due to the thin boundary layer and flow split. As shown in the figure below

Figure 5.

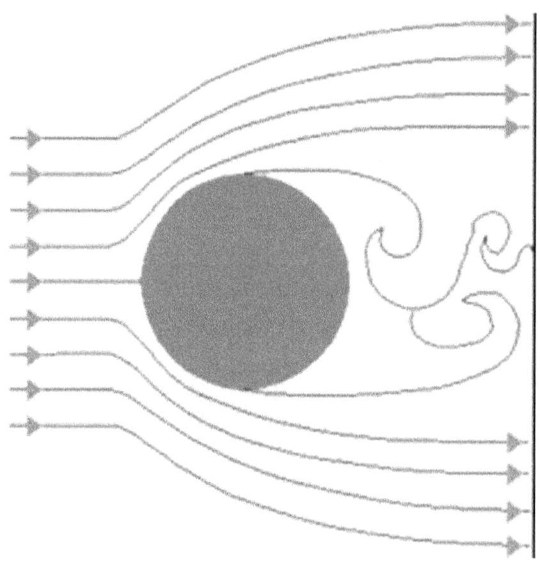

Prandtl's theory of thin boundary layers is a breakthrough in fluid mechanics and aerodynamics. Surprisingly, the thin layer is actually responsible for significant body resistance, while the flow out of this layer does not affect resistance. Thus, Prandtl was able to solve a paradox that lasted more than a century and revolutionized fluid mechanics, making it the father of modern aerodynamics.

A uniform stream of incompressible inviscid fluid flowing in a steady irrotational flow past a solid body exerts no force on it. We offer an instructive derivation of this finding, d'Alembert's paradox (de Sá, 2021) Propose a solution to d'Alembert's paradox by comparing the observation of significant drag/lift in fluids with low viscosity, such as air and water, with the mathematical prediction of zero drag/lift of stationary irrotational solutions of the incompressible inviscid Euler equations, known as potential flow (Hoffman and Johnson, 2010)

KUTTA CONDITION

The Kutta Condition is a principle of constant flow hydrodynamics that can be applied to a variety of scenarios. The term "aerodynamics" refers to stiff bodies with sharp corners, such as the trailing edges of aerodynamic surfaces.

Irrotational, inviscid, incompressible flow (potential flow) over an airfoil is a Kutta condition. This method is not based on the panel method, contrary to popular belief. It uses an O-type elliptic grid generation technique to formulate a finite difference scheme on a boundary-fitted grid (Mohebbi and Sellier, 2014)

The Kutta Condition Applied to Airfoils

If a clean symmetrical body, including a cylinder with an oval cross-section, movementsthru the fluid without an attitude of attack, it does now no longer create carry. There are factors of stagnation withinside the body: one withinside thefront and one withinside the back. When the oval cylinder movementsthru the liquid at a non-0attitude of attack, there are nonetheless stagnation factorsat the body: one at the lowest of the cylinder, close to the main edge; and the opposite is on thepinnacle of the cylinder, close to the trailing edge. The circulateround this clean cylinder is 0 and no carry is generated in spite of the advantageousattitude of attack.

If a profile with a sharp trailing edge moves through the air with a positive angle of attack, the two points of stagnation are on the underside near the leading edge, as in the case of an oval cylinder; and up near the trailing edge. When the air passing through the underside of the wing reaches the trailing edge, it must pass around the trailing edge and along the top of the wing until it stagnates at the top of the wing At the trailing edge, a vortex flow forms, and because the radius of the sharp trailing edge is zero, the air speed travelling around the trailing edge must be infinitely high! Real fluids can't go indefinitely fast, although they can go very fast. Because of the high airspeed near the trailing edge of the wing, strong viscous forces in the air near the trailing edge cause a powerful vortex to form on the top of the wing near the trailing edge. This vortex, known as the beginning vortex, is carried away by the wing profile as it moves.

To confirm the existence of embryonic eddies in fluids, aerodynamic pioneers were able to picture them. The initial vortex, which is on the top of the wing towards the trailing edge, attracts air via the top of the wing, and the core of the vortex has a very low pressure. The top of the profile's stagnation points are also drawn in the direction of the initial vortex. The top stagnation point reaches the trailing edge after the airfoil has travelled little through the air, and the original vortex is ejected away from the airfoil and continues spinning in the air where the airfoil left it. Due to viscous forces, the original vortex dissipates quickly.

The Kutta Condition in Aerodynamics

One result of the Kutta circumstance is that the airflow thru the pinnacle of the wing is a good dealquicker than the airflow below the bottom. An air packet drawing near the wing alongside the stagnation go with the drift line will break up into elementson the stagnation factor: one 1/2 of will by skipthru the pinnacle and the alternative1/2 ofalongside the bottom. The go with the driftthru the pinnacle is a good dealquicker than the go with the driftalongside the bottom, so the 2 halves will by no means meet once more. They do not even reconnect after the profile has long gonethru. This is now and againknown as a "break up. There is a famousfalse impressionreferred to as the "sametour time" fallacy, which states that the 2 halves come collectivelyonce moreon the trailing fringe of the profile. This false impression contradicts the phenomenon of separation, which has turn out to becleanbecause the discovery of Martin Kutta.

When the wing's velocity or attack angle changes, a vulnerable preliminary vortex forms above or beneath the trailing section. This vulnerable preparatory vortex combines with the Kutta condition's internal cooperation to create a new attack speed or attitude. This alters the airfoil stream, causing it to carry in response to changes in the rate or inclination of the assault. The Kutta situation explains why the wings always have sharp trailing edges, despite the fact that this is undesirable from a layout and production standpoint. A plane with a trailing wing that is softly rounded produces relatively little carry.

Figure 6.

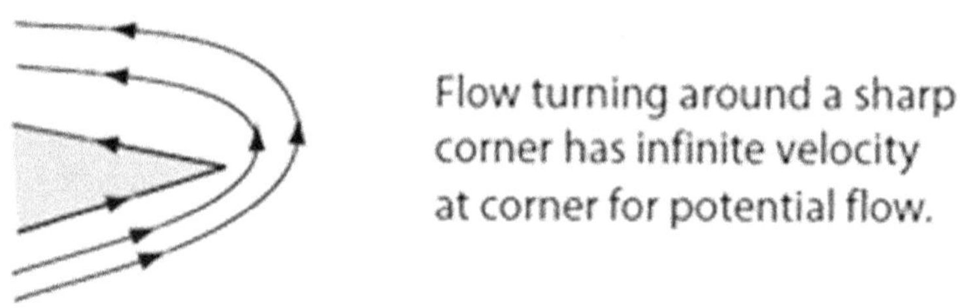

Figure 7.

CONTINUUM MECHANICS

The study of liquid elements depicts the movements of fluids and gases and their Association with strong bodies. There are numerous ways of assisting partition liquid element into exceptional subjects. Here made a separation into compressible and incompressible streams. Compressible streams are those in which there are changes in the liquid thickness are significant. A significant strength worried about compressible streams, gas elements, manages fast streams where thickness changes are enormous and wave peculiarities happen every now and again.

Incompressible streams, of either gases or fluids, are streams where thickness changes in the liquid are not a significant piece of the material science. The investigation of incompressible stream incorporates such subjects as water power, hydrodynamics, grease hypothesis, optimal design, and limit layer hypothesis. It likewise contains foundation data for such unique subjects as hydrology, delineated streams, disturbance, turning streams, and organic liquid mechanics. Incompressible stream not just involves the focal situation in liquid elements but on the other hand is principal to the commonsense subjects of hotness and mass exchange.

Continuum Assumption

Liquid mechanics, strong mechanics, electrodynamics, and thermodynamics are for the most part models of actual sciences in which the world is considered to be a continuum. The continuum presumption just implies that actual properties are envisioned to be disseminated all through space.

Each point in space has a limited set of attributes, such as speed, temperature, stress, and the strength of the electric field. The attributes may change esteem as you move from one point to the next, and there may even be surfaces where a few properties jump spasmodically. The interface between a solid and a liquid, for example, is envisioned as a surface where the thickness bounces from one value to the next. The continuum presupposition, on the other hand, prevents attributes from becoming unbounded or unclear at a single isolated point.

Sciences that hypothesize the presence of a continuum are basically perceptible, sciences and arrangement, generally talking, with occasions that might be seen with the independent eye. Occasions in the minuscule universe of atoms, cores, and rudimentary particles are not represented by continuum laws, nor are they depicted as far as continuum thoughts.

Nonetheless, there is an association between the two perspectives. Continuum properties may be deciphered as midpoints of occasions including an incredible number of infinitesimal particles. The development of such a translation falls into the disciplines of factual thermodynamics (measurable mechanics) and dynamic hypothesis. Now and again we will examine some of the more straightforward tiny models that are utilized for continuum occasions. This guides in a more profound comprehension of continuum properties, yet not the slightest bit does it make the thoughts ''more genuine.'' The basic suppositions of continuum mechanics remain without anyone else without reference to the minute world.

WHAT IS THE MEANING OF INCOMPREHENSIBLE FLOW?

Incompressible stream is a primary region of liquid mechanics. It incorporates inside its limits a considerable number of issues and peculiarities that are found in designing and nature.

Streams of gases, just as those of fluids, are often incompressible. Laypeople are normally astonished to discover that the example of the progression of air can be like that of water.

From a thermodynamic point of view, gases and fluids have very various attributes. As we probably are aware, fluids are frequently displayed as incompressible liquids. In any case, incompressible liquid is a thermodynamic term, while incompressible stream is a liquid mechanical term. We can have an incompressible progression of a compressible liquid. The principle model for incompressible stream is that the Mach number below (M ® 0).

This is a fundamental condition. Also, different conditions concerning heat move should be fulfilled. There are a few distinct circumstances of hotness move under which incompressible stream can happen. In this part a definite investigation of these circumstances will be made.

As part of this review we infer the conditions that oversee incompressible stream. Then, at that point, we will be in a situation to notice a portion of the overall elements and qualities of incompressible stream.

Incompressible flow refers to a flow in which the density in each liquid packet remains constant; H. in every infinitesimal volume of fluid moving in the flow. This type of flow is also known as isochoric flow, which means "equal space / area". It is important to highlight the difference between an "incompressible flow" and an "incompressible fluid": while the former is a property of the flow, the latter is a property of the material. Liquids are compressible from the start, but many are considered incompressible because the density variation is negligible for common applications. Incompatibility is a property that every liquid exhibits under certain conditions.

Dependence on Material Property

The conditions for which an incompressible stream happens shift contingent upon the material properties of the liquid. Any liquid can be considered as incompressible under specific conditions on speed and strain. The propensity of a liquid to change volume when exposed to an outside load relies upon the actual material

Dimensional Analysis

Although the presence of estimation as a strategy and methodology for assigning a number to a variable was assumed, we investigated the estimation and layered structure of actual components in this study. The significant reality used in layered investigation is that the actual factors have no standard or central units of measurement

The charge of an electron, Planck's constant, the gravitational constant between drawing in masses, the speed of light, and so on are all well-known constants in material science. These constants, however, are not applicable to every single actual transaction. The charge on an electron is not a necessary unit for measuring flow in an electric engine. The speed of light is unquestionably not the most important unit for determining the speed of a water wave.

We are forced to create our own scales since we lack any universally important estimation units in the real world. Because our estimating scales are self-assertive advances, they cannot play a significant role in actual cycles. If we modify the size of the length unit, all factors, including length, should increase or decrease in a timely manner. It should be feasible to work on an issue and track down substantial data only by considering the layered part of it. This should be possible regardless of whether the topic is too complex to even consider allowing us to study it thoroughly.

COMPRESSIBLE FLOW

To arrive at our objective of getting when a stream might be considered as incompressible, we need to know a portion of the attributes of compressible streams. Our principle intention is to concentrate on the progression of a liquid that has totally subjective thermodynamic conditions

We will concentrate on three issues. Compressible, Couette stream, a cylinder swaying toward the finish of an extremely long cylinder, and compressible stream with little cross-segment region changes. The compressible Couette stream issue shows how a shear speed profile produces nuclear power by gooey scattering. The temperature slopes from this interaction then, at that point, lead to thickness varieties. It is critical to know the scales and boundaries that influence this cycle.

From a pragmatic outlook, the compressible Couette issue is amodel of how hotness is produced in a course. A similar impact happens on rapid planes and rockets in light of the fact that there is a layer of shear stream close to the surface. It is some of the time noticed that this issue is a precise answer for the Navier–Stokes condition that is, an answer for the full conditions without expecting any vehicle properties or thermodynamic capacities are consistent

Stream Functions (Velocity Potential)

The stream function and the velocity potential are two highly helpful concepts introduced in this chapter. These numbers have physical meanings, and they're frequently utilised as dependent variables in flow issue solutions. They take the place of the velocity components in this job. Only some sorts of flows that match certain kinematic constraints have a stream function and a velocity potential. When a flow possesses symmetry with regard to a coordinate system, that is, when it does not depend on one coordinate and has a zero rate of expansion, it is called a stream function; • v = 0. This second requirement is always met because the rest of the book deals with incompressible fluids. Analyzing the equation regulating the stream function yields a plethora of accurate answers.

The velocity potential, on the other hand, does not require symmetry in the flow field but does impose a considerably more stringent constraint on particle motion, specifically that the vorticity be zero throughout the flow; = v = 0. In irrotational flows, viscous forces can be overlooked. The velocity potential is extremely useful in inviscid flows as a result of this.

Vorticity Dynamics

In the force conditions for incompressible stream we centre our consideration around speed and strain as the significant things of interest. Collaborations that happen in a stream field are clarified as far as inactivity, pressure powers, gravity powers, and thick powers. These fundamental ideas are the components available to us in deciphering liquid unique occasions. In many examples it is invaluable to decipher the stream occasions as far as the vorticity and the powerful occasions that are communicating to give a specific vorticity appropriation. The presence of vorticity for the most part demonstrates that gooey impacts are significant. This happens on the grounds that liquid particles must be set into pivot by an unequal shear pressure. Vorticity elements, generally talking, offer a strategy to isolate a stream into thick and inviscid impacts. It is particularly significant in situations where there is just a feeble connection among gooey and inviscid impacts.

Moderate Reynolds Numbers

Moderate Reynolds numbers are found in streams when delay, pressure powers, and gooey powers are all important. Moderate Reynolds numbers cannot be described as having a precise range of values. The mathematical properties of a certain object are determined by the specific stream scenario and the way in which the Reynolds numbers are defined. Moderate Reynolds numbers for a stream passing through a circular chamber range from VD/v = 0.5 to roughly 200. Moderate Reynolds values for streams entering an opening might range from 0.5 to 50. Another issue with showing a cautious reach for modest Reynolds numbers is that there is no sudden shift in the stream on either end of the range.

Idleness affects are minor in low-Reynolds-number streams; therefore they develop smoothly into moderate-Reynolds-number streams. Furthermore, gooey impacts are logically imprisoned to partition regions at the upper end of the moderate-Reynolds-number system. Individual judgement is required to assign a specific number to these advancements. Despite the difficulties in defining them categorically, moderate Reynolds-number streams are ones in which both tension and gooey powers contribute significantly to increases in liquid speed over a significant portion of the stream. Limit layers are something we really want to avoid with this last stipulation.

High Reynolds Number

Many designing streams are at high Reynolds numbers. It is generally typical for the stream in a line to have a Reynolds number of 10^5, or the stream over the wing of a little plane to havea Reynolds number of 10^6. We will observe that the stream field can be isolated into two sections: an inviscid stream in the significant piece of the stream area, and limit layers closethe dividers. Limit layer standards likewise apply to thin districts of high (shear layers) inside the principle stream locale. Inviscid stream and limit layers, is to stress that these subjects are not unmistakable however that they stand firm on integral footings in the hypothesis of liquid mechanics.

Ideal Flows

At high Reynolds numbers, UL/v, inviscid streams occur. Inviscid streams, as the reader may have heard, are situations in which the liquid is frictionless or has no consistency. These explanations are not meant to be taken seriously. The questions are as follows: Does the stream design possess any critical unequal gooey powers? If this is not the case, the conditions that govern the stream are those obtained by setting = 0 in the Navier–Stokes conditions. This is equivalent to allowing the Reynolds number to become infinite with a proper speed scale U and a reasonable length scale L. The ideal stream is a special type of inviscid stream with no vorticity. The particles in an ideal stream have never been subjected to uneven shear pressure and thus do not turn. The strain and gravity forces that create the stream cannot cause any molecule to turn.

Allow us to consider a few common circumstances where ideal stream is a sensible supposition. Presently, consider a wing mounted in a uniform airstream. The approaching stream is without vorticity, and along these lines the significant piece of the stream contains irrotational movement.

The vorticity produced at the wing's outer layer is bound to a slender thick wake and two vortices emanating from the wing's tips. Another type of inside stream is one in which a large repository provides an irrotational stream to a pipe or channel. Once again, the stream confines the vorticity to the dividers, and the principle stream is an optimal stream. In this case, for some random Reynolds number UD/v, the pipe may be so long that the vorticity diffuses away from the divider, defiling a large portion of the cross segment. When this occurs, the stream is currently in poor condition. If a stream divides, either on an outer stream or within an inward stream, the wake or distribution district only affects a portion of the stream. As a result, the stream is currently not completely irrotational.

In these cases, a heavy reliance on the inviscid stream limit layer hypothesis fails. After a while, there is a large section of the stream where the movement is inviscid and irrotational. In these districts, the stream actually submits to the best stream conditions. Because these are elliptic situations, the wake locale has an upstream impact. From one point of view, the inviscid stream is over a compelling body shape, which includes the genuine body as well as the impact of the wake. The problem with this method is that the size and state of the wake are unknown in advance. Several attempts have been made to display such streams with PC configurations. A fundamental fix is a hypothesis about the concept of the wake, as well as how it interacts with the primary stream

WHAT IS CIRCULATION?

When considering this theory, frictional force, which has a negative impact on the efficiency of most mechanical devices, turns out to be very important for the creation of lift. Further reading will reveal that lift cannot be produced in the absence of friction.

Elementary Flow

Any complicated incompressible flow may beunified via way of means of superposition of collection of fundamental flows. Primarily, four sorts of fundamental flows are categorized namely, uniform flow, Source flow, Doublet flow and Vortex flow. With the mixture of this kind of fundamental flow, any range of glidesituationsmay bederived. In this sectioncirculatecharacteristic and capacitycharacteristic for these kinds of fundamental flows are in shortdescribed. Velocity Potential characteristic and Steam characteristic are scalar function that assist look at whether or not the given fluid float is rotational or irrotational. Both the function offers a particular Laplace equation. The fluid float may be rotational or irrotational float primarily based totally on whether or not it satisfies the Laplace equation or not.

Potential Flow

In this Phase, we can introduce a number of "primary best flows." These flows will form the foundation upon which we will build more complex flows. "Superposition" is the primary precept we're relying on. This allows for the linear addition of multiple flows, resulting in more complex flows. This is possible because the primary underlying equations governing the flows are linear.

We have shown that for perfect (irrotational and incompressible with 0 viscous forces) flows, there is an orthogonal dating between variables that characterise the waft, stream function, and velocity potential. The orthogonal situation demonstrates that if we comprehend one, it is far more straightforward to determine the other. We'll use each of these waft descriptors to some extent, but we'll mostly use the stream function representation of the many flows we can explore. We can also restrict our effects to two-dimensional scenarios, albeit this isn't always necessary.

We have the simplified continuity equation for ideal flows, which considers the density as a constant and allows the density to be removed without delay inside the equation. This has the impact of enticing a few of the pace additives that must retain genuine in order to achieve mass conservation. Furthermore, the irrotational float condition necessitates a vorticity of 0, resulting in additional circumstances on pace derivatives with inside the float.

The boundary situations for such flows must be distinct, and they must solve the governing equations. We no longer have the no-slip boundary case to help establish the rate fee because those flows are inviscid. This means that the rate can be any (fixed) fee at a floor (which is no longer the same as the floor rate). We can, however, specify that a floor be impermeable if no float crosses the barrier. As a result, the rate ordinary to the floor must have a factor of zero. So, at the very least, we can say something quantifiable about a pace element.

The viscosity of a fluid is the most significant attribute since it determines the flow qualities. Because viscosity is a collective feature, it is difficult to calculate accurately when simulating a fluid using a particle model. We describe a new method in this article that has a higher signal-to-noise ratio than previous methods. It works by simulating counter-flowing Poiseuille flows without utilising explicit boundaries by employing periodic boundary conditions. The viscosity of the two flows is therefore related to their mean flow velocity. The approach is applied to two very distinct systems. First, we'll look at a simple generic fluid model called dissipative particle dynamics, which requires precise viscosity values to characterise the model fluid. The more lifelike the second. Lennard-Jones (Backer et al., 2005) The linear instability and nonlinear stability theories are used to investigate Poiseuille flow with pair stresses in a fluid layer. The eigenvalue issues for nonlinear stability for x,z and y,z disturbances are also derived. The Chebyshev collocation method is used to arrive at the eigenvalue equation, which is then solved numerically. The Chebyshev collocation method is used to solve the equivalent of the Orr-Sommerfeld eigenvalue problem. The challenges that arise while computing the Orr-Sommerfeld equation's spectrum are explained. For a wide range of the pair stresses coefficient M, the critical Reynolds number Rc, critical wave number ac, and critical wave speed cc are determined (Harfash and Meften, 2020)

Steady Flow

A consistent flow is the only wherein the amount of liquid flowing according to 2nd via any section, is constant. This is the definition for the suitable case. True consistent flow is gift handiest in Laminar flow. In turbulent flow, there are chronic fluctuations in pace. Pressure additionally varies at every point. But if this fee of alternate of strain and pace are same on each aspects of a constant common value, the glide is consistent flow. The specific time period use for that is suggest consistent glide. Steady flow can be uniform or non-uniform

Uniform Flow

The go with the drift is described as uniform go with the driftwhilstwith inside thego with the driftdisciplinethe speed and different hydrodynamic parameters do now no longerextrude

From factor to factor at any immediate of time. For a uniform go with the drift, the velocity is a feature of time only.

No spatial distribution of hydrodynamic and different parameters is possible for a uniform flow.

Regardless of whether it adjusts with time-unsteady uniform flow, any hydrodynamic parameter can have a completely unique price within the entire field.

Non-Uniform Flow

The float is described as non-uniform when the velocity and certain hydrodynamic characteristics change from one factor to the next.

The variations in position for a non-uniform flow can be found either in the flow direction or in directions perpendicular to it.

Near solid barriers through which the fluid travels, non-uniformity in a direction perpendicular to the flow is constantly encountered. There are four different possible combinations:

- Steady Uniform Flow
- Steady Non-Uniform Flow
- Unsteady Uniform Flow
- Unsteady Non-Uniform Flow

Steady Uniform Flow

Flow at a constant rate through a uniform cross-sectional duct

Steady Non-Uniform Flow

Flow at a constant rate through a non-uniform cross-sectional duct

Unsteady Uniform Flow

Inconsistent Uniform Flow, Flow at different rates through a long straight pipe with a uniform cross-section.

Unsteady Non-Uniform Flow

Uneven, erratic flow, Flow at varying rates through a duct with a non-uniform cross-section.

Turbulent Flow

Turbulent glide is a term used in hydrodynamics to describe irregular flows that include eddies, eddies, and glide instabilities. Excessive pulsed convection and occasional pulsed diffusion are used to determine this. This contrasts with laminar glide, which occurs when fluid travels in parallel layers without breaking between them. The upward push of cigarette smoke, waterfalls, blood slide in arteries, and most of the recirculation of the Earth's atmosphere are only a few instances of turbulence in natural phenomena and manmade packages. Turbulence occurs in human packaging such as automotive aerodynamics, but also in many corporate packages such as heat exchangers, quenching operations, or continuous steel casting.

What is Turbulent Flow?

Go with the flow of the turbulence. is a phrase used in hydrodynamics to characterise eddies, eddies, and drift instabilities, among other abnormal flows. It is regulated by excessive momentum convection and coffee momentum diffusion. Laminar drift, on the other hand, happens when fluid flows in parallel layers without breaking between them. Cigarette smoke rising into the atmosphere, waterfalls, blood flow with artery drift, and much of the earth's ecosystem recirculation are just a few examples of turbulence in natural occurrences and human undertakings. Turbulence is seen in a range of commercial undertakings, including heat exchangers, quenching operations, and continuous steel casting.

Turbulence Structure

In 1920 Lewis Fry Richardson summed up his paintings of turbulence shapes for weather packets with a well-known poem published in Weather Prediction using Numerical Process1: "Large vortices have small vortices which depend on their speed, and small vortices have large vortices. Small viscosity, thisprecept was dictated with the help of considerations on the use of electricity; large vortices are very progressive and tend to be unstable. Their movement feeds smaller vortices due to the Neighbourhood Kinetic Electric Switch These smaller vortices go through the same process, inflicting even smaller vortices that inherit electricity from their parent's vortex, and so on. This electrical switch is generally known as a "cascade of electricity" and is generally inertial, where a small dissipation of electricity occurs, however a sufficiently small scale is achieved that the viscosity of the fluid can use with it. This final turbulence scale offers a laminar proximity regime and is characterized using a low Re value.

Turbulence Scale

Due to the complexity of turbulence and its unpredictable behaviour, scientists exploited statistical patterns to better comprehend turbulent flows. In 1941, Kolmogorov expanded on Richardson's theory. For a sufficiently high Reynolds number, Kolmogorov claimed that small-scale eddies are isotropic, whereas large eddies may be anisotropic (or anyway rely upon the topology of a selected region). This assumption is critical because it assures that statistical analysis of tiny eddies is unaffected by geometry and, as a result, is representative of all turbulent flows. In the context of this theory, Kolmogorov quantitatively described the fundamental properties of the lower scale of turbulence.

Applications

Many natural occurrences (such as river currents, atmospheric currents, and herbal convection) as well as human programmes (such as city wind float, aerodynamic assessment, non-stop casting, and cooling process, cooling / heating systems) contain turbulent currents. Sim Scale was used to create little programmes in this time. The first is a measurement of float in a cylinder. This case can be used to simulate the float of a river across a bridge column, but it also has instructional value because it is a regularly used testing reference. It's also widely used in troubleshooting settings to illustrate particular modes (related with the presence, magnitude, and frequency of vortexes).

CONCLUSION

In this Phase we made a clean rationalization approximately the subsequent title, Magnus Effect and little bit defined extra detailly approximately their applications, like Base Ball Golf, Tennis in Sports and in Modern software like, The Magnus Spherical Airship, Magenn Air Rotor System (MARS) and approximately Kutta situation& The Kutta situation had been implemented at the air foil floor and their aerodynamics because of kutta conditions, provide an explanation for approximately the Kutta-Joukowski" theorem and Bernoulli's equation, and clean rationalization approximately circulation, Turbulence. In the Turbulence waft we made rationalization approximately the turbulence structure, what is supposed via way of means of Turbulences scale, and their software.

REFERENCES

Backer, J. A., Lowe, C. P., Hoefsloot, H. C. J., & Iedema, P. D. (2005). Poiseuille flow to measure the viscosity of particle model fluids. *The Journal of Chemical Physics*, *122*(15), 154503. doi:10.1063/1.1883163 PMID:15945641

Cayzac, R., Carette, E., Denis, P., & Guillen, P. (2011). Magnus effect: Physical origins and numerical prediction. *Journal of Applied Mechanics*, *78*(5), 051005. doi:10.1115/1.4004330

de Sá, C. V. S., & von Borries Lopes, A. (2021). An instructive derivation of d'Alembert's paradox. *European Journal of Physics*, *43*(1), 015004. doi:10.1088/1361-6404/ac329a

Harfash, A. J., & Meften, G. A. (2020). Poiseuille Flow with Couple Stresses Effect and No-slip Boundary Conditions. *Journal of Applied and Computational Mechanics*, *6*(Special Issue), 1069–1083.

Hoffman, J., & Johnson, C. (2010). Resolution of d'Alembert's paradox. *Journal of Mathematical Fluid Mechanics*, *12*(3), 321–334. doi:10.100700021-008-0290-1

Johnson, W. (1986). The Magnus Effect—Early investigations and a question of priority. *International Journal of Mechanical Sciences*, *28*(12), 859–872. doi:10.1016/0020-7403(86)90032-9

Kenyon, K. E. (2016). On the Magnus Effect. *Nature and Science*, *8*(02), 49–52. doi:10.4236/ns.2016.82006

Mohebbi, F., & Sellier, M. (2014). On the Kutta condition in potential flow over airfoil. *Journal of Aerodynamics*.

Tong, F., Duan, J., & Li, X. (2021). Characteristics of wall-shear stress fluctuations in shock wave and turbulent boundary layer interaction. *Journal of Turbulence*, *22*(12), 761–783. doi:10.1080/14685248.2021.1974466

Virk, D., & Virk K. (2018). A Review on Magnus Effect and its Applications. *International Journal of Information and Computing Science, 5*(8).

Chapter 4
Airfoil Theories and Their Applications

Kumaran T.
Vel Tech Rangarajan Dr. Sagunthala R&D Institute of Science and Technology, India

Sivarasan E. N.
National Chung Hsing University, Taiwan

ABSTRACT

One of the most important design impacts in aircraft is an airfoil. This airfoil is also considered the cross-section of the wing, so the flow characteristics of air depend on the airfoil's shape. There are some airfoil theories we can consider for the design consideration while designing the wing. The airfoil theories describe the nature of fluid flow over the wing based on the angle of attack for variable speed conditions. This chapter deals with airfoil terminology and the theories of airfoil like the methodology of conformal transformation, Cauchy-Riemann relations, complex potential, Kutta-Joukowski, thin airfoil theory, and their applications. The theories mentioned above explain transformation and its applications, Kutta condition, Kelvin's circulation theorem, starting vortex creation of airfoil for variable speed considered for aircraft, and the advantages of the optimization.

INTRODUCTION

Airfoil is the basic geometrical constraint to design a wing or blades in aerodynamic applications. The airfoil nomenclature is shown in figure 1 and the airfoil has a leading edge and trailing edge which is the initial and end portion of the airfoil. The shape of the airfoil is defined based on the airfoil theories and conditions to satisfy the aerodynamic performance characteristic and the nature of the aircraft is considered. The essential characteristics like the angle of attack are one of the important constraints which influence the aerodynamic performance of the aircraft. The angle of attack depends on the chord of the wing (Gudmundsson, 2013). The wing chord is the connecting line between the leading edge and the trailing edge of the airfoil. Since the start of the flight, a great variety of unique airfoils have been developed. These have a variety of features, some of which are excellent and others which are less so.

DOI: 10.4018/978-1-6684-4230-2.ch004

Figure 1. Airfoil nomenclature of a conventional aircraft
Source:(Snorri Gudmundsson, Chapter 8 - The Anatomy of the Airfoil)

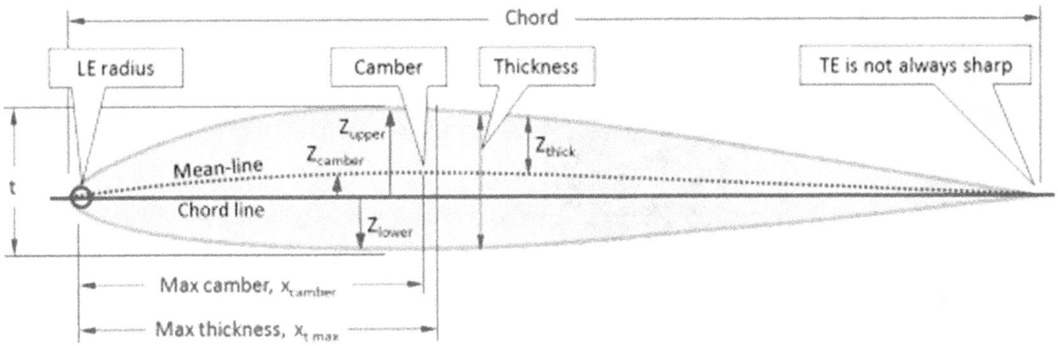

Typically, airfoils are designed using direct analysis and inverse design. Every design is based on the properties of airfoil (Houghton and Carpenter, 2003; Arfken et al., 2013). The properties of the airfoil were also similar to the aerodynamic characteristic of a body since the airfoil is used to design the wing. So the properties like lift coefficient (CL), Angle of attack, Minimum Drag Coefficient (Cdmin), Lift Curve Slope (CLα), and The pressure coefficient (CP) are very important parameters also influencing the performance of the wing. Based on the properties and the shape the airfoil is designed in a variety of configurations. The design of airflow over a wing is complex often for various geometrical aspects like an asymmetrical, asymmetrical wing. Conformal mapping is a critical technique that is used to solve complex airfoil flow conditions with the fewest geometric constraints. This method is used to represent an airfoil by comparing the solution for a sphere to that for an airfoil using the Joukowski transformation. Through the use of an inviscid, incompressible viscous fluid model, the flow around a cylinder was calculated by superimposing elementary potential flows. The modified flow equations and basic theories of meteorology were used to calculate lift as a function of the angle of attack for each airfoil. Mathematical equations were used by the NACA to develop its early airfoil series, which included the 4-digit, 5-digit, and modified 4-/5-digit designs. These equations define the camber (curvature) and thickness variation along the length of the mean-line (geometric centerline) of the airfoil section. For example, the shape of the 6-Series was developed using theoretical rather than geometrical methods, which results in a more intricate appearance. Before the National Advisory Committee for Aeronautics (NACA) produced this series, the shape of an airfoil was determined by historical experience with known shapes and experimental alterations (Catwell, 2014).

Conformal Mapping

A conformal map is a way to transform a complex-valued function from one coordinate system to another. Applying a transformation function to the original complex function accomplishes this objective. Consider the complex plane Z in figure 2 as an example. The complex function Z, Z=x+iy is used to define coordinates in this plane. The x-direction streamlines are given by and the equipotential curves are given by $\varphi = x$ in this illustration of a basic uniform fluid flow.

Figure 2. Conformal mapping transformation
Source:(https://www.grc.nasa.gov/www/k-12/airplane/map.html)

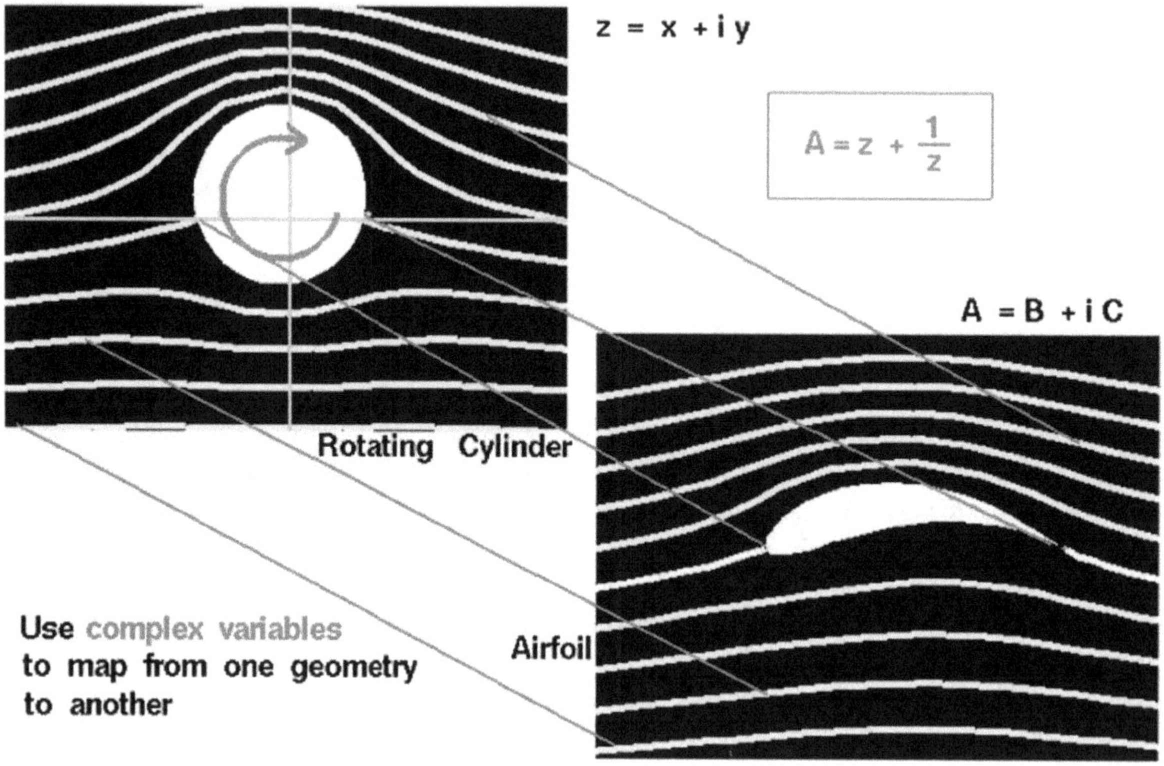

A CM may be utilised to transform this complex plane Z into a new complex plane w = f(K), where f(Z) is the transformation function w = ÖZ, as shown in the Z plane, the variables x and y have been changed into the new variables u and v (Selig and Maughmer, 1992). While the relative geometry of the streamlines and equipotential curves has altered as a result of this transformation, the collection of curves remains perpendicular. CM isn't complete without this angle-preserving feature. It is evident that the flow of air and the curves remain the same in the converted W plane and the harmonic function in the Z plane is also converted. Then the streamlines of the airfoil section have to be generated by analysing the airflow in the Z plane and using a particular CM function to convert this method of an airfoil in the W plane.

Applications of CM

- CMs are extremely useful for solving engineering and physics issues that may be stated as functions of a complicated variable but have unpleasant geometries. The analyst can change the problematic geometry into a lot more convenient one by selecting the proper mapping.
- In scattering and diffraction difficulties, CM can be applied. The mathematical challenge for the scattering and diffraction of planar electromagnetic waves is to find a solution to the scalar wave function that satisfies both the boundary and radiation conditions at infinity. Only in a few circumstances are exact answers accessible for such challenges. For broader scenarios, CMs are

employed to investigate far-field representations of scattered and diffracted waves (DeLillo and Sahraei, 2019).

Modelling the Airfoils Using CM

Because of the intricate, frequently non-symmetric geometries involved, fluid modelling interactions around airfoils is difficult. CM with complex variables is a valuable intermediary step that permits difficult Issues involving airfoil flow should be approached as problems with simpler geometry. To relate the flow problem for a cylinder to that of an airfoil, use the Joukowsky transformation, the CM approach was used to describe the fluid flow around the NACA airfoils. To use an incompressible, inviscid fluid model, the flow around a cylinder was calculated by superimposing elementary potential flows. For every airfoil, updated flow solutions and fundamental aerodynamics theories were applied to determine lift as a function of angle of attack. The thin airfoil approach's lift calculations are compared to these calculations. CM is a mathematical approach that allows complicated geometries to be translated into simpler geometries while maintaining the original geometry's angles and orientation. When investigating the flow around an airfoil, this technique can be used to simulate the flow around a cylinder, which makes computations much easier. The CM approach allows lift measurements on the cylinder to be compared to those on the equivalent airfoil since the functions that regulate fluid flow fulfil Laplace's equation. (DeLillo and Sahraei, 2019) The CM technique is utilised to simulate the two-dimensional fluid flow around airfoils in this work. We'll start by going over how airfoils are classified geometrically. After that, we'll go over the physical model we used to illustrate the incompressible, inviscid fluid flow over an airfoil and the theory behind CM. This will be accomplished by connecting the flow over a cylindrical surface to the flow around an aerofoil using the Joukowsky transformation, which is a form of CM. When we opted to portray three NACA airfoils using the CM transformation, we utilized numerical tools to determine fluid flow and lift around them. Finally, we'll compare our lift results to those produced using thin airfoil theory, a different method of modelling airfoil aerodynamics. To explain any inconsistencies between the two models, these findings will be complemented by quantitative error analysis.

In modelling, the airflow properties of an airfoil are divided into two types of branches. The airflow through an airfoil is treated as a single branch with distinct and important fluid characteristics. Because every fluid has viscous qualities, it is more desirable in this branch, but it entails a level of physical value and theoretical issues that are beyond the scope of this approach of conformal mapping implementation. However, we may replicate the theoretical flow around an airfoil using a process similar to that employed in the surrounding fluid and referred to as a fluid model. These fluids are viscous, and taking into account the fluid flow allows for the creation of the desired conditions.

The first of these requirements is that the airfoil must be travelling at subsonic speeds through the fluid. This is crucial because shock waves form at velocities approaching the speed of sound, causing the fluid flow to stop and the perfect fluid idealisation to fail. Airfoils flowing through flow zones between Mach 0.0 and 0.4 will be the focus of our models since this is the level of fluid constraint in the fluid flow that may be regarded as minimal.

The Kutta criteria explained the flow over the selected airfoil for subsequent assumptions. At the trailing edge, the airflow from the airfoil's upper and lower portions merges. The shear between an airfoil border and the fluid can grow gradually, allowing the flow to be pretty smooth to the trailing edge of each wing. An airfoil's boundary conditions must be met for it to generate an aerodynamic force lift (DeLillo and Sahraei, 2019).

The condition of Kutta imposes two extra restrictions on flow over an airfoil. To start with the theory explains the mandatory of the airfoil leading and trailing edges to be stagnated where airstream velocity vanishes. First and foremost, the airfoil's angle of attack hasn't crossed a vital threshold called the Stalling angle. At this point, the stalling angle of an airfoil is no longer physically relevant, and the airflow over the airfoil is no longer reasonably smooth.

Cauchy-Riemann Relations

To evaluating the problem, if a function w = f(K) is analytic at a point for K=x + iy and w = u + iv. The Cauchy-Riemann equations supply this.

w = f(K) states that it is differentiable at a point K = Ko if, and only if,

$$\frac{\partial u}{\partial x} = \frac{\partial v}{\partial y} \text{ and } \frac{\partial u}{\partial y} = -\frac{\partial v}{\partial x}$$

When these equations exist, it can be proved that the complex derivative can be calculated using any of these methods.

$$\frac{df}{dK} = \frac{\partial f}{\partial x} \text{ or df /dK} = \frac{df}{dK} = -\mathrm{i}\frac{\partial f}{\partial y}$$

If the function f(K) = K2 = x2 – y2 + 2ixy then u = x2 − y 2 and v = 2xy; therefore

$$\frac{\partial u}{\partial x} = 2x,\ \frac{\partial u}{\partial y} = 2y,\ \frac{\partial v}{\partial x} = 2x,\ \frac{\partial v}{\partial y} = 2y.$$

the C-R equations are always fulfilled in this case, and hence the function is analytic everywhere.

Finally

$$\frac{df}{dz} = \frac{\partial f}{\partial x} = 2\mathrm{x} + 2\mathrm{iy} = 2\mathrm{z}$$

or, equivalently,

$$\frac{df}{dz} = -\mathrm{i}\frac{\partial f}{\partial x} = -\mathrm{i}(-2\mathrm{y} + 2\mathrm{ix}) = 2\mathrm{z}.$$

By just differentiating f(K) as if it were a real function, we should get this result. This will always be the case for analytic functions, i.e., an analytic function F (K) may be obtained using the criteria for differentiating real functions (Ganguli, 2008; Benson and Benson, 1996; Kapania et al., 2008).

Kutta-Joukowski Theorem

The circulation of air and the interaction between pressure change and velocity is one of the most well-known definitions of lift on the wing. The Kutta Jankowski theorem explains these estimates of the wing's lift production. As a result, we must define the terms "circulation" and "boundary layer" When this idea is put to the test, the frictional force that limits the effectiveness of most moving parts turns out to be quite important for the lift's formation. We'll see how the lift can't be created without friction in the next section.

Consider an airfoil (a chopped portion of a wing) travelling through the air to better comprehend lift creation. Let's take a closer look at what's going on on the wing's surface with this image. Due to surface friction, the air at the airfoil's surface has no relative velocity (due to Van der Waals forces) (Riso et al., 2016). Due to the frictional effect, the air layer above this minimal liquid layer slows down. The lower-velocity air layer, for example, tries to slow down the higher-velocity air layer, and so on. This procedure is repeated until the airflow approaches that of a free stream. A 'Boundary Layer' is defined as a layer of air where viscosity has a major influence near the airfoil surface shown in figure 3. The wing's capacity to create lift is dependent on this boundary layer.

Figure 3. Boundary layer over an airfoil surface
Source: (De paula, Adson. (2016)

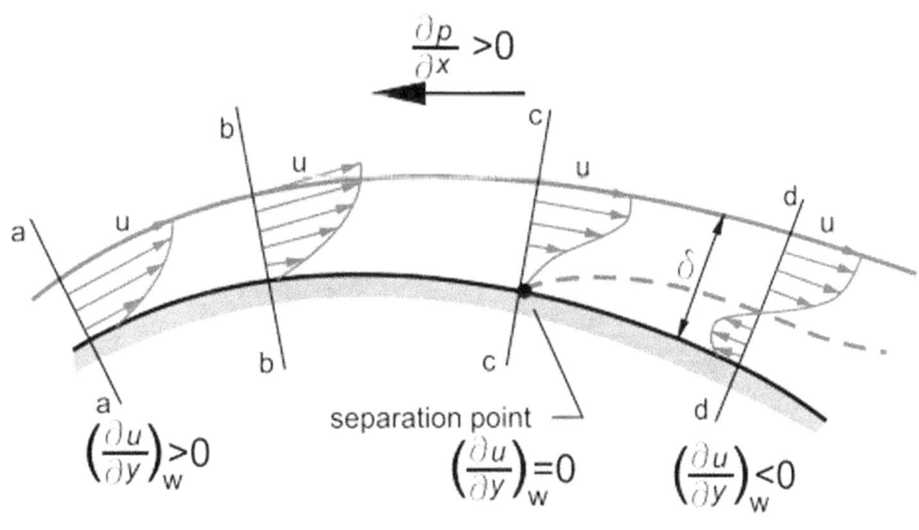

In the case of an airfoil, at the very beginning of the flow, two stagnation spots are established. On the top surface of the airfoil, one is close to the leading edge and the other is near the trailing edge. The stagnation point travels toward the bottom surface of an airfoil as the AoA increases. The trailing edge stagnation point, which was initially on the top airfoil surface, was seen to migrate precisely at the trailing edge. This circumstance is referred to as the Kutta condition. This indicates that there is movement around an airfoil, which helps to dislodge the stagnation point. The circulation of equal and opposing strength sets and departs from the end portion of the airfoil is required to ensure force conservation (Mohebbi et al., 2019; Sun and Wu, 2022). The vortex that forms is referred to as a beginning vortex.

We can now relate how "circulation on the wing" causes lift now that we have a picture of what it entails. The graphic on the left depicts airflow around the wing, while the diagram in the centre depicts the circulation caused by the vortex that we discussed before. The resulting diagram is obtained by adding the two flows (Vector). The length of the arrows represents the magnitude of the airflow velocity. The top surface's flow rises, while the bottom surface's flow drops, resulting in higher pressure on the lower layer than on the upper layer. As a result, Bernoulli's concept is re-applied, resulting in a net upward force called Lift. By increasing our knowledge of high and low-pressure generation, " Bernoulli's high-low pressure justification for lift formation is completed by the Kutta-Joukowski theorem.

Kutta-Joukowski Condition

An extra condition on the flow field is required to calculate the airfoil's circulation. We may anticipate the circulation to adapt until the overall flow exiting the trailing edge of the airfoil is smooth if we consider the entire stream to be made up of a homogeneous contribution (with no circulation) and a circulatory contribution. The Kutta-Joukowski condition is so named because it regulates the circulation and, as a result, the lift on the airfoil in a unique way. The flow over the symmetrical Joukowski airfoil is shown in figure.4

Figure 4. Symmetrical Joukowski airfoil.
Source:(https://www.grc.nasa.gov/www/k-12/airplane/short.html)

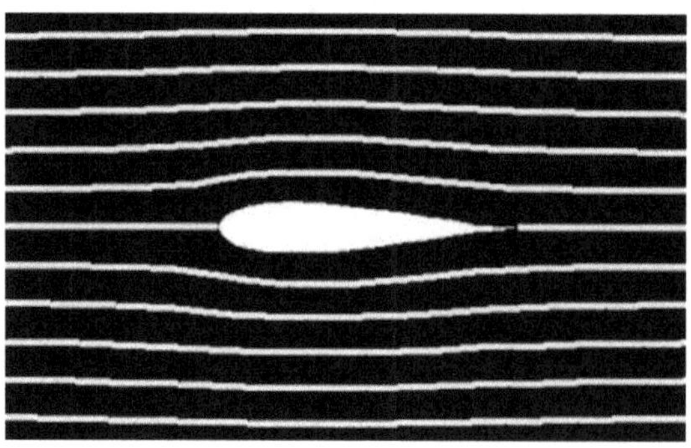

So get flow without circulation at first with two stagnation sites on the upper portion and lower portion of the airfoil (Paula, 2016). The fluid on the bottom surface of the airfoil must speed around the sharp trailing edge to approach the rearward stagnation point on the top surface. This necessitates an infinite fluid velocity at the end of the airfoil, which is implausible in viscous air. In this flow configuration, the rear stagnation point is pushed downward until it reaches the trailing edge of the airfoil, which is unstable. At this point, net circulation is present around the airfoil, with the zero velocity stagnation point cancelling out the limitless velocity at the back edge, resulting in a smooth flow. The Kelvin circulation theorem states that in a nonviscous fluid, circulation is preserved. Because viscosity is only genuinely relevant in the boundary layer at high Reynolds numbers (assuming it stays attached), for the viscous incompressible fluid around an airfoil at high Re, the circulation should be roughly retained. A beginning vortex with the opposite circulation is created downstream of the airfoil to compensate for the circulation formed around the airfoil. This is depicted in Prandtl's classic flow visualizations, as shown in the adjacent illustrations (Wahidi and Bridges, 2009; Benson and Benson, 1996). The take-off spacing is limited by this initial vortex (also known as the downwash). The downwash is frequently chaotic, and tiny planes can be flipped by a large jet's downwash.

Importance of Kutta Condition

The importance of this condition is to allow the physicist to consider the effect of the viscosity of the air in the momentum equation. In practice, it's critical for estimating lift on a wing. There are an infinite number of valid solutions when the conservation of energy and momentum equations are applied to an inviscid fluid flow around a solid body, such as a potential flow. Applying the governing equations, in the form of the N–S equations is one technique to get the proper answer. However, in most cases, these do not lead to a solution.

The Kutta condition is a method of including certain frictional effects while excluding others, such as shear and other flow separation effects. A multitude of terms has been used to characterise the condition. One is that the trailing edge's velocity cannot fluctuate indefinitely. Even though the viscous air may change in speed increasing the variations of flow separation and the trailing edge has a zero flow velocity. The Kutta condition has not been applied to time-dependent flow. the stagnated point initiated on the top surface of the airfoil and the rear velocity flow accelerates.

Kutta Condition in Potential Flow Over Airfoil

The numerical treatment of fluid dynamics issues has been transformed by the introduction of high-speed digital computers. Numerical approaches have now become a standard tool for studying fluid flows over bodies such as airfoils. Incompressible potential flows are one of the most important types of fluid flows to investigate while investigating low-speed aerodynamics difficulties. Fluid dynamicists developed numerical ways to address incompressible potential flow problems (the Laplace's equation) over an airfoil because of the restrictions associated with accurate (analytical) solutions using complex variables methods (CM). Panel techniques have been the main aerodynamic tools to quantitatively address such flows since the late 1960s. Panel approaches can be used to solve the dynamic problem of fluids governed by Laplace's equation. These techniques split the aerodynamic surface into regularly spaced panels, scatter such as source, doublet, and vorticity of uncertain flow strength on every panel. In this method of an ideal gas potential flow over an airfoil, the vorticity panel strength and circular quantities are important and their review leads to the method of calculating the aerodynamic surface flow velocity as a result of pressure coefficients (Selig and Maughmer, 1992) thus they will not be described further here. For more information, the reader might look up the references listed above. Dealing with the panels and their qualities, on the other hand, is quantitatively much more difficult than the approach given in recent research publications, and requires a significant amount of programming work. The Kutta condition has to be initiated in an analytical loop using the desired equation for vorticity panel strength. It provides a new method for numerically addressing ideal gas potential flow over an airfoil that takes into account factors such as vorticity panel strength and circulation. This method employs a mesh grid generation in elliptical equations to generate the mesh and estimate the aerodynamic parameters. An equation is created for the airfoil trailing edges, at the end of the results were compared to the recommended approach test situations. (both analytical and numerical). The Velocity Potential the Kutta Condition, the suggested solution is simple to implement. The velocity potential ϕ, $(\partial\phi/\partial n = 0)$

or the stream function ψ, $(\partial\psi/\partial s = 0)$ can be used to n and s are the scale parameter to the airfoil and the distance along the aerofoil section, respectively, to determine the wall boundary condition. To convert the physical realm onto the computational domain, we can use transformation connections.$\partial\phi/\ \partial n$airfoil surface $= -1\ J\ddot{O}\alpha\ (\alpha\phi\xi - \beta\phi\eta) = 0$

The Kutta condition is used to solve numerical simulation of a 2D incompressible flow field over an airfoil. To solve Laplace's solution for the ψ at each grid point, the proposed method uses the elliptic grid generating methodology (O-type). As a result, the panels and values used in the panel approach, such as vorticity panel strength and circulation, are not addressed. The trailing edges of finite-angle and cusped trailing edges are both impacted. A novel and relatively simple to implement stream function formula is developed for finite-angle and cusped trailing edges. The precise answers achieved in both circumstances demonstrate the numerical scheme's correctness and precision (Collicott et al., 2016).

Kutta Condition in Compressible Flow Over Isolated Airfoils

To calculate subsonic flow over isolated airfoils, the condition is utilised in the team equation. The proposed technique solves the stream function problem for fluid flow in the computing domain using a body-fitted mesh and the finite-difference method. For airfoils with both finite angles and cusped trailing edges, an equation is developed for applying the Kutta condition. A comparison of the suggested numerical method's findings with those obtained through experimental and other numerical techniques demonstrates that they are in great agreement, confirming the proposed method's accuracy and correctness. To solve 2D stable, irrotational, subsonic (subcritical) compressible flow over isolated airfoils, the research proposes using the stream function equation and a unique technique to perform the Kutta condition. To propose and apply an exact Kutta condition scheme into the analytical loop, the precise derived formula for the ψ at the airfoil trailing edge is employed. The exact phrase is broad enough to include both finite-angle and cusped trailing edges. Various test scenarios employing physical and other mathematical approaches were used to validate the proposed numerical methodology. The results revealed that the proposed strategy is exceptionally accurate and trustworthy.

AIRFOIL DESIGN METHODS

Airfoil design involves an analysis of turbulent flow characteristics and the link between pressure and form. Designing an airfoil can have a series of objectives. With a low drag coefficient in mind, the airfoils are designed to provide a certain amount of lift with little drag. There are some situations where the drag does not matter; what matters is how high you can go. This performance may be necessary despite the fact that the thickness, the stability moment, the off-design performance, or other unforeseen restrictions are all present and accounted for using an airfoil that has previously been created by someone who knows what they're doing is one method of airfoil design. When the aims of a particular design challenge meet with the objectives of the primary airfoil design, "design by authority" works effectively. The benefit of this approach is that testing results are already available. There aren't going to be any shocks, such as an early stall. On the other hand, accessible tools have now been developed to the point where one may be pretty certain that the expected performance will be realised. It is currently increasingly usual to utilise "designer airfoils" that are individually customised to the demands of a certain project. The method of bespoke airfoil design is discussed in this section of the notes. The direct and inverse design approach for airfoil design may be divided into two types.

Direct Methods for Airfoil Design

Direct airfoil design processes include defining section geometry and estimating pressures and performance. When the supplied shape is examined for performance, a modification is made to improve that shape's properties. In this sort of technique, there are two key sub-problems:

- The determination of the performance metric
- The method of changing the shape

Direct airfoil design starts with a given airfoil form (such as a NACA airfoil), determines which problem is the most problematic, and fixes it. As a result, the performance will be increased. The process of fixing the most obvious problems in a particular airfoil is repeated until the segment is devoid of significant flaws. Such airfoils may not necessitate a detailed specification of an objective function, but it does demand considerable expertise in spotting possible difficulties and, in many cases, many experiences with solving them.

The pressure peak at low Cl can be reduced by lowering the surface "bump" at the leading edge and decreasing the lower surface wall thickness aft of the hump. Until Cl reaches roughly 0.2, the lower surface stream will be linked. Double-check that we haven't harmed the Clmax too severely.

Inverse Design

The target pressure distribution is another form of the objective function. Sometimes, it is possible to supply a distribution of relevant Cps and to strive for a difference between the actual and estimated Cps. Several inverted design techniques are based on this principle. For example, the camber line geometry that generates a specific pressure differential on an airfoil in potential flow may be solved using the thin airfoil theory. The second step of airfoil design occurs when an objective has been established. The aerofoil's shape is altered in this part of the design in order to improve performance. There are numerous methods for accomplishing this:

- Using the information on the effects of changes in geometry on Cp and changes in Cp on performance.
- Using approximate solutions to define the airfoil geometry and letting the computer decide on the order in which the changes should be made to improve the design.

Thick Airfoil Design

Thick airfoils have the disadvantage of lowering the minimum pressure. As a result, the undesirable pressure gradient worsens, requiring an earlier start to recovery. If a maximum thickness point is specified, the maximum thickness section must rebound from the steepest gradient point that is possible. This is exactly the kind of difficulty that Liebeck discusses when it comes to maximum lift. A boundary layer exists in the thickest possible section throughout the recovery, almost to the point of separation.

High Lift Airfoil Design

The high lift aerodynamic design in airfoils is required to produce aerodynamic characteristics like lift coefficient, and negative or significant pressure on the airfoil shown in figure 5. The limit of this approach may be influenced by the compressibility of the fluid and the pressure recovery due to the boundary layer. It has been shown that keeping the boundary layer on the separation boundary maximized lift starting at a certain recovery height and position (Jaganrai et al., 2015; Lecture notes, n.d.). For a Re of 5 million, such distributions are depicted below. The contrast between laminar and turbulent findings should be noted. At Re = 10 million, the thickest part is 57 per cent thick, yet it will split abruptly with any angle of attack.

Figure 5. CP versus laminar and turbulent flow conditions
Source: (http://aero-comlab.stanford.edu/aa200b/lect_notes/AirfoilDesign.pdf)

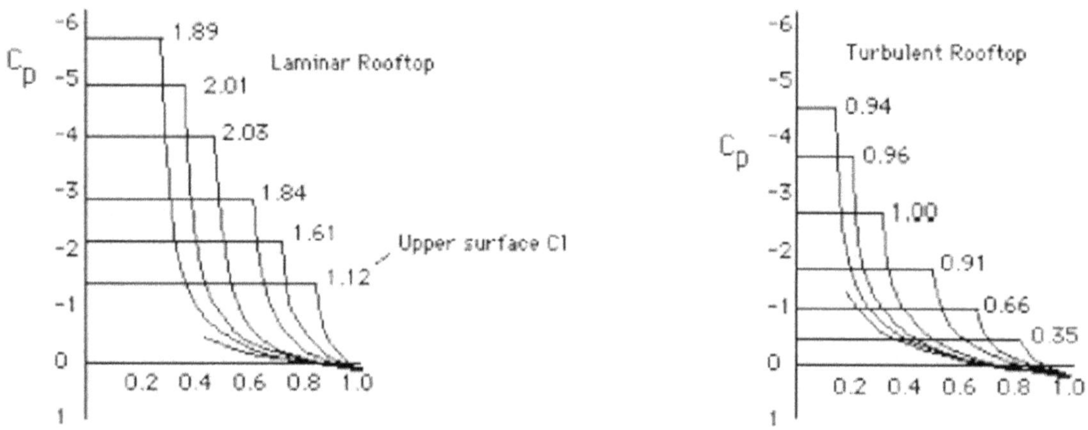

Laminar Airfoil Design

Laminar flow can aid in the reduction of skin friction drag, the enhancement of maximum lift, and the reduction of heat transfer. At low Reynolds numbers, the flow is good and also a smooth surface utilizing an airfoil with a beneficial pressure gradient may be done with minimum effort. The next section shows how to alter pressures to create long laminar flow runs over the top and lower surfaces.

Transonic Airfoil Design

If we want to minimise shock losses at a desired transonic speed, which leads to the transonic airfoil design challenge. As a result, the minimal pressure coefficient that may be tolerated is essentially limited. The challenge in transonic design is the creation of an airfoil section with significant lift and thickness while avoiding strong shock waves because both lift and thickness reduce the minimum Cp. Because they can tolerate some supersonic flow even without increasing drag, most sections can serve well as "supercritical airfoils." The highest local Mach numbers over a well supercritical airfoil really shouldn't exceed 1.2 to 1.3, according to a simple rule of thumb. When compared to completely subcritical devices, this results in a significant increase of available Cl. Supercritical sections generally relate to a certain type of airfoil that is intended to work effectively in supersonic flow regimes. Many of the following design principles are commonly used in such sections to optimise aerodynamic forces or thickness at a given Mach no.

Low Reynolds Number Airfoil Design

Airfoil design has piqued the curiosity of applied aerodynamics practitioners for more than a century. The discipline is being fuelled by an ever-increasing mix of airfoil design needs for novel applications like UAVs, and this trend is expected to continue. When all conceivable permutations of the plethora of airfoil design criteria are considered, it becomes clear that the number of distinct sets of requirements considerably outnumbers the number of current airfoils. As a result, the innovation and use of airfoil design methodologies will continue to be the most cost-effective option.

Enriching airfoil "catalogues" for the sake of enrichment, on the other hand, is thought to be of minimal benefit. The goal of this lecture is to discuss the design of low-Reynolds-number airfoils, which has been the topic of many studies and has been published in numerous important conferences. First, numerous techniques for toes airfoil design is examined, with inverse methods emerging as the preferred way over the others. Second, various portions of these notes discuss the significant effect of laminar separation bubbles on low Reynolds number airfoils, necessitating the development of an inverse approach that can more or less directly regulate transition movement with the angle of attack.

The intended transition behaviour is the foundation of a design philosophy that has been incorporated into an inverted airfoil design technique. Finally, various sample airfoils are shown in this lecture to demonstrate the general approach to low Reynolds number airfoil design. In each case, cutting-edge airfoil design and analysis software was applied. The systematic and parametric analyses demonstrate relevant performance patterns and trade-offs in airfoil design at low Reynolds numbers, even though each airfoil was developed for a distinct purpose. The complete design process has been confirmed by wind tunnel experiments, as will be shown, and the results are shown alongside the projections. Airfoil design becomes more complicated when the boundary layer becomes very low in controlling desirable pressure values without splitting at lower Reynolds numbers. As a result, these designs have low pressure and limited lift capacity. It may be challenging to maintain a turbulent flow boundary layer and the lowest pressure zones. The boundary layer can only withstand modest pressure recovery in such situations. An all-laminar segment can yield a CL of around 0.4 or have a thickness of about 7.5 per cent, according to the equations for laminar separation (Lecture notes, n.d.).

Laminar Separation Bubbles and Transition

Airfoil flows with low Reynolds numbers can be characterised primarily by the presence of laminar separation bubbles. A lot of emphases has been paid to laminar separation bubbles, as they are the principal source of performance degradation in comparison to airfoil sections with a higher Reynolds number On an airfoil, laminar separation bubbles are formed due to a lack of turbulent flow in the connected boundary layer. Before the transition, the laminar flow separates. A "flow separation sphere" is formed when the turbulence reattaches to the airfoil's surface below the transition. The pressure drag over the laminar separation region is the primary cause of the airfoil's high drag when Reynolds numbers are low. The presence and size of a flow separation bubble can be discovered using surface oil flow visualisation. Specifically, in low Reynolds number airfoils, the length of the flow separation sphere and, therefore, the magnitude of the drag increase caused by this bubble are strongly impacted by their transition site. Laminar separation bubbles can have a deleterious influence on small Reynolds airfoils if the transition is not controlled.

Low Moment Airfoil Design

Extending lift as far back on the airfoil as necessary is not always possible when the airfoil pitching moment is constrained. Such conditions occur in the building of tailless aircraft, helicopter rotor blades, and even sails, kites, and massive pterosaurs. This particular airfoil is a Lie-back section with a positive Cmo. that performs well at low Reynolds numbers. Its performance isn't horrible, but it pales in comparison to other parts without a Cm limit in Clmax. (At Re = 500,000, Clmax =1.35 compared. 1.60 for typical sections.)

CONCLUSION

This paper discusses the airfoil characteristics and various design methods involved to validate the results for various optimized solutions using Cauchy Reimann solutions. The mathematical expressions are used to evaluate complex geometry problems. Various airfoil design theories are discussed in this chapter to understand the design criteria of the wing concerning aerodynamic characteristics. The transformation of the 3D model with a similar geometric transformation is explained in conformal mapping and the Kutta-Joukowsky condition explains the transformation of the cylindrical surface to the airfoil.

REFERENCES

Arfken, G. B., Weber, H. J., & Frank, F. E. (2013). *Mathematical Methods for Physicists* (7th ed.). Academic Press.

Benson, T., & Benson, T. (1996). Interactive educational tool for classical airfoil theory. In *35th Aerospace Sciences Meeting and Exhibit* (p. 849). Academic Press.

Cantwell, B. (2014, April 6). *The NACA airfoil series* [Lecture notes]. Department of Aeronautics and Astronautics, Stanford University.

Collicott, S. H., Valentine, D. T., Houghton, E. L., & Carpenter, P. W. (2016). *Aerodynamics for Engineering Students* (7th ed.). Butterworth-Heinemann.

DeLillo, T. K., & Sahraei, S. (2019). Computation of plane potential flow past multi-element airfoils using conformal mapping, revisited. *Journal of Computational and Applied Mathematics*, *362*, 246–261. doi:10.1016/j.cam.2018.08.031

Ganguli, S. (2008). *Conformal Mapping and its Applications.* Academic Press.

Gudmundsson, S. (2013). *General aviation aircraft design: Applied Methods and Procedures.* Butterworth-Heinemann.

Houghton, E. L., & Carpenter, P. W. (2003). *Aerodynamics for engineering students*. Elsevier.

Jaganraj, R., Boopathy, G., & Varun, V. (2015). Vigneshwar, Experimental Investigation of Passive Flow Control on Bluff Bodies. *International Journal of Applied Engineering Research, 10*(8), 19793–19798.

Kapania, N. R., Terracciano, K., & Taylor, S. (2008). Modeling the fluid flow around airfoils using conformal mapping. *SIAM Undergraduate Research Online*, *1*(2), 70–99. doi:10.1137/08S010104

Mohebbi, F., Evans, B., & Sellier, M. (2019). On the Kutta condition in compressible flow over isolated airfoils. *Fluids*, *4*(2), 102. doi:10.3390/fluids4020102

Paula, A. A. D. (2016). *The airfoil thickness effects on wavy leading edge phenomena at low Reynolds number regime* (Doctoral dissertation). Universidade de São Paulo).

Riso, C., Riccardi, G., & Mastroddi, F. (2016). Nonlinear aeroelastic modeling via conformal mapping and vortex method for a flat-plate airfoil in arbitrary motion. *Journal of Fluids and Structures*, *62*, 230–251. doi:10.1016/j.jfluidstructs.2016.02.002

Selig, M. S., & Maughmer, M. D. (1992). Multipoint inverse airfoil design method based on conformal mapping. *AIAA Journal*, *30*(5), 1162–1170. doi:10.2514/3.11046

Sun, S. Y., & Wu, G. X. (2022). Inviscid flow passing a lifting body with a higher order boundary element method. *Engineering Analysis with Boundary Elements*, *136*, 144–157. doi:10.1016/j.enganabound.2021.12.012

Wahidi, R., & Bridges, D. (2009, June). Experimental investigation of the boundary layer and pressure measurements on airfoils with laminar separation bubbles. In *39th AIAA Fluid Dynamics Conference* (p. 4278). Academic Press.

Chapter 5
Finite Wing Theory

Madhankumar G.
https://orcid.org/0000-0003-0931-1860
Sathyabama Institute of Science and Technology, India

Mothilal T.
KCG College of Technology, India

Kumar K. M.
https://orcid.org/0000-0002-4268-6309
St. Joseph's College of Engineering, India

Muralidharan G.
Surya Group of Institutions, India

Mala D.
University College of Engineering, Panruti, India

ABSTRACT

Wing design is a very complicated and intricate issue. It is not feasible to cover everything in this chapter; however, it is possible to discuss some of the essential ideas that underpin design for high lift and low drag. Lift may be increased in four ways for fixed air characteristics and free-stream speed: increased wing area, increased angle of attack, increased camber, increased circulation through the use of high-momentum fluid. One of the most important applications of potential flow theory was the study of lifting surfaces such as aircraft wings, since the boundary conditions on a complex geometry can significantly complicate any attempt to tackle the problem via analytical techniques, which involves some simplification assumptions in order to arrive at a solution. These assumptions will be related to the concept of three-dimensional thin wing issues in this chapter.

DOI: 10.4018/978-1-6684-4230-2.ch005

1 INTRODUCTION

Wing design is a very complicated and intricate issue. It is not feasible to cover everything in this article, however, it is possible to discuss some of the essential ideas that underpin design for high lift and low drag. The lift may be increased in four ways for fixed air characteristics and free-stream speed: Increased wing area, Increased angle of attack, Increased camber, Increased circulation through the use of high-momentum fluid. One of the most important applications of potential flow theory was the study of lifting surfaces such as aircraft wings since the boundary conditions on a complex geometry can significantly complicate any attempt to tackle the problem via analytical techniques, which involves some simplification assumptions in order to arrive at a solution. These assumptions will be related to the concept of three-dimensional thin wing issues in this chapter.

2 LIFT AND DRAG ESTIMATION IN A FINITE WING

We know an aeroplane is in the air because of the lift created by its wings as a result of higher pressure on the lower (bottom) surface and lower pressure on the suction (top) surface. The lift is produced by the pressure differential between the lower and higher surfaces. It also causes the flow to curve around at the wingtips as it is driven from the pressure surface to the suction surface. As a result, there will be a spanwise component of flow from the wingtip to the wing root, causing the streamlines on the upper surface to bend toward the wing root, as seen in Figure 1. Similarly, the spanwise component of flow on the bottom surface will be in the direction of the wing root to the wingtip. As a result, it is clear that the flow across the wings is three-dimensional, and their aerodynamic characteristics differ significantly from those of their airfoil sections (Figure 2).

Figure 1. Front and top views of a three-dimensional wing's flow pattern

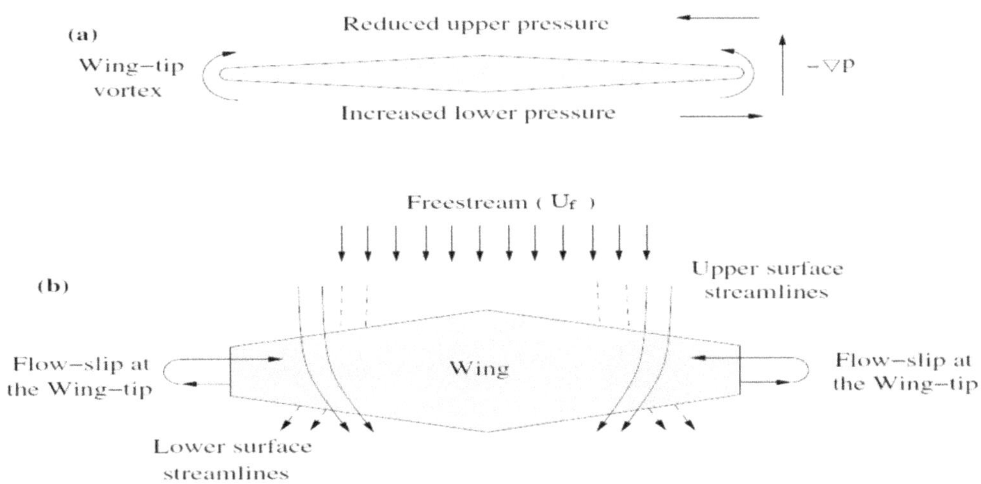

Figure 2. Diagram showing the shedding of wingtip vortices and the downwash

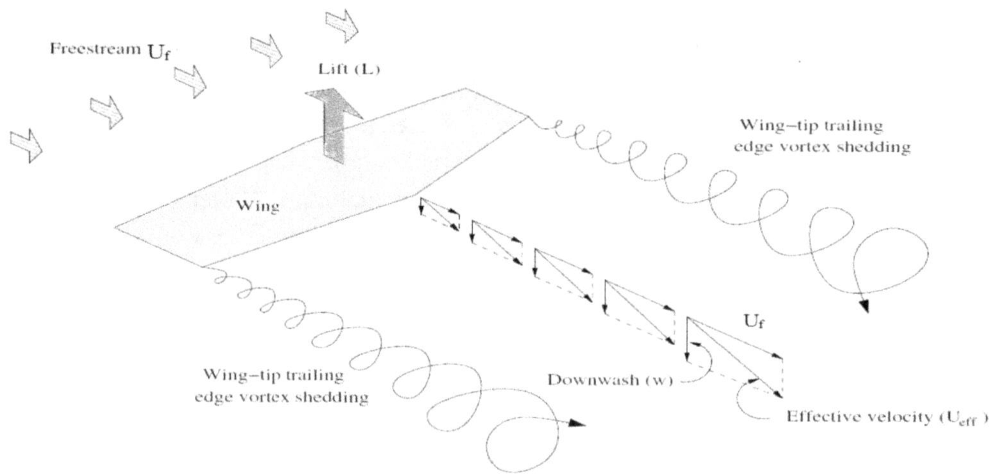

The curling of the flow at the wing tips has additional effect on the wing's aerodynamic properties. This "flow slip" creates a circulatory motion that trails downstream of the wing, resulting in a trailing vortex at each wing tip, shedding downstream of the wing. These trailing vortices, which are emitted at the wing tip, cause a minor downward component of air velocity (opposite to the direction of lift) in the region of the wing. This downward component of air velocity is known as downwash, and it is generally represented by the symbol 'w'. The downwash may alternatively be interpreted as the effect of the lifting wing "pressing down" on the air, causing the air to have more downward motion in the vicinity of the wing. As seen in Figure 3, downwash mixes with freestream velocity (U_f) to generate a local relative wind that is canted downward in the area of each airfoil segment.

Figure 3. a) Two-dimensional airfoil section; b) three-dimensional wing are shown in a schematic diagram.

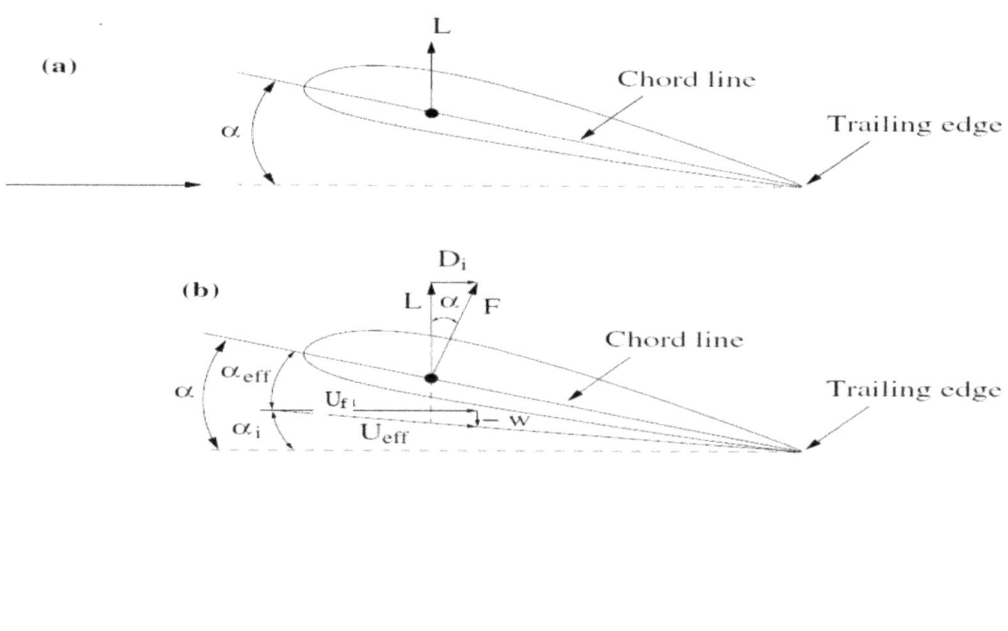

The geometric angle of attack (α) is defined as the angle formed by the chord line and the freestream direction. However, as seen in Figure 3, the local wind speed is inclined below the freestream velocity (U_f) by the angle (α_i) which is referred to as the induced angle of attack. When the induced downwash is w_i and the freestream velocity is U_f, the induced angle of attack (α_i) is expressed as

$$\alpha_i = \tan^{-1}\left(-\frac{w_i}{U_f}\right) \tag{1.1}$$

The negative sign in the above statement indicates that the downwash is in the downward direction, i.e., the opposite direction of the lift force. The induced angle can alternatively be stated as using the small-angle approximation as

$$\alpha_i \approx \left(-\frac{w_i}{U_f}\right) \tag{1.2}$$

The downwash created by shedding trailing edge vortices from the wing tips, as well as its influence on the inclination of the local relative airflow, has two primary effects on the local airfoil section. In reality, the actual angle of attack as observed by the airfoil locally is less than the geometric angle of attack (α). This angle of attack is referred to as the effective angle of attack (α_{eff}) which is written for a three-dimensional wing in Figure 3.

$$\alpha_{eff} = \alpha - \alpha_i \tag{1.3}$$

In addition, the effective freestream velocity (U_{eff}) will now be calculated as

$$U_{eff} = \sqrt{U_f^2 + w_i^2} \tag{1.4}$$

However, given minor downwash (w ≈ 0),

$$U_{eff} \approx U_f \tag{1.5}$$

Because the effective angle of attack is smaller than the geometric angle of attack, the lift created by a three-dimensional wing is less than that generated by a two-dimensional wing (airfoil section) in the absence of downwash. As a result, as compared to its airfoil section, a finite wing must operate at a larger geometric angle of attack to obtain the same lift per unit span. Downwash is connected with an extra drag penalty in addition to reducing the angle of attack. We know that the local lift force functions normally in relation to the local relative wind speed, and that the tilt generated by downwash tilts the lift force by the same angle αi as shown in Figure 3. As a result, the local lift has a component along the freestream direction, implying that there will be drag caused by downwash. This increased drag is referred to as induced drag (Di). Furthermore, the lift created by the wing per unit span is

$$L' = \rho U_f \Gamma(y)\cos(\alpha_i) \tag{1.6}$$

Introducing minor-angle approximation $\cos(\alpha_i) \approx 1$, then the lift per unit span becomes

$$L' = \rho U_f \Gamma(y) \tag{1.7}$$

Similarly, the wing's drag per unit span will be

$$D_i' = \rho U_f \Gamma(y)\sin(\alpha_i) \tag{1.8}$$

Again, from minor-angle approximation $\sin(\alpha_i) \approx \alpha_i$, then the drag per unit span becomes

$$D_i' = -\rho U_f \left(\frac{w_i}{U_f}\right)\Gamma(y) = L'\alpha_i \tag{1.9}$$

If the wing has a varied distribution of circulation throughout its full span, as shown in Figure 4, the total lift (L) on the wing by semi-span (b) may be stated as

$$L = \int_{-b}^{b} \rho U_f \Gamma(y)\,dy \tag{1.10}$$

And the induced drag will be

$$D_i = -\int_{-b}^{b} \rho U_f \left(\frac{w_i}{U_f}\right)\Gamma(y)\,dy = \rho U_f \int_{-b}^{b} \Gamma(y)\alpha_i(y)\,dy \tag{1.11}$$

This is the necessary formulation for induced drag (D_i) in terms of induced downwash (w_i) and circulation (Γ).

2.1 The Relationship Between Trailing Edge Vortices and Load Distribution

The strength of the vortices shed between these portions equals the change in circulation from section to section. The circulation around the section y is Γ shown in Figure 4, and it is decreased to $\Gamma - d\Gamma$ at the section y + dy. Thus, the trailing edge vortices of strength $d\Gamma$ will be shed between the sections y and y + dy. If γ(y) is the vortices' strength per unit length, then

$$d\Gamma = -\gamma(y)dy \tag{1.12}$$

Figure 4. Load distribution spanning a straight wing

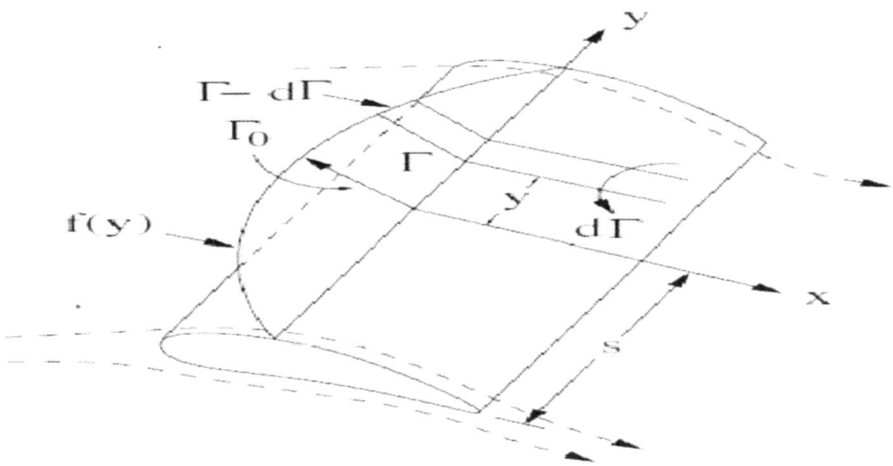

Consider the impact of the trailing vortex filaments of strength *dΓ* that are shed from the section at y. The downwash generated at some other point y_1 along the span will be provided by

$$dw'(y) = -\frac{\gamma(y)dy}{4\pi(y - y_1)} \tag{1.13}$$

We obtain by integrating the preceding equation between the limits -b and b

$$w(y) = -\frac{1}{4\pi}\int_{-b}^{b}\frac{\gamma(y)dy}{(y - y_1)} \tag{1.14}$$

This is the load distribution throughout the span of the wing.

3 VORTEX IDEA

Humans have been fascinated with vortices since the dawn of time. Everyone has seen the spinning of leaves when a strong wind blows or the eddies at the surface of a water body in their daily activities. Until recently, people believed that supernatural forces were responsible for their production and harmful consequences. Despite the fact that this myth has been dispelled, many vortex flows continue to be dangerous and mysterious. Furthermore, important astronomical mysteries like as the genesis of the solar system and the structure of our galaxy (Milky Way) remain unsolved vortex issues.

So, what exactly is a vortex? The answer to this question is neither simple nor original. A vortex, on the other hand, is commonly defined in two ways. The first definition indicates that a vortex is the rotating motion of a large number of fluid particles around a common centre; however, the pathways of rotation of these particles do not have to be circular and can also be asymmetrical. However, most vortices in nature have pathlines that are not perpendicular, but rather skewed to the axis of rotation, forming a spiral pattern. Because the fluid particles contain a component of velocity parallel to the axis of rotation, these vortical flows are referred to as spiral vortices.

Cauchy and *Stokes* provided the second definition of the vortex. The vorticity was defined as the angular velocity of any substance (fluid particles in our example) at a location in space (flow field). Though each vortex contains vorticity, this is not always the case. For example, the parallel shear flow possesses vorticity but is not a vortex.

3.1 Forced Vortex

We know that in rigid body rotation, all of the components are in coupled translational and rotational movements, with no relative motion between them. When a fluid is rotated at a constant angular velocity along a common axis, the entire fluid mass moves as if it were a solid body. Individual particles are in motion, but they are not deformed because there are no shear forces operating between the fluid layers. This form of fluid spinning is referred to as a forced vortex.

When the viscous forces occurring between the fluid layers are stronger than the inertial forces, the forced vortex behaviour takes over. Because the inertia force to viscous force ratio is represented in terms of the Reynolds number $Re = \frac{\rho U_f l}{\mu}$. A low Reynolds number flow suggests significantly higher viscous stresses, indicating a proclivity to act more like a solid body, i.e., like a forced vortex. As a result, in a forced vortex flow, the angular velocity (ω) of rotation is constant but the tangential velocity (v_θ) grows linearly with the radius of rotation (r). Thus,

$$v_\theta = \omega_r \tag{1.15}$$

The fluid particles in this vortical flow rotate on their own axis in addition to translating about a shared axis. As a result, a forced vortex is a combination of translational and rotating movements. A constant source of energy or torque is required to sustain the solid body-like spinning of a forced vortex.

3.2 Free Vortex

The influence of viscosity becomes little in a flow with a high Reynolds number. In the limit, for extremely high Reynolds numbers, the flow can be considered to be inviscid. A free vortex is a form of fluid spinning. The momentum of a body will be preserved in the absence of any influencing factors, according to Newton's second law of motion. If no influencing forces are present, the angular momentum associated with a free vortex flow will stay invariant under steady-state circumstances. It is simple to demonstrate that in a free vortex flow, the tangential velocity (v_θ) is inversely proportional to the radius (r). In other words,

$$v_\theta = \frac{\Gamma}{2\pi r} \tag{1.16}$$

where Γ is the counter-clockwise circulation around a closed contour obtained. It is worth noting that the tangential velocity approaches to infinity at the centre (r = 0). As a result, at the centre of a free vortex, a singularity exists.

3.3 Vortex Line, Vortex Tube, and the Vortex Filament

A vortex line, also known as a vorticity line, is defined as a line whose tangent is parallel to the local vorticity vector ($\vec{\zeta}$) everywhere. As a result, it is effectively a line pointing in the direction of a vortex vector. In the vorticity field, a vortex line is comparable to a streamline in the velocity field. Furthermore, a vortex line connected to a fluid is orthogonal to a streamline connected to a fluid particle. The following relationship defines the vortex lines:

$$\frac{dx}{\zeta_x} = \frac{dy}{\zeta_y} = \frac{dz}{\zeta_z} \tag{1.17}$$

Where ζ_x, ζ_y, and ζ_z are the components of the vorticity vector along x, y and z directions in Cartesian space. A vortex tube is a bundle of vortex lines that is comparable to a streamtube, which is a bundle of streamlines. It is, in fact, a cylindrical tube in space with surface components made up of vortex lines travelling around a closed curve specified by the diameter of a randomly chosen tube's cross section. The circulation around this curve is proportional to the number of vortex lines that travel through the curve's enclosed region.

The vortex tube strength, also known as the vortex flux, is the integral of the vorticity over the tube's cross section. The vortex flow is the same at every cross section along the tube because the vorticity is divergence free. A vortex tube in an inviscid flow with the fluid; that is, a vortex tube associated with some fluid components will remain associated with those elements. Finally, when the cross-sectional area of the tube falls to zero, a vortex filament is the limiting case of a vortex tube. In other words, a vortex filament is a vortex tube with an infinitesimal cross section.

4 VORTEX MOTION THEOREMS OF HELMHOLTZ

Helmholtz's theorems, named after scientist Hermann von Helmholtz (1821–1894), explain the three-dimensional motion of fluid in the vicinity of vortex filaments in aerodynamics. They are valid for inviscid and incompressible flows, as well as flows without nonconservative forces. These theorems did, in fact, pave the way for the legendary Prandtl (1921) to develop the traditional lifting line theory, which remains one of the most important contributions to aerospace research.

It has been found that the strength of a vortex eventually decreases owing to viscous factors. Because all actual fluids have viscosity, the Helmholtz theorems do not apply to them. The following are descriptions of the four Helmholtz theorems.

4.1 The First Theorem of Helmholtz

A vortex tube's strength, measured by the circulation around the perimeter of any tube cross section, remains constant along its length. This theorem applies to any fluid in which the vorticity field is divergence free, i.e., there is no divergence.

$$\vec{\nabla}.\vec{\zeta} = \vec{\nabla}.\left(\vec{\nabla}\times\vec{v}\right) = 0 \tag{1.18}$$

Proof: The Gauss divergence theorem can be used to establish this statement.

$$\iint_S \vec{\zeta}.\overrightarrow{dS} = \iiint_\vartheta \vec{\nabla}.\vec{\zeta}\, d\vartheta \tag{1.19}$$

where ϑ indicates the volume of a vortex tube subsection and S is its bounding surface. When we plug Eq. (1.18) into Eq. (1.19), we obtain

$$\iint_S \vec{\zeta}.\overrightarrow{dS} = \iiint_\vartheta \vec{\nabla}.\vec{\zeta}\, d\vartheta = 0 \tag{1.20}$$

The vorticity vector, by definition, is directed normal to its bounding surface, i.e., S, and along the sides of the tube. Surprisingly, the only nonzero components in Eq. (1.20) emanate from the top and bottom surfaces of the vortex tube. Thus,

$$\iint_S \vec{\zeta}.\overrightarrow{dS} = \int_{S_1} \vec{\zeta}.\overrightarrow{dS} + \int_{S_2} \vec{\zeta}.\overrightarrow{dS} = 0 \tag{1.21}$$

where S_1 and S_2 are the areas of the surface at the bottom and top of the volume ϑ, respectively. Also, at S_1, the area vector ($\hat{n}$) and the vorticity vector ($\vec{\zeta}$) are pointing in opposing directions, but they are pointing in the same way at S_2. As a result, Eq. (1.21) may be written as

$$\iint_S \vec{\zeta}.\overrightarrow{dS} = -\iint_{S_1} \vec{\zeta}.\hat{n}dS + \iint_{S_2} \vec{\zeta}.\hat{n}dS = 0 \tag{1.22}$$

Alternatively,

$$\iint_{S_1} \vec{\zeta}.\hat{n}dS = \iint_{S_2} \vec{\zeta}.\hat{n}dS \tag{1.23}$$

Furthermore, we may use the Stokes curl theorem to get

$$\iint_S \vec{\zeta}.\hat{n}dS = \oint_C \vec{v}.\overrightarrow{dS} \tag{1.24}$$

However, by definition $\oint_C \vec{v}.\overrightarrow{dS} = \Gamma$, where Γ is the circulation around the closed curve C bounded by the region S. As a result, if C_1 and C_2 are the closed contours that bound the surfaces S_1 and S_2, respectively, then Eq. (1.24) yields the required conclusion.

$$\Gamma_{C_1} = \Gamma_{C_2} \tag{1.25}$$

In other words, the circulation (the intensity of the vortex tube) stays constant along its length.

4.2 The Second Theorem of Helmholtz

This theorem states that a vortex line can never cease abruptly in a fluid. The vortex lines and vortex tubes should be connected to make a closed route. They must be infinite in length and begin (or finish) at solid limits.

4.3 The Third Theorem of Helmholtz

This theorem states that an inviscid, incompressible, and barotropic [$\rho=\rho(P)$] fluid that is initially irrotational will stay such indefinitely. As a result, such a fluid cannot generate vorticity. The inviscid and incompressible flow over a bow-shock, which produces vortices, is an exception to this rule.

Proof: This theorem is easily shown using Kelvin's circulation theorem. It says that for a barotropic fluid travelling under conservative body forces, circulation (Γ) around an arbitrary closed curve moving with the fluid is time independent, i.e., $\frac{D\Gamma}{Dt} = 0$. We may derive the following from this theorem:

$$\frac{D\Gamma}{Dt} = \iint_S \left[\frac{\partial \vec{\zeta}}{\partial t} + \vec{\nabla} \times \left(\vec{\zeta} \times \vec{v} \right) \right] . \overrightarrow{dS} = 0 \tag{1.26}$$

Because the above statement holds true for any arbitrary area S. To maintain the equality, the integrand must be zero. Consequently, we have

$$\frac{\partial \vec{\zeta}}{\partial t} = -\left[\vec{\nabla} \times \left(\vec{\zeta} \times \vec{v} \right) \right] \tag{1.27}$$

Alternatively,

$$\frac{\partial \vec{\zeta}}{\partial t} = \left[\vec{\nabla} \times \left(\vec{v} \times \vec{\zeta} \right) \right] \tag{1.28}$$

As a result, if vorticity is 0 at the start i.e., $\vec{\zeta} = 0$, it will remain zero throughout.

4.4 The Fourth Theorem of Helmholtz

This theorem asserts that for a vortex line with a unique identity, the ratio of vorticity to the product of fluid density and line length remains constant throughout time. In other words,

$$\frac{\zeta}{\rho l} = \text{constant} \tag{1.29}$$

As a result, stretching a vortex line increases the vorticity.

5 BIOT AND SAVART LAW OF VORTEX MOTION

Many aerodynamic applications need the calculation of the velocity caused by a concentrated distribution of vorticity (a vortex) with any direction. We must invert the equation obtain the velocity $v(r,t)$ in terms of the vorticity $\zeta(r,t)$ given by

$$\vec{\zeta} = \vec{\nabla} \times \vec{v} \tag{1.30}$$

The selected step-by-step technique will be outlined next. We know that the continuity equation for an incompressible fluid is given by

$$\vec{\nabla}.\vec{v} = 0 \tag{1.31}$$

Because the divergence of velocity for an incompressible fluid is zero, the velocity ($\vec{v}$) may be written as the curl of another vector field, say of $\vec{g}$ (r, t). So, we write

$$\vec{v} = curl\ g = \vec{\nabla} \times \vec{g} \tag{1.32}$$

Because the curl of any vector's gradient is always zero, the vector $\vec{g}$ is undefined to the extent of the gradient of the scalar function of location and time. It follows from Eq. (1.32) that

$$\vec{\nabla} \times \vec{v} = \vec{\nabla} \times \left(\vec{\nabla} \times \vec{g}\right) = \vec{\nabla}\left(\vec{\nabla}.\vec{g}\right) - \nabla^2 \vec{g} \tag{1.33}$$

Let us also suppose that

$$\vec{\nabla}.\vec{g} = 0 \tag{1.34}$$

It should be noted that the above assumption is permissible since the function $\vec{g}(\vec{r},t)$. To the extent of a gradient vector, is indeterminate. Eqs. (1.30), (1.33), and (1.34) yield

$$\nabla^2\vec{g} = -\vec{\nabla}\times\vec{v} = -\vec{\zeta} \tag{1.35}$$

This is the Poisson's equation for vector $\vec{g}$, which may be thought of as a vector potential. Once $\vec{g}$ the velocity field is calculated as a solution of equation, it may be derived using Eq. (6.33). (6.36). If the vector $\vec{g}(\vec{r},t)$ and vorticity vector $\vec{\zeta}(\vec{s},t)$ are represented in Cartesian space in terms of their respective components, as

$$\vec{g} = (g_x, g_y, g_z) \tag{1.36}$$

$$\vec{\zeta} = (\zeta_x, \zeta_y, \zeta_z) \tag{1.37}$$

Eq. (1.35) can thus be expressed as

$$\begin{aligned}\nabla^2 g_x &= -\zeta_x \\ \nabla^2 g_y &= -\zeta_y \\ \nabla^2 g_z &= -\zeta_x\end{aligned} \tag{1.38}$$

The answer to Eq. (1.35) may be written as

$$\vec{g}(\vec{r},t) = \frac{1}{4\pi}\iiint_R \frac{\vec{\zeta}(\vec{s},t)}{|\vec{r}-\vec{s}|}\,d\vartheta \tag{1.39}$$

where $\vec{\zeta}(\vec{s},t)d\vartheta$ is a vortex distribution element at the points s and R in the area where the vorticity is present scattered, as seen in Figure 5.

Proof: We wish to obtain the solution for the following equation.

$$\nabla^2\vec{g} = -\vec{\zeta}(\vec{r})$$

or in Cartesian space, the following set of equations:

$$\nabla^2 g_x = -\zeta_x(\vec{r})$$

$$\nabla^2 g_y = -\zeta_y(\vec{r})$$

$$\nabla^2 g_z = -\zeta_z(\vec{r})$$

Figure 5. Coordinates used in calculating velocity from a vortex distribution.

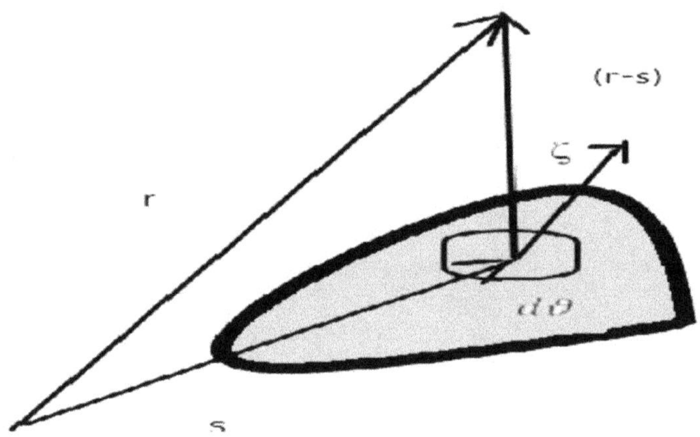

Figure 6. Coordinates used in obtaining the solution of Poisson's equation for vector $\vec{g}$

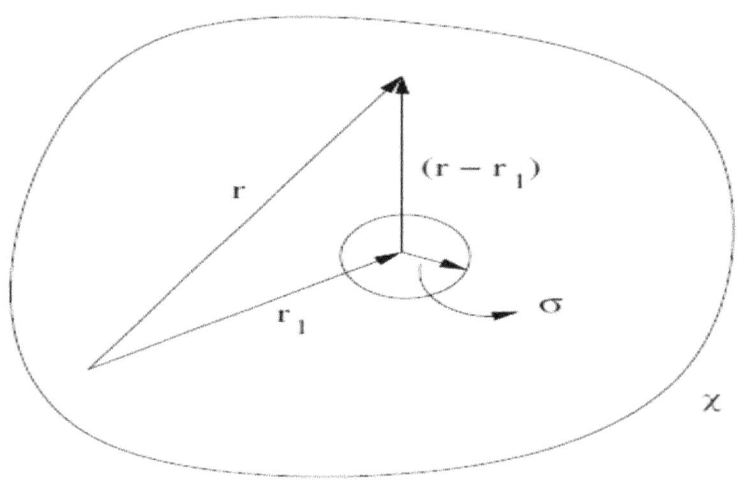

Consider the whole universe and assume that the field $\vec{g}(\vec{r},t)$. At infinity, it disappears sufficiently strongly (Figure 6). Furthermore, it is necessary to consider the solution to one of the system's equations, as provided in Eq (1.38). To build the answer, we will utilise the nomenclature presented in Figure 5 and the Green's theorem. Thus,

$$\iiint_{\vartheta}\left(\phi\nabla^{2}\psi-\psi\nabla^{2}\phi\right)d\vartheta=\oiint_{S}(\phi\vec{\nabla}\psi-\psi\vec{\nabla}\phi).\hat{n}dS \tag{1.40}$$

where ϑ represents the volume and S signifies the boundary surface area. In Eq. (1.40), we may identify using $g_x(r)$ for the aforementioned equation, as provided in Eq (1.38). Furthermore, consider the function ψ as

$$\psi\left(r\right)=\left(\vec{r}-\vec{r_1}\right)^{-1} \tag{1.41}$$

Therefore, we write

$$\nabla^2\phi = -\zeta_x(r)$$

$$\nabla\phi = \nabla g_x$$

$$\nabla\psi=\frac{r-r_1}{\left(\left|\vec{r}-\vec{r_1}\right|\right)^{3}}$$

$$\nabla^2\psi=0$$

everywhere except at $r=r_1$ where it becomes infinite

Because the point $r=r_1$ is a singular point in the sense that, ψ,$\nabla\psi$ and $\nabla 2\psi$ become infinite at that point, we now set a tiny sphere of radius σ around that point r1 an_d apply Green's theorem in the region encompassed between the sphere and an arbitrarily drawn big surface χ. To fill the whole space, we subtract to infinity and decrease the sphere to the point r1. *As* a result, we receive

$$\iiint_{R}\frac{\zeta_x\left(r\right)}{\left|\vec{r}-\vec{s}\right|}d\vartheta=-\lim_{\chi\to\infty}\oiint_{\chi}\left(g_x\frac{r-r_1}{\left(\left|\vec{r}-\vec{r_1}\right|\right)^{3}}+\frac{\nabla g_x}{\left|\vec{r}-\vec{r_1}\right|}\right).\hat{n}dS-\lim_{\sigma\to\infty}\oiint_{\sigma}\left(g_x\frac{r-r_1}{\left(\left|\vec{r}-\vec{r_1}\right|\right)^{3}}+\frac{\nabla g_x}{\left|\vec{r}-\vec{r_1}\right|}\right).\hat{n}dS \tag{1.42}$$

Because we previously assumed that the field dies out sufficiently strongly at infinity, we require that both g_x and ∇g_x vanish as we approach infinity, such that the limit of the surface integral over χ vanishes as $\chi\to\infty$. As a result, the limit of integral over χ is set to zero.

Consider the integral over the sphere with the origin at the point r1 *in* spherical coordinates. In other words,

$$-\iint\left(g_x+\frac{\partial g_x}{\partial x}\right)\sin\theta d\theta d\phi$$

In the limiting case, for σ=0 from the above integral gives $-4\pi g_x(r_1)$. Then we get,

$$\iiint_R \frac{\zeta_x(\vec{r})}{|\vec{r}-\vec{r}_1|} d\vartheta = 4\pi g_x(\vec{r}_1)$$

It is worth noting that the integration variable in the preceding integral is r, and if we swap the roles of r and r_1, the solution of equation (1.35) may be written as

$$g_x(r) = \frac{1}{4\pi} \iiint_R \frac{\zeta_x(\vec{r}_1)}{|\vec{r}-\vec{r}_1|} d\vartheta \tag{1.43}$$

The variable of integration is now r_1. Similarly, the additional components, $g_y(\vec{r})$ and $g_z(\vec{r})$, are found, and their solutions are comparable to those of $g_x(\vec{r})$. Based on these findings, we may deduce that the answer for $\vec{g}$ is

$$\vec{g}(r) = \frac{1}{4\pi} \iiint_R \frac{\zeta_x(\vec{r}_1)}{|\vec{r}-\vec{r}_1|} d\vartheta$$

Replacing r_1 = s, we obtain the same solution as given by Eq.(1.39).

Now, by substituting Eq.(1.39) into (1.32), the velocity field is obtained as

$$\vec{v} = \vec{\nabla} \times \vec{g} = \frac{1}{4\pi}\left[\vec{\nabla} \times \iiint_R \frac{\zeta(\vec{s},t)}{|\vec{r}-\vec{s}|} d\vartheta\right] \tag{1.44}$$

Suppose elemental vector $\overrightarrow{dg}$ is the contribution to $\vec{g}$ at $\vec{r}$ due to vortex element $\vec{\zeta} d\vartheta$ located at $\vec{s}$.

$$\overrightarrow{dg}(\vec{r},t) = \frac{1}{4\pi} \frac{\zeta(\vec{s},t)}{|\vec{r}-\vec{s}|} d\vartheta \tag{1.45}$$

Similarly, let the elemental velocity $\overrightarrow{dv}(\vec{r},t)$ is the contribution to $\vec{v}$ at $\vec{r}$.

$$\overrightarrow{dv}(\vec{r},t) = \frac{1}{4\pi} curl_r \frac{\zeta(\vec{s},t)}{|\vec{r}-\vec{s}|} d\vartheta \tag{1.46}$$

The subscript r on the curl indicates that the curl is to be taken in relation to the coordinates of the point r.

Figure 7. Biot and Savart law was derived using coordinates.

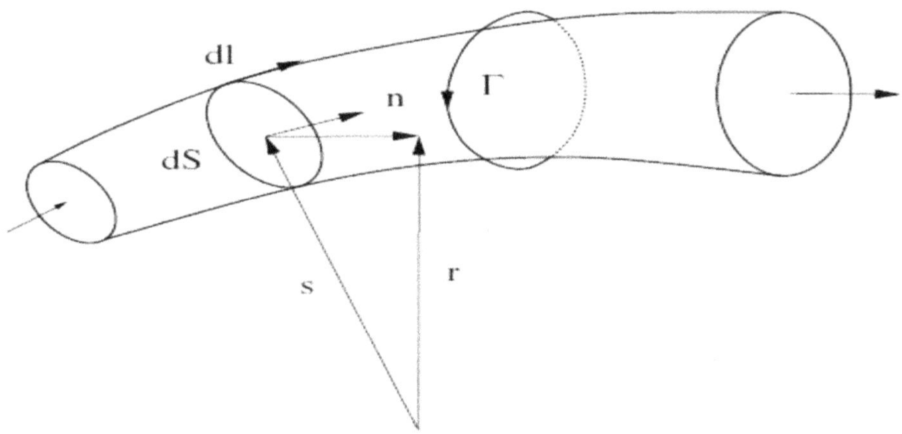

Consider the strength Γ of a vortex filament (circulation) seen in Figure 8. Choose a volume element $d\vartheta$ of this filament as the cylinder created by a cross-sectional surface $\hat{n}dS$, and an element's length is dl along the filament (Figure 7). The contribution of the vortex element at $\vec{s}$ to the vector potential $\vec{g}$ at a field point $\vec{r}$ is given by

$$\overrightarrow{dg}(\vec{r}) = \frac{1}{4\pi}\frac{\zeta(\vec{s})}{|\vec{r}-\vec{s}|}\left(\hat{n}dS.\overrightarrow{dl}\right) \tag{1.47}$$

The elemental length vector $\overrightarrow{dl}$, on the other hand, may be expressed as

$$\overrightarrow{dl} = \frac{\vec{\zeta}}{|\vec{\zeta}|}dl \tag{1.48}$$

Along with

$$\vec{\zeta}.\hat{n}dS = \Gamma$$

Eq. (1.47) can be expressed as

$$\overrightarrow{dg}(\vec{r}) = \frac{\Gamma}{4\pi}\frac{\overrightarrow{dl}}{|\vec{r}-\vec{s}|} \tag{1.49}$$

The contribution of the filament element to the velocity at point r is given by

$$\overrightarrow{dv}(\vec{r}) = \vec{\nabla}_r \times \frac{\Gamma}{4\pi} \frac{\overrightarrow{dl}}{|\vec{r} - \vec{s}|} \tag{1.50}$$

When performing the curl, the letters $\overrightarrow{dl}$ and $\vec{s}$ are considered fixed. As a result, the above equation becomes

$$\overrightarrow{dv}(\vec{r}) = \frac{\Gamma}{4\pi} \frac{\overrightarrow{dl} \times (\vec{r} - \vec{s})}{|\vec{r} - \vec{s}|^3} \tag{1.51}$$

This is the well-known Vortex Motion Law of Biot and Savart. Integrating Eq. (1.51) across the whole length of the filament yields the velocity at position $\vec{r}$ due to the complete vortex filament. Thus,

$$\vec{v}(r) = \frac{\Gamma}{4\pi} \int \left(\frac{\overrightarrow{dl} \times (\vec{r} - \vec{s})}{|\vec{r} - \vec{s}|^3} \right) \tag{1.52}$$

5.1 Biot and Savart Law: Velocity Induced by a Straight Vortex Filament

In this part, we will use Biot and Savart's law (Eq. (1.52)) to compute the velocity caused by a straight line segment. We know from Helmholtz's first theorem that a vortex line cannot begin or stop in a fluid except at the solid boundaries. In this case, we want to compute the induced velocity by a segment that is a portion of a continuous vortex line. This vortex segment exhibits continuous circulation (Γ) over the length of the line and may be oriented in Cartesian space in any direction, as shown in Figure 8. It is worth noting that the direction of induced velocity is derived by using the right-hand curl rule.[1] Also, suppose the distance $(\vec{r} - \vec{s})$, between the vortex segment and the point Q in Eq. (1.52) is $\vec{q}$. Thus, the velocity imparted $\overrightarrow{dv}$ by a segment $\overrightarrow{dl}$ on this line at point Q may be represented as

$$\overrightarrow{dv} = \frac{\Gamma}{4\pi} \left(\frac{\overrightarrow{dl} \times \vec{q}}{|q|^3} \right) \tag{1.53}$$

where $q = |\vec{q}|$. The following equation may be expressed in scalar form using the geometry given in Figure 8 as

$$dv_i = \frac{\Gamma}{4\pi} \frac{\sin\theta}{q^2} dl \tag{1.54}$$

$dl = \left|\overrightarrow{dl}\right|$ in this case as well. We may derive the following relationships from the figure:

$$d = q \sin\theta \tag{1.55}$$

Figure 8. A straight vortex filament induces velocity at point P.

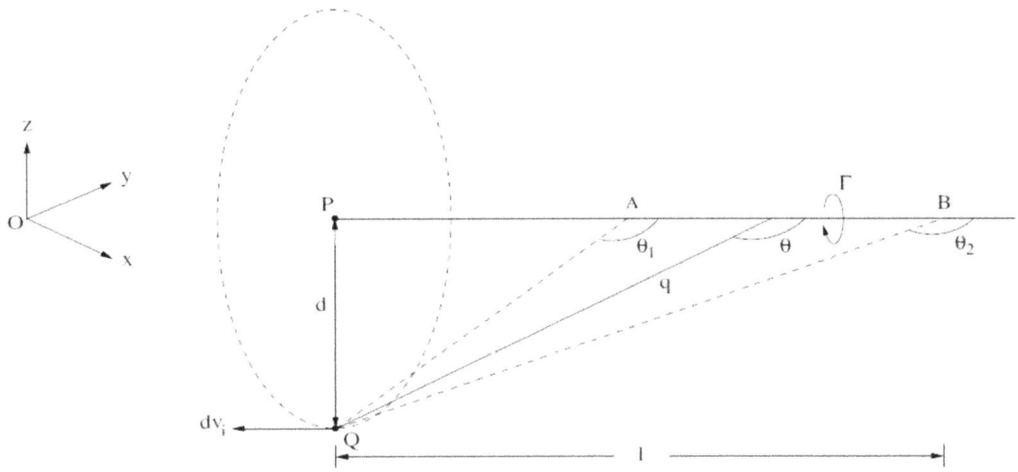

From the Figure 8,

$$\frac{d}{l} = \tan\left(\pi - \theta\right) = -\tan\theta$$

On rearranging,

$$l = -\frac{d}{\tan\theta} \tag{1.56}$$

Differentiating the preceding phrase with regard to yields

$$dl = \left(\frac{d}{sin^2\theta}\right)d\theta \tag{1.57}$$

When we plug the aforementioned relationships into Eq. (1.54), we get

$$dv_i = \frac{\Gamma}{4\pi}\left(\frac{\sin^2\theta}{d^2}\right)\times\sin\theta\times\left(\frac{d^2}{\sin^2\theta}\right)d\theta$$

On rearranging,

$$dv_i = \frac{\Gamma}{4\pi d}\sin\theta d\theta \tag{1.58}$$

Assume that the line segment endpoints, designated by A and B, intersect the angles θ_1 and θ_2, respectively, at the point Q. The aforementioned equation may be integrated across the straight vortex line segment (A to B) as follows:

$$v_i = \frac{\Gamma}{4\pi d}\int_{\theta_1}^{\theta_2} \sin\theta d\theta$$

We ultimately get by solving the above integral

$$v_i = \frac{\Gamma}{4\pi d}\left(\cos\theta_1 - \cos\theta_2\right) \tag{1.59}$$

This finding is critical in aerodynamics since it indicates that the velocity produced by a straight vortex line segment is a function of its strength Γ, distance d, and view angles θ_1 and θ_2. Using this approach, we will now investigate two specific examples to calculate the induced velocity near the vortex line.

5.1.1 Induced Velocity by an Infinite Vortex Segment

When both ends of a vortex stretch to infinity, it is referred to as an endless vortex. It is a two-dimensional example in which θ_1 =0 and θ_2=π. As a result of Eq. (6.60), the velocity produced at a location near an infinite vortex becomes

$$v_i = \frac{\Gamma}{2\pi d} \tag{1.60}$$

5.1.2 A Semi-Infinite Vortex Segment Causes Velocity

A semi-infinite vortex, as the name implies, is a vortex in which one of its ends spreads to infinity. Assume, for example, that the vortex line segment end B in Figure 8, extends to infinity θ_2=π. As a result of Eq. (1.59), the velocity produced at a location precisely opposite the segment end A, such that $\theta_1 = \frac{\pi}{2}$, will be

$$v_i = \frac{\Gamma}{4\pi d} \tag{1.61}$$

This is precisely half the velocity caused by an infinite vortex section.

Example Problem 1:

Consider a vortex sheet, where the velocity above is 6 ms^{-1}, respectively. The element of this vortex sheet is 0.8 m wide, which rolled up into a line vortex. Calculate the strength of the vortex.

Given vortex sheet are rolled up into a line vortex. Therefore, the radial velocity given for a vortex sheet $V_r = 6\ ms^{-1}$

Vortex sheet width, d = 0.8 m
Radius of the vortex sheet, r = d/2 =0.8/2 = 0.4 m

The strength of the vortex sheet is given by

$$\Gamma = 2\pi r V_r = 2\pi X0.4X6 = 15.08m2s^{-1}$$

6 THE EVOLUTION OF PRANDTL'S LIFTING LINE THEORY AND THE VORTEX SYSTEM

The idea of vorticity is critical in explaining a wide range of fluid flow phenomena. Remember that a vorticity is basically a measure of the fluid particle's moment of momentum about its own centre of mass. Assume a tiny sphere of fluid moving along an axis going through its centre freezes and solidifies. Because the moment of momentum is preserved during freezing, the angular velocity of the solid sphere at the time of creation is exactly half of the vorticity of the fluid before freezing. Furthermore, the Kelvin and Helmholtz dynamical theorems connect changes in the vorticity of the fluid particle to the moments of the forces operating on the fluid particle.

As seen in Figure 9, an airfoil begins at rest and travels through the air. As illustrated in Figure 9a, the circulation around any curve, say C, in the flow field is initially and remains zero (Γ_C=0) . As a result, the flow stays irrotational everywhere except in the thin boundary layer near to the airfoil, where viscosity produces vorticity.

It's an odd truth that viscosity aids in the generation of lift on the airfoil. To illustrate this, consider an airfoil at rest with no circulation in the flow field, as shown in Figure 9a. When the airfoil starts moving, the viscous action at the trailing edge results in the shedding of a vortex known as the starting vortex (Figure 9b), and the circulation around a closed curve C_1 containing the starting vortex is non-zero $\left(\Gamma_{C_1} = 0\right)$, as shown in Figure 9c. If we examine a bigger curve (C_2) in the inviscid zone that encloses both the vortex and the airfoil, Kelvin's circulation theorem informs us that the circulation around the curve C_2 will be zero as well, i.e., $\left(\Gamma_{C_2} = 0\right)$.

Figure 9. The bound vortex on an airfoil and the beginning vortex downstream are depicted schematically.

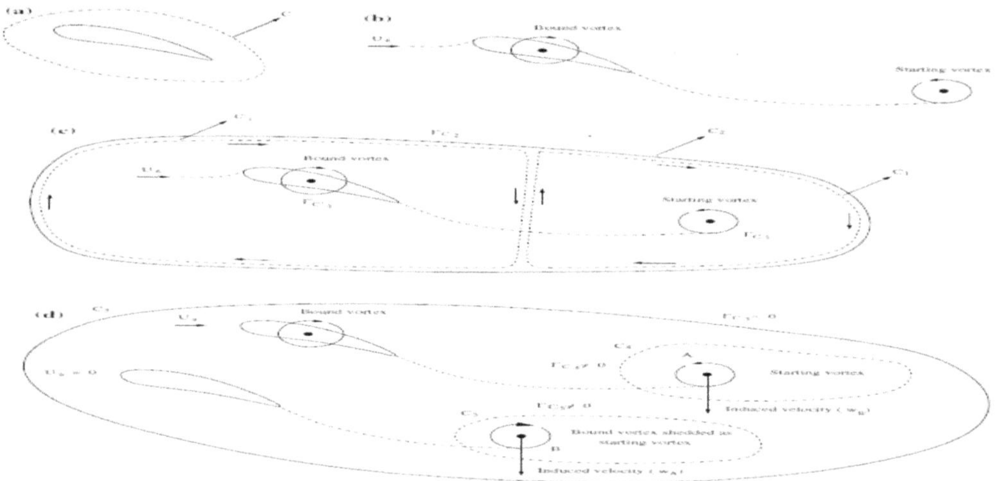

If a curve $C_1^{'}$ encircles the airfoil, the circulations around the vortex and the airfoil require that

$$\Gamma_{C_1} + \Gamma_{C_1^{'}} = \Gamma_{C_2} \tag{1.62}$$

Since $\Gamma_{C_2} = 0$, thus

$$\Gamma_{C_1^{'}} = -\Gamma_{C_1} \tag{1.63}$$

This is a significant aerodynamic result. It asserts that circulation around the beginning vortex, indicated by A, necessitates equal and opposite circulation around the airfoil. This circulation around the airfoil can be attributed to a fictional vortex, which is required for lift production. This artificial vortex is known as a bound vortex because it remains linked to the airfoil while in motion. It is important to note that a confined vortex differs from a free vortex, which may move freely across the flow area. Furthermore, if the airfoil comes to a halt, the bound vortex, shown by B in Figure 9d, sheds downstream again due to viscous action at the sharp trailing edge. Because the viscosity has acted on either of these vortices, the circulations around curves C4 and C5, respectively, have occurred.

Surrounding the initial vortex and the shed bound vortex are discovered to be nonzero in an anticlockwise orientation. However, the curvature (C_3) encircles both vortices and passes through the fluid, where there has never been any friction, the circulation stays zero. As a result of the equal and opposing circulations of vortices. The entire bounded region C_3 has zero net vorticity. The convected vortices A and B of equal and opposing strength the fluid carries it downstream. As seen in Figure 9d, the vortex A causes downwash w_A on the vortex B, which induces downwash w_B on the vortex A.

Furthermore, when all of the torque generating elements in Kelvin's theorem (pressure forces, body forces, and viscous forces) are missing, the geometrical interpretation of vortex motion dynamics may be expressed in terms of Helmholtz's theorem. Readers are urged to refer to Sect. 1.4 for further information on these theorems. However, for ease of reference, the substance of these four Helmholtz's theorems is as follows

- The strength (i.e., circulation) of a closed-loop filament remains constant over its length, according to the first theorem.
- The second theorem argues that a vortex line can never abruptly cease in a fluid; instead, they form complete loops.
- The third theorem asserts that a fluid line that is a vortex line at any moment stays a vortex line.
- The fourth theorem states that the ratio of vorticity to the product of the fluid density and the length of the vortex line remains constant as time passes on a vortex line of fixed identity.

The preceding rules may be validated by using the example of smoke rings, which are actually vortex filaments. These filaments create complete loops because they can never terminate. Furthermore, along a vortex line, each vortex induces a component of velocity on the other vortices (downwash), allowing these smoke rings to convect away on their own. As a result, these smoke rings, which are in fact fluid masses, are transported away by the vorticity. Furthermore, these vorticities (smoke rings) never die, but fade away owing to viscosity.

So far in our study of vorticity, we've looked at an airfoil that is believed to be an infinite-span wing (i.e., two-dimensional wing). However, the wings of actual aircraft are finite in length and made up of a sequence of airfoils with varying cross sections. That is, a genuine wing has a three-dimensional form.

When an aeroplane is in motion, the lower surface (or pressure surface) of its wings experiences a greater pressure zone, while the upper surface experiences a lower pressure region (or suction surface). This pressure differential does, in fact, cause lift on the wing. It's worth noting that the wings have a pressure greater than the freestream static pressure at the bottom surface, but a pressure lower than the freestream static pressure at the top surface. Furthermore, because pressure is a continuous function, it must be the same at the wing tips. In turn, the fluid components have a propensity to travel from the bottom surface of the wing around the wing tips to the top surface, resulting in equal pressure on both sides.

Figure 10. The shedding of wing tip vortices, as well as the upwash and downwash behind a flying aircraft.

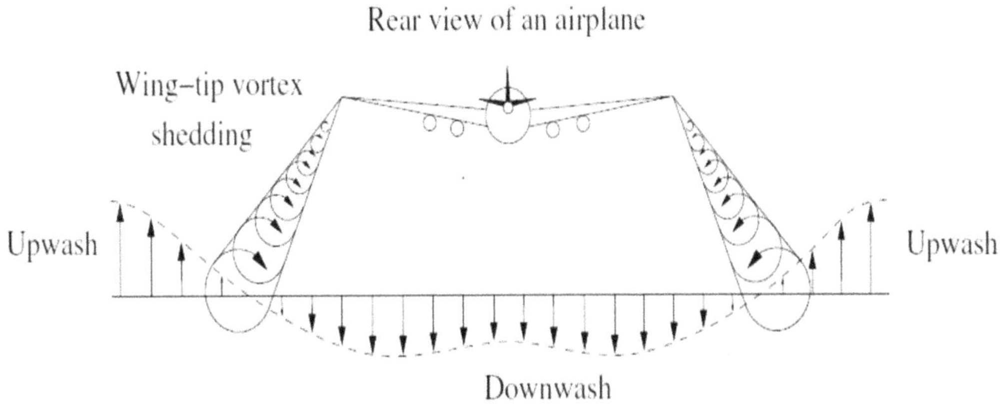

Flow slip refers to the movement of fluid components from a high-pressure zone to a low-pressure region around the wing tips. As a result, there is a spanwise component of flow on the wing's top surface in the direction of tip to root. Similarly, the flow on the lower surface has a spanwise component in the direction of root to tip. In addition, there is an oncoming freestream of air travelling toward the wing. When these two movements are combined, the resulting motion causes the flow to be slanted inward, toward the wing root on the top surface, and outward, toward the wing tip on the bottom side.

Figure 11. A horseshoe vortex model that has been simplified

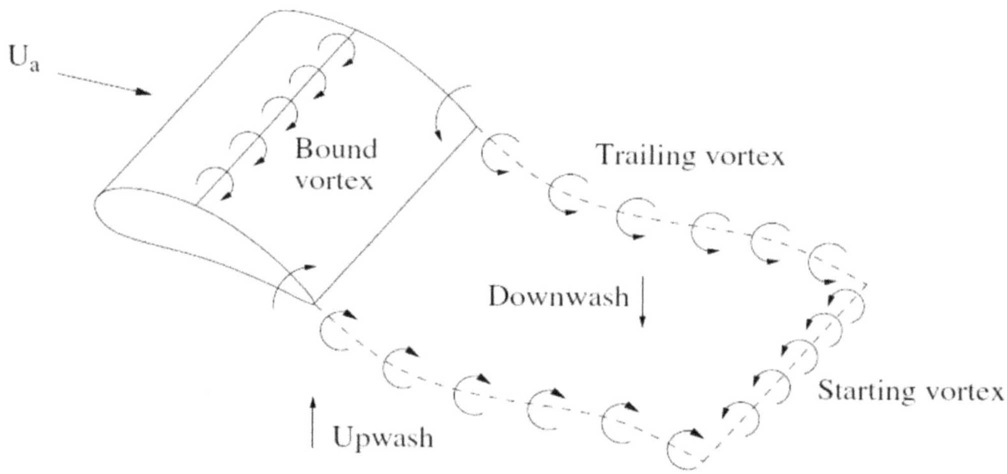

As a result, for the flow exiting the trailing edge of the wings, the upper surface pathlines are certainly inclined to the lower surface pathlines, resulting in helical pathways or vortices. The wings shed a complete line of vortices, with the vortex strength highest at the wing tip and rapidly decreasing to zero at mid-span. After a short distance downstream, these vortices combine to produce two unique helical vortices. As depicted in Figure 10, these vortices shed from the trailing edge of the wings are known as trailing edge vortices or trailing vortices. The figure indicates that the rotations of vortices shed from either wing are polar opposites, one clockwise and the other anticlockwise. However, regardless of their the direction of vortex rotation in these orientations is such that the flow is always oriented toward the wake region (at the rear of the aircraft). Furthermore, these trailing vortices generate further vortices. The downward component of velocity behind the wing (inside the wingspan). If we remain still and look at these vortices from the same angle. When we look at the back of a moving aeroplane, we notice that all of the fluid downward movement inside the vortex system (referred to as referred to as downwash), but all fluid outside of this system will progress upward (known as upwash). The gradient of velocity is occasionally necessary to build a connection between the upwash and the downwash. Quite massive and cause tremendous movements, which might be devastating for an aeroplane flying through it, as it could result in structural deterioration.

Although all of the vortices outlined above may be addressed independently, in real flows, they all coexist and constitute a complex vortices system. For a finite-span wing, the horseshoe vortex model offers a simplified depiction of this complicated vortex system. As illustrated in Figure 11, it is a closed loop composed of a line of bound vortices of continuous circulation going with the wing and important for creating lift, and two lines of trailing wing tips vortices, each with a form resembling roughly a horseshoe. When the wing begins to travel through the fluid, the vortex loop is finally completed by the line of initial vortices shed. This initial vortex is dissolved as a result of the action of viscosity.

Furthermore, despite the fact that the simplified vortex model is unrealistic because it expects uniform circulation at all sections of a finite wing, with a reduced effective wingspan and the same mid-plane circulation, a satisfactory model for the flows induced sufficiently far away from the aircraft is provided.

6.1 Lanchester–Prandtl Wing Theory

The lifting line theory is a mathematical model that predicts the aerodynamic characteristics of a limited span wing. Because Lanchester (1907) and Prandtl (1918) separately developed this theory, it is also known as the Lanchester–Prandtl lifting line theory or the Lanchester–Prandtl wing theory. The intensity of vortices was considered to decrease over the whole wing in this model because they continually shed downstream a trail of vortices producing a vortex sheet from the entire trailing edge, rather than just from the wing tips. By dispersing an unlimited number of horseshoe vortices throughout the whole width, this theory replicates a finite wing.

A lifting line is formed by the bound vortices put on a line going through the aerodynamic centres of each airfoil segment of the wing. In order to account for the change in lift (or circulation) over the span, it was also assumed that the lifting line, which consists of distributed bound vortices, has the circulation variable. As a result, free vortices (trailing vortices) emerge and are shed downstream into infinity. These trailing vortices are responsible for creating downwash at the lifting line, which alters the local angle of attack at each airfoil segment. Because of the addition of the aspect ratio, AR, Lifting Line Theory correlates significantly better with Lightning wind tunnel data than Thin Airfoil Theory. This variable allows you to calculate the effect of trailing vortices and their downwash on wing lift. This similar ratio may be used to approximate the induced drag caused by downwash on the wing.

Furthermore, the aerodynamic forces are more accurately characterised at the aerodynamic centre than at any other place on the chord line because it is the point at which the pitching moment coefficients do not vary with angle of attack, i.e., $\frac{dC_M}{d\alpha} = 0$. Furthermore, an airfoil may be statistically simulated by distributing vortices continuously throughout the surface of the airfoil section, forming a vortex sheet. The vortex panel is a sheet that may be discretized into a succession of straight pieces of varying strengths. This panel, which is a two-dimensional straight-line segment, does, in fact, constitute a quadrilateral panel for a three-dimensional wing, with each element of the vortex sheet seen as a straight vortex segment travelling down the y-axis.

Consider a quadrilateral panel with dimensions c (y) along the x-axis and 2β along the y-axis, as illustrated in Figure 12, to determine the downwash caused by the trailing vortices at some position downstream of the wing. Consider a spanwise portion of a standard horseshoe vortex system with strength $\Delta\Gamma(\beta) = -\left(\frac{d\Gamma(\beta)}{dy}\right)d\beta$ is chosen at the quarter-chord position $\frac{c(y)}{4}$ and its influence investigated at the control point $\frac{3c(y)}{4}$ in the quadrilateral panel. Furthermore, a continuous distribution of bound vortices on this vortex segment generates a lifting line that induces downwash in its vicinity. Using Eq.

(1.59), calculate the downwash (Δw_b) at the control point $\left[\frac{c(y)}{2}, y\right]$ because of the fact that this portion is provided by

$$\Delta w_b = -\frac{\Delta\Gamma(\beta)}{4\pi\frac{c(y)}{2}}(\cos\theta_1 - \cos\theta_2) \tag{1.64}$$

We can deduce the geometry of the horseshoe vortex system from the geometry depicted in Figure 12,

$$\cos\theta_1 = \frac{(y+\beta)}{\left[\left(\frac{c}{2}\right)^2 + (y+\beta)^2\right]^{1/2}}$$

And

$$\cos\theta_2 = \frac{(y-\beta)}{\left[\left(\frac{c}{2}\right)^2 + (y-\beta)^2\right]^{1/2}}$$

Figure 12. Replacement of a finite wing with the system of horseshoe vortices

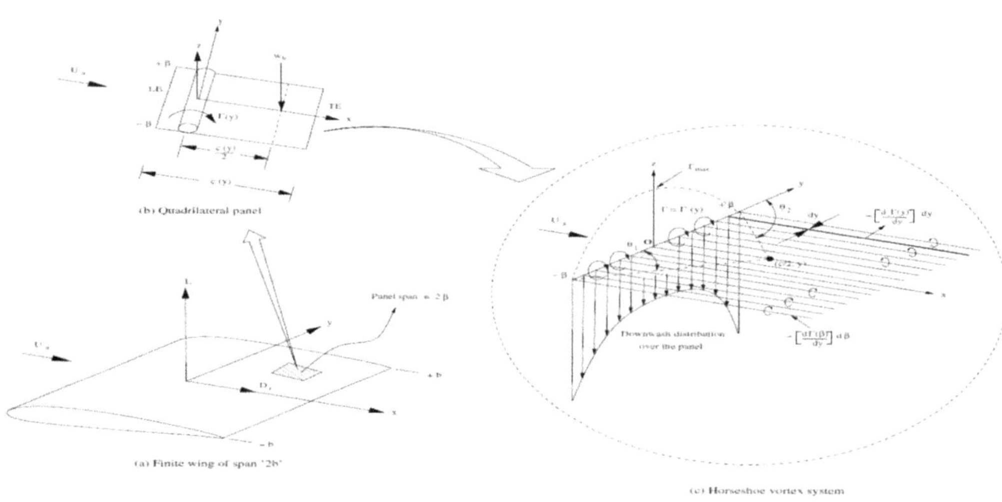

From aforementioned substitutions, Δw_b becomes

$$\mathrm{d}w_b = -\frac{\Delta\Gamma(\beta)}{4\pi\frac{c(y)}{2}}\left(\frac{(y+\beta)}{\left[\left(\frac{c}{2}\right)^2+(y+\beta)^2\right]^{1/2}}-\frac{(y-\beta)}{\left[\left(\frac{c}{2}\right)^2+(y-\beta)^2\right]^{1/2}}\right) \tag{1.65}$$

However, given a wing with a high aspect ratio, the factor $\left(\frac{c}{2}\right)^2$ can be ignored, and the preceding equation is reduced to

$$\mathrm{d}w_b = -\frac{\Delta\Gamma(\beta)}{4\pi\frac{c(y)}{2}}(2) = -\frac{\mathrm{d}\Gamma(\beta)}{2\pi\frac{c(y)}{2}} \tag{1.66}$$

It is important to note that this is the downwash produced by a single horseshoe vortex system. The lifting line, on the other hand, is made up of an unlimited number of horseshoe vortices, each with an arbitrarily tiny strength d. Taking into account all of these horseshoe vortices' contributions, the induced downwash for the whole lifting line, assessed at y, will become

$$w_b = -\frac{\Gamma(y)}{2\pi\frac{c(y)}{2}} \tag{1.67}$$

Next, we want to look at the downwash caused by the wing's trailing vortices. Because the intensity of bound vortices varies along the spanwise direction, and because Helmholtz's theorem stipulates that no vortex filament may begin or stop in a flow suddenly, except at solid barriers, any local alteration in this circulation must be shed downstream. Indeed, a change in circulation along a horseshoe vortex system's lifting line will shed downstream in the form of trailing vortex, and the power of each following vortex is equal to the change in circulation along the lifting line.

Furthermore, for a finite-span wing, the trailing vortex shedding is considered to be shed from a semi-infinite vortex line with the strength $\Delta\Gamma(y) = \left(\frac{d\Gamma(y)}{dy}\right)dy$. Thus, the intensity of a vortex line beginning at $y=\beta$ will be $\Delta\Gamma(\beta) = \left(\frac{d\Gamma(\beta)}{dy}\right)dy$. The downwash at the control point may be expressed as using Eq. (1.61) and for $\theta_1 \approx \frac{\pi}{2}$ and $\theta_2 \approx \pi$,

$$\Delta w(y) = \frac{\Delta\Gamma(y)}{4\pi}\frac{1}{(y-\beta)} \tag{1.68}$$

This is the result of a single trailing vortex filament. To compute the downwash caused by the entire wing, we must integrate equation (1.68) across the entire length of the wingspan, i.e. within the constraints $-b \leq y \leq b$, gives

$$w_i(y) = \frac{1}{4\pi} \int_{-b}^{b} \frac{\left[-\frac{d\Gamma(y)}{dy} \right] dy}{(y-\beta)} \qquad (1.69)$$

Furthermore, we may deduce from the boundary condition that no flow can penetrate the solid surface of a wing. In other words, the flow velocity normal to the wing surface will be nearly zero. Clearly, on the solid surface of the wing, the total of normal velocity components produced by the wing (w_b), trailing vortices (w_i), and the normal velocity component of the freestream flow ($U_a \propto$) must be zero.

$$w_b + w_i + U_a \propto = 0 \qquad (1.70)$$

The subscripts 'b' and 'i' represent the bound and induced effects, respectively. Furthermore, by combining Eqs. (1.67) and (1.69), we get (1.70).

$$-\frac{\Gamma(y)}{2\pi \frac{c(y)}{2}} + \frac{1}{4\pi} \int_{-b}^{b} \frac{\left[-\frac{d\Gamma(y)}{dy} \right] dy}{(y-\beta)} + U_a \propto = 0 \qquad (1.71)$$

Taking the preceding equation and dividing it by U_a yields

$$-\frac{\Gamma(y)}{2\pi \frac{c(y)}{2} U_a} + \frac{1}{4\pi} \int_{-b}^{b} \frac{\left[-\frac{d\Gamma(y)}{dy} \right] dy}{(y-\beta) U_a} + \propto = 0 \qquad (1.72)$$

This equation is known as Prandtl's lifting line integrodifferential equation for the spanwise load distribution (y), and it is a mixture of several angles, as shown in Figure 2. When the terms of Eqs. (1.2) and (1.72) are compared, the induced angle of attack αi and effective angle of attack (α_{eff}) are obtained.

$$\alpha_i = \frac{1}{4\pi} \int_{-b}^{b} \frac{\left[-\frac{d\Gamma(y)}{dy} \right] dy}{(y-\beta) U_a} \qquad (1.73)$$

$$\alpha_{eff} = \frac{\Gamma(y)}{2\pi \frac{c(y)}{2} U_a} \tag{1.74}$$

Furthermore, differences in lift distribution across the wing span modify the downwash produced at each airfoil section, which alters the induced angle of attack α_i. As a result, the effective angle of attack (α_{eff}) changes over the wingspan, i.e., $\alpha_{eff}=\alpha_{eff}(y)$. Thus, an airfoil at $y=\beta$ (Note that this is also the position of a trailing vortex filament for the quadrilateral panel under consideration, as illustrated in Figure 12b) will have an effective angle of attack eff ($\alpha_{eff}(\beta)$). Furthermore, for a cambered airfoil section, α_{eff} is measured from the zero-lift angle ($\alpha_{L=0}$), but for a symmetric airfoil, it is measured from the origin $C_{L(\alpha=0)}$. The lift curve slope (a_0) for a cambered airfoil section is given by

$$a_0 = \frac{dC_{L'}}{d\alpha} = 2\pi$$

The local lift coefficient ($C_{L'}$) for an airfoil section situated at $y=\beta$ may be calculated using the above relation for an airfoil section as

$$C_{L'}\left(\alpha_{eff(\beta)}\right) - C_{L'}\left(\alpha_{L=0}\right) = a_0\left(\alpha_{eff(\beta)} - \alpha_{L=0}\right) = 2\pi\left(\alpha_{eff(\beta)} - \alpha_{L=0}\right) \tag{1.75}$$

However, since $C_{L'}\left(\alpha_{L=0}\right) = 0$, the above equation reduces to

$$C_{L'}\left(\alpha_{eff(\beta)}\right) = 2\pi\left(\alpha_{eff(\beta)} - \alpha_{L=0}\right) \tag{1.76}$$

Furthermore, the local lift coefficient ($C_{L'}$) for the airfoil section at β is defined by

$$C_{L'}\left(\alpha_{eff(\beta)}\right) = \frac{L'}{q_a c(\beta)} \tag{1.77}$$

where L' is the lift per unit span, $q_a = \frac{1}{2}\rho U_a^2$ an is the freestream dynamic pressure, and $c(\beta)$ is the panel's chord length. The preceding formula may be rewritten using the Kutta–Joukowski theorem as

$$C_{L'}\left(\alpha_{eff(\beta)}\right) = \frac{\rho U_a \Gamma(\beta)}{\frac{1}{2}\rho U_a^2 c(\beta)} = \frac{2\Gamma(\beta)}{U_a c(\beta)} \tag{1.78}$$

Assume a two-dimensional airfoil section with a local lift curve slope of a_0 and a local effective angle of attack of α_{eff} as observed by it. By plugging Eq. (1.78) into (1.76), we get

$$\alpha_{eff(\beta)} = \frac{\Gamma(\beta)}{\pi U_a c(\beta)} + \alpha_{L=0} \tag{1.79}$$

Remembering the generalised form of the effective angle of attack from Eq. (1.2) and assuming the airfoil section is positioned at $y=\beta$, the local effective angle of attack becomes

$$\alpha_{eff(\beta)} = \alpha(\beta) - \alpha_i(\beta) \tag{1.80}$$

The following equation, using Eqs. (1.79) and (1.73), produces the geometric angle of attack at $y=\beta$ as

$$\alpha(\beta) = \frac{\Gamma(\beta)}{\pi U_a c(\beta)} + \alpha_{L=0}(\beta) + \frac{1}{4\pi}\int_{-b}^{b} \frac{\left[-\frac{d\Gamma(y)}{dy}\right]dy}{(y-\beta)U_a} \tag{1.81}$$

This is the basic equation of Prandtl's classical lifting line theory. It denotes that the local geometric angle of attack for a particular cambered airfoil section is equal to the sum of the effective angle of attack and the induced angle experienced by the section locally. $\Gamma(\beta)$ may be easily estimated from Eq. 1.81 for a given finite wing with known geometric angle of attack in a freestream of specified velocity. Also offered is a generic boundary condition that fulfils the solution of equation (1.81) is

$$\Gamma(y=\pm b)=0 \tag{1.82}$$

6.2 Symmetric Elliptical Aerodynamic Load Distribution

Consider the circulation distribution illustrated in Figure 12c. The circulation $\Gamma(y)$ across the finite wing can be represented as shown in the image.

$$\left(\frac{\Gamma(y)}{\Gamma_{\max}}\right)^2 + \left(\frac{y}{b}\right)^2 = 1 \tag{1.83}$$

This is simply an ellipse equation. As a result, Eq. (1.83) is more properly known as the elliptical distribution of circulation. However, because the circulation curve only forms a half ellipse across the wingspan, the area under this curve is $\frac{\pi b \Gamma_{max}}{2}$, or half of the overall ellipse size.

To derive the total downwash at a particular position $y=\beta$, by the entire trailing vortex sheet, let us differentiate equation (1.83) with regard to y.

$$\frac{d\Gamma(y)}{dy} = \frac{\Gamma_{max}}{2}\left[1-\left(\frac{y}{b}\right)^2\right]^{-1/2}\left(-2\frac{y}{b^2}\right) \tag{1.84}$$

Substituting Eq.(1.84) into (1.69) yields

$$w_i(y) = \frac{\Gamma_{max}}{4\pi b^2}\int_{-b}^{b}\frac{\left[1-\left(\frac{y}{b}\right)^2\right]^{-1/2} ydy}{(y-\beta)} \tag{1.85}$$

Note that, the above integral has a singularity which can be removed from Cauchy principle value which can be solved by using the following substitution:

$$y = b\cos\theta \tag{1.86}$$

On differentiating the above, we get

$$dy = -b\sin\theta d\theta \tag{1.87}$$

The elliptic distribution of circulation, given by Eq. (1.83), in transformed coordinates is obtained from Eq. (1.86).

$$\Gamma(\theta) = \Gamma_{max}\sin \tag{1.88}$$

Putting Eq. (1.86) into (1.85) and observing that at the wing tips, $\theta=\pi(y=-b)$ and $\theta=0(y=-b)$, obtains

$$w_i(\theta) = \frac{\Gamma_{max}}{4\pi b}\int_{-b}^{b}\frac{\cos\theta d\theta}{(\cos\theta - cos\beta)} \tag{1.89}$$

It is worth noting that the above integral has a similar shape to a Glauert identity, i.e.,

$$\int_{0}^{\pi}\frac{\cos n\theta d\theta}{(\cos\theta-\cos\theta_0)} = \frac{\pi\sin n\theta}{\sin\theta} \tag{1.90}$$

Where n = integers (including "0"). Relating Eq.(1.89) with (1.90), we find n=1. Thus,

$$\int_{0}^{\pi}\frac{\cos\theta d\theta}{(\cos\theta-\cos\beta)} = \frac{\pi\sin\theta}{\sin\theta} = \pi \tag{1.91}$$

Eq. (6.91) thus gives the total induced downwash for the whole wingspan as

$$w_i(\theta) = -\frac{\Gamma_{max}}{4b} \tag{1.92}$$

Finally, by plugging Eq. (1.92) into (6.1), the induced angle of attack αi is obtained.

$$\alpha_i = -\frac{\Gamma_{max}}{4bU_a} \tag{1.93}$$

It is worth noting that Eqs. (1.92) and (1.93) are crucial conclusions of finite wing theory. They claim that given a symmetric elliptical lift distribution throughout the wingspan, both induced downwash ($w_i(\theta)$) and induced angle α_i are constant along the span rather than a function of "y." Furthermore, both $w_i(\theta)$ and α_i approach 0 if 2b→∞, i.e. when the wingspan is endlessly enormous.

6.2.1 Aerodynamic Loads and Coefficients for an Elliptic Lift Distribution

According to the Kutta–Joukowski theorem, the lift per unit length (L') is written as

$$L'(y) = \rho U_a \Gamma(y)$$

Substituting the expression for $\Gamma(y)$ from Eq. (1.83), we get

$$L'(y) = \rho U_a \Gamma_{max}\left[1-\left(\frac{y}{b}\right)^2\right]^{1/2} \tag{1.94}$$

To find the total lift, given by L, for the full wingspan, integrate the previous equation between the limitations; $-b \leq y \leq b$. In other words,

$$L = \frac{\pi b}{4}\rho U_a \Gamma_{max} \tag{1.95}$$

For a symmetric elliptic lift distribution, this is the total lift operating on the whole wingspan. The total induced drag (D_i) on the wing will be calculated next. Remember Eq (1.12)

$$D_i = \rho U_a \int_{-b}^{b} \Gamma(y)\alpha_i(y)dy$$

Substituting Eqs.(1.83), (1.93) and $\frac{y}{b} = \cos\theta$ into above integral and by changing the limits of integration yields

$$D_i = \frac{\rho^{\text{“}2}_{max}}{4}\int_0^{\pi} sin^2\theta d\theta$$

On solving, the overall induced drag for a symmetric elliptical loading is given by

$$D_i = \frac{\pi\rho\Gamma_{\max}^2}{8} \tag{1.96}$$

Further, by definition, the lift coefficient is expressed as

$$C_L = \frac{L}{\frac{1}{2}\rho U_a^2 S} \tag{1.97}$$

where S denotes the wing's planform area The formula for C_L is obtained by substituting the total lift L from Eq. (1.95) in the aforementioned equation to get

$$C_L = \frac{\pi}{2}\frac{b}{S}\frac{\Gamma_{\max}}{U_a} \tag{1.98}$$

Simplifying the above equation for $\Gamma_{\max}$, we get

$$\Gamma_{\max} = \frac{2U_a S C_L}{\pi b} \tag{1.99}$$

In addition, the coefficient of induced drag may be expressed as

$$C_{D_i} = \frac{D_i}{\frac{1}{2}\rho U_a^2 S} \tag{1.100}$$

In the previous equation, using the value for total induced drag (D_i) provided by Eq. (1.96) yields C_{Di} as

$$C_{D_i} = \frac{\pi}{4}\frac{\Gamma_{\max}^2}{U_a^2 S} \tag{1.101}$$

Substituting Eq.(1.99) into (1.101), we get

$$C_{D_i} = \frac{1}{\pi}\frac{S}{b^2}C_L^2 \tag{1.102}$$

The aspect ratio (AR), defined as the ratio of the square of the semi-wingspan to the planform area of the wing, is a very essential and useful geometric parameter for a finite wing. Thus,

$$AR = \frac{b^2}{S} \tag{1.103}$$

In turn, Eq.(1.102) takes the form

$$C_{D_i} = \frac{C_L^2}{\pi AR} \tag{1.104}$$

This is an essential conclusion of finite wing theory, which demonstrates that the coefficient of generated drag is exactly proportional to the square of the coefficient of lift and inversely proportional to the wing's aspect ratio. In other words, the coefficient of generated drag is determined by the lift coefficient as well as the aspect ratio. In other words,

$$C_{D_i} = C_{D_i}\left(C_L, AR\right) \tag{1.105}$$

However, because the induced drag is simply a result of the lift produced on a limited wing, this dependency of the induced drag on the lift is not unique. We previously described how the formation of wing tip vortices caused by pressure differences between the suction and pressure surfaces generates changes in geometric angles of attack (α) at each airfoil section. A reduction in α tilts the lift downstream in the freestream direction, introducing a component of the lift along the freestream direction. This extra force component is referred to as induced drag or, more colloquially, drag owing to lift. Because the same pressure imbalance exists between the top and bottom surfaces of a wing, it is the process that generates both lift and wing tip vortices (i.e., induced drag); consequently, induced drag and lift on a finite wing are inextricably related.

Furthermore, because the coefficient of induced drag is inversely proportional to the aspect ratio of the wing, we must have the maximum feasible aspect ratio in order to minimise the induced drag. The employment of very large aspect ratio wings, on the other hand, is impossible due to significant structural restrictions.

6.3 Lifting Characteristics of Wing

The maximum lift coefficient of the airplane $C_{L_{max}}$ depends upon many factors. Only the most important of these will be considered here and they are listed below.

- Airfoil maximum lift coefficient $c_{l_{max}}$.
- Wing aspect ratio A, taper ratio λ, and sweepback angle Λ.
- Trailing edge flap design and deflection angle.
- Leading edge flap design and deflection angle.

The methods that will be used to estimate $C_{L_{max}}$ for the various configurations of a wing are taken mainly from the USAF Stability and Control DATCOM, Hoak et al. (1978) and are empirical in nature. There are other, more sophisticated, approaches based on different computational fluid dynamics (CFD) schemes which will be discussed in varying degrees of detail. Such scientifically richer methods will generally have been covered in the fluid dynamics analysis courses of an engineering degree program and may be implemented, if desired. However, it is important to develop some familiarity with empirical and approximate techniques rarely covered in academic courses. The rapid turnaround they provide is of great value in preliminary design situations in industry.

6.3.1 Determination of the Wing Lift Curve Slope for Swept Back Wing

The airfoil characteristics described thus far are based on two-dimensional flow whereas wings have finite span and three-dimensional effects must be considered. This lifting line theory on wings should be reviewed to complement the mainly empirical approaches presented here. The three-dimensional lift curve slope of conventional wings $C_{L\alpha}$ is given, per radian, by the following equation:

$$\frac{C_{L\alpha}}{AR} = \frac{2\pi}{2+\left[\frac{AR^2\beta^2}{k^2}\left(1+\frac{\tan^2\wedge_{\frac{c}{2}}}{\beta^2}\right)+4\right]^{1/2}} \tag{1.106}$$

Thus $C_{L\alpha}$ is a function of wing aspect ratio, mid-chord sweep angle $\wedge_{\frac{c}{2}}$, Mach number, and airfoil section (defined parallel to the free stream) lift curve slope. The factor k in equation (1.) is the ratio of the experimental two-dimensional (i.e., airfoil) lift curve slope (per radian) at the appropriate Mach number $(C_{l\alpha})_M$ to the theoretical value at that Mach number, $\frac{2\pi}{\beta}$ or $k=(C_{l\alpha})_M/2\pi\beta$. Note that the theoretical (Prandtl-Glauert) correction for subsonic compressibility is $(C_{l\alpha})_M=(C_{l\alpha})/\beta$, so in the absence of an experimental value for $(C_{l\alpha})_M$ the value for $k=(C_{l\alpha})/2\pi$, that is the ratio of the actual low-speed airfoil lift curve slope to that of the airfoil in ideal incompressible flow will suffice. The section lift curve slope (per degree) is obtained from thin airfoil theory and $\beta=\sqrt{1-M^2}$ is the Prandtl-Glauert factor.

The sweep-conversion formula, may be used to find the mid-chord sweep for any straight-tapered wing as follows:

$$\tan\wedge_{1/2c} = \tan\wedge_{LE} - \frac{2}{AR}\left(\frac{1-\lambda}{1+\lambda}\right) \tag{1.107}$$

Recall that λ is the taper ratio, $\frac{c_t}{c_r}$. To find the sweepback angle of the leading edge from that at any other constant percent chord line $\left(n=\%\frac{c}{100}\right)$, for trapezoidal wing planforms yields

$$\tan\wedge_{LE} = \tan\wedge_{nc} - \frac{4n}{AR}\left(\frac{1-\lambda}{1+\lambda}\right) \tag{1.108}$$

For example, if the quarter chord sweepback angle is known (n=1/4=0.25), the sweepback angle of the leading edge is easily determined. In a similar fashion, once the sweepback angle of the leading edge is known, the sweepback angle of any other constant percent chord line can be easily found:

$$\tan\wedge_{nc} = \tan\wedge_{LE} - \frac{4n}{AR}\left(\frac{1-\lambda}{1+\lambda}\right) \tag{1.109}$$

6.3.1.1 Sample Calculation of the Wing Lift Curve Slope

Consider the 63-009 airfoil, a symmetric section of thickness ratio t/c = 9% being used in a wing with an aspect ratio AR = 6, a leading-edge sweepback Ù=45.6°, and a taper ratio λ=0.55. To find the lift curve slope of this wing at a Mach number M=0.45, we may use equation (1.106). First we must determine the sweepback of the half-chord line, which may be found using equation (1.109):

$$\tan\wedge_{c/2} = \tan\wedge_{LE} - \frac{2}{AR}\left(\frac{1-\lambda}{1+\lambda}\right) = \tan(45.6) - \frac{2}{6}\left(\frac{1-0.55}{1+0.55}\right) = 0.924$$

Table 1. Basic characteristics of NACA airfoil at Reynolds number $Re_c = 9 \times 10^6$

Airfoil	α_0(deg)	$C_{m,0}$	$C_{l\alpha}$	a.c	$\alpha_{cl,max}$	$C_{l,max}$	$\alpha^0 w$
63-006	0	0.005	0.112	0.258	10.0	0.87	7.7
63-009	0	0	0.111	0.258	11.0	1.15	10.7
63_1-012	0	0	0.116	0.265	14.0	1.45	12.3
63_2-015	0	0	0.117	0.271	14.5	1.47	11.0
63_3-018	0	0	0.115	0.271	15.5	1.54	11.2
63-210	-1.2	-0.035	0.113	0.261	14.5	1.56	9.6
63_1-212	-2.0	-0.035	0.114	0.263	14.5	1.63	11.4
63_2-215	-1.0	-0.03	0.116	0.267	15.0	1.60	8.8
63_3-218	-1.4	-0.033	0.118	0.271	14.5	1.85	8.0

We also require the value of $k=(C_{l\alpha})_M/2\pi\beta$. Because $(C_{l\alpha})_{M,}$ the experimental value for $C_{l\alpha}$ at M=0.45, is not provided here, we approximate it by $C_{l\alpha}/\beta$, where $C_{l\alpha} = 0.111/^0\left(6.363\frac{1}{rad}\right)$ is the low-speed value given in Table 1. Then equation (1.106) yields

$$C_{L\alpha} = \frac{2\pi AR}{2+\left[\frac{AR^2\beta^2}{k^2}\left(1+\frac{\tan^2 \wedge_{\frac{c}{2}}}{\beta^2}\right)+4\right]^{1/2}}$$

$$= \frac{2\pi(6)}{2+\left[\frac{6^2\left(1-0.45^2\right)^2}{\left(\frac{6.363}{2\pi}\right)^2}\left(1+\frac{0.924^2}{\left(1-0.45^2\right)^2}\right)+4\right]^{1/2}} = \frac{37.68}{9.50} = 3.965$$

This result, $C_{L\alpha}$=3.965 per radian, may also be written as $C_{L\alpha}$=0.069 deg^{-1}. We see that the lift curve slope of the finite wing is substantially less than that of the airfoil of which it is comprised. Note that doubling the Mach number from 0.45 to 0.9 increases the lift curve slope of the wing by almost 10%.

SUMMARY

The attributes of an airfoil section that are, in fact, the same as the properties of an infinite-span wing. These characteristics differ for finite-span wings mounted to an actual aircraft. This is because, unlike an airfoil, which is a two-dimensional object, a wing is fundamentally a three-dimensional entity, with flow in the spanwise direction. That is, the flowover the wings is three-dimensional in nature, and hence its aerodynamic characteristics change greatly from those of their airfoil sections.

The downwash generated by trailing edge vortices shedding from the wing tips, as well as its influence on the inclination of the local relative airflow, has two primary effects on the local airfoil section.

In reality, the actual angle of attack as observed by the airfoil locally is less than the geometric angle of attack (α). For a three-dimensional wing, this angle of attack is known as the effective angle of attack (α_{eff}). In other words,

$$\alpha_{eff} = \alpha - \alpha_i$$

In addition, the effective freestream velocity (U_{eff}) will now be calculated as

$$U_{eff} = \sqrt{U_f^2 + w^2}$$

However, given minor downwash ($w \approx 0$),

$$U_{eff} \approx U_f$$

Furthermore, the downwash produced by these trailing edge vortices from the wing tips creates an extra drag component known as induced drag.

The ideas of vortex sheets and vortex filaments are useful in analysing the aerodynamic properties of limited span wings. The induced velocity of a vortex filament may be computed using the Biot–Savart law from a directed line segment (dl).

$$\vec{v}(r)=\frac{\Gamma}{4\pi}\int\left(\frac{\vec{dl}\times(\vec{r}-\vec{s})}{|\vec{r}-\vec{s}|^{3}}\right)$$

The velocity produced by a finite-length straight vortex filament is given as

$$v_i=\frac{\Gamma}{4\pi d}(\cos\theta_1-\cos\theta_2)$$

Both generated downwash and induced angle are constant along the wingspan for a symmetric elliptical lift distribution.

$$w_i(\theta)=-\frac{\Gamma_{max}}{4b}$$

$$\alpha_i=-\frac{\Gamma_{max}}{4bU_a}$$

For a symmetric elliptic lift distribution, the total lift and induced drag operating on the whole wingspan.

$$L=\frac{\pi b}{4}\rho U_a\Gamma_{max}$$

$$D_i=\frac{\pi\rho\Gamma_{max}^{2}}{8}$$

The three-dimensional lift curve slope of conventional wings $C_{L\alpha}$ is given, per radian, by the following equation:

$$\frac{C_{L\alpha}}{AR}=\frac{2\pi}{2+\left[\frac{AR^2\beta^2}{k^2}\left(1+\frac{\tan^2\wedge_{\frac{c}{2}}}{\beta^2}\right)+4\right]^{1/2}}$$

REFERENCES

Lanchester, F. W. (1907) Aerial flight, Vol. 1: Aerodynamics. Archibald Constable & Co.

Prandtl, L. (1918). *Tragftii geltheorie, 1. Mit-teilung*. Nachr Ges Wiss Gottingen.

Prandtl, L. (1921). Uber die Eindringungsfestigkeit (Harte) plastischer Baustoffeund die Festigkeit von Schneiden. *Zeitschrift für Angewandte Mathematik und Mechanik, 1*, 15–21. doi:10.1002/zamm.19210010102

Von Helmholtz, H. (1868). Über discontinuierliche Flüssigkeits-Bewegungen. Monatsberichte derKöniglichen Preussische Akademieder Wissenschaften zu Berlin, 23, 215–228.

ENDNOTE

1 In the Biot and Savart law formulation for calculating induced velocity, if we curl our fingers from the line segment $\overrightarrow{dl}$ toward $\hat{r}$, the thumb illustrates the direction of induced velocity.

APPENDIX A

Descriptive Type Questions

1. An airplane is flying at 100 m/s in a steady level flight. If the aircraft weighs 60kN and have elliptical wing of span 20 m, determine the induced drag.
2. Determine the wing loading of an aircraft weighing 1000 kg, if the wing area is 25 m^2.
3. Derive the lift and drag equation for a finite wing which experiences freestream velocity parallel to chord length and perpendicular to the wing span.
4. Show that the integral on the right-hand side of Kelvin's circulation theorem vanishes, if the fluid is barotropic.
5. Prove that the elliptical spanwise lift distribution leads to a constant downwash along the span.
6. Consider a vortex sheet, where the velocities above and below the sheet are 8 and $4ms^{-1}$, respectively. The element of this vortex sheet is 0.2m wide, which rolled up into a line vortex after some time. Calculate the strength of the vortex.
7. Plot the curve between the induced drag coefficient and the lift coefficient for an elliptical load distribution over the wing of aspect ratio 7.
8. If the aspect ratio of a glider, having elliptical planform wing, is 6. Calculate the change in minimum angle of glide, if the aspect ratio is doubled. Assume $C_D = 0.03 + 0.07C_L^2$
9. An airplane is flying at $170ms^{-1}$ in a steady level flight. If the aircraft weighs 90kN and have elliptical wing of span 15m, determine the induced drag.
10. Describe the motion of a vortex pair (a) when their circulations are equal and in the same direction, and (b) when their circulations are equal but in opposite directions.
11. Determine the flow field due to (a) a vortex filament, which is in the form of a circular ring (vortex ring), (b) a plane vortex sheet, which is formed by a distribution of horseshoe vortex filament, and (c) an infinite row of point vortices of equal strength distributed along a straight line at equal intervals.
12. Find the path of a vortex bounded by two walls perpendicular to each other.

APPENDIX B

Multiple Choice Questions

1. For a wing of finite span, if the lift coefficient is doubled then the induced drag coefficient
   - (a) will remain same
   - (b) will be increased two times
   - (c) will be increased three times
   - (d) will be increased four times

2. For an aircraft with tapered straight wing, root chord and tip chord are 0.9 and 0.5 m, respectively. If the total wingspan is 14 m, then the aspect ratio of wing will be
   (a) 10
   (b) 20
   (c) 0.1
   (d) 30
3. Which of the following wing shapes has the lowest induced drag according to Prandtl's lifting line theory?
   (a) elliptical
   (b) straight rectangular
   (c) straight tapered
   (d) tapered sweptback
4. Which form of static stability is provided by an aircraft's wing sweep back?
   (a) longitudinal stability
   (b) lateral stability
   (c) directional stability
   (d) both (b) and (c)
5. When the aspect ratio of a finite span wing is raised, the lift curve slope increases.
   (a) decrease
   (b) increase
   (c) remain same
   (d) cannot say
6. When a wing's "flap" is deflected, the lift curve slope changes.
   (a) increase
   (b) decrease
   (c) remain same
   (d) increase or decrease depending upon the shape of the flap.
7. The swept back wings experience_____________ lift than the straight wings for any given length.
   (a) less
   (b) more
   (c) same
   (d) cannot say
8. Stall characteristics vary depending on the kind of aircraft. In a "benign stall," which of the following is TRUE?
   (a) One wing stalls slightly before the other.
   (b) The nose drops gently and the wings remain level throughout.
   (c) The nose rises, pushing the wing deeper into the stalled state.
   (d) All the above are FALSE.
9. If the cruise speed of an aeroplane is raised three times during typical cruise motion (straight level flight), the produced drag is___________.
   (a) increased by three times
   (b) increased by nine times
   (c) 1/3rd of the original
   (d) 1/9th of the original

10. If the coefficient of lift of a finite wing is doubled while the aspect ratio is lowered to half, the produced drag is_______________.
    (a) doubled
    (b) four times the original
    (c) eight times the original
    (d) 1/8th of the original

Answer Key

1. (c)
2. (b)
3. (a)
4. (d)
5. (b)
6. (c)
7. (a)
8. (b)
9. (d)
10. (c)

Chapter 6
Viscous Flow and Its Effect

Kaliappan S.
https://orcid.org/0000-0002-5021-8759
Velammal Institute of Technology, Chennai, India

Raj Kamal M. D.
Velammal Institute of Technology, Chennai, India

Joseph Manuel D.
Velammal Institute of Technology, Chennai, India

Balaji V.
Loyola Institute of Technology, India

Murugan P.
Jimma Institute of Technology, Ethiopia

ABSTRACT

Viscosity is a property that expresses the internal drag of a fluid to motion; impact of viscosity states the statics and flows. Statics means whenever fluids at zero velocity have no relative movements between layers of fluid and thus du/dy = 0. At the time there is no shear stress and viscosity of the fluid is free. Fluid viscosity plays a major role on the fluid floating in it. The authors focused on solids and fluids and the no slip condition, momentum transfer through molecular motion, shear stress and viscosity, Couette flow, and Poiseuille flow. Here the authors made a discussion the Newtonian viscous flow, and the statement of Newton's law of viscosity was examined. The discussion has been extended up to viscosity and the effect of their temperature and impact of increasing in temperature has been explained along with surface tension.

DOI: 10.4018/978-1-6684-4230-2.ch006

INTRODUCTION

The Incompressible and Compressible has a vital phenomenon in fluid mechanics, (Yuan et al., 2019) the compressible flow regime refers to the supersonic/subsonic flow scenario. The incompressible technique, on the other hand, could not accurately depict a number of significant instances with low Mach numbers. In a poppet valve, a cavitating jet is in a similar situation To address the disparity between incompressible and compressible approaches in a thorough manner, (Anand and Christoy, 2021). We address the challenge of stable low Reynolds number flow of a generalised Newtonian fluid through a narrow elastic tube by coupling fluid lubrication theory to a structural problem expressed in terms of Donnell's shell theory (Ahmad et al., 2019). In constant two-dimensional boundary layer flow, the flow and heat transfer of a Cassonian fluid across an extended plate with a Newtonian heating boundary condition are investigated. The control partial differential equations are transformed to nonlinear ordinary differential equations using appropriate similarity transformations before being numerically solved using an implicit finite difference scheme. The flow and heat transfer properties of several integrated parameters, such as Casson's fluid parameter, Eckert's number Ec, Prandtl's number Pr, and conjugate Newtonian heating parameter, are examined and explained (Goldstein, 1948). Singularities may occur in the solution of non-linear parabolic equations due to their non-linearity, according to the research. The equations discovered may aid computers working with laminar boundary layers who require more than a rough answer to have a target to aim for (Dhawan and Narasimha, 1958). On a flat plate, the transition in the boundary layer is investigated from the perspective of intermittent turbulent spot formation. Emmons' probability calculations can be expanded to explain the apparent statistical resemblance of transition zones based on the idea of localised laminar breakdown, which has some experimental evidence. Using these concepts, comprehensive computations of boundary layer characteristics such as mean velocity profiles and skin friction during transition are possible. The intermittency factor is the parameter in the mean velocity profiles, which correspond to a universal one-parameter family. The existence of a relationship between the transition Reynolds number and the rate of generation of turbulent spots is inferred from experimental data (Mureithi et al., 2013) the flow of a boundary layer over a constantly moving heated flat surface with velocity in a streaming flow with velocity and temperature dependent viscosity is investigated in this work. Through the viscous dissipation term, the momentum and energy equations are intertwined. A suitable similarity variable is used to translate the coupled boundary layer equations into a self-similar form. To solve the self-similar boundary layer equations, an effective numerical approach is applied. It is proven that when the velocity ratio is low enough, increasing viscous dissipation dramatically improves local heat transmission, resulting in temperature overshoots near the wall. The temperature-dependent viscosity, as well as the velocity and temperature distribution within the boundary layer, is all demonstrated to be affected by the viscosity variation parameter.

2. VISCOUS FLOW

Viscous flow is a flow in which frictional effects are dominating. The fluid is flowing in layers. The layer that sticks to the wall has no velocity, which is known as the no slip condition, and it causes friction at the wall. Frictional effects exist between two layers of fluid, with the slower layer attempting to slow down the quicker layer, which can be measured using viscosity. We discover that for a wide range of total wedge angles, local solutions exist, and that a class of individual wedge angles and stress exponents is chosen. For all combinations of individual wedge angles, partial local solutions exist, and the stress exponents are calculated as functions of these angles (Anderson and Davis, 1993).

2.1 Solids are Rigid and Fluids Flow

Imagine you're holding a brick between your palms. The energies from your hand are transmitted through the brick as you move your right hand away from you and your left hand towards you. Because the molecules in the brick have definite places, this is the case. The intermolecular connections resist in a spring-like manner when gently moved. Each layer in static equilibrium experiences the same shear stress shown in the figure 1.

Figure 1.

Imagine the brick has been replaced with treacle. The treacle pours into a different shape as you move your hands. Because the molecules have no fixed locations, they cannot withstand a shear stress in static equilibrium. When one layer is moved, the other layers flow over it to accommodate the shift. However, there is a shear stress in the moments leading up to mechanical equilibrium. We'll find soon that it's related to the strain rate as be shown in the figure 2.

Figure 2.

What if we replaced the treacle with the in viscid fluid we've been considering so far? The slide of an in viscid fluid is excellent. It is unable to withstand any shear stress. It would be ideal as a lubricant (if you could keep it in the desired place) it has been shown in the figure 3.

Figure 3.

Minimizing the rate of energy dissipation for a class of trial stress distributions is used to estimate the resistance of a porous media to a fluid flowing through it. In terms of certain two- and three-point averages that characterise the medium, rigorously valid lower bounds on the resistance are determined. The results are reported for both flow with and without pore wall slide (Prager, 1961).

2.2 No Slip Condition and Momentum Transfer

Consider the molecules in a gas that are trapped between two plates that move from left to right. Experiments show that the molecules adhere to surfaces long enough to achieve thermal equilibrium before returning to the gas. As a result, when they leave the surface, they have the same x-velocity and temperature as the surface on average. This is a no-slip condition has been shown in the figure 4.

Figure 4.

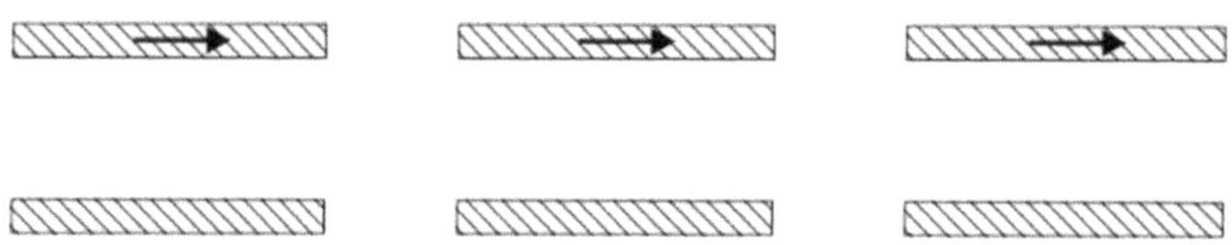

Nearby molecules clash with molecules that have just left the surface. The excess x-momentum of the molecules arriving from the top surface has been diffused into adjacent layers of fluid after many encounters. These, in turn, jostle with molecules nearby, transferring x-momentum further into the fluid. Eventually, x-momentum diffuses all the way to the bottom plate, and a linear velocity profile is obtained by averaging the velocities of all the molecules. This is similar to heat diffusion, in which a linear temperature profile is obtained and it has been shown in the figure 5.

We explain how Prandtl's boundary layer equations are affected by the viscosity of a viscous fluid in prospective flow away from the boundary layers. For sound waves that perturb the Navier–Stokes equations linearized on a state of rest, potential flow equations for viscous compressible fluids are derived. These linearized equations support a potential flow with the interesting characteristic that the Bernoulli equation, potential, and stress are all affected by viscosity. The impact of viscosity is to cause spatially periodic waves to decay in time or time-periodic waves to decay and expand in space. It is neither essential nor sufficient that potential flows meet the Navier–Stokes equations, which encompasses all potential flows of incompressible fluids as well as potential flows in the acoustic approximation derived here (Joseph, 2003).

2.3. Shear Stress and Viscosity

We treat the fluid as a continuum because there are too many molecules to consider each one individually. The momentum transfer caused by molecular motion must therefore be simulated in some way. Fluid layers adjacent to each other exchange momentum at a constant rate. Proportional to the gradient in velocity the rate of change is determined by Newton's second law (f = ma).A force is essentially a change in momentum across a certain area. When multiplied by the area. This is the shear stress area τ and it has been shown in the figure 6.

Figure 5.

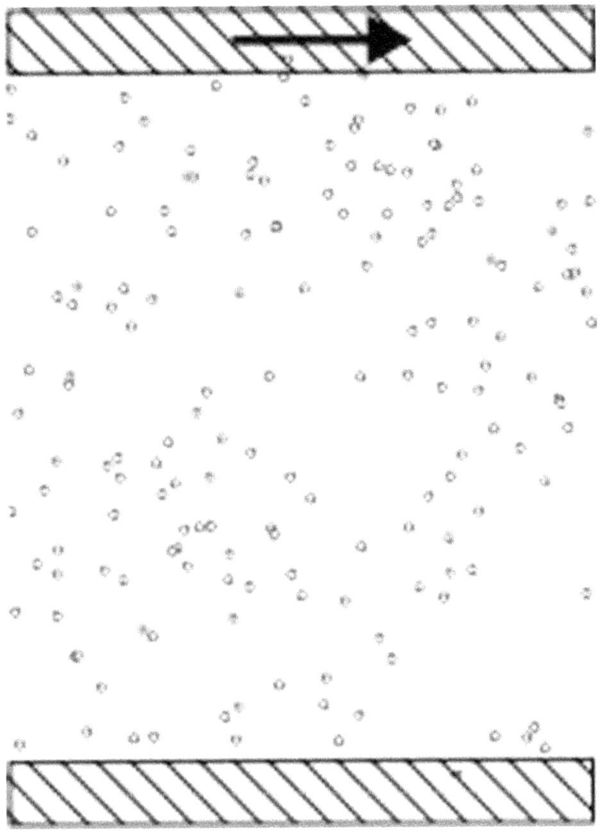

Figure 6.

In a certain time and over a certain area, some molecules swap places

The coefficient of proportionality is the viscosity, μ:

$$\tau = \mu \frac{dv_x}{dy}$$

Because viscosity is tightly tied to molecular motion, it fluctuates dramatically with temperature. Because the average molecular speed increases with temperature, the momentum transfer per unit time also increases, viscosity increases in gases.

In liquids, however, it decreases as the temperature rises, as you can see when pouring boiling water from a kettle. This is due to the fact that liquid molecules do not simply bounce off one another. Instead, they form brief bonds with one another, enhancing momentum transfer. As the temperature rises, this bond energy becomes less relevant in comparison to their kinetic energy.

Momentum diffuses at a rate proportional to the velocity gradient in most fluids, just as heat diffuses at a rate related to the temperature gradient. These are known as "Newtonian fluids." Because long chains of molecules and small suspended particles can align or distort with the flow direction, the viscosity is affected by the velocity gradient. Non-Newtonian fluids are the name for these fluids and it has been shown in the figure 7.

Figure 7.

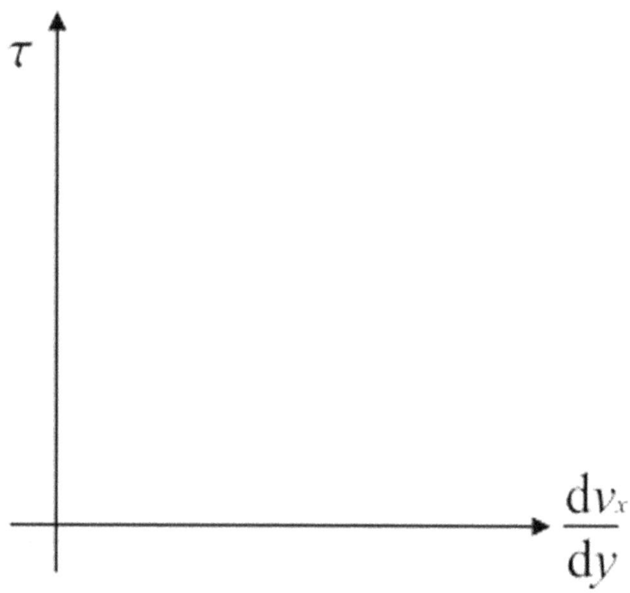

A novel viscosity-as-a-function-of-shear-rate model has been developed. This model includes three fluid properties: zero-shear rate viscosity, infinite-shear rate viscosity, and non-Newtonian fluid indication. The model can be reconfigured to simulate fluid that acts as shear-thinning, shear-thickening, or Newtonian behaviour with only a few assumptions. By connecting the viscosity to the shear rate, it was applied to a range of drilling fluids in this study. When compared to other well-known models, it predicts with a higher R-squared value (0.99) than the power-law and Herschel-Bulkley models (Goh and Wan Nik, 2018).

2.4 Couette Flow

Couette flow is the constant incompressible viscous flow between moving plates described in the previous section. Because there is no acceleration (see bottom of page), the forces on a control volume must accumulate to zero using f = ma.It has been shown in the figure 8.

Figure 8.

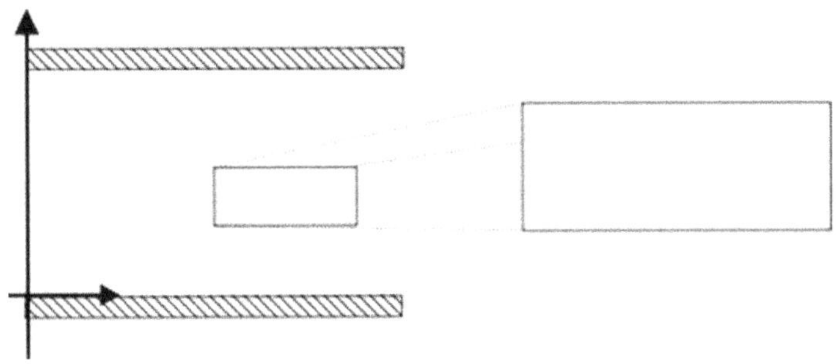

Because the pressure gradient in Couette flow is zero, we can simulate the shear stress using the viscosity of the fluid and the velocity gradient from our molecular argument. As a result, we insert =dv_x/ dy into this expression: v_x = B y +C is the solution. The boundary conditions are used to calculate the constants B and C: We should have looked at the material derivative of velocity (i.e. the acceleration of a fluid blob) before starting to demonstrate that the flow is not accelerating)

$$\frac{dv}{dt}=\frac{Dv}{Dt}=\left(\frac{\partial}{\partial t}+v\cdot\nabla\right)v=\frac{\partial v}{\partial t}+(v\cdot\nabla)v=\frac{\partial v}{\partial t}+\left(v_x\frac{\partial}{\partial x}+v_y\frac{\partial}{\partial y}\right)v$$

The constant generalised planar Couette flow of viscous incompressible fluid between two porous parallel plates through a porous media with a magnetic field has been researched by Awe. The velocity, average velocity, shear stress, skin frictions, volumetric flow, drag coefficients, and stream lines have all been investigated (Sharma, 2020). The water channel in the Couette flow was an infinite-belt kind with counter-moving walls. Because the belt and channel walls were transparent, fluid-suspended reflective flakes could be used to see the flow pattern in the streamwise-spanwise plane. A high-amplitude pointwise disruption prompted the transition. Based on half-channel height and half the velocity differential between the walls, the transitional Reynolds number, i.e. the lowest Reynolds number at which turbulence can be sustained, was determined to be 360 10. A large enough amplitude of the initial disturbance caused a growing turbulent patch for Reynolds numbers above this amount. For Reynolds numbers up to 1000, its form and spreading rate were established (Tillmark and Alfredsson, 1992).

2.5 Poiseuille Flow

The flow between stationary plates with a pressure gradient is now considered. A force balance on the same control volume yields the velocity profile it has been shown in the figure 9.

The viscosity of a fluid is the most significant attribute since it determines the flow qualities. Because viscosity is a collective feature, it is difficult to calculate accurately when simulating a fluid using a particle model. We describe a new method in this article that has a higher signal-to-noise ratio than previous methods. It works by simulating counter-flowing Poiseuille flows without utilising explicit boundaries by employing periodic boundary conditions. The viscosity of the two flows is therefore related to their mean flow velocity. The approach is applied to two very distinct systems. First, we'll look at a simple generic fluid model called dissipative particle dynamics, which requires precise viscosity values to characterise the model fluid. The more lifelike the second. Lennard-Jones (Backer et al., 2005). The linear instability and nonlinear stability theories are used to investigate Poiseuille flow with pair stresses in a fluid layer. The eigenvalue issues for nonlinear stability for x,z and y,z disturbances are also derived. The Chebyshev collocation method is used to arrive at the eigenvalue equation, which is then solved numerically. The Chebyshev collocation method is used to solve the equivalent of the Orr-Sommerfeld eigenvalue problem. The challenges that arise while computing the Orr-Sommerfeld equation's spectrum are explained. For a wide range of the pair stresses coefficient M, the critical Reynolds number Rc, critical wave number ac, and critical wave speed cc are determined (Harfash and Meften, 2020).

Figure 9.

a.

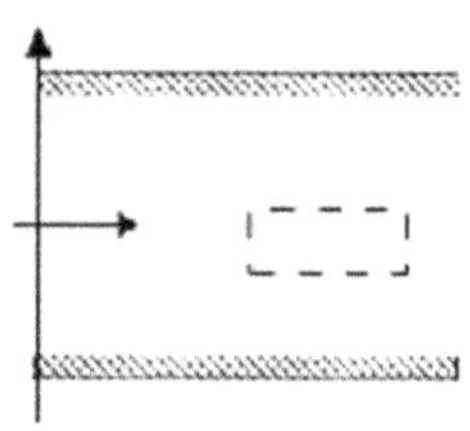

b.

$$\frac{d\tau}{dy}\delta y\delta x - \frac{dp}{dx}\delta x\delta y = 0$$

$$\Rightarrow \frac{d\tau}{dy} = \frac{dp}{dx}$$

$$\Rightarrow v_x = \left(\frac{1}{2\mu}\frac{dp}{dx}\right) y^2 + By + C$$

2.6 Viscous Flow Down a Slope

On a steady viscous flow down a hill, a comparable force balance can be accomplished and it has been shown in the figure 10.

Figure 10.

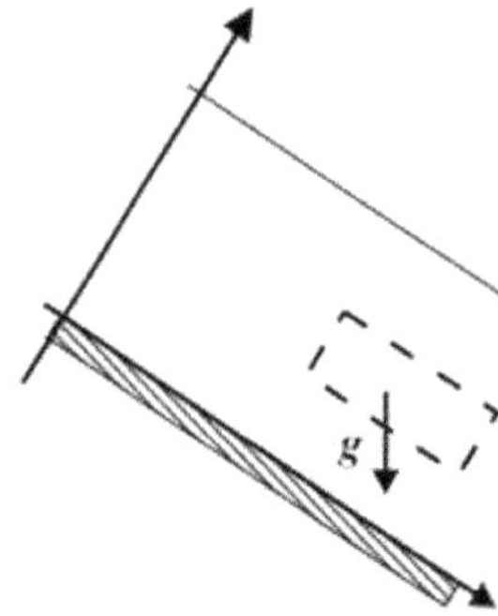

3. NEWTONIAN VISCOUS FLOW

Newtonian fluids are named after Sir Isaac Newton (1642-1726), who described the flow behaviour of fluids using a simple linear relationship between shear stress [mPa] and shear rate [1/s]. The viscosity of the fluid [mPas] is the proportionality constant, and the relationship is now known as Newton's Law of Viscosity. Newtonian and non-Newtonian were shown in the below figure 11.

Figure 11.

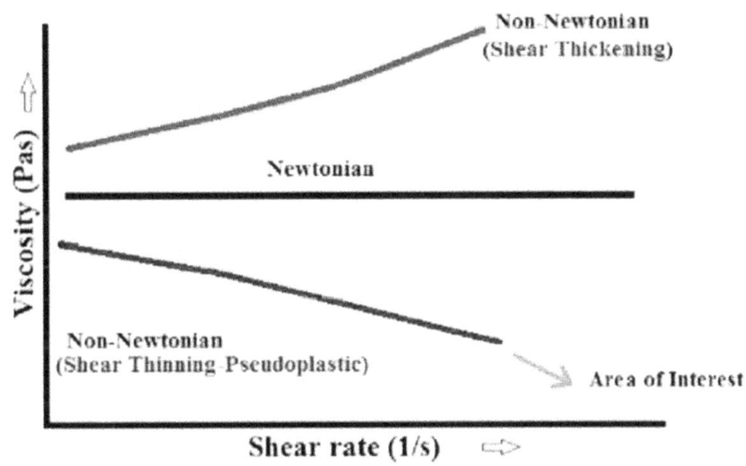

Even at vanishing Reynolds numbers, a flow vessel with an elastic wall can flex dramatically due to viscous fluid flow within it (no fluid inertia). Because of the change in cross-sectional area, deformation increases throughput. In contrast to the Hagen–Poiseuille law for a rigid tube, the latter results in a non-constant pressure gradient in the flow-wise direction and, as a result, a nonlinear flow rate–pressure drop connection. Many biofluids are non-Newtonian, and generalised Newtonian (say, power-law) rheological models are good approximations (Anand and Christoy, 2021) the majority of food industries throughout the world use the fluid movement process. In these industries, viscous fluids are frequently employed to produce large quantities of useful food products. These industries must guarantee that the necessary equipment, such as pipelines, is kept clean and clear in order to transfer viscous liquids smoothly. Fluids are pumped through these pipes, causing friction against the pipe walls. The fundamental foundation of mass transfers in the food industry is flow characteristics. These fluid flow parameters are largely responsible for the product's quality (Sarow, 2020)

4. SURFACE TENSION

Surface tension is defined as a fluid's tendency to shrink into the smallest possible surface area. This liquid property is based on the fact that the molecules of the liquid at the surface are in a different state than those in the liquid's interior. The molecule at the liquid surface is surrounded by other molecules and is attracted in all directions equally. As a result, the molecule's net attraction force is zero. Surface tension is determined not only by the forces of attraction between particles, but also by the forces of attraction of solids, liquids, and gases in close proximity to it.

The work or energy required to remove the surface layer of the molecules in a unit area is equal to the energy required for surface tension. Surface tension is measured in joules per unit area (square meters). It is frequently expressed as a force exerted on a surface perpendicular to a line of unit length. At 20 °C, water has a surface tension of 0.07275 joules per square metre. Other organic liquids with lower surface tensions, such as benzene and alcohols, have higher surface tensions than mercury. As the temperature rises, the net force of attraction between molecules decreases, as does surface tension. The most common example in our daily lives is when we add soap to the water. Because the surface tension of the liquid with soap decreases, it easily mixes with the dirt on hand and cleans it.

Surface tension liberates the energy required to increase the liquid's surface area by unit area. Surface tension is also a fundamental property of the liquid surface that resists force. It serves as a barrier between foreign materials and liquid, as well as the force that holds liquid molecules together.

Some other examples: liquid drops are spherical because a sphere has the smallest surface area for a given volume. (Surface tension attempts to obtain the liquid's minimum surface area.) Cohesive and adhesive forces are the driving forces behind the formation of surface tension.

5. VISCOSITY

Viscosity is the property of a fluid that opposes the relative motion of two surfaces of the fluid moving at different speeds, or in simpler terms, viscosity means friction between the molecules of fluid or opposition to flow. Liquids that flow quickly have lower internal resistance, and vice versa.

This is due to the strong intermolecular forces that exist between the molecules. These liquids have a low viscosity and are less viscous. Internal resistance is high in liquids that flow slowly. This is due to the weak intermolecular forces that exist between them. As a result, they are less viscous or have a low viscosity.

For a given fluid at a given temperature, the tangential, or shearing, stress that causes flow is directly proportional to the rate of shear strain, or rate of deformation; the shear stress divided by the rate of shear strain is constant. The constant is referred to as absolute, dynamic, or simply viscosity. Newtonian fluids are fluids that behave in this manner. The viscosity unit is newton-second per square metre.

When the temperature rises, the viscosity of liquids decreases while the viscosity of gases increases. As a result, when heated, liquids flow more easily, whereas gases flow more slowly. Viscosity is also known as frictional forces that act between fluid and the surface of contact in some cases. The surface could be solid, like a pipe, with water as the fluid.

Viscosity is defined as the resistance provided by the pipe to the flow of water. Viscosity can, however, be created between two liquids moving at different speeds. For example, oil has a high viscosity, but when we heat it up in the car, it becomes less viscous and thus flows smoothly through an engine.

5.1 Specific Viscosity

A measure of a fluid's resistance to flow, expressed as the ratio of the fluid's absolute viscosity to that of a reference fluid (usually water in the case of liquids)

5.2 Kinematic Viscosity

The kinematic viscosity is an atmospheric variable defined as the ratio between the dynamic viscosity μ and the density ρ of the fluid, i.e. and depends on both air temperature and pressure. For air at sea level pressure and 20 °C, ρ = 1.205 × 10−3 g cm−3 and v = 0.15 cm2 s−1

5.3 Effects of Temperature on Viscosity

The viscosity of a gas increases as the temperature rises. Although viscosity should be proportional to the square root of absolute temperature, according to the kinetic theory of gases, it grows more rapidly in practise.

There will be molecular interchange in a liquid comparable to that which occurs in a gas, but there will be additional attractive, cohesive forces between the molecules of a liquid (which are much closer together than those of a gas). Liquid viscosity is influenced by both cohesion and molecular exchange.

Increasing the temperature of a liquid has the effect of reducing cohesive forces while increasing the rate of molecule interchange.

Shear stress is reduced in the first effect, whereas it is increased in the second. As a result, the viscosity of liquids decreases as the temperature rises. The drag force increases as viscosity increases in gases and decreases in liquids at high temperatures.

5.4 Impact on Increase in Temperature

In gases, rising temperature causes the sphere to slow down, but in liquids it accelerates. At normal temperature, the molecules of a liquid are strongly bonded together by attractive intermolecular forces (e.g. Van der Waal forces).

Individual molecules are difficult to move because they are closely connected to their neighbours; hence these attractive forces are responsible for viscosity.

The kinetic or thermal energy of the molecules increases as the temperature rises, making them more mobile.

The viscosity is reduced as the attractive binding energy is reduced. If you keep heating the liquid, the kinetic energy will surpass the binding energy, causing molecules to escape and the liquid to evaporate.

5.5 Vortices Dynamics

The revolving motion of fluid around a common centreline is frequently referred to as a vortex. The vorticity of the fluid, which quantifies the rate of local fluid rotation, defines it. The fluid typically loops around the vortex, increasing in speed as the vortex approaches and decreasing in pressure. The Wing tip vorticity is shown in the figure 12.

A vortex is a vorticity structure that keeps a degree of coherence with time while it is adverted by the flow, as defined by us. A vortex, however, has no distinct boundary due to viscosity. Plotting the vorticity contours reveals the presence of vorticity but not the location of a vortex's boundary.

The dynamics of three point vortices flowing in an ideal fluid in the plane are well known to be represented in Hamiltonian form, with the resulting equations of motion being totally integrable in the sense of Liouville and Arnold. For particular (admissible) sorts of Hamiltonian perturbations of the three vortex system in a plane, the persistence of regular behaviour (specifically periodic motion) associated with totally integrable systems is the subject of this work. Following a brief review of the dynamics of the integrable planar three vortex system, it is demonstrated that the admissible class of disturbed systems includes three vortices in a half plane, three coaxial slender vortex rings in three space, and "limited" four vortex dynamics in three space (Blackmore et al., 2007).

Figure 12.

6. CONCLUSION

In this chapter, we had discussed about the Viscous Flow and its Effect by the following section, first we made introduction about the Viscous and explain the topic called Solids are Rigid and Fluids Flow then explain the No slip condition and momentum transfer, Shear Stress and we had focused to detailed explanation on the Couette Flow, Poiseuille Flow and about the Viscous Flow Down a Slope, Newtonian Viscous Flow, and also explanation is given for the Surface Tension, and finally we made to discussion on the Viscosity and their effects of temperature on Viscosity, Impact on increase in temperature Vortices dynamics and concluded

REFERENCES

Ahmad, K., Wahid, Z., & Hanouf, Z. (2019). Heat transfer analysis for Casson fluid flow over stretching sheet with Newtonian heating and viscous dissipation. *Journal of Physics: Conference Series, 1127*(1), 012028.

Anand, V., & Christov, I. C. (2021). Revisiting steady viscous flow of a generalized Newtonian fluid through a slender elastic tube using shell theory. *ZAMM-Journal of Applied Mathematics and Mechanics. Zeitschrift für Angewandte Mathematik und Mechanik, 101*(2), e201900309.

Anderson, D. M., & Davis, S. H. (1993). Two-fluid viscous flow in a corner. *Journal of Fluid Mechanics*, *257*, 1–31.

Backer, J. A., Lowe, C. P., Hoefsloot, H. C. J., & Iedema, P. D. (2005). Poiseuille flow to measure the viscosity of particle model fluids. *The Journal of Chemical Physics*, *122*(15), 154503.

Blackmore, D., Ting, L., & Knio, O. (2007). Studies of perturbed three vortex dynamics. *Journal of Mathematical Physics*, *48*(6), 065402.

Dhawan, S. J., & Narasimha, R. (1958). Some properties of boundary layer flow during the transition from laminar to turbulent motion. *Journal of Fluid Mechanics*, *3*(4), 418–436.

Goh, E. G., & Wan Nik, W. B. (2018). A generalized model for viscosity as a function of shear rate. *Journal of Engineering and Applied Sciences (Asian Research Publishing Network)*, *13*(9), 3219–3223.

Goldstein, S. (1948). On laminar boundary-layer flow near a position of separation. *The Quarterly Journal of Mechanics and Applied Mathematics*, *1*(1), 43–69.

Harfash, A. J., & Meften, G. A. (2020). Poiseuille Flow with Couple Stresses Effect and No-slip Boundary Conditions. *Journal of Applied and Computational Mechanics*, *6*(Special Issue), 1069–1083.

Joseph, D. D. (2003). Viscous potential flow. *Journal of Fluid Mechanics*, *479*, 191–197.

Mureithi, E. W., Mwaonanji, J. J., & Makinde, O. D. (2013). *On the boundary layer flow over a moving surface in a fluid with temperature-dependent viscosity*. Academic Press.

Prager, S. (1961). Viscous flow through porous media. *Physics of Fluids*, *4*(12), 1477–1482.

Sarow, S. A. (2020, June). Flows of Viscous Fluids in Food Processing Industries: A review. *IOP Conference Series. Materials Science and Engineering*, *870*(1), 012032.

Sharma, A.S. (2020). Steady Generalized Plane Couette Flow of Viscous Incompressible Fluid between Two Porous Parallel Plates through Porous Medium with Magnetic Field. *International Journal of Creative Research Thoughts, 8*(12).

Tillmark, N., & Alfredsson, P. H. (1992). Experiments on transition in plane Couette flow. *Journal of Fluid Mechanics*, *235*, 89–102.

Yuan, C., Song, J., & Liu, M. (2019). Comparison of compressible and incompressible numerical methods in simulation of a cavitating jet through a poppet valve. *Engineering Applications of Computational Fluid Mechanics*, *13*(1), 67–90. doi:10.1080/19942060.2018.1552202

Chapter 7
Aerodynamic Force Measurements Using Blower Balance Tunnel at Low Reynolds Number

Suresh Chinnasamy
ACS College of Engineering, India

Paramaguru Venugopal
ACS College of Engineering, India

Ramesh Kasimani
Government College of Technology, Coimbatore, India

ABSTRACT

This chapter describes the basic concepts of aerodynamics, evolution of lift and drag, types of drag, reduction of wing tip vortices, non-planar wing concepts for increased aerodynamic efficiency, various methods for determination of aerodynamic forces of an airplane, classification of wind tunnels, blower balance tunnels, and a case study report on aerodynamic force measurement of the non-planar wing systems. To increase the aerodynamic efficiency of the monoplane configuration, the 'C-wing' configuration is presented in this chapter. The aim is to prove, at all angles of attack, C-wing produces a higher (L/D) ratio than straight wing for the same wetted surface area. The aerodynamic characteristics of three different wing models with NACA-64215 aerofoil such as straight wing, C-wing, and inverted C-wing at different angles of attack and low Reynolds number are shown. The inverted C-wing created more lift but produced more vibration, which may lead to lesser structural integrity.

DOI: 10.4018/978-1-6684-4230-2.ch007

INTRODUCTION

Fluid mechanics is the branch of physics deals with the mechanics of fluids (liquids, gases, and plasmas) and the forces on them. It can be divided into fluid statics, the study of fluids at rest; and fluid dynamics, the study of the effect of forces on fluid motion.

Properties of Fluid

Any characteristic of a system is called *property*. It may either be *intensive* (mass independent) or *extensive* (that depends on size of system). The state of a system is described by its properties. The number of properties required to fix the state of the system is given by *state postulates*. Most common properties of the fluid are:

- **Pressure (p)**: It is the normal force exerted by a fluid per unit area. In SI system the unit and dimension of pressure can be written as, N/m^2 and $M\ L^{-1}\ T^{-2}$, respectively.
- **Density (ρ)**: The density of a substance is the quantity of matter contained in unit volume of the substance. For mass density; Unit: kg/m3, Dimension: M L-3.
- **Temperature (T)**: It is the measure of hotness and coldness of a system. In thermodynamic sense, it is the measure of internal energy of a system. Many a times, the temperature is expressed in centigrade scale (degree Celsius) where the freezing and boiling point of water is taken as 0 degree Celsius and 100 degree Celsius, respectively. In SI system, the temperature is expressed in terms of absolute value in Kelvin scale (K = degree Celsius + 273).
- **Viscosity (μ)**: When two solid bodies in contact, move relative to each other, a friction force develops at the contact surface in the direction opposite to motion. The situation is similar when a fluid moves relative to a solid or when two fluids move relative to each other. The property that represents the internal resistance of a fluid to motion (i.e. *fluidity*) is called as *viscosity*.
- **Specific Weight (υ)**: In fluid mechanics, specific weight represents the force exerted by gravity on a unit volume of a fluid. For this reason, units are expressed as force per unit volume (e.g., N/m^3 or lbf/ft^3). Specific weight can be used as a characteristic property of a fluid.

Classifications of Fluid Flows

Some of the general categories of fluid flow problems are as follows:

1. **Viscous and Inviscid flow**: The fluid flow in which frictional effects become signification, are treated as viscous flow. When two fluid layers move relatively to each other, frictional force develops between them which is quantified by the fluid property 'viscosity'. Boundary layer flows are the example viscous flow. Neglecting the viscous terms in the governing equation, the flow can be treated as inviscid flow.
2. **Internal and External flow:** The flow of an unbounded fluid over a surface is treated as 'external flow' and if the fluid is completely bounded by the surface, then it is called as 'internal flow'. For example, flow over a flat plate is considered as external flow and flow through a pipe/duct is internal flow. However, in special cases, if the duct is partially filled and there is free surface, then

it is called as open channel flow. Internal flows are dominated by viscosity whereas the viscous effects are limited to boundary layers in the solid surface for external flows.

3. **Compressible and Incompressible flow:** The flow is said to be 'incompressible' if the density remains nearly constant throughout. When the density variation during a flow is more than 5% then it is treated as 'compressible'. This corresponds to a flow Mach number of 0.3 at room temperature.
4. **Laminar and Turbulent flow:** The highly ordered fluid motion characterized by smooth layers of fluid is called 'Laminar Flow', e.g. flow of highly viscous fluids at low velocities. The fluid motion that typically occurs at high velocities is characterized by velocity fluctuations are called as 'turbulent.' The flow that alternates between being laminar & turbulent is called 'transitional'. The dimensionless number i.e. Reynolds number is the key parameter that determines whether the flow is laminar or turbulent.
5. **Steady and Unsteady flow:** When there is no change in fluid property at point with time, then it implies as steady flow. However, the fluid property at a point can also vary with time which means the flow is unsteady/ transient. The term 'periodic' refers to the kind of unsteady flows in which the flow oscillates about a steady mean.
6. **Natural and Forced flow:** In a forced flow, the fluid is forced to flow over a surface by external means such as a pump or a fan. In other case (natural flow), density difference is the driving factor of the fluid flow. Here, the buoyancy plays an important role. For example, a warmer fluid rises in a container due to density difference.
7. **One/Two/Three dimensional flow:** A flow field is best characterized by the velocity distribution, and thus can be treated as one/two/three dimensional flow if velocity varies in the respective directions.

Reynolds Number

The Reynolds number (Re) helps predict flow patterns in different fluid flow situations. At low Reynolds numbers, flows tend to be dominated by laminar (sheet-like) flow, while at high Reynolds numbers flows tend to be turbulent (Deakin, D 2011). The turbulence results from differences in the fluid's speed and direction, which may sometimes intersect or even move counter to the overall direction of the flow (eddy currents). These eddy currents begin to churn the flow, using up energy in the process, which for liquids increases the chances of cavitation. Reynolds numbers are an important dimensionless quantity in fluid mechanics.

The Reynolds number is the ratio of inertial forces to viscous forces within a fluid which is subjected to relative internal movement due to different fluid velocities. A region where these forces change behavior is known as a boundary layer.

With respect to laminar and turbulent flow regimes:

- laminar flow occurs at low Reynolds numbers, where viscous forces are dominant, and is characterized by smooth, constant fluid motion;
- turbulent flow occurs at high Reynolds numbers and is dominated by inertial forces, which tend to produce chaotic eddies, vortices and other flow instabilities.

The Reynolds number is defined as $Re = \rho\, u\, L / \mu$ or $u\, L / \nu$ where:

- ρ is the density of the fluid (SI units: kg/m^3)
- u is the flow speed (m/s)
- L is a characteristic linear dimension (m)
- μ is the dynamic viscosity of the fluid (Pa·s or N·s/m^2 or kg/(m·s))
- ν is the kinematic viscosity of the fluid (m^2/s).

BASICS OF AERODYNAMICS

Aerodynamics is the study of objects (cars, airplanes, etc.) moving relative to a fluid, such as air. This also includes stationary objects where the fluid is moving, such as a hair dryer or wind tunnel. Lift and drag forces are arguably the two most important performance measures studied in aerodynamics.

Wind tunnels are large tubes with air moving inside. The tunnels are used to copy the actions of an object in flight. Researchers use wind tunnels to learn more about how an aircraft will fly.

Evolution of Aerodynamic Forces

Aerodynamics describes the motion of air and the forces acting on the solid bodies moving through air or remaining stationary in flow of air stream. Four aerodynamic forces acting on an aircraft are weight, lift, thrust, and drag. Aerodynamics concentrates on Lift and Drag, which are created by air passing over and around the aircraft.

- **Weight** is the force exerted on an aircraft by gravity. The pull of gravity acts through the aircraft's centre of gravity, which is the point at which an aircraft would balance if suspended. The magnitude of this force changes only with a change in mass of the aircraft.
- **Lift** is produced by air passing over the wing of an airplane or over the rotor blades of a helicopter. Lift is the force that overcomes the weight of an aircraft so that it can rise in the air.
- **Thrust** is the force that moves an aircraft through the air. In a conventional fixed-wing aircraft, thrust is provided by the power plants while the wings supply the lift. In a helicopter both thrust and lift are supplied by the main rotor blades.
- **Drag** is the force of resistance by the air to the passage of an aircraft through it. Thrust force sets an aircraft in motion and keeps it in motion against drag force. The aerodynamic parameter that quantifies the performance of wing is the ratio of Lift to Drag known as aerodynamic efficiency (Hossain et. al 2011).

Types of Drag

There are three basic types of drag, namely parasite drag, wave drag and induced drag.

Parasite Drag

In aeronautical engineering, the parasite drag is the total drag that is not associated with lift induced drag. It is the drag acts on the solid object when it is moving through air. Parasite drag is combination of form drag and skin friction drag(Abdel-Motaleb et.al 2019)

Wave Drag

It is the drag produced at transonic and supersonic speeds due to the formation of shock waves.

Induced Drag

It is the drag due to lift that arises when the moving object redirects the air flow coming round it. It is also known as vortex drag or lift induced drag (Sidairi et al., 2016).

Induced Drag Reduction Methods

1. Increasing Aspect ratio of the wing (Bushnell, DM 2003)
2. Increase the span of the wing
3. Increase the span efficiency of wing
4. Decrease the airfoil thickness.(Bushnell et.al 1979)
5. Use of non planar wing concepts such as Winglets, Ring wing, C wing etc (Harish, V & Kumar, C 2016).

From the above list, the four methods (1) to (4) are limited for a given finite wing geometry (Streit et al., 2010). Induced drag may be easily reduced by increasing the span of a planar wing. Higher structural weight and cost make such efforts counterproductive (Jahanmiri, 2011).

In the last method (5), the Nonplanar wings (Kroo 2001) offer the possibility of reduced drag compared with planar wings of the same span and area. Hence the researchers focus on implementing nonplanar wings for improving the aerodynamic efficiency of an aircraft.

Non-Planar Wing Systems

Non-planar wing systems are optimally loaded wings with higher aerodynamic efficiency than planar wings having the same aspect-ratios (Dakka, S & Johnson, O 2019). Methods for the analysis of such wings have evolved due to the requirements of minimum induced drag (or drag due to lift) for powered aircraft, and later for endurance flight, flight altitude flight, and human-powered flight (Ning and Kroo, 2008).

The wing systems include wings with spanwise camber, C-wings, ring wings, and various combinations of basic lifting surfaces (Kroo, I 2005). Wings with endplates, winglets, and other tip devices belong to the same category of non-planar systems.

Winglets for Drag Reduction

Recent attention in winglets spans the preceding 25 years. Conventional winglets, defined as small fins or vertical protrusion at the wingtips, improve the overall performance of the airplane by reducing the induced drag thus increasing the lift-to-drag ratio (L/D) known as aerodynamic efficiency (Janardhanan *et al.* 2019). The applications of winglets to commercial transport aircraft were first tested in KC-135A aircraft in the late 19th century (Lappas &Ikenaga 2019). The winglets could increase an aircraft's range by as much as seven percent at cruise speeds. Wingtip devices (McLean 2005)(winglets, feathers, sails, etc.) can improve drag due-to-lift efficiency by 10 to 15% if they are designed as an integral part of the wing(Abdelghany et al. 2016). Winglets that are 20% of the semi-span in height reduce the drag by 17% (Kroo 2001). As add on devices, however, they are detrimental to the overall performance of the wing(Salahuddin *et al.* 2013). As aircraft gets larger and fly farther, the purpose of wingtip devices technology will become more remarkable (Cosin et.al 2010). The problem of the aerodynamics of the wingtip region and wingtip devices arises when studying ways of increasing aerodynamic efficiency.

There are many wingtip devices with each having dissimilar functions, their intended consequence is forever to reduce the aircraft's drag by a particular revival of the tip vortex energy(Cerón-Muñoz *et al.* 2008). It can also progress aircraft managing personality and enhance safety for aircraft. Such devices raise the effective aspect ratio of a wing devoid of materially escalating the wingspan. An addition of span would lower induced drag (Henderson et.al 1989). Wingtip devices boost the lift generated at the wingtip (by smoothing the airflow across the upper wing near the tip) and diminish the lift-induced drag caused by wingtip vortices, improving the lift-to-drag ratio(Khan *et al.* 2019). Winglets function by increasing the effective aspect ratio of the wing without contributing significantly towards the structural loads. The winglet theory was first developed in the late 1800s by Frederick W. Lanchester. His investigations proved that under high lift conditions, wingtip drag can be condensed by introducing a vertical surface at the wingtip. This increases fuel efficiency in powered aircraft and amplifies the cross-country speed in a glider, in both cases increasing range.

An important advantage of using wingtip devices as additional aerodynamic means is the fact that it can be mounted on existing aircraft without serious wing structure modifications (Bourdin, P 2002). At the same time, it can be considered as alternatives to wing extensions made for enhancing the lift-to-drag ratio both in aircraft design and in improving existing aircraft (Lam and Maull, 1993).

Limitations of Conventional Winglets

Conventional winglets provide the utmost drag hold back and get a better lift to drag ratio only under cruise condition (Amendola et al., 2018). All through non-cruise conditions, these winglets are less likely to develop aircraft performance, and subsequently, they do not deliver the best possible fuel effectiveness during take-off, landing, and climb (Beechook and Wang, 2013). Non-cruise flight setting adds up to a considerably bulky portion of a flight and consequently, winglet designs must be optimized (Pfeiffer, 2004) to be effective during both cruise and non-cruise flight conditions (Cerón-Muñoz et al., 2008). Make inquiries on conventional winglets enhancement methods have been more dominant compared to any other types of the winglet. In modern days, extensive investigation has been amplified to boost aircraft performance during flight.

C-Wing Aerodynamics

Various wingtip devices have been proposed and implemented. The most widely adopted wingtip design is the wing with a winglet (Fazle Rabbi et.al 2015). This vertical extension of the wingspan reduces the wingtip vortices and hence downwash ultimately minimizing induced drag. Now introducing the concept of C-wing (Smith et al. 1996)for aircraft configuration has been exciting in terms of its aerodynamic characteristics. The winglet decreases the total vortex-induced drag by scheduling the loads on each of the lifting surfaces. The vortex flow of the main plane is carried onto the winglet to load the winglet inward. If the horizontal extensions in top and bottom of the conventional plate winglet, that forms the "C" shape

(Bahumanyam. 2014) at the wing tip, then the vorticity is further extended and creates downward load with constant over-all lift. In the C-wing configuration there will be the cushion effect during ascent and descent which may cause the drag to increase radically, to impede this effect the slots can be introduced.

In the succeeding sections, the different types of aerodynamic force measurement systems, blower balance tunnel and how the C-wing configuration is effective in reduction of vortex drag will be discussed.

AERODYNAMIC MEASUREMENT USING WIND TUNNEL

Experimental testing is an important tool required to validate all computational results. With current technology, vortex lattice methods and computational fluid dynamics remain attractive (Bravo et al., 2019) and relatively inexpensive tools to use for the analysis of aircraft wings. However, they still retain weakness concerning the simulation of real-life flow conditions, such as separation, turbulence, and boundary layer growth. Experimental testing, therefore, remains a primary analysis tool for capturing such effects. In itself, experimental wind tunnel testing also contains sources of error, with regard to the effects of the tunnel and model support on the model and airflow as well as potential calibration and Reynolds Number discrepancies (Smith et al., 2001). The low-speed subsonic wind tunnel hada 1m by 1m section and a maximum airspeed of 50 m/s. The scale-down aircraft wing wind tunnel models were fabricated and tested with 30 m/s airspeed.

FORCE MEASUREMENTS IN BLOWER BALANCE TUNNEL

A wind tunnel is a tool used in aerodynamic research to study the effects of air moving past solid objects (Dinesh et.al 2014). Wind tunnel balance is said to have an open-air circuit. Here the air is drawn directly from the surroundings into the wind tunnel and rejected back into the surroundings. The tunnel balance is a three-component type (three forces) designed using the electrical strain gauges to indicate separately on the digital indicator(Fan 2010). The balance is intended for indicating the lift, drag, and side force in the case of aerofoils, and drag force only in the case of bluff bodies, spherical, flat disc, Hemi-spherical. These models are mounted on the string (vertical square rod) situated exactly beneath the test section. The output from the lift, drag & side force (strain gauges) is connected to the respective multi-pin sockets provided at the control panel.

AERODYNAMIC INVESTIGATION ON DIFFERENT C-WING MODELS

In this experimentation, the flow quality and aerodynamic forces were found over three wing models such as Straight wing, C wing, and Inverted C wing. The flow Force measurements using tunnel balance were described in this section.

The aircraft wing model was designed by referring to the dimension of an aircraft Airbus A300 -600 R wing with airfoil NACA 64215 (Kaushik *et al.* 2018). There were three models such as straight wing, C wing,and inverted C wing having the same lifting area had been fabricated. The design procedure of the wing model and experimental investigation such as Flow visualization (Ristić and Eng 2007)and Blower Balance test have been presented here.

Selection of Airfoil

An airfoil is the shape of a wing. An airfoil-shaped body moved through a fluid produces an aerodynamic force (Azlin et.al 2011). Subsonic flight airfoils have a characteristic shape with a rounded leading edge, followed by a sharp trailing edge, often with asymmetric curvature of upper and lower surfaces. The lift on an airfoil is primarily the result of its angle of attack and shape. For the construction purpose, NACA 64215 was the airfoil taken as a reference from the aircraft Airbus A300-600R. NACA 64215 is the supercritical airfoil(Stryczniewicz 2019). They have a higher drag divergence Mach number, they develop shock waves further aft than traditional airfoils, they greatly reduce shock-induced boundary layer separation, and their geometry allows for more efficient wing design.

- 6 –A single digit representing the theoretical optimum lift coefficient at the ideal angle of attack (C_{LI}=0.15*6).
- 4 – A single digit for the X-coordinate of the point of the maximum chamber at X= 0.05*4.
- 2 – A single-digit indicating whether the chamber is simple.
- 15 – The maximum thickness in percent of chord.

Design and Fabrication of Model

The wing geometry of commercial aircraft Airbus A300-600R was selected for the design. It is a wide-body twin-engine jet airliner that was developed and manufactured by Airbus. The model fabricated was a scaled-down of the original prototype. Since the wind tunnel test section available for the study was limited. The geometry of the wing had been taken as follows,

The X coordinates and Y coordinates were taken from the literature used to generate airfoil in the CATIA V5 R20 workbench then extrude the airfoil into the wing (Djojodihardjo et al., 2013). Then, the vertical and horizontal extrusions at the wingtip were drawn with an airfoil of required geometry to generate the C surface. The inverted C winglet (Guerrero et.al 2018) had been drawn by extruding the vertical surface to the negative coordinate. Based on CATIA V5 R20, the three models such as the commercial wing, the wing with C-winglet, and the wing with inverted C-winglet were fabricated by compression molding(Janardhanan et al., 2019).

Table 1. Wing geometry

Scaling ratio	1:50
The span of the wing	448 mm
Root chord length	180 mm
Tip chord length	54 mm
The height of the vertical winglet	103 mm
The height of the horizontal winglet	116 mm
The sweep angle of the wing	19°
Cant angle (angle between upper wing and vertical winglet)	90°

Figure 1. Fabricated models of a straight wing, c-wing, and inverted c-wing

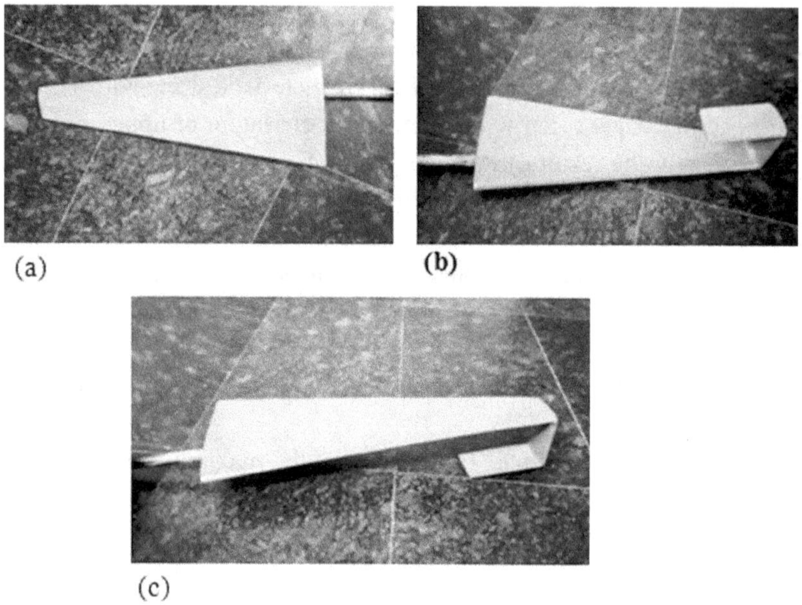

The fabricated wing models of (a) straight wing, (b) C wing, and (c) Inverted C wing respectively is shown in figure 1. All the wings have the same planar area and fabricated in the compression molding process (Tanaka &Wood 2010).

Blower Balance Tunnel

The test was carried out at Hindusthan College of Engineering and Technology, Coimbatore. The low-speed subsonic wind tunnel balance is shown in figure 2. The specifications are:

- Blower type: a) Axial with a circular duct for (delivery and suction) b) Operated at 50 Hz
- Motor: AC motor with thyristor speed controller (AC drive)
- Drive: AC type
- Electrical input: Single phase, 220V, AC supply with neutral and earth.
- Test section speed: 5-10 m/s

The aerodynamic force measurements had been investigated on three fabricated models such as commercial wing, C-wing, and inverted C-wing. Lift and Drag force were determined in three-component balance in blower tunnel balance.

Due to the test section flow speed limitation, the test section speed was set as 5 m/s. However, this could be useful in predicting the aerodynamic characteristics of wings in very speed flight i.e less than stall speed(Richardson & Power 1996).Stall speed is the slowest speed that an airplane can fly. The typical value for commercial airplanes is ranging from 8 – 15 m/s.

Figure 2. Blower balance tunnel

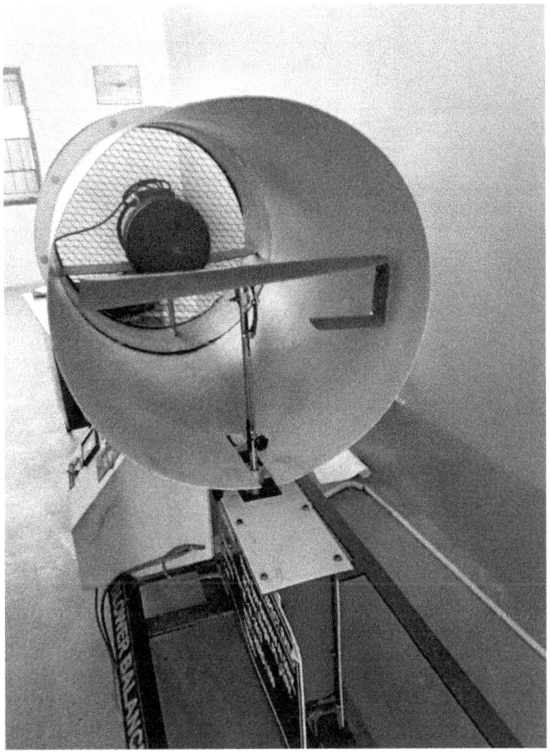

Results of Experimentation

The three different wind tunnel models based on commercial aircraft wing have been fabricated for experimental investigation. They are conventional straight wing, C wing, and Inverted C wing. The flow over the wings has been studied using the flow visualization technique (Pollock, 1980) and aerodynamic force measurements were determined by the blower balance tunnel.

The experiments were conducted at a low subsonic speed such as for flow visualization 50 m/s has been considered (Eller, D & Heinze, S 2005). And the test section speed of 5 m/s was maintained for force measurement. The intention was to understand the behavior of the nonplanar C wing system at low speed (low Reynolds number) flow i.e Re = 54,000.

Blower Balance Force Measurements

The aerodynamic forces namely lift and drag had been determined at various angles of attack such as 0^0, 5^0 and10^0 using wind tunnel balance. The results are tabulated in table 1 and 2.

It is inferred from table 2, that the Lift and Drag increase with an increase in the angle of attack (Chattot, JJ 2004, 2006). The Coefficient of Lift is higher for the inverted C wing over other wings. Likewise, the coefficient of drag is lesser for the inverted C wing. But during the investigation, it was seen that the inverted C wing underwent a heavy vibration. The magnitude of vibration in the tunnel itself was very high, which may lead to structural failure of the wing due to flutter in cruise flight at maximum flight speed. Hence it could not be recommended to incorporate into the straight wing.

Table 2. Lift and drag for different wing models

Sl. No	Alpha	Lift (N)			Drag (N)		
		Conventional wing	C wing	Inverted C wing	Conventional wing	C wing	Inverted C wing
1	0°	0	0.1962	0.4905	0.00981	0.00981	0.00981
2	5°	0.1962	0.1962	1.4715	0.01962	0.00981	0.02943
3	10°	0.4905	0.8829	1.177	0.02943	0.02943	0.04905

Figure 3. Coefficient of lift versus angle of attack for three wing model

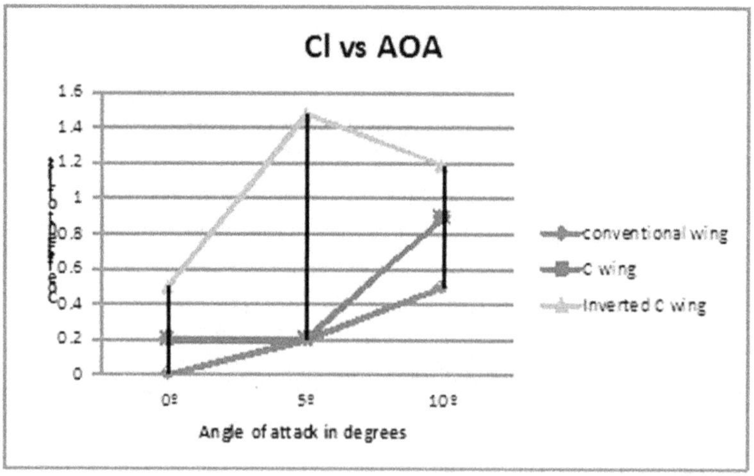

Figure 4. Coefficient of drag versus angle of attack for three wing model

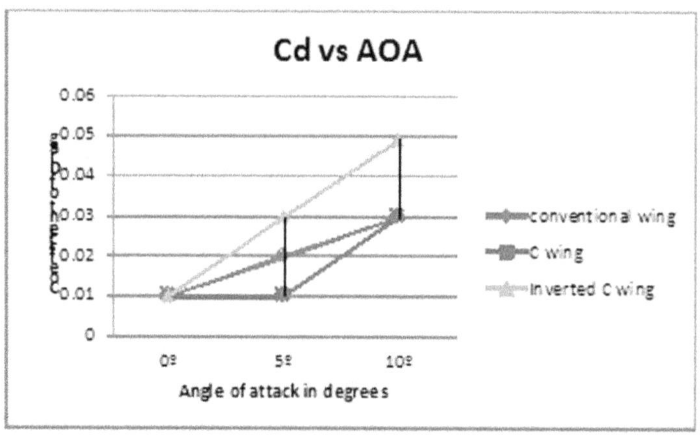

It is observed from the Figure 3 and 4, the Coefficient of Lift is higher for the inverted C wing over other wings. Likewise, the coefficient of drag is lesser for the inverted C wing.

Figure 5. Cl / Cd versus angle of attack for three wing model

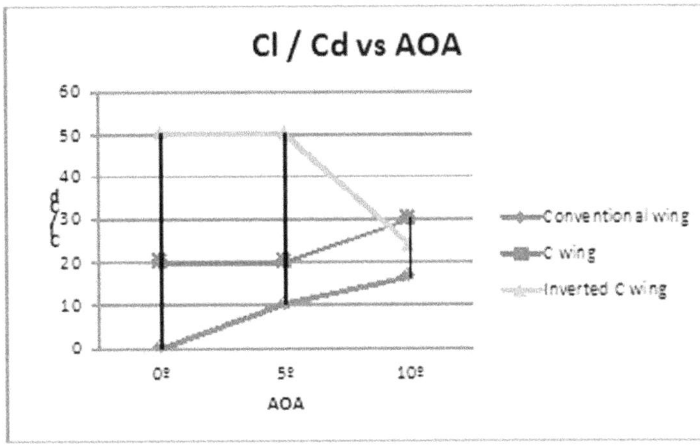

Table 3. Lift to drag ratio for different wing models

Sl. No	Alpha	L/D		
		Conventional wing	C wing	Inverted C wing
1	0°	0	20	50
2	5°	10	20	50
3	10°	16.67	30	23.99

From table 3 and Figure 5, it is evident that the Lift / Drag ratio is higher in the inverted C wing than in other wing models (Raj Arora *et al.* 2005). The inverted C wing (Bourdin, P 2006) was undergoing very high vibration and hence it could not be incorporated in the commercial wing due to the structural limitation. Hence, the C wing configuration could be preferred to the conventional straight wing so as to get enhanced aerodynamic efficiency.

Hence, with all the results obtained from the various studies undertaken, it could be observed that the C wing provides better aerodynamic characteristics without compensating much structural integrity and design complexity (Bourdin, P 2008).

CONCLUSION

Following were the conclusions drawn from the experimental investigation:

1. Aerodynamic characteristics for the aircraft model with NACA 64-215 have been presented,
2. Lift curve slope increases more with the addition of the wing with C-winglet and wing with inverted C-winglet at the same time drag decreases more for the wing with C-winglet and increases for the inverted C-winglet.

3. The winglet design is capable to reduce induced drag force and convert wingtip vortices to additional thrust which will save cost by reducing the usage of fuel, noise level reduction and increase the efficiency of the aircraft engine.
4. From the results and discussions, it was understood that the optimum angle of attack is 5°. At that angle, the C-wing has an optimum lift and minimum drag and also has the optimum lift to drag ratio. The inverted C-wing creates more lift when compared to the C-wing. But inverted C-wing produces more vibration at an angle of attack of 5° which causes the structural damage(Jansen *et al.* 2010). Finally, it can be concluded from the experimental investigation of the commercial wing, the C-wing, and inverted C-wing, the C-winglet has the optimum performance at a 5° angle of attack.

This chapter outlined the basics of aerodynamics, drag force and its effects, importance of induced drag reduction in commercial airplanes, non-planar wing concepts and blower balance tunnel for aerodynamic force measurements.

REFERENCES

Abdel-Motaleb, S. A., Taylor, J. D., Hunsaker, D. F., & Coopmans, C. (2019). Comparison of induced and parasitic drag on wings with minimum induced drag. AIAA Scitech 2019 Forum. doi:10.2514/6.2019-2120

Abdelghany, E., Khalil, E., Abdellatif, O., & El Harriri, G. (2016). Winglet Cant And Sweep Angles Effect On Aircraft Wing Performance. *The International Conference on Applied Mechanics and Mechanical Engineering*, 17, 1–17. 10.21608/amme.2016.35282

Al Sidairi, K.A., & Rameshkumar, G.R. (2016). Design of Winglet Device for Aircraft. *International Journal of Multidisciplinary Sciences and Engineering*, 23–30.

Amendola, G., Dimino, I., Concilio, A., Andreutti, G., Pecora, R., & Casciọ, M. L. (2018). Preliminary design process for an adaptive winglet. *International Journal of Mechanical Engineering and Robotics Research*, *7*(1), 83–92.

Analysis, W., & Deakin, D. (2011). Simulate To Innovate. *Aerospace (Basel, Switzerland)*, (July), 32–33.

Azlin, M. A., Taib, C. M., Kasolang, S., & Muhammad, F. H. (2011, July). CFD analysis of winglets at low subsonic flow. In *Proceedings of the World Congress on Engineering* (*Vol. 1*, pp. 6-8). Academic Press.

Bahumanyam, P. K. (2014). Evaluation of Novel Wing design for UAV. *Proceeding of Comsol Conference*.

Beechook, A., & Wang, J. (2013, September). Aerodynamic analysis of variable cant angle winglets for improved aircraft performance. In *2013 19th International Conference on Automation and Computing* (pp. 1-6). IEEE.

Bourdin, P. (2002). Planform effects on lift-induced drag. In *20th AIAA Applied Aerodynamics Conference*. American Institute of Aeronautics and Astronautics Inc. 10.2514/6.2002-3151

Bourdin, P., Gatto, A., & Friswell, M. I. (2006). The application of variable cant angle winglets for morphing aircraft control. In *Collection of Technical Papers - AIAA Applied Aerodynamics Conference.* American Institute of Aeronautics and Astronautics Inc. 10.2514/6.2006-3660

Bourdin, P., Gatto, A., & Friswell, M. I. (2008). Aircraft control via variable cant-angle winglets. Journal of Aircraft, 414–423. doi:10.2514/1.27720

Bravo-Mosquera, Cerón Muñoz, & Catalano. (2019). Design and computational analysis of a closed non-planar wing aircraft coupled to a boundary layer ingestion propulsion system. In AIAA Propulsion and Energy Forum and Exposition. American Institute of Aeronautics and Astronautics Inc.

Bushnell, D. M. (2003). Aircraft drag reduction - A review. In *Proceedings of the Institution of Mechanical Engineers, Part G: Journal of Aerospace Engineering*. Professional Engineering Publishing Ltd. 10.1243/095441003763031789

Bushnell, D. M., & Tuttle, M. H. (1979). Survey and Bibliography on Attainment of Laminar Flow Control. In *Air Using Pressure Gradient And Suction.* NASA Reference Publication.

Cerón-Muñoz. Catalano, & Coimbra. (2008). Passive, active, and adaptative systems for wing vortex drag reduction. *ICAS Secretariat - 26th Congress of International Council of the Aeronautical Sciences 2008, 4*, 1537–1548.

Chattot, J. J. (2004). Analysis and design of wings and wing/winglet combinations at low speeds. *AIAA Paper, 13*(October), 2256–2265. doi:10.2514/6.2004-220

Chattot, J. J. (2006). Low-speed design and analysis of wing/winglet combinations including viscous effects. *Journal of Aircraft, 43*(2), 386–389. doi:10.2514/1.15349

Cosin, R., Catalano, F. M., Correa, L. G. N., & Entz, R. M. U. (2010). Aerodynamic analysis of multi-winglets for low speed aircraft. *27th Congress of the International Council of the Aeronautical Sciences 2010*, 1622–1631.

Dakka, S., & Johnson, O. (2019). Aerodynamic design and exploration of a blended wing body aircraft at subsonic speed. *International Journal of Aviation, Aeronautics, and Aerospace*, *6*(5). doi:10.15394/ijaaa.2019.1411

Dinesh, Venkatesan, DV, SK, & Kumar. (2014). Diagnostic Investigation of Aircraft Performance at Different Winglet Cant Angles. *International Journal of Mechanical, Aerospace, Industrial and Mechatronics Engineering, 8*(12), 1987–1996.

Eller, D., & Heinze, S. (2005). Approach to Induced Drag Reduction with Experimental Evaluation. *Journal of Aircraft, 42*(6), 1478–1485. doi:10.2514/1.11713

Fan, Z. (2010). Measurement of Aerodynamic Forces and Moments in Wind Tunnels. In *Encyclopedia of Aerospace Engineering*. John Wiley & Sons, Ltd. doi:10.1002/9780470686652.eae078

Fazle Rabbi, M., Nandi, R., & Mashud, M. (2015). Induce Drag Reduction of an Airplane Wing. *American Journal of Engineering Research, 4*(6), 219–223.

Guerrero, J., Sanguineti, M., & Wittkowski, K. (2018). CFD Study of the Impact of Variable Cant Angle Winglets on Total Drag Reduction. *Aerospace (Basel, Switzerland)*, *5*(4), 126. doi:10.3390/aerospace5040126

Harish, V., & Kumar, C. (2016). International Journal of Current Engineering and Technology Drag Reduction using Suction Slit to the Main Wing. *International Journal of Current Engineering and Technology, 6*(4).

Henderson, W. P., & Holmes, B. J. (1989). Induced drag - Historical perspective. SAE Technical Papers. doi:10.4271/892341

Hossain, A., Rahman, A., Hossen, J., Iqbal, P., Shaari, N., & Sivaraj, G. K. (2011a). Drag reduction in a wing model using a bird feather like winglet. *Jordan Journal of Mechanical and Industrial Engineering*, *5*(3), 267–272.

Kaushik, V., Mahore, M., & Patil, S. (2018). Analysis of Dimpled Wing of an Aircraft. *International Journal of Engineering Development and Research, 6*.

Kroo, I. (2001). Drag due to Lift: Concepts for Prediction and Reduction. *Annual Review of Fluid Mechanics*, *33*(1), 587–617. doi:10.1146/annurev.fluid.33.1.587

Kroo, I. (2005b). Nonplanar wing concepts for increased aircraft efficiency. Advanced Concepts for Future Civil Aircraft, 1–29.

Ning, A., & Kroo, I. (2008, August). Tip extensions, winglets, and c-wings: conceptual design and optimization. In *26th AIAA Applied Aerodynamics Conference* (p. 7052). 10.2514/6.2008-7052

Smith, M. J., Komerath, N., Ames, R., Wong, O., & Pearson, J. (2001). Performance analysis of a wing with multiple winglets. *19th AIAA Applied Aerodynamics Conference*. 10.2514/6.2001-2407

Smith, S. C. (1996). *A Computational and Experimental Study of Nonlinear Aspects of Induced Drag*. NASA Report No 3598.

Section 3

Compressible Flow Aerodynamics

Chapter 8
Thermodynamics of Fluids in Motion

Rathinavel S.
Nehru Institute of Engineering and Technology, Coimbatore, India

Senthil Kumar S.
SRM Institute of Science and Technology, Tiruchirappalli, India

Senthilkumar T. S.
SreeSowdambika College of Engineering, Aruppukkottai, India

Vignesh Kumar V.
St. Joseph College of Engineering, Chennai, India

ABSTRACT

Thermodynamics is a science that deals with energy (heat and work) transformations and properties of substance that are affected by these transformations. Thermodynamics has been a significant part for a long time in engineering field with broad application areas such as power plants, transportation vehicles, and some direct energy conversion devices. Generally, the chemical scientist focuses on chemical reactions, phase equilibrium, and catalysis. Also, it is necessary to check whether the reactions were completed or proceed to a precise limit only. Thermodynamics also contributes to the above-mentioned chemical arena. This chapter discusses the introductory part of thermodynamics, work and energy transfer, laws of thermodynamics, principles of energy conversion, combined forms of first law and second law of thermodynamics, thermodynamic relations, and fluid compressibility. The above-mentioned issues have been well presented with inclusion of short notes. Also, illustrative example problems are solved.

DOI: 10.4018/978-1-6684-4230-2.ch008

INTRODUCTION

Thermodynamics has its own set of terms, just like any other science. We begin with a review of thermodynamics and unit systems, then move on to a discussion of basic concepts such as system, state, state postulate, equilibrium, process, and cycle. Density, specific gravity, and specific weight, as well as the system's intensive and extensive properties, are all defined. Finally, we show how to solve engineering problems using an intuitive systematic problem-solving technique that can be used as a model.

THERMODYNAMICS AND ENERGY

Thermodynamics is a science that deals with energy. Also, it examines how it affects the physical properties of a substance. Although everyone has an intuitive understanding of what energy is, it is difficult to define it precisely. The ability to change can be viewed as energy. One of nature's most fundamental laws is the conservation of energy principle. It simply states that during an interaction, energy can take on different forms, but the total amount of energy remains constant. To put it another way, energy cannot be created or destroyed. As shown in Figure 1, A rock falling off a cliff accelerates because its potential energy is converted to kinetic energy.

Figure 1.

CLASSICAL AND STATISTICAL THERMODYNAMICS

A substance made of many particles is known as a molecule. The behaviour of these particles naturally affects the properties of the substance. The behaviour of matter can be studied from two different perspectives:

1. Classical thermodynamics (macroscopic approach) and
2. Statistical Thermodynamics (microscopic approach).

In the macroscopic approach, a certain amount of matter is considered, but the events that occur at the molecular level are ignored. The macroscopic approach is only interested in the consequences of many molecules acting together, and these consequences can be realized by the human senses. In the microscopic approach, the behaviour of the molecules is described by adding the results of each molecule's behaviour.

THERMODYNAMICS UNITS

Dimensions can be used to describe any physical quantity. Units are the magnitudes designated to the dimensions. Some basic dimensions are chosen as primary or fundamental dimensions, such as mass m, length L, time t, and temperature T, while others, such as velocity V, energy E, and volume V, are represented in terms of the primary dimensions and are referred to as secondary or derived dimensions. At present, two sets of units that are used widely are the English system (also known as the United States Customary System) and the metric SI (from Le Système International d' Unités), which is also known as the International system. The SI is a simple and logical system based on a decimal relationship between the various units that are used in most industrialized countries for scientific and engineering work.

THERMODYNAMIC SYSTEM AND CONTROL VOLUME

A system is a quantity of matter or a region in space that has been chosen for study. The surroundings or environment refers to everything outside of the system. The system boundary separates the system from its surroundings as shown in Figure 2. The boundary is the real or imaginary surface that separates the system from its surroundings. A system's boundary can be fixed or movable. In the view mathematically, the boundary has zero thickness and thus it cannot contain any mass or occupy any volume in space.

Figure 2.

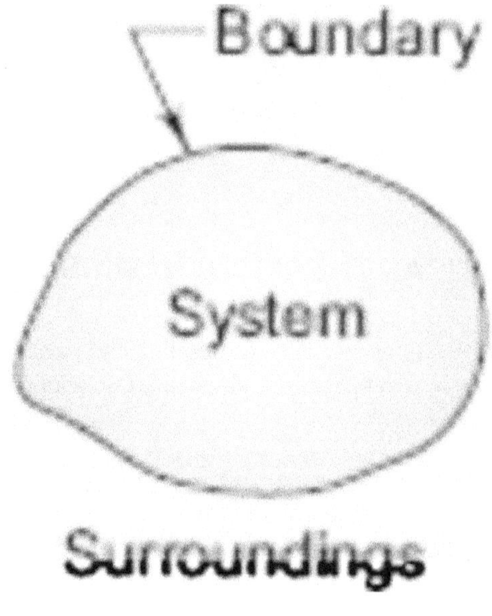

Figure 3.

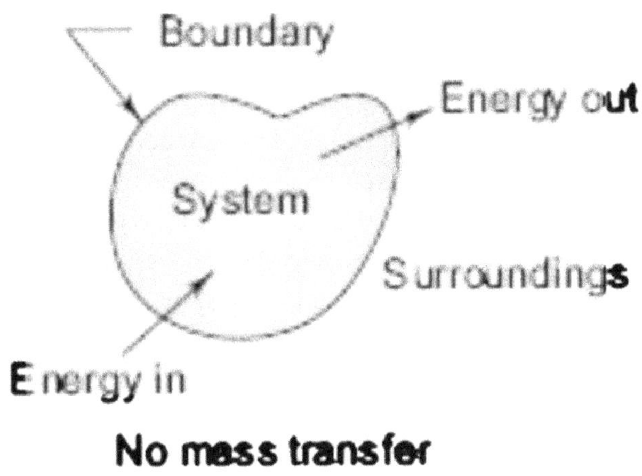

Figure 4.

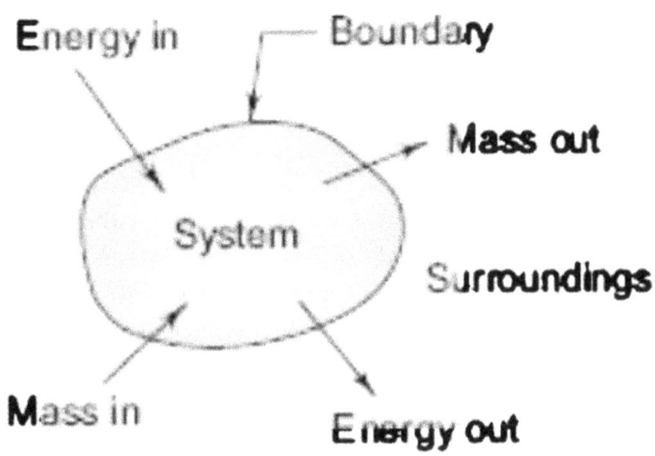

Figure 5.

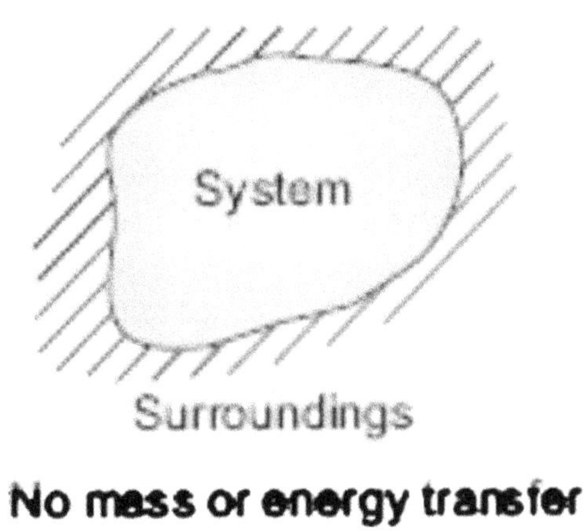

Depending on whether a fixed mass or a fixed volume in space is studied, systems can be classified as closed or open or isolated. A closed system (also known as a control mass or simply system depending on the context) is made up of a fixed amount of mass that cannot cross its boundary as shown in Figure 3. A closed system is one in which no mass can enter or leave. However, energy can cross the boundary in the form of heat or work, and the volume of a closed system does not have to be fixed.

An open system, also known as a control volume, is a properly selected region in space. A device that involves mass flow, such as a compressor, turbine, or nozzle, is usually enclosed as shown in Fig.4. The best way to study flow through these devices is to choose a region within the device as the control volume. A control volume's boundary can be crossed by both mass and energy.

Isolated systems exist when there is no interaction between the system and its surroundings as shown in Figure 5. There is no mass or energy transfer across the system boundary, and it has a fixed mass and energy.

STATE

Consider a system that isn't changing. At this point, we can measure or calculate all of the properties across the entire system, giving us a set of properties that completely describes the system's condition or state. All the system's properties have fixed values at any given state. If even one property's value changes, the state will shift to a new one.

Thermodynamics is concerned with states of equilibrium. The word equilibrium denotes a state of equilibrium. There are no unbalanced potentials (or driving forces) within the system in an equilibrium state. When a system in equilibrium is isolated from its surroundings, it does not change.

There are different types of equilibrium, and a system isn't in thermodynamic equilibrium until all the relevant types of equilibrium are fulfilled. For instance, if the temperature is the same throughout the system, the system is in thermal equilibrium. That is, no temperature differential, which is the driving force for heat flow, is present in the system. Mechanical equilibrium is linked to pressure, and a system is in mechanical equilibrium if the pressure at any point in the system does not change over time.

When the mass of each phase of a system reaches and stays at an equilibrium level, the system is said to be in phase equilibrium. Finally, a system is said to be in chemical equilibrium if its chemical composition does not change over time, implying that no chemical reactions take place. A system is not in equilibrium unless it meets all the relevant equilibrium criteria.

State Postulate

Once an adequate number of properties have been specified, the rest of the properties will automatically assume certain values. To put it another way, specifying a set of properties is enough to fix a state. The state postulate specifies the number of properties required to fix a system's state: "Two independent, intensive properties completely specify the state of a simple compressible system".

PROCESSES AND CYCLES

A process is any change in a system's equilibrium state from one to another, and the path of the process is the series of states through which the system passes during the process. To identify a process, one must include the process's initial and final states, as well as the path it takes and its interactions with the environment.

A quasi-static, or quasi-equilibrium, the process proceeds in such a way that the system always remains infinitesimally close to an equilibrium state. A quasi-equilibrium process is sufficiently slow to allow the system to adjust itself internally so that properties in one part of the system do not change faster than properties in other parts.

Figure 6 illustrate that when a gas is compressed quickly in a piston-cylinder device, the molecules near the face of the piston don't have enough time to escape, causing them to pile up in a small region in front of the piston, resulting in a high-pressure region. Because of the pressure difference, the system is no longer in equilibrium, rendering the entire process non-quasi-equilibrium. If the piston is moved slowly enough, the molecules will have enough time to redistribute, and there will be no molecule pileup in front of the piston. As a result, the pressure inside the cylinder will be nearly constant and always rise at the same rate. Because equilibrium is always maintained, this is a quasi-equilibrium process.

Figure 6.

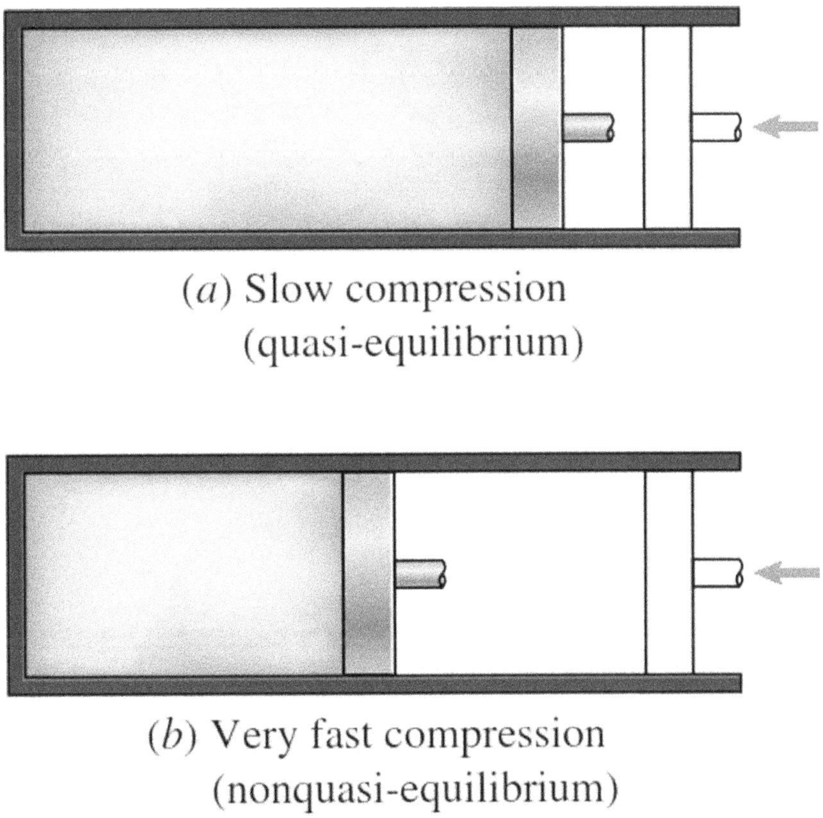

It should be noted that a quasi-equilibrium process is an idealized version of a real process, not a true representation. However, many real processes come close to it, and they can be modelled as quasi-equilibrium with little error. For two reasons, engineers are interested in quasi-equilibrium processes. First, they are simple to analyze; second, work-producing devices that operate on quasi-equilibrium processes produce the most work. As a result, quasi-equilibrium processes serve as metrics against which real processes can compete.

The use of thermodynamic properties as coordinates in process diagrams is extremely helpful in visualizing the processes. Temperature T, pressure P, and volume V are some of the most common properties used as coordinates (or specific volume v). The P-V diagram of a gas compression process is shown in Figure 7. The process path denotes a series of equilibrium states that the system passes through during a process and is only relevant for quasi-equilibrium processes. We can't characterize a non-quasi-equilibrium process by a single state, so we can't talk about a process path for the whole system. Instead of a solid line, a dashed line between the initial and final states denotes a non-quasi-equilibrium process.

Figure 7.

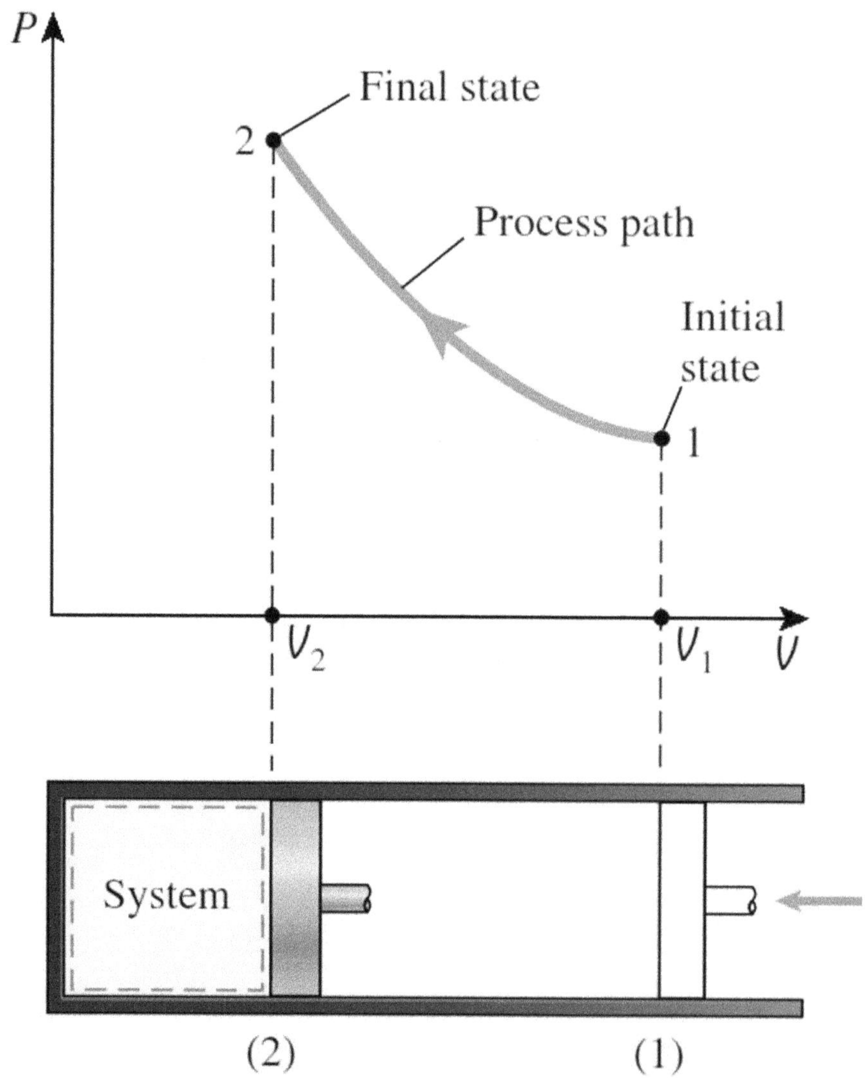

When a system returns to its initial state at the end of a process, it is said to have completed a cycle. That is, the initial and final states of a cycle are the same.

PROPERTIES OF A SYSTEM

Property refers to any characteristic of a system. Pressure P, temperature T, volume V, and mass m are some of the more common properties. Properties are classified as intensive or extensive. Temperature, pressure, and density are examples of intensive properties that are independent of a system's mass. Extensive properties are those whose values are determined by the system's size (or extent). Extensive properties include total mass, total volume, and total momentum, to name a few. Uppercase letters are used to denote extensive properties (except mass m), while lowercase letters are used to denote intensive properties (with pressure P and temperature T being the obvious exceptions). Extensive properties per unit mass are called specific properties. Specific volume (v = V/m) and specific total energy (e = E/m) are two examples of specific properties.

Continuum

In the gas phase, the matter is made up of atoms that are widely spaced. However, it is very convenient to ignore a substance's atomic nature and think of it as a continuous, homogeneous matter with no holes, or a continuum. We can treat properties as point functions and assume that they vary continuously in space with no jump discontinuities using the continuum idealization.

DENSITY AND SPECIFIC GRAVITY

Density (ρ) is defined as mass per unit volume

$$\rho = \frac{m}{V}\ (\mathrm{kg/m^3}) \tag{1}$$

The reciprocal of density is the specific volume v, which is defined as volume per unit mass. That is,

$$\mathrm{v} = \frac{V}{m} = \frac{1}{\rho} \tag{2}$$

For a differential volume element of mass dm and volume dV, density can be expressed as $\rho = dm/dV$.

Temperature and pressure affect the density of a substance in general. Most gases' density is proportional to their pressure but inversely proportional to their temperature. Liquids and solids, on the other hand, are essentially incompressible substances with little variation in density as a function of pressure.

The density of a substance is sometimes expressed as a percentage of the density of a well-known substance. Then it's known as specific gravity, or relative density, which is defined as the ratio of a substance's density to that of a standard substance at a given temperature.

$$Specific\ Gravity = \frac{\rho}{\rho_{H_2O}} \tag{3}$$

ZEROTH LAW OF THERMODYNAMICS

If two bodies are in thermal equilibrium with a third body, they are also in thermal equilibrium with each other, according to the zeroth law of thermodynamics. It may appear preposterous that such a self-evident fact is referred to as one of the fundamental laws of thermodynamics. It can't be deduced from the other laws of thermodynamics, but it does serve as a foundation for temperature measurement accuracy. The zeroth law can be restated as two bodies are in thermal equilibrium if both have the same temperature reading, even if they are not in contact, by replacing the third body with a thermometer.

FORMS OF ENERGY

Energy can take many forms, including thermal, mechanical, kinetic, potential, electric, magnetic, chemical, and nuclear. The total energy E of a system is the sum of these forms. On a unit mass basis, the total energy of a system is denoted by e and expressed as

$$e = \frac{E}{m} kJ / kg \tag{4}$$

The macroscopic forms of energy, such as kinetic and potential energies, are those that a system possesses with some outside reference frame. The microscopic forms of energy are those that are related to a system's molecular structure and the degree of molecular activity and they are unaffected by external reference frames. The internal energy of a system is denoted by U and is the sum of all microscopic forms of energy.

Kinetic energy (KE) is the energy that a system has as a result of its motion relative to some reference frame. The kinetic energy is expressed as

$$KE = m\frac{V^2}{2}(kJ) \tag{5}$$

or, on a unit mass basis,

$$KE = \frac{V^2}{2}(kJ / kg) \tag{6}$$

Potential energy (PE) is the energy that a system has because of its elevation in a gravitational field. The potential energy is expressed as

$$PE = mgz\ (kJ) \tag{7}$$

or, on a unit mass basis,

$$PE = gz\ (kJ/kg) \tag{8}$$

The total energy of a system consists of the kinetic, potential, and internal energies and is expressed as

$$E = U + KE + PE = U + m\frac{V^2}{2} + mgz\ (kJ) \tag{9}$$

or, on a unit mass basis,

$$e = u + ke + pe = u + \frac{V^2}{2} + gz\ (kJ/kg) \tag{10}$$

ENERGY TRANSFER BY HEAT

Heat and work are two types of energy that can cross the boundary of a closed system. It's essential to understand the difference between these two types of energy.

Heat is a type of energy that is transferred between two systems (or between a system and its surroundings) due to a temperature difference. That is, an energy interaction is only considered heat if it occurs due to a temperature difference. As a result, no heat transfer can occur between two systems that are at the same temperature. Heat can be easily identified because it is caused by a temperature difference between the system and its surroundings. Heat transfer per unit mass of a system is denoted q and is determined from

$$q = \frac{Q}{m}(kJ / kg) \tag{11}$$

ENERGY TRANSFER BY WORK

Work is an energy interaction between a system and its surroundings, similar to heat. As previously stated, energy can cross a closed system's boundary in the form of heat or work. As a result, if the energy crossing a closed system's boundary is not heat, it must be working. Work is defined as the energy transfer associated with a force acting over a distance. The work done per unit mass of a system is denoted by w and is expressed as

$$w = \frac{W}{m}(kJ / kg) \tag{12}$$

Because heat and work are directional quantities, a complete description of a heat or work interaction requires both the magnitude and direction to be specified. Adopting a sign convention is one way to do so. The widely accepted formal sign convention for heat and work interactions are as follow: Heat transfer to a system and work done by a system are positive; heat transfer from a system and work done on a system is negative.

LAW OF CLASSICAL THERMODYNAMICS

0^2 Relation among properties
0 Thermal equilibrium
1 Conservation of energy
2 Degradation of energy (irreversibilities)

The formal law of thermodynamics does not often include the 0^2 law. The existence of relationships among the properties is an equation of state which is mentioned by the 0^2 law. It can be either in tabular or graphical information. To define a state we require only three independent properties of the system for a single component or pure substance. Properties like, temperature and pressure are not independent if the substance exists in more than one phase (as in a liquid together with its vapour). So selection of property must be taken care of. For the unit mass, we require only two independent properties to formulate the state. So we can express any property with two known independent properties.

$$P = f(x,y) \tag{13}$$

If the systems are separated by a non-adiabatic or insulated wall, there will be a continuous change in the state of the system until it reaches equilibrium. The two systems are then said to be in thermal equilibrium with each other and will then have one property in common which we call the temperature.

Property Relations

Property relations are by combing the first and second laws.

Consider the first law for a stationary system that executes an infinitesimal process:

$$\delta q = \delta w + du$$

If it is a reversible process,

$$\delta w = p\,dv \tag{14}$$

$$\delta q = T\,ds \tag{15}$$

Substitute above (14) & (15) into the first law yields

$$T\,ds = du + p\,dv \tag{16}$$

On Differentiating the enthalpy, we obtained

$$dh = du + p\,dv + v\,dp \tag{17}$$

Combining equations (16) and (17) produces

$$T\,ds = dh - v\,dp \tag{18}$$

Even though the assumptions of a reversible process were done to derive equations (18) and (18), the results of the equations contain only properties and thus are valid relations to use between any end states, whether reached reversibly or not.

$$T\,ds = du + p\,dv$$

$$T\,ds = dh - v\,dp$$

The first law in an alternative form

$$\delta q - \delta w = du \tag{19}$$

Since internal energy is a property, changes in internal energy depend only on the end states of a process. Let's now substitute an irreversible process between the same endpoints as our reversible process. Then du must remain the same for both the reversible and irreversible cases, with thc following rcsult:

$(\delta q - \delta w)$ rev $= du = (\delta q - \delta w)$ irreversible.

For example, the extra work that would be involved in an irreversible compression process must be compensated by the same amount of heat released (an equivalent argument applies to an expansion). In this fashion, irreversible effects will appear to be "washed out" in equations (3) and (6) and we cannot tell from them whether a particular process is reversible or irreversible.

THE FIRST LAW OF THERMODYNAMICS

So far, we've looked at different types of energy separately, such as heat Q, work W, and total energy E, with no attempt to relate them during a process. The conservation of energy principle, also known as the first law of thermodynamics, provides a solid foundation for studying the relationships between various forms of energy and energy interactions. The first law of thermodynamics states that energy cannot be created or destroyed during a process; it can only change forms, based on experimental observations. As a result, during a process, every ounce of energy should be considered.

During these experiments, careful measurements revealed the following: Regardless of the nature of the closed system or the details of the process, the network done for all adiabatic processes between two specified states of a closed system is the same. This statement appears to be very powerful, with the potential for far-reaching implications, given that there is an infinite number of ways to perform work interactions under adiabatic conditions. This principle is called the first law of thermodynamics or just the first law.

The existence and definition of the property total energy E is a major consequence of the first law. Given that the network is the same for all adiabatic processes of a closed system between two specified states, the net work's value must be determined solely by the system's end states and thus must correspond to a change in a system property. The total energy is this property. It's worth noting that the first law makes no mention of the total energy of a closed system at a given state. It simply states that during an adiabatic process, the change in total energy must be equal to the network done. As a result, total energy at a given state can be assigned any convenient arbitrary value to serve as a reference point.

First Law for a Closed System Undergoing a Change of State

The algebraic summation of all energy transfer across system boundaries is zero, and the expression $(\Sigma W) = (\Sigma Q)$ only applies to systems undergoing cycles. However, if a system goes through a state change that involves both heat and work transfer, the net energy transfer will be stored or accumulated within the system. The net energy transfer $(Q - W)$ will be stored in the system if Q is the amount of heat transferred to the system and W is the amount of work transferred from the system during the process. Internal energy, or simply the energy of the system, is the name given to energy in storage that is neither heat nor work. Therefore

$$Q - W = \Delta E \tag{20}$$

where ΔE is the increase in the energy of the system, or

$$Q = \Delta E + W \tag{21}$$

Here Q, W, and ΔE are all expressed in the same units (in joules). Energy may be stored by a system in different modes.

First Law for an Open System (Mass Balance and Energy Balance Equation)

A steady flow system is depicted in Fig. 8, in which one stream of fluid enters and another leaves the control volume. Within the control volume, there is no accumulation of mass or energy, and the properties of any location within the control volume remain constant over time.

The following quantities are defined in Figure 8.

A_1, A_2 – a cross-section of stream, m^2
w_1, w_2 – mass flow rate, kg/s
p_1, p_2 – pressure, absolute, N/m^2
v_1, v_2 – specific volume, m^3/kg
u_1, u_2 – specific internal energy, J/kg
V_1, V_2 – velocity, m/s
$\frac{dQ}{dT}$ – net rate of heat transfer through the control surface, J/s

$\frac{dW}{dT}$ – net rate of work transfer through the control surface, J/s

Figure 8.

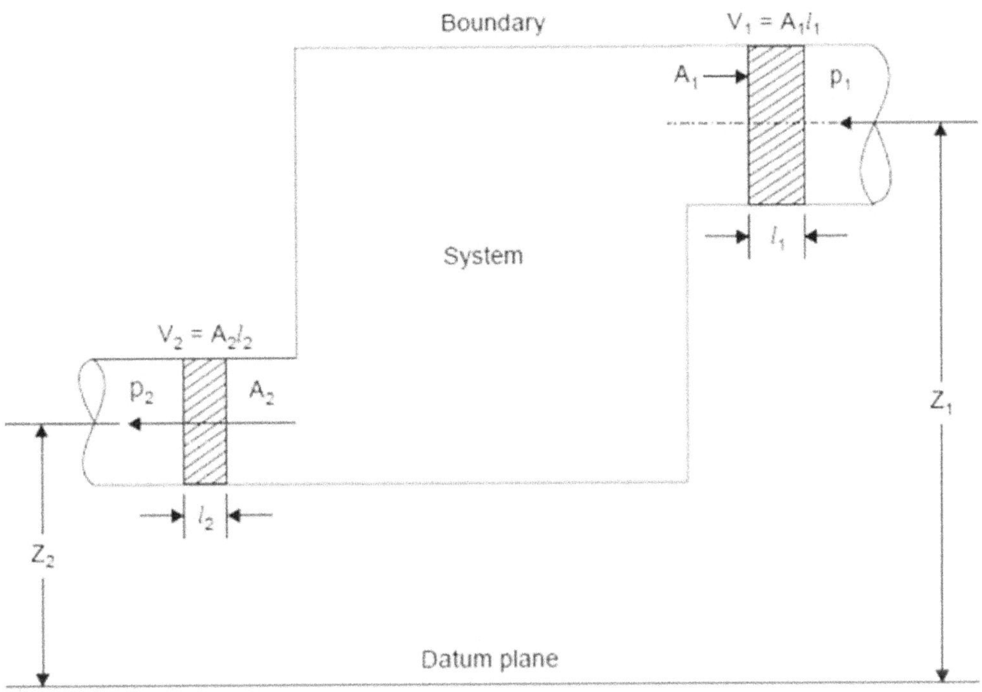

First Law for an Open System (Mass Balance)

If there is no accumulation of mass within the control volume, the mass flow rate entering must equal the mass flow rate leaving, according to mass conservation.

$w_1 = w_2$

$$\frac{A_1 V_1}{v_1} = \frac{A_2 V_2}{v_2} \tag{22}$$

This equation is known as the equation of continuity

First Law for an Open System (Energy Balance)

Work can be transferred in two ways in a flow process: external work and flow work. All work transfer across the control surface other than that caused by normal fluid forces is referred to as external work.

Shear (shaft or stirring) work and electrical work are the only types of external work that matter in engineering thermodynamics.

THE SECOND LAW OF THERMODYNAMICS

Previously, the conservation of energy principle, or the first law of thermodynamics, was applied to processes involving closed and open systems. Energy is a conserved property, as stated several times in those chapters, and no process is known to have violated the first law of thermodynamics. As a result, it is reasonable to conclude that a process must satisfy the first law for it to take place. However, as previously stated, satisfying the first law does not guarantee that the process will take place.

The first law does not impose any restrictions on a process's direction, but it does not guarantee that the process will take place. The first law's inadequacy in determining whether a process can occur is remedied by introducing a second general principle, the second law of thermodynamics. The two most important forms of the second law are given below.

- **Kelvin-plank statement** – It is impossible to construct a continuously operating device that generates no other effect than the removal of heat from a single thermal reservoir and produces an equivalent amount of work.
- **Clausius Statement** – It's impossible to build a device that runs in a cycle and has no effect other than transferring heat from a low-temperature body to a high-temperature body.

The first statement is about a heat engine, while the second is about a heat pump. Even though these statements lack mathematical proof, their violation has yet to be documented.

The second law of thermodynamics can be used for more than just determining the direction of processes. The second law also states that energy has both quantity and quality.

ENTROPY

The second law results in the definition of entropy, a new property. Entropy is a somewhat abstract property that is difficult to describe physically without considering the system's microscopic state. The best way to understand and appreciate entropy is to look at how it's used in everyday engineering processes, which is exactly what we'll be doing.

The second law of thermodynamics frequently leads to inequalities in expressions. For example, a reversible heat engine operating between the same two thermal energy reservoirs is more efficient than an irreversible (i.e., actual) heat engine. Similarly, an irreversible refrigerator or heat pump with the same temperature limits has a lower coefficient of performance (COP) than a reversible one.

$$\oint \frac{\delta Q}{T} \leq 0 \tag{23}$$

That is, the cyclic integral of dQ/T is always less than or equal to zero. This inequality is valid for all cycles, reversible or irreversible. The symbol $\oint$ (integral symbol with a circle in the middle) denotes that the integration will be carried out throughout the cycle. Differential amounts of heat transfer can be considered in any heat transfer to or from a system. The sum of all these differential amounts of heat transfer divided by the temperature at the boundary is then the cyclic integral of Q/T.

The above cyclic integral equation is known as Clausius inequality, which is valid for or all thermodynamic cycles, reversible or irreversible, including the refrigeration cycles.

In 1865, Clausius recognized that he had discovered a new thermodynamic property, which he named entropy. Its designation is S, and its definition is

$$dS \geq \frac{\delta Q}{T} \tag{24}$$

Entropy Principle

If any insignificant process is undergone by a system, the above equation for the total mass.

$$dS \geq \frac{\delta Q}{T} \tag{25}$$

For an isolated system that does not interact with the environment in any way, $\delta Q=0$.
Therefore, for an isolated system

$$dS_{iso} \geq 0 \tag{26}$$

For a reversible process,

$$dS_{iso}=0 \tag{27}$$

or S = Constant

For an irreversible process

$$dS_{iso}>0 \tag{28}$$

As a result, it is demonstrated that the entropy of a closed system can never decrease. When the process is reversible, it always increases and only stays constant. This is referred to as the entropy principle, or simply the entropy principle.

What is Entropy?

From the preceding discussion, it is clear that entropy is a useful property that can be used in the second-law analysis of engineering devices. However, this does not imply that we fully comprehend entropy. Because we don't have any. We are unable to provide an adequate response to the question.

What exactly is entropy? However, the fact that we can't fully describe entropy doesn't negate its usefulness. We couldn't define energy either, but that didn't stop us from understanding energy transformations and the principle of conservation of energy. Granted, entropy isn't as well-known as energy. However, as we use entropy more, we will gain a better understanding of it and a greater appreciation for it. By considering the microscopic nature of matter, the next discussion should shed some light on the physical meaning of entropy.

Entropy can be thought of as a metric for molecular disorder or randomness. The positions of molecules become less predictable as a system becomes more disordered, and entropy rises. As a result, it's not surprising that a substance's entropy is lowest in the solid phase and highest in the gas phase.

MAXWELL'S THERMODYNAMIC RELATIONS

Maxwell's relations are deduced by considering four different and important useful thermodynamic functions like internal energy U, enthalpy H, Helmholtz free energy (or simply the free energy) F, and Gibbs free energy (or the Gibbs function) G.

1. Internal Energy

The internal energy (U) is a thermodynamic potential, which means that it provides information through differentiation.

$$dU = TdS - pd\forall \tag{29}$$

$$\left(\frac{\partial U}{\partial \forall}\right)_S = -p \tag{30}$$

and

$$\left(\frac{\partial U}{\partial S}\right)_\forall = T \tag{31}$$

since dU is an exact differential, thus

$$\left(\frac{\partial T}{\partial \forall}\right)_S = -\left(\frac{\partial p}{\partial S}\right)_\forall \tag{32}$$

2. Enthalpy

$$dH = TdS + \forall dp \tag{33}$$

$$\left(\frac{\partial H}{\partial p}\right)_S = \forall \tag{34}$$

and

$$\left(\frac{\partial H}{\partial S}\right)_p = T \tag{35}$$

Again, if dH is an exact differential, hence

$$\left(\frac{\partial T}{\partial p}\right)_S = \left(\frac{\partial \forall}{\partial S}\right)_p \tag{36}$$

3. Helmholtz Free Energy

It is defined as

$$F = U - TS \tag{37}$$

Upon differentiating both sides,

$$dF = dU - SdT - TdS \tag{38}$$

or

$$dF = -pd\forall - SdT \tag{39}$$

since dF is an exact differential,

$$\left(\frac{\partial p}{\partial T}\right)_\forall = \left(\frac{\partial s}{\partial \forall}\right)_T \tag{40}$$

4. Gibbs Free Energy

It is defined as $G = F + p\forall$. Using Eq. (37), we get

$$G = U - TS + p\forall \tag{41}$$

and

$$dG = dU - TdS - SdT + pd\forall + \forall dp \tag{42}$$

or

$$dG = \forall dp - SdT \tag{43}$$

since dG is also an exact differential

$$\left(\frac{\partial \forall}{\partial T}\right)_p = -\left(\frac{\partial s}{\partial p}\right)_T \tag{44}$$

Equations (32), (36), (40) and (44) are known as Maxwell's relations.

BOSE-EINSTEIN STATISTICS AND BOSE-EINSTEIN DISTRIBUTION FUNCTION

In particles that follow Bose-Einstein statistics, the probability of a microstate of energy E in an n-particle system is $\rho^{BE}_{MS,N,E}$. In an isolated system of Bose-Einstein particles, the entire probability sum is given below

$$1 = \sum_{\{Ni\}} W^{BE}\left(N_i, g_i\right)\rho^{BE}_{MS,N,E} \tag{45}$$

The number of ways to assign indistinguishable particles to quantum states, if any number of particles can occupy the same state to find the W^{BE} (N_i,J_i),

By considering the number of ways that N_i particles can be assigned to g_i quantum states associated with the energy level ε_i, it is observed that the least number of quantum states that can be utilized is one; then all of the particles can be put into one quantum state. At the other extreme, it is not possible to use more than the N_i quantum states that are used, when it's been given each particle its quantum state. This problem can be viewed as finding the number of ways that can draw as many as g_i boxes around the N_i points.

Firstly, create a scheme to draw such boxes. Suppose in a linear frame on which there is a row of locations, then each location can hold one particle. This frame is closed at both the ends and between each successive pair of particle-holding locations, there is a slot into which a wall can be inserted. This frame is sketched in figure 9 below.

When (g_i) walls are inserted into these slots, the frame which contains g_i boxes can also be inserted into the walls so that the N_i particles are distributed among the g_i boxes in such a way that we can have any desired number of particles in any desired number of boxes. (Of course, placement of the walls is subject to the constraints that we use at most g_i boxes and exactly N_i particles). This can be achieved by constructing the frame to have ($N_i + g_i$ -1) particle-holding locations. To see this, we think about the case that requires the largest number of particle-holding locations. This is the case in which all N_i particles are in one box. (see figure 10.) For this case, we need N_i occupied locations and (g_i - 1) unoccupied locations.

Figure 9.

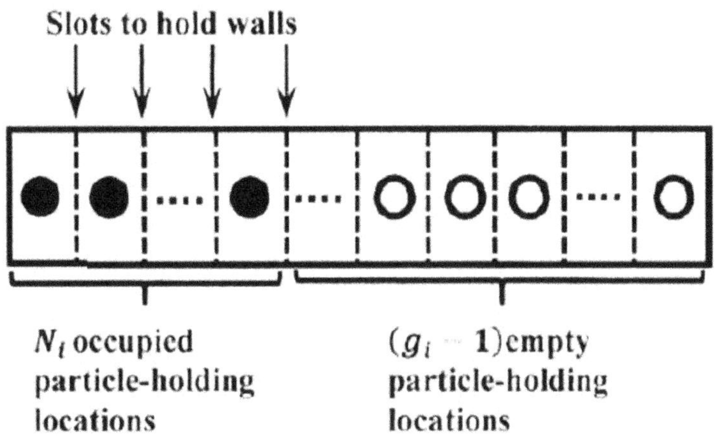

Figure 10.

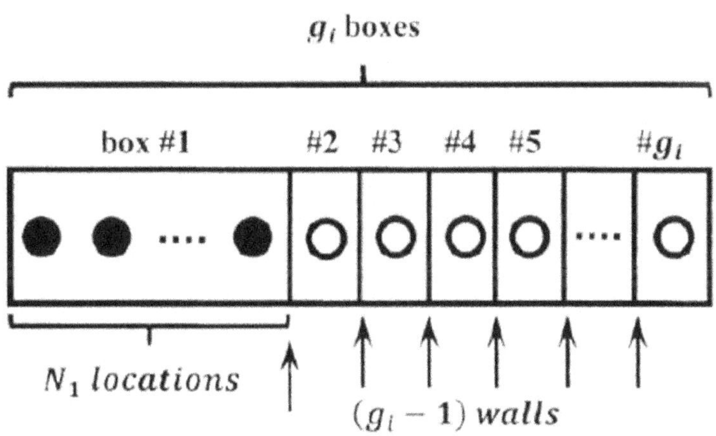

Now we consider the number of ways that we can insert (g_i-1) walls into the (N_i + g_i -1) slots. The first wall can go into any of (N_i + g_i -1). The second can go into any of (N_i + gi-1-(-1)) or (N_i + g_i – 2) slots. The last wall can go into any of (N_i + g_i -1 – (g_i -2)) or (N_i +1) slots. The number of ways of inserting the (g_i -1) walls is therefore

$$(Ni+gi-1)(Ni+gi-2)\ldots(Ni+1)$$
$$=\frac{(Ni+gi-1)(Ni+gi-2)\ldots(Ni+1)(Ni)\ldots(2)(1)}{Ni!}=\frac{(Ni+gi-1)!}{Ni!} \tag{46}$$

This total is greater than the answer we seek because it includes all permutations of the walls. It does not matter whether the first, the second, or the last wall occupies a given slot. Therefore, the expression we have obtained over-counts the quantity we seek by the factor $(g_i\text{-}1)!$ which is the number of ways of permuting the $(g_i\text{-}1)$ walls. We have therefore that the Ni particles can be assigned to g_i quantum states in

$$\frac{(Ni+gi-1)!}{Ni!(gi-1)!}$$

ways, and hence

$W^{BE}(N_i,g_i)$

$$=\prod_{i=1}^{\infty}\left[\frac{(Ni+gi-1)!}{Ni!(gi-1)!}\right]=\left[\frac{(N1+g1-1)!}{N1!(g1-1)!}\right]*\left[\frac{(N2+g2-1)!}{N1!(g2-1)!}\right]*\ldots*\left[\frac{(Ni+gi-1)!}{Ni!(gi-1)!}\right]*\ldots \tag{47}$$

So that the total probability sum for a Bose-Einstein system becomes (via. Equation):

$$1=\sum_{\{Ni\}}\prod_{i=1}^{\infty}\left[\frac{(Ni+gi-1)!}{Ni!(gi-1)!}\right]\left[\rho^{BE}(\varepsilon i)\right]^{Ni} \tag{48}$$

To find the Bose-Einstein distribution function, we seek the population set $\{N_1, N_2, \ldots, N_i, \ldots\}$ for which W^{BE} is a maximum, subject to constraints

$$\mathrm{N}=\sum_{i=1}^{\infty}N_i \tag{49}$$

$$E=\sum_{i=1}^{\infty}N_i\varepsilon i \tag{50}$$

The mnemonic function is

$$\begin{aligned}F_{mn}^{BE}=\sum_{i=1}^{\infty}[(N_i+gi-1)\ln(Ni+gi-1)-(Ni+gi-1)-NilnN_1\\ +Ni-(gi-1)\ln(gi-1)+(gi-1)]+\alpha\left(N-\sum_{i=1}^{\infty}Ni\right)+\beta\left(E-\sum_{i=1}^{\infty}\varepsilon i\right)\end{aligned} \tag{51}$$

We seek the Ni for which F_{mn}^{BE} is an extremum; that is, the Ni satisfying

$$0=\frac{\partial_{mn}^{BE}}{\partial Ni}=\frac{Ni+gi-1}{Ni+gi-1}+\ln(Ni+gi-1)-1-\frac{Ni}{Ni}-\ln Ni+1-\alpha-\beta\varepsilon i-ln\frac{Ni}{Ni+gi-1}-\alpha-\beta\varepsilon i \tag{52}$$

Solving for Ni, we find

$$\text{Ni} = \frac{(gi-1)e^{-\alpha}e^{-\beta\varepsilon i}}{1-e^{-\alpha}e^{-\beta\varepsilon i}} \approx \frac{gie^{-\alpha}e^{-\beta\varepsilon i}}{1-e^{-\alpha}e^{-\beta\varepsilon i}} \tag{53}$$

Where the last expression takes advantage of the fact that gi is usually a very large number, the error introduced by replacing (gi-1) with gi is usually negligible. If $1 >> e^{-\alpha}e^{-\beta\varepsilon i}$, the Bose-Einstein distribution function reduces to the Boltzman distribution function.

Problems With Thermal and Calorific Properties

1. Air flows over a square plate having a dimension of 0.5 m x 0.5 m. The free stream temperature of the air is 280°C. At a steady state, the plate temperature is 55°C. If the convective heat transfer coefficient is 250 $W/m^2.K$. Determine the heat transfer rate from the air to one side of the plate.
2. A steel pipe having a diameter of 15 cm has hot air at 900°C flowing through it. The pipe is shielded with two layers of protecting materials each having a thickness of 3 cm and having thermal conductivities of 0.3 W/mK and 0.8 W/mK. The inlet and exit heat transfer coefficients are 80 and 30 W/m2 K respectively. The atmosphere is at 280°C. Calculate the rate of heat loss from an 80 m long pipe. Neglect the pipe thickness.
3. A 20mm diameter rod is to be heated from 800°C to 1600°C. The density of the rod is 800 kg/m^3 and specific heat of 600 J/kgK. It is placed concentrically inside a long cylindrical furnace having an internal diameter of 120 mm. The inner side of the furnace is at a temperature of 9000°C and has an emissivity of 0.5. If the surface of the rod has an emissivity of 0.3, find the time required to heat the rod.

Problems With Perfect Gas

1. In the closed stationary system, an air expands reversible adiabatic process from 0.8 MPa, 25°C to 0.6 MPa. Find out the exit temperature, and per kg of air, the change in enthalpy, the heat transferred, and the work done.
2. Natural gas is required on a site 1500 m above storage level. The gas is steadily pumped at – 160°C, 1.6 bar from the storage to a delivery point on the site where its pressure is 2.2 bar, its temperature 32°C, and its flow rate 1150 m^3/hr. The work transfer to the gas at the pump is 15 kW, find the heat transfer to the gas between the two points. Neglect the change in K.E. and assume that the gas properties of methane (CH4) may be treated as an ideal gas having $\gamma = 1.33$ ($g = 9.75\ m/s^2$).
3. The closed system consists of 1kg of air, initially at 10°C and occupying 0.8 m^3 volume of the system, undergoes a constant pressure heating process to 180°C. There is no work other than pdv work. Find (a) the work done during the process, (b) the heat transferred, (c) the change in the gas.
4. A gas occupies 0.024 m^3 at 700 kPa and 95°C. It is expanded in the non-flow process according to the law PV 1.2 = constant to a pressure of 70 kPa after which it is heated at constant pressure back to its original temperature. Sketch the process on the p-v and T-s diagrams, and calculate for the whole process the work done, the heat transferred, and the change of entropy. Take cp = 1.047 and cv = 0.775 kJ/kg K for the gas.
5. A heat engine uses 1 kg of air as the working fluid. The maximum and minimum temperatures of the cycle are 880 K and 520 K. The maximum pressure is 1 MPa and the volume of the gas

doubles during the isothermal heating process. Determine the total work and heat supplied so that the efficiency is the maximum possible for the given maximum and minimum temperatures.

6. A two-stage air compressor receives 0.238 m^3/s of air at 2.5 bar and 55°C and discharges at 16 bar. The polytropic compression is 1.35. Determine (a) the minimum power needed for compression, (b) the power needed for single-stage compression to the same pressure, (c) the maximum temperature for (a) and (b), and (d) the heat removed in the intercooler.
7. The air contained in a cylinder fitted with a piston is compressed reversibly according to the law pv1.25 = const. The mass of air in the cylinder is 0.1 kg. The initial pressure is 100 kPa and the initial temperature is 20°C. The final volume is 1/8 of the initial volume. Determine the work and the heat transfer.

Problems With Internal Energy

1. A vertical cylinder chamber gadget comprises water and is warmed on top of a range. During the interaction,65 Btu of heat is moved to the water, and heat losses from the side walls amount to 8 Btu. The cylinder ascends because of evaporation, and 5 Btu of work is finished by the vapour. Determine the alter within the vitality of the water for this handle.
2. The pump increases the water pressure from 15 psia to 70 psia. Find out the power contribution essential, in hp, to pump 0.8 ft3/s of water. Ensures the water temperature at the inlet has any substantial outcome on the required drift control?
3. Study a room that is at first at the outside temperature of 20°C. The room consists of a 40-W lightbulb, a 110-W TV set, a 300-W refrigerator, and a 1200-W iron. Expecting no heat transfer over the walls, Regulate the rate of increase of the energy content of the room when all of these electric devices are on.
4. A geothermal pump is used to pump the saline water. The density of saline is 1050 kg/m^3 at a rate of 0.3 m^3/s from the depth of 200 m. The pump effectiveness is 74% and determines the input power required for the pump. Terminate the frictional losses in pipes, and assume the seawater at 200 m depth to be showing to the atmosphere.
5. Lake water is impelled to a 15 m above storage tank at the rate of 70 L/s while using 15.4 kW electricity. Neglecting frictional losses in the pipes and any dissimilarities in kinetic energy, determine (a) the overall efficiency of the pump–motor and (b) the Inlet and exit pressure difference of the pump.
6. The wind turbine rotating speed is 15 rpm under stable winds flowing through the turbine at the rate of 42,000 kg/s. The turbine blade tip velocity is 250 km/h. If the turbine blade produces 180 kW power, determine (a) the velocity of the air and (b) turbine efficiency. Assume the density of air is to be 1.31 kg/m^3.
7. An oil pump consumes 44 kW of electricity while impelling oil with $\rho = 860$ kg/m^3 at the rate of 0.1 m^3/s. The inlet and exit dimensions of the circular pipe are 8 cm and 12 cm, correspondingly. If the oil pressure is raised to 500 kPa and the efficiency of the motor is 90%, determine the mechanical efficiency of the oil pump.

Problems With the First Law of Thermodynamics

1. An engine is tested using a water brake at 1500 rpm. The measured torque of the engine is 11000 mN and the water consumption of the brake is 0.5 m^3/s, its inlet temperature being 30°C. Calculate the exit water temperature, assuming that the engine power is ultimately transformed into heat which is absorbed by the cooling water.
2. An engine engages in two work interactions during the single cycle: 25 kJ to the fluid and 35 kJ from the fluid, and three heat interactions, two of which are known: 65 kJ to the fluid and 35 kJ from the fluid. Evaluate the magnitude and direction of the third heat transfer.
3. Water is pumped from a lower reservoir to a higher reservoir by providing 20 kW of shaft power. The free surface of the upper reservoir is 55 m higher than that of the lower reservoir. If the flow rate of water is measured to be 0.04 m^3/s, determine mechanical power that is converted into thermal energy during this process due to frictional effects.
4. A system composed of 2 kg of fluid expands in a frictionless engine from an initial state of 2 MPa, 160°C to a final state of 25°C. If there is no heat transfer, find out the suitable process.
5. An electric generator attached to a windmill produces an average electrical power output of 8 kW. The power is used to charge a storage battery. 0.75 kW rate of heat transfer occurs from the battery to the surroundings. Determine the total amount of energy stored in the battery in 10h of operation.
6. Air is contained in a rigid wall insulated tank with a volume of 0.3 m^3. The tank is personalized with a paddle wheel which transfers energy to the air at a constant rate of 5 W for 30 min. The initial density of the air is 1.2 kg/m^3. If no changes in KE or PE occur, determine (a) specific volume at the final state of air, and (b) change in specific internal energy of the air.
7. A mass of gas 1.5 kg undergoes a quasi-static expansion which follows a relationship $p = a + bV$, where a and b are constants. The initial and final pressures are 800 kPa and 150 kPa and the corresponding volumes are 0.50 m^3 and 1.80 m^3. The specific internal energy of the gas is given by the relation $u = 1.3\ PV - 63$ kJ/kg. where p is the kPa and v is in m^3/kg. Determine the total heat transfer and the maximum internal energy of the gas attained during expansion.

Problems With the Second Law of Thermodynamics

1. During the heat addition process of a Carnot cycle, 850 kJ of heat is added to the working liquid from a source at 300°C. Determine (a) the change in entropy of the working fluid, (b) the entropy change of the source, and (c) the total entropy change for the process
2. A Refrigerant-134a enters the cooling coils of the refrigeration system as a saturated liquid–vapour mixture at a pressure of 240 kPa. The refrigerant absorbs 160 kJ of heat from the cooled space, which is maintained at –16°C and leaves as saturated vapour at the same pressure. Determine (a) the change in entropy of the refrigerant, (b) the entropy change of the cooled space, and (c) the total entropy change for this process.
3. The radiator heating system has a volume of 35 L and is filled with superheated water vapour at the pressure of 180 kPa and 165°C. At this moment both the inlet and the exit valves of the radiator are closed. After a while, the temperature of the steam drops to 28°C as a result of heat transfer to the room air. Determine the entropy change of the steam during this process.
4. A rigid tank contains refrigerant-134a and its volume 0.5-m^3. Initially at 220 kPa and 60% quality. Now the heat is transferred to the refrigerant from a source at 35°C until the pressure expands to

385 kPa. Determine (a) the entropy change of the refrigerant, (b) the entropy change of the heat source, and (c) the total entropy change for this process.

5. An engine that takes in 110 MKJ at a temperature of 485 K and rejects 62 MJ at a temperature of 263 K and delivers 18 kWh of mechanical work. What would you advise investing money to put this engine on the market?
6. The heat transfers from the heat engine and the heat pump are used to heat the water circulating through the radiators. The efficiency of the heat engine is 32% and the coefficient of performance of the heat pump is 3. Determine the ratio of the heat transfer to the circulating water and heat transfer to the heat engine.
7. A heat pump working on the Carnot cycle takes heat from a reservoir at 8°C and delivers heat to a reservoir at 55°C. A reversible heat engine is used to drive the heat pump which takes in heat from a reservoir at 750°C and rejects heat to a reservoir at 72°C. The reversible heat engine also drives a machine that absorbs 25 kW. If the heat pump extracts 22 kJ/s from the 7°C reservoirs, Determine (a) the rate of heat supply from the 750°C sources, and (b) the rate of heat rejection to the 72°C sinks.

Problems With Energy Equations

1. In a steam boiler, hot gases from fire heat transfer to water which vaporizes at a constant temperature. The gases are cooled from 1100°C to 550°C while the water evaporates at 220°C. The specific heat of gases is 1.005 kJ/kgK, and the latent heat of water at 220°C, is 1858.5 kJ/kg. All the heat transferred from the gases goes to the water. How much does the total entropy of the combined system of gas and water increase as a result of the irreversible heat transfer between the result based on 1 kg of water evaporated? If the temperature of the surroundings is 30°C, find the increase in unavailable energy due to irreversible heat transfer.
2. Exhaust gases leave an internal combustion engine at 800°C and 1 atm, after having done 1050 kJ of work per kg of gas in the engine (cp of gas =1.1 kJ/kg K). The temperature of the surroundings is 30°C. (a) How much available energy per kg of gas is lost by throwing away the exhaust gases (b) What is the ratio of the lost available energy to the engine work.
3. A pressure vessel has a volume of 1 m^3 and contains air pressure at 1.4 MPa and a temperature is 175°C. The air is cooled to 25°C by heat transfer to the surroundings at 25°C. Calculate the accessibility in the initial and final conditions of the irreversibility process. Take P_0 = 100 kPa.
4. In a rotary compressor, air enters at 1.5 bar and 25°C where it is adiabatically compressed to 5.5 bar and 300°C. Calculate the irreversibility and the entropy production for mass flow rate. The atmosphere is at 1.5 bar, 30°C. Neglect the K.E. changes
5. A 2 kg mass of air in a pressure vessel expands from 5 bar, 60°C to 1.05 bar, 30°C while receiving 1.3 kJ of heat from a reservoir at 240°C. The environment is at 1.02 bar, 18°C. Calculate the maximum work and the work done on the atmosphere.
6. A mass of 5 kg of air is in a vessel at 250 kPa and 25°C. Heat is transferred to the air from a reservoir at 650°C. until the temperature of the air rises to 350°C. The environment is at 150 kPa, 30°C. Determine (a) the initial and final availability of air, and (b) the maximum useful work associated with the process.

Problems With Fluid Compressibility

1. The piston and cylinder device, whose piston is at the ideal position, initially embraces 3 kg of air at 200 kPa and 27°C. A 400 kPa pressure is required to move the piston. Heat is transferred to the air until the volume is doubled. Determine the work done by the air and the total heat transferred to the air.
2. Determine the change in the entropy of air, in kJ/kg×K, as it undergoes a change of state from 100 kPa and 20°C to 600 kPa and 300°C using the equation of state $P(v - a) = RT$ where $a = 0.10$ m^3/kg, and compare the result to the value obtained by using the ideal gas equation of state.
3. A single-acting 4 stroke engine of 0.15 m bore develops an indicated power of 4 kW when running at 216 rpm. Determine the area of the indicator diagram that would be obtained with an indicator having a spring constant of 25 × 106 N/m^3. The length of the indicator diagram is 0.1 times the length of the stroke of the engine.
4. Determine the change in the internal energy of air, in kJ/kg, as it undergoes a change of state from 100 kPa and 20°C to 600 kPa and 300°C using the equation of state P(v - a) = RT where a = 0.10 m^3/kg, and compare the result to the value obtained by using the ideal gas equation of state.
5. A horizontal pump forces the water of 1 m^3/min from an open well to a closed tank where the pressure is 0.9 MPa. Compute the work done by the pump, upon the water in an hour just to force the water into the tank against the pressure. Sketch the system upon which the work is done before and after the process.
6. A cylinder has a piston of an area of 0.12 m^2 and contains gas at a pressure of 1.5 MPa. The gas expands accordingly, due to the process which is denoted by a straight line on a p-v diagram. The final pressure is 0.15 MPa. Calculate the work done by the gas on the piston if the stroke is 0.30 m.

Problems With Combined 1st and 2nd Law of Thermodynamics

1. Water is being warmed in a closed pan on top of a range while being stimulated by a blade type of wheel. During the interaction, 30 kJ of heat is transferred to the water, and 5 kJ of heat is lost to the surrounding air. The wheel has work done of 500 Nm. Determine the final energy of the system if its initial energy is 12.5 kJ.
2. A vertical piston and cylinder device comprise water and it is heated up to the maximum level., 65 Btu of heat is transferred to the water during this process, and 8 Btu amount of heat losses from the side walls. The piston increases as a consequence of evaporation and 5 Btu of work is done by the vapour. Determine the change in the energy of the water for this process.
3. The proprietor is considering these heating systems to warm his house: Electric resistance heating with $0.12/kWh and 1 kWh = 3600 kJ, gas heating with $1.24/therm and 1 therm = 105,500 kJ, and oil heating with $2.3/gal and 1 gal of oil = 138,500 kJ. Assuming efficiencies of 100% for the electric furnace and 87% for the gas and oil furnaces, determine the suitable heating system with the lowest energy cost.
4. A steam power plant produces 300 MW of total power with an overall thermal efficiency of 32%. The actual gravimetric air-fuel ratio in the furnace is calculated to be 12 kg air/kg fuel. The heating value of the coal is 28,000 kJ/kg. Determine (a) the amount of coal consumed during 24 hours and (b) the rate of air flowing through the furnace.

5. A refrigerator removes heat from the food compartment at a rate of 5040 kJ/h for each kW of power it consumes. Determine the coefficient of performance of the refrigerator and the rate of heat rejection to the outside air.
6. A refrigerator is used to reduce the water temperature from 23 to 5°C in a continuous manner. The heat rejected in the condenser is 570 kJ/min and the power is 2.65 kW. Determine the rate at which water is cooled in L/min and the COP of the refrigerator. The specific heat of water is 4.18 kJ/kg. °C and its density is 1 kg/L.
7. A Carnot engine works between a source at 1000 K to 300 K. If the heat engine is supplied with heat at a rate of 800 kJ/min, determine (a) the thermal efficiency and (b) the power output of this heat engine.

CONCLUSION

A brief discussion on the classical and statistical thermodynamics along with the thermodynamic system has been made. Beyond that laws of thermodynamics along with the postulates from various Maxwell's relations, Gibbs equation and Bose-Einstein statistics have also been discussed. At the same time that many of the important problems and aspects of the theory of thermodynamic fluids have been dealt with in the chapter.

Chapter 9
Supersonic Nozzle Flow

Nithya Subramani
Vel Tech Rangarajan Dr. Sagunthala R&D Institute of Science and Technology, India

Manigandan Sekar
Sathyabama Institute of Science and Technology, Symbiosis International University, India

Yasin Sohret
Anadolu University, Turkey

Gowtham Gajapathy
Vel Tech Rangarajan Dr. Sagunthala R&D Institute of Science and Technology, India

ABSTRACT

The nozzle flow is considered to be much more important in aerospace applications. Based on the pressure, velocity and temperature, flow velocity, Mach number and area, the nozzle shape will vary to meet the required condition. Nozzle is a part where the potential energy is converted into kinetic energy. The high pressure and temperature combustion product of gas is converted into high velocity and low-pressure gas as exhaust. In aviation, the main force of thrust is generated due to high velocity exhaust. This chapter gives an explanation and mathematical expression of various nozzles in both subsonic and supersonic flow along with the flow associated issues. The 3D model of convergent chevron nozzle was analyzed for the characteristics of jet mixing and acoustic effect at the exit. These results were compared with the same dimension chevron nozzle with wedges. The added wedges enhanced the jet mixing.

INTRODUCTION

Characteristics of Flow

Based on Mach number, the fluid flow is characterized in different regions and also the speed of the object is identified. It is an important quantity which has no dimension because of the ratio of two velocities with same dimension and it plays an important role in getting the proper idea of the motion of vehicles especially airplanes and rockets moving in an air medium. Mach number is the ratio between the speed

DOI: 10.4018/978-1-6684-4230-2.ch009

of the object in air medium and the speed of sound. The speed of sound is 343 m/s (meter/second) at the temperature of 20 degrees Celsius in the considered medium.

$$M = \frac{Speed\ of\ object}{Speed\ of\ sound} = \frac{v}{c}$$

V = Speed of object
C = Speed of sound

Mach number is generally represented with M. When an object is flying in the sky, the air particles around them are disturbed by the speed of the object. If the Mach number is greater than the speed of sound, then the air particles are compressed thus the compressibility effects are taken into account. As a result of identifying the Mach number, it is possible to find whether the flow is compressible or not.

Classification of Flow Regions

According to the Mach number (M) variations, the flow is classified in different regimes. If the Mach number remains less than 0.3 then the flow is said to be Incompressible. If it varies from 0.3 to 0.8, then the flow is called Subsonic. The flow is sonic when the Mach number is equal to one. The flow is said to be transonic when the Mach number is greater than 0.8 and till 1.2. If the Mach number is greater than 1.2 then it is called as supersonic. Hypersonic flows are greater than Mach number 5. The different flow regime according to the Mach number is listen in the table 1

Table 1. Different flow regime according to Mach number

Flow type	Mach number Limits
Incompressible	0 to 0.3
Subsonic	0.3 to 0.8
Transonic	0.8 to 1.2
Sonic	1
Supersonic	1 to 5
Hypersonic	5 to 8

If the Mach number is high, then the fluid is considered as compressible and compressibility effects are significant. Incompressible fluid is the constant density flow. Even if the pressure is changed there is not much difference in volume with the constant temperature too. The change in pressure is related to change in volume with very small amount when the temperature is constant and lies within the regime of incompressible flow. From Bernoulli's principle, the pressure variation is related to velocity change. The flow is said to be compressible when there is a volume change with the respective pressure variation. For aircrafts, the height is also considered for performance and it changes the Mach number. When the aircraft flies at different altitude, the pressure varies and speeds of sound changes. In that particular, the change of height resulting in change of Mach number creates necessity of the aircraft to adjust itself.

The flow variation is closely considered and monitored since the flow dominates and decides the performance of the engine. The converging nozzle is used when the flow is subsonic. Converging nozzle is the duct in which the area is decreased at certain inclination constantly. When the incoming flow is supersonic, pressure and temperature of the incoming flow are high while the velocity and Mach number of the inlet flow are lower. When the subsonic flow is passed through the convergent nozzle, the pressure and temperature gt reduced while the velocity and Mach number get increased. With the same purpose of increasing the exit velocity of the supersonic incoming flow, the divergent nozzle can be used. For the divergent nozzle with the supersonic flow, the pressure and temperature of the fluid get reduced and conversely the velocity and Mach number are increased. For subsonic flow, the divergent nozzle is acted as diffuser to reduce the speed of the flow. In the same way, the convergent nozzle is acted as diffuser for supersonic flow. When the compressibility effects are considered in the nature of the flow, the shock wave problems are also considered.

Shock Waves

Shock waves are strong pressure waves happen in any elastic medium as air, water or any solid material that is ejected from explosions at the speed greater than the speed of sound. The rapid pressure changes happen across the shock waves.

Normal Shock Wave

The normal shock waves are the strongest pressure waves that are able to change the supersonic flow into subsonic flow across it. These waves are perpendicular to the incoming flow.

Oblique Shock Waves

Oblique shock waves are at an angle to the incoming flow and the flow behind oblique shocks is still supersonic unlike normal shock waves.

The isentropic flow condition is assumed to find out the relationships between various parameters. The most important isentropic relation includes area- velocity relation and area - Mach number relation. This proposed chapter gives the explanation and mathematical expression of various nozzles in both sub-sonic and supersonic flow along with the flow associated issues in nozzle. The 3D model of convergent chevron nozzle was analyzed for the characteristics of jet mixing and also the acoustic qualities at the exit. These results were compared with the same dimension chevron nozzle with the formation wedge design. The addition of wedge reduced the jet noise and enhanced the jet mixing.

Governing Equations

The basic governing equations are the continuity equation, momentum equation and energy equation.

Continuity equation

$$\rho_1 u_1 A_1 = \rho_1 u_1 A_1 \tag{1}$$

Momentum equation

$$P_1 A_1 + \rho_1 u_1^2 A_1 + \int_{A1}^{A2} P\, dA = P_2 A_2 + \rho_2 u_2^2 A_2 \quad (2)$$

Energy equation

$$h_1 + u_1^2 / 2 = h_2 + u_2^2 / 2 \quad (3)$$

Area – Velocity Relation

The nozzle characteristics can be found by deriving the follow Area-Velocity Relation. It gives the relationship between area of the nozzle and velocity of the flow that decides the nature and limits of the flow passing through the nozzle. To get the relationship derivation, the flow has to be considered as quasi one dimensional flow. In such flow, all flow properties are varied in only one direction that is X axis. The area change is in effect in quasi one dimensional while the area is considered to be constant in one dimensional flow.

In such flow $A = A(x)$, $P = P(x)$, $\rho = \rho(x)$, $V = u = u(x)$ and all properties are changed in the direction of X. The differential form governing equations are used to get the final equation. The assumptions of the flow are adiabatic and Inviscid and no dissipation mechanisms are followed as there is no friction, thermal conduction or diffusion on the flow.

$$\frac{d\rho}{\rho} + \frac{du}{u} + \frac{dA}{\rho a} = 0 \quad (5)$$

To eliminate the term of $\frac{d\rho}{\rho}$, consider pressure form of the same expression

$$\frac{dP}{\rho} = \frac{dP}{d\rho} \cdot \frac{d\rho}{\rho} = -\, u\, du \quad (6)$$

Since the flow is isentropic, any change in pressure dp is creating the respective isentropic change in density dρ.

$$\frac{dp}{d\rho} = \left(\frac{\partial p}{\partial \rho}\right)_2 = a^2$$ (7) Substitute equation 7 in 6

$$a^2 \frac{d\rho}{\rho} = -\, u\, du \quad (8)$$

$\frac{d\rho}{\rho}$ = - u du / a^2 . (9) Multiply and divide the equation 9 with u

$$\frac{d\rho}{\rho} = - u^2\, du / u\, a^2 \quad (10)$$

Substitute the Mach number representation for the value of u/a

$$\frac{d\rho}{\rho} = - M^2\, (du / u) \quad (11)$$

Substitute the equation 11 in 12 then it comes as

$$\frac{d\rho}{\rho} + \frac{du}{u} + \frac{dA}{\rho a} = 0 \quad (12)$$

$$- M^2\, (du / u) + \frac{du}{u} + \frac{dA}{\rho a} = 0 \quad (13)$$

$$\frac{dA}{A} = (M^2 - 1)\, \frac{du}{u} \quad (14)$$

The equation 14 is called as Area-Velocity relation. From this equation, the observed relations are

Figure 1. Area – velocity relation (John.D. Anderson (2017). Modern Compressible Flow)

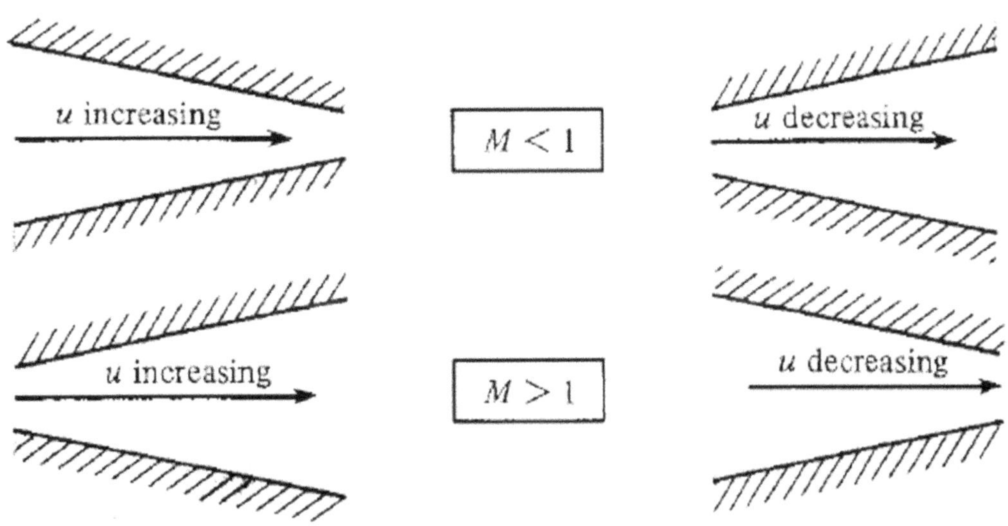

1. When M®0, this flow is incompressible flow and after the limit substitution it comes as Au = Constant which is the continuity equation of incompressible flow.
2. When M = 1, it comes as $\frac{dA}{A} = 0$, the area can be minimum or maximum with the respective results. The minimum area of the nozzle where the M = 1 is called throat area.
3. When 0 £ M < 1, it is the subsonic flow and results as if the there is a velocity increment then that is because of the area decrement and vice versa is also the same. It is true for incompressible flow as the velocity gets increased with the converging duct while the opposite of this as the velocity decrement in diverging duct is noticed.
4. When M > 1, the flow is supersonic and the velocity increment is observed in area increment. For a supersonic flow, the converging duct decreases the velocity and the diverging duct increases the velocity.

Figure 1 gives the relationship between Area- Velocity relation at the subsonic and supersonic speed for convergent and divergent ducts. For any gas in incompressible limit to expand it isentropically, It should pass through the convergent – divergent nozzle as subsonic to supersonic. For any gas to compress isentropically, it also must flow through the convergent – divergent nozzle as supersonic to subsonic. The application of C-D nozzle is in seen in rocket where the combustion chamber is attached with the bell shaped convergent – divergent nozzle and also in supersonic wind tunnel that has to produce supersonic velocity in the test section. The flow in C-D nozzle is shown in the below figure 2 with the velocity variations in two segments for both subsonic and supersonic flow.

Figure 2. C-D nozzle with different Mach numbers (John.D. Anderson (2017). Modern Compressible Flow)

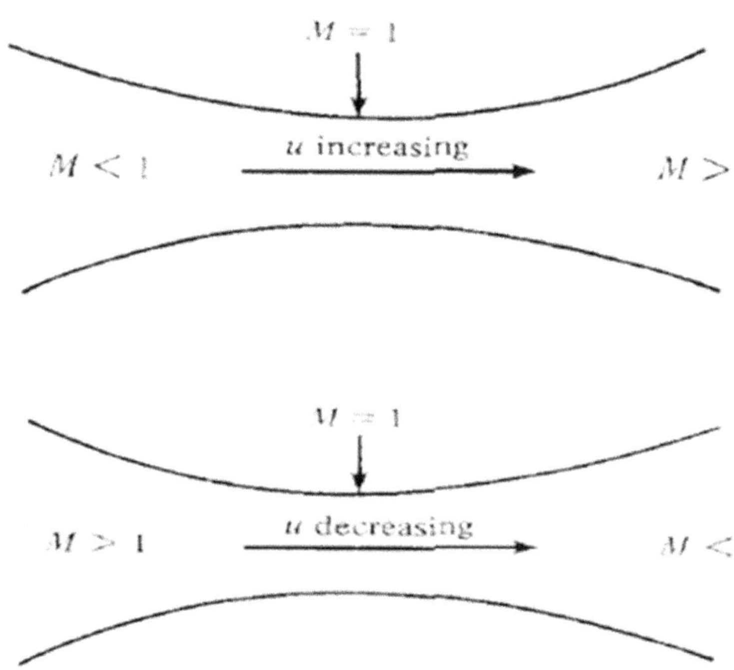

Besides the area velocity relation, it is also much more important to find the Area-Mach number relation for nozzles.

Area-Mach Number Relation

To derive the subsonic-supersonic flow relations, the isentropic flow is assumed and the Mach number, velocity of object, area, density and speed of sound at the throat section are M^*, u^*, A^*, ρ^* and a^* respectively. At the throat the special superscript * is used for all parameters while at any location other than the throat is expressed with A, M and u for the local area, local Mach number and local velocity. Figure 3 gives the Geometry for Area-Mach number relation.

Figure 3. Geometry for Area-Mach number relation (John.D. Anderson (2017). Modern Compressible Flow)

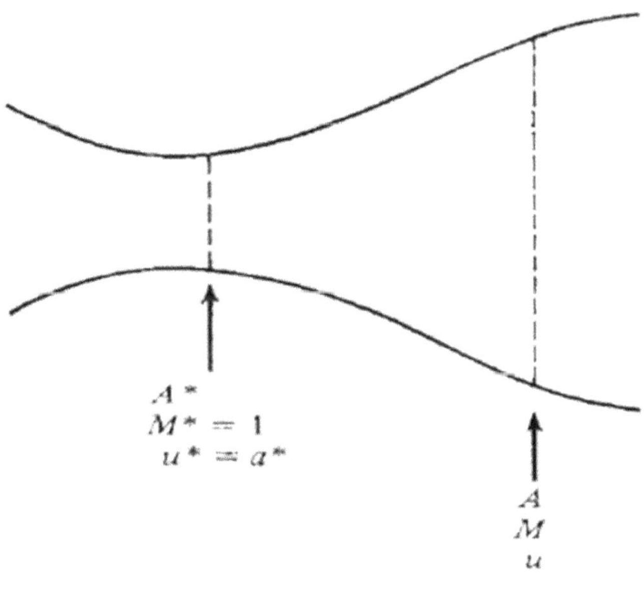

By applying the continuity equation to the selected local and throat sections, the equation comes as

$$\rho^* u^* A^* = \rho u A \quad (15)$$

Equation 15 can be rewritten as

$$A / A^* = (\rho^* / \rho) (u^* / u) \quad (16)$$

Since at the throat $u^* = a^*$

$$A / A^* = (\rho^* / \rho) (a^* / u) . \quad (17)$$

The stagnation density is multiplied and divided with the equation of 17

$$A / A^* = (\rho^* / \rho^o) (\rho^o / \rho)(a^* / u) \tag{18}$$

The ratio of stagnation density to local density and the ratio of stagnation density to throat density are given below

$$\rho^o / \rho = \{ 1 + \frac{\gamma - 1}{2} M^2 \}^{1/\gamma-1} \tag{19}$$

$$\rho^o / \rho^* = \{ \frac{\gamma + 1}{2} \}^{1/\gamma-1} \tag{20}$$

$$(u / a^*)^2 = M^{*2} = (\frac{\gamma + 1}{2} M^2 / 1 + \frac{\gamma - 1}{2} M^2) \tag{21}$$

$$(A / A^*)^2 = (\rho^* / \rho^o)^2 (\rho^o / \rho)^2 (a^* / u)^2 \tag{22}$$

Substituting the equations of 5, 6 and 7 in the equation of 8 the equation becomes

$$(A / A^*)^2 = (\frac{2}{\gamma + 1})^{2/\gamma-1} (1 + \frac{\gamma - 1}{2} M^2)^{2/\gamma-1} (1 + \frac{\gamma - 1}{2} M^2 / \frac{\gamma + 1}{2} M^2) \tag{23}$$

$$(A / A^*)^2 = 1 / M^2 [\frac{2}{\gamma + 1} (1 + \frac{\gamma - 1}{2} M^2)]^{(\gamma+1/\gamma-1)} \tag{24}$$

This relation is called as Area – Mach number relation. From the final result it is understood that the local area must be greater than or at least equal to that of throat area. Figure 4 represents the Area-Mach Number relation for both supersonic and subsonic.

When Mach number is less than one, the converging duct acts as a nozzle and diverging duct acts as a diffuser. For the Mach umber greater than 1, the diverging duct act as a nozzle and converging duct act as a diffuser. The pressure, temperature, velocity and Mach number variation in different area ducts with respect to various Mach number is shown in figure 5. The temperature and density are inversely proportional to each other. As the temperature decreases across the nozzle, the density of the fluid is increased with the increased velocity.

Figure 4. Area – Mach Number relation (John.D. Anderson (2017). Modern Compressible Flow)

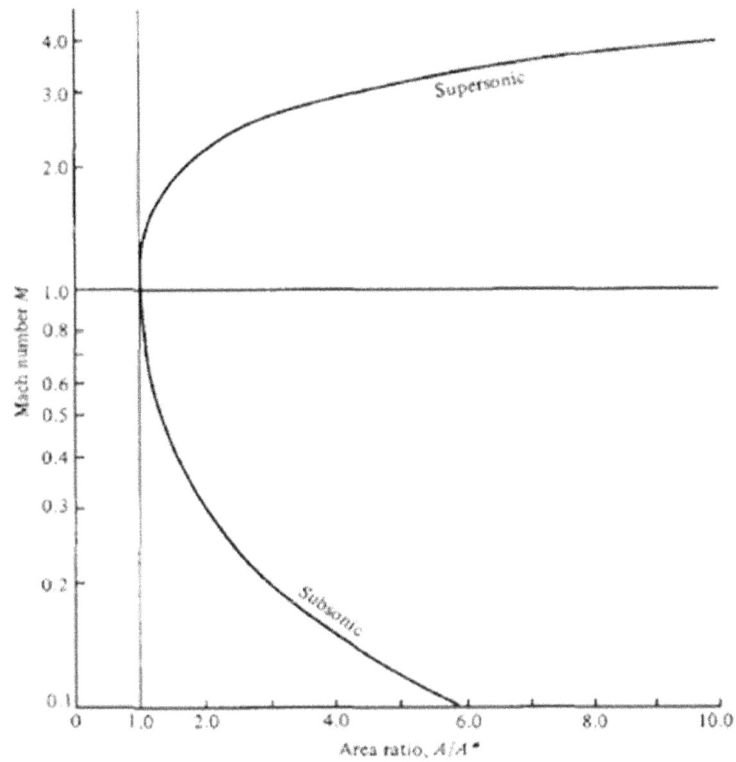

Figure 5. Fluid properties at different ducts (John.D. Anderson (2017). Modern Compressible Flow)

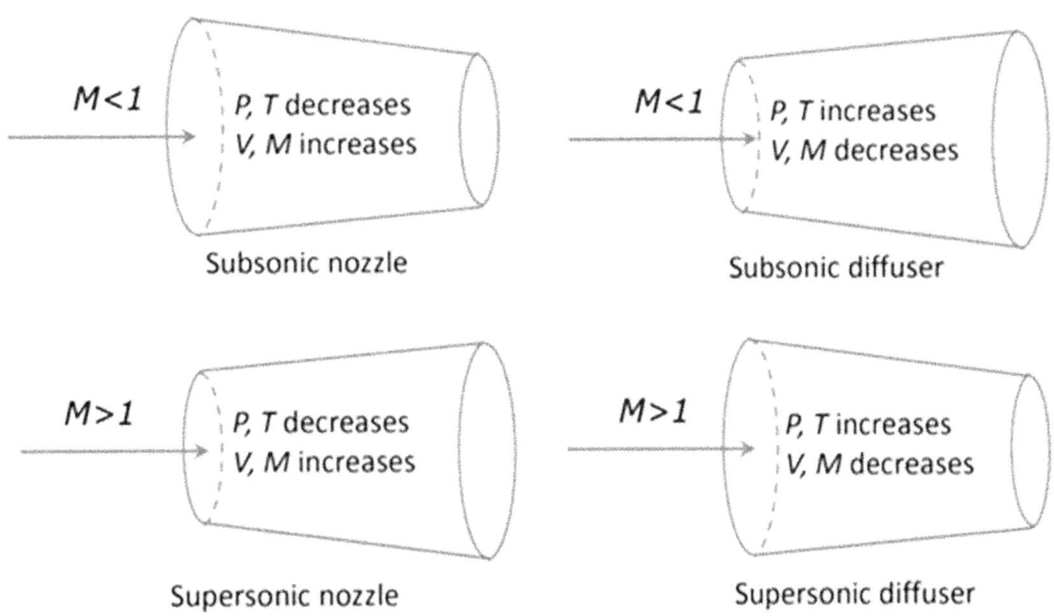

OPERATING PRINCIPLE NOZZLE

The nozzle is a part of the engine that is usually fixed at the exit of the aircraft engine to convert the internal energy into propulsive energy. The main distinction of the aircraft is its jet at the nozzle. The nozzle can produce the jet at various speeds based on the shape of the nozzle and incoming fluid properties. The nozzle purpose is to increase the exit velocity and then brings it to the ambient pressure and also makes the coherent jet.

Classification of Nozzles

The mission requirements vary from one type of aircraft to another aircraft. There many types to expand the gases and mix it with the atmospheric air. The nozzle types are

Convergent Nozzles

Convergent nozzles are used in most of the jet engines. It has converging area as the distance increases in x axis. This nozzle is particularly used for subsonic and sonic flow. The convergent nozzle will choke the flow when it passes the convergent duct. If the nozzle pressure ratio is above the critical level of 1.8:1. The expansion of the gas to the atmospheric pressure is obtained downstream of the duct. Most of the thrust is produced by the jet momentum but still there is some amount of pressure because of the pressure variation between the throat static pressure and atmospheric pressure.

Divergent Nozzles

The divergent nozzle is a duct where the passage area increases as the length of the passage increases. This type of nozzle is used in scramjet where the nozzle incoming air is supersonic to expand further in the nozzle section.

Convergent – Divergent Nozzle

Convergent – divergent nozzle is a combination of both the convergent and divergent nozzle. When the convergent reaches the critical value, the flow will become choked and the flow cannot be further improved and it reduces the thrust producing efficiency of the nozzle. To overcome this problem, the addition of diverging part will further decrease the pressure downstream of the nozzle. This concept is applied in all rocket nozzles.

Choked flow is the resultant of the compressible effect. The flow velocity is limited once the flow is reached its chocked condition with the further decrement in the pressure variation at the downstream. When the flow passes through the convergent (constriction or throat of C-D nozzle) area to the lower pressure environment, the velocity is increased. Initially, the velocity increases for the mass conversion but because of the venture effect, the static pressure reaches at certain point and after this point there will not be any further mass flow increment because of any decrement in the downstream for the fixed upstream condition. Thus it is called as limiting condition.

Fixed Area Nozzle

On a jet engine or rocket engine, the exit area of the nozzle remains constant. This type of nozzle is called as fixed area nozzle. this nozzle is used for only at subsonic speeds because it is only acceptable in the aircraft performance that changes with respect to altitude variation and the respective fluid properties variation. It cannot be used in afterburning nozzle and also in supersonic flight nozzle where the extended expansion is required. All these basic details are explained clearly in modern incompressible flow book. (John, D. Anderson 2017)

Variable Area Nozzle

The variable area nozzle is used in combat aircrafts that is using the afterburner setup. This variable area nozzle has numbers of moving and extendable plates and it form nearly circular cross sectional area. When the aircraft uses the afterburner and it is used for supersonic flight, the variable area nozzle is used t the divergent section of the nozzle configuration.

Variable Area Convergent –Divergent Nozzle

Turbofan engine use the convergent – divergent nozzle where any additional airflow is not required to be bumped to the C-D nozzle. The divergent nozzle portion may be the internal part of the afterburner setup with certain angled extension after throat.

Hypersonic Speed Nozzle

The hypersonic speed engine uses the single ramp expansion nozzle that is utilized in the application of scramjet engine. In these types of engines, the airframe integration with engine plays an important role. The amount of air wetted by the lower body of the aircraft is used to produce propulsive energy.

Nozzle Operating Condition

The pressure at the exit of the nozzle differs from the ambient pressure. The nozzle operation can be classified into three types as correctly expanded, overly expanded and under expanded. Figure 6 provides the three operating conditions of nozzle.

Correctly Expanded Nozzle

If the nozzle is in optimum condition, the exit pressure of the nozzle is equal to ambient pressure of the atmosphere. Under this condition, the flow inside the nozzle is perfectly expanded and produces more thrust.

P_e = Pressure at the nozzle exit
P_∞ = Ambient pressure
$P_e = P_\infty$

Figure 6. a) over expanded, b) correctly expanded and c) under expanded conditions
(http://www.aerospaceweb.org/question/propulsion/q0220.shtml)

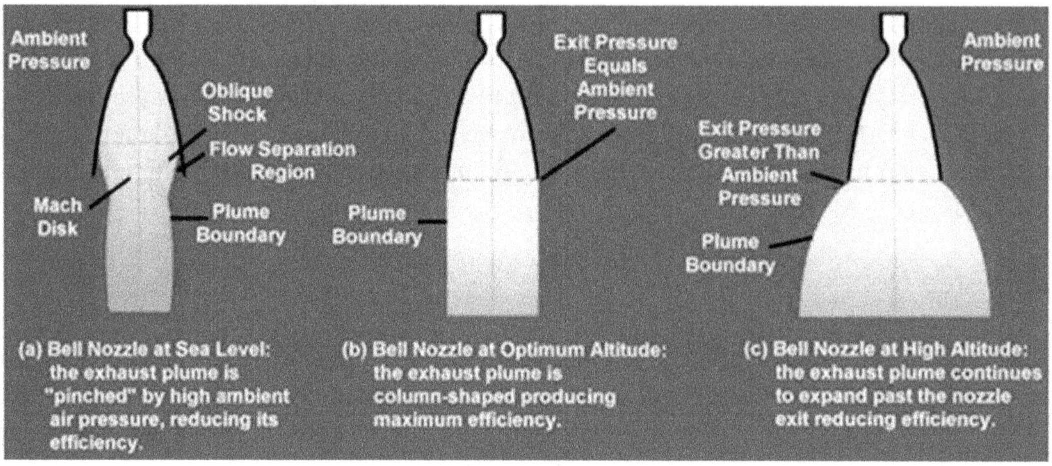

Under Expanded Nozzle

In an under expanded nozzle, the gas expansion still continues even the exit of the nozzle because the nozzle exit pressure is higher than the ambient pressure. Consequently, the nozzle exit gas is expanded in the outer atmosphere that reduces the nozzle efficiency and the expansion happens outside the nozzle will be wasted and cannot be used in the form of useful thrust.

$P_e > P_\infty$

Over Expanded Nozzle

When the exit pressure is lower than the ambient pressure, the higher atmospheric pressure squeezes the exit pressure inside the nozzle thus the portion exit are in the nozzle is wasted. As such it is called as over expanded nozzle.

$P_e < P_\infty$

REYNOLDS NUMBER

Reynolds number is a dimensionless metric that aids in flow behavior prediction. Laminar flow has a low Reynolds number, while turbulent flow has a high Reynolds number. The building of an optimal flow model for effective fluid system design is aided by the identification of flow regime with Reynolds number. Reynolds number is the ratio between inertia force and viscous force.

$R_e = \rho \, v \, D / \mu$
ρ – Density of the working fluid

v – velocity of the working fluid
D- Diameter of the pipe or duct
μ - Viscosity of the fluid

The Reynolds number has the advantage of being considerably easier to compute for each flow and to visualize the fluid flow. The Reynolds number is used to anticipate flow and to establish computational or experimental methodologies for observing flow dynamics.

When the assumption is stated in terms of Reynolds number, the fluid has laminar characteristics up to 2300. The viscous force is more apparent in contrast to the inertial force because laminar flow is often observed when the fluid is moving at a low velocity or has a smaller cross-section and has a smoother flow path. When the Reynolds number is 3500 or greater, the flow is called turbulent. Because of the larger flow rate and uneven route, the inertial force is enhanced. When the Reynolds number is between 2300 and 3500, the flow is in the transition phase, meaning it has just gained velocity but isn't irregular enough to be classified as turbulent.

ADDITIONAL APPLICATIONS OF NOZZLE

Nozzle Vectors (Thrust Vectoring)

Nozzle vectors are also called as thrust vectoring. It is the additional feature or ability of an aircraft to control thrusts in different directions at various altitudes. The thrust is manipulated by the movements created by the nozzle. The rockets cannot use the thrust vectoring outside the earth atmosphere. By using four methods it is done successfully

1. Gimbal engine or nozzles
2. Reactive fluid injection
3. Auxiliary vernier thrusters
4. Exhaust vanes or jet vanes

Gimbaled Thrust

It is done by deflecting both the combustion chamber nozzle to achieve thrust vectoring. In liquid rocket engine, the thrust produced and direction change can be done by deflecting the entire system while it is done by deflecting the nozzle alone in solid rockets by means of electronic actuators or hydraulic actuators. It is called as gimbal method because the combustion chamber and the nozzle is connected by the use of ballpoint with a hole in the center or it also can be done by using the flexible seal.

Reaction Fluid Injection

Another method of obtaining the thrust vectoring result is reaction fluid injection during which the liquid is injected at the exit of the nozzle. The liquid injectors are placed at the aft of the nozzle. The whole engine combustion chamber or the nozzle doesn't have to be deflected. The exhaust plume is adjusted by injecting the liquid wherever it is required resulting the different thrust on the side of the injection and the net thrust was changed and produced the respective control.

Vernier Thrusters

It is a small thruster that is connected to the small combustion chamber by a gimbal point and deflect its own axis. This is particularly used for space applications.

Jet Vanes

This is the earliest method used in rocket engines where the jet vanes are placed at the exhaust system. These parts are deflected without having to move any other parts of the engine like combustion chamber and nozzle. However, this system reduces the mechanical operational difficulty, it reduces the rockets efficiency. The design material requirement for the vanes is also there that it has to be made by using refractive materials or coolant material to avoid melting due to excessive temperature of the exhaust. Graphite and copper are the known materials used for vanes. Graphite has high melting point and the copper material has high heat capacity and both having the quality of thermal conductivity. However separate cooling system has to be used unless the vanes will undergo melting.

Difficulties in Nozzle

The nozzle instability is characterized because of many possible reasons. The main reasons are the flow inside the nozzle and the external flow pattern outside the nozzle. The internal flows are affected highly by the shock wave formation and separation of boundary layers. The external flows are also affected by the extension of the shock wave and pattern of it. In addition to it, compression and expansion waves produced by the wall also decide the performance of the nozzle. The shock wave formation and its location in the nozzle is controlled by the shape of the nozzle and its upstream conditions. The nozzle performance can be decided by the factors of the exhaust jet mixing with the ambient air and the noise produced at the exhaust.

Flow Mixing at the Exit

The jet comes out at high velocity from the nozzle exit compared to the ambient air. It is important to get the proper and smooth mixing of it with the atmospheric air thus improving the performance of the nozzle. Since the exit flow is supersonic, the mixing of exhaust with the ambient air is much more difficult and makes tremendous noise unless it is smooth. There are many methods to promote the efficient mixing such as active and passive methods. The thrust can be improved by making the proper mixing. The active methods are the secondary fluid injection. The reason why it is called as active is the use of additional energy to activate the used mechanism. The passive methods are placing different size and shapes of the tabs and lobes at the exit. By changing the exit shape of the nozzle also the effective mixing can be achieved for example, the rectangle and elliptical throat and nozzle. When the tabs or any other secondary devices are attached with the nozzle exit, the vortices will be created at the exit and it gives energy to the shear layer and the mixing will be improved by reducing the potential core length. The potential core length is the supersonic center line from the exit of the nozzle. The length of the shock train will be reduced thus the mixing will be improved. The streamwise vortices created by chevron cut and the effect of geometric parameters were measured and justified (Malay Suvagiya & S. D. Sharma 2018)

Noise

The noise created by the aircraft and rocket nozzle is the major unfavorable thing due to the mixing of exhaust air and atmospheric outer air. This jet noise can be effectively controlled by the modifying the nozzle design parameters. Experimental and statistical analysis was carried out to reduce the noise in chevron nozzle (Kaleeswaran.P & P. Shanmughasundaram 2016).

Over the past decade, the demand and the attraction on air transport has been increased and aircraft industry is growing day by day consistently. Over all in India, there are 464 airports and out of them 125 airports belong to Airport Authority in India. The 78% of the domestic national passenger traffic and 22% of the international traffic is covered. The reason behind this preference on the mode of air transport is the advantages which possess over the other mode of transport such as short time and high-speed travel, easy access to overseas even uninhabited and remote places where other mode of transport is not possible, economical, time saving and presence of natural obstacles are not there in the route of air transport path. However, apart from high capital investment and risky nature, the noise pollution associated with aircraft is a growing concern. During the operation of aircraft, an enormous amount of nose is created because of many reasons. Noise is created by the aircraft and its components. When the aircraft is moving, the aircraft and its propeller compress or reduce the density of air, thus, increasing the movement of the air molecules. This air molecules movement create the pressure waves. When the pressure waves are strong enough, it produces the noise frequency within the level of hearing frequency spectrum. For example, because of auxiliary power units when the aircraft is parked. Not only during parking but also during takeoff, landing, taxiing and flying. The noise created by the aircraft is classified into three types as noise from engine and other mechanical parts, aerodynamic noise and aircraft system noise. Source of the mechanical noise is the movement of the mechanical bearings and transmission chain systems. Among all other types of sources, higher intensity sound peaks are created because of the rotational speed of the engine and movement of the engine parts. While takeoff and landing, engines are the primary source of most of the noise created. Especially, in turbofan engines, buzzsaw noise is created when the tip of the fan blade reaches supersonic speed. The acoustic characteristics were found for different nozzles and validated (Daniel, C & Petre, C 2013). Apart from turbofan engines, huge amount of jet noise is created in all types of jet engines. In fact, major part of engine noise is jet noise. It includes turbulence at the nozzle exit and jet mixing. When high velocity exhaust gas is leaving the nozzle exit, it creates vortices which later turns into turbulence within certain distance. As this hot turbulent air mixes with the cold atmospheric air, the jet mixing noise is created. This turbulence mixing happens in an annular way initially then moves considerable length normal to the exhaust flow direction. This can be multiple times of the diameter of the nozzle. The mixing region length is generally four to five diameters of the nozzle. If the jet is supersonic flow or chocked flow, series of shock cell is created. Within the shock cell, the is expanded and contracted and the length of this shock cell is taken up to 10 diameters from the nozzle exit. Screech tone and sound created by the shock waves are also associated with the shock cells.

In addition to it, noise from aircraft systems and aerodynamic noise are relatively lower than the engine jet noise. A few aircraft systems that create noise for both military and civilian are cockpit display system, conditioning system and mainly pressurization system. The APU (Auxiliary power unit) used in aircraft is also creating considerable noise. The aerodynamic noise involves the air movement around the aircraft and its control surfaces. The main parameter to consider is the shape. By narrowing the nose and shape of the nose and other control surfaces, much of the sound is created. In propeller aircrafts, the moving air around the propeller blade creates much of the noise. Similarly, in helicopters, the main and tail rotor generates more noise. Aerodynamic noise is also called as airframe noise. It can be classified further into two types as bluff body noise and edge noise. The bluff body noise is created because of the vortex shedding on both sides of the bluff bodies. In the form of edge noise, when the turbulent wave is moved to the edge of the surface, the pressure fluctuations create noise.

Aircraft noise adversely affects both environment and living beings. The level of sound is measured in decibel (dB). Generally during takeoff and flying, the noise range is more than 140 dB. This is mainly because of jet noise and turbulence mixing. The maximum level of noise can be heard by human beings is 60 dB. The hearing of 70 dB for a long time can affect the hearing capacity 160 dB immediately affect and create issues with ears. A few serious health issues are also created by noise such as hypertension, hearing impairment, cardiovascular disease and sleep disturbance. Thus, reducing noise becomes much more important in aviation sector. There two basic methods to control noise such as active methods and passive methods. In active methods, separate power source is required to suppress the sound. The particular frequency of anti-noise is created as the frequency of the noise created by the aircraft. As a result, the noise will be cancelled. In the passive methods, some mechanical tabs or modification of the nozzle is used to reduce the noise. Along with the noise reduction, the thrust enhancement was also achieved for the conventional chevron design with a few dimensional changes (Sasi kumar M., Abirami K., Sandhiya K., Vijay G. and Vishnu Varthan S.2018). In chevron nozzle, the lobes shape was also tried to mitigate the noise created by the jet mixing (Sivasankar.G.A, & S. Balamurugan 2017).

Chevron Nozzle

To minimize the noise produced by the mixing of air, there have been many tests conducted experimentally on small scale chevrons at the exit of the nozzle. On the circumference of the exhaust nozzle, the triangle tooth pattern is there to enhance the jet mixing thus reducing the jet noise. Chevron nozzles are employed mainly in turbofan engines where the cool bypass air get mixed with the hot engine core air at the exit of the nozzle. Most of the turbofan engines have the serrations at the trailing edge of the nozzle. By providing the cut in the nozzle exit, the velocity of the outgoing air is reduced and therefore the temperature is also reduced. As a result of increasing the air interaction, the improved mixing is obtained. There are different types of chevron nozzles in use such as

1. Baseline chevron nozzle
2. Conventional chevron nozzle
3. Sinusoidal chevron nozzle
4. Asymmetric chevron nozzle

In baseline chevron nozzle, the trailing edge is simple without corrugations while in conventional chevron, the saw-tooth pattern is there in the exit tip of the nozzle. Sinusoidal chevron removes the sharp serrations with the smooth pattern and the asymmetric chevron consists the irregular triangle tooth in dimension. The numerical analysis was done for different configurations of the chevron nozzle and it was found that the length and shape of the serrations on the nozzle exit plays an important role in both noise reduction and jet mixing (Jenifer N & Preethi S 2016). The chevron nozzle was tested at the speed of transonic and the respective effects were identified (Usha Bharathi.A, & Vikash et al. 2015). Separate vortex stabilizers were also used to minimize the jet mixing noise at the exit of the nozzle(Vance Dippold 2008).

METHODOLOGY

The investigation on the chevron with wedge and without wedge was done in this study. This combined shape proved its efficiency in controlling the supersonic jet at the exit of the nozzle. The jet mixing of plain cold jet with the nozzle exhaust hot jet was promoted significantly using this shape rather than plain saw-tooth cut at the lip of the nozzle. A simple 3D analysis was carried out to find the ability of the modified chevron. The nozzle taken for the tests was convergent – divergent nozzle with the Mach number of 3. The 3D models were created by CATIA and analyzed by ANSYS Fluent software. Two sets of nozzle designs were analyzed by using the acoustic method. By measuring the noise frequency, it was possible to measure the effectiveness of each nozzle lip design. By placing wedge at the lip of the chevron nozzle, jet mixing was promoted further. The used wedge inside the saw-tooth was expected to generate vortex at the tip of it and provide additional energy to the mixing. 3D compressible flow modeling was done by ANSYS FLUENT. The RANS (Reynolds-Averaged Navier-Stokes) equations combined with k-ω SST (Shear-Stress Transport) turbulence model was created to find the flow characteristics. The solver used was 3D density-based solver and compressibility effects were considered with curvature corrections. The fluid was considered to be air and ideal condition was used for analysis. The meshed model was created and analyzed further. The details of the designs are shown in table. Two different nozzle models were created with chevron and chevron with wedge. The obtained results were compared to each other to find the efficient one in the jet control and noise reduction. The details of the chevron nozzle with and without wedge are listed in table 2. Figure 7 and 8 illustrates the design of nozzle without wedge and with wedge. The chevron nozzle was analyzed with triangular shape (Grigore,C, Virgil, S, & Daniel-Eugeniu C 2014).

Table 2. Details of chevron with and without wedge

S.NO	Nozzle	Inlet Diameter	Exit diameter	Number of saw-tooth	Saw-tooth length (mm)	Wedge length (mm)
1	Chevron	42	30	8	6	
2	Chevron with wedge	42	30	8	6	3

RESULTS

From the obtained results, it is clear that the jet mixing is smoother in chevron with wedge than the plain chevron nozzle. It is because of the small size vortices created by the wedge present in the second model of chevron. The mixing ability of the small size vortices is more stable rather than big vortices. The maximum pressure level reached by the chevron nozzle is 205 dB while the maximum pressure level reached by the wedge nozzle is 201 dB. From this result, it is evident that a significant amount of sound is reduced in the case of wedge chevron.

Figure 7. Chevron nozzle without wedge

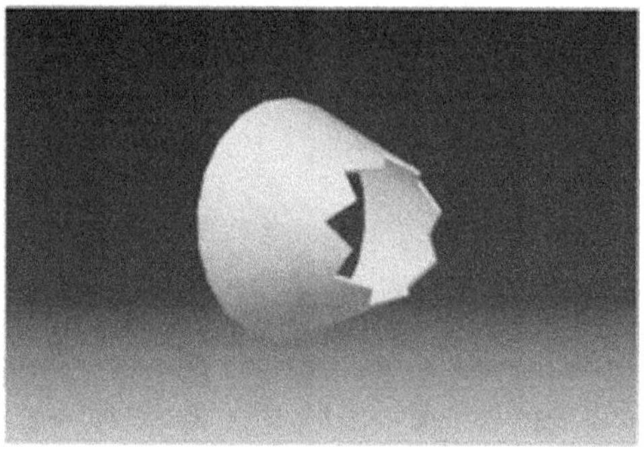

Figure 8. Chevron nozzle with wedge

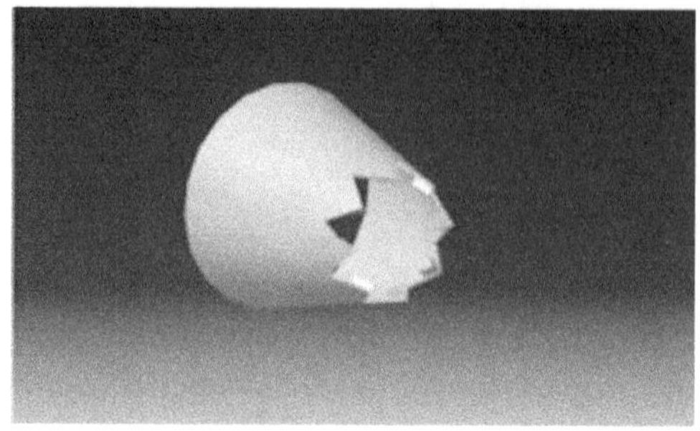

The potential core length is the measure of the jet mixing of exhaust at the exit of the nozzle. From the tip of the nozzle to the end of the supersonic core jet is called potential core length. As long as the supersonic jet extents and creates shocks, it makes noise when it gets mixed with the ambient air. The potential core length can be measured in terms of diameter of the nozzle exit. The potential core length of the chevron nozzle without the wedge was 3.93D while the potential core length of the chevron nozzle with wedge was 3.05D and the respective figure was 9 and 10.

Figure 9. Potential core length for chevron nozzle without wedge

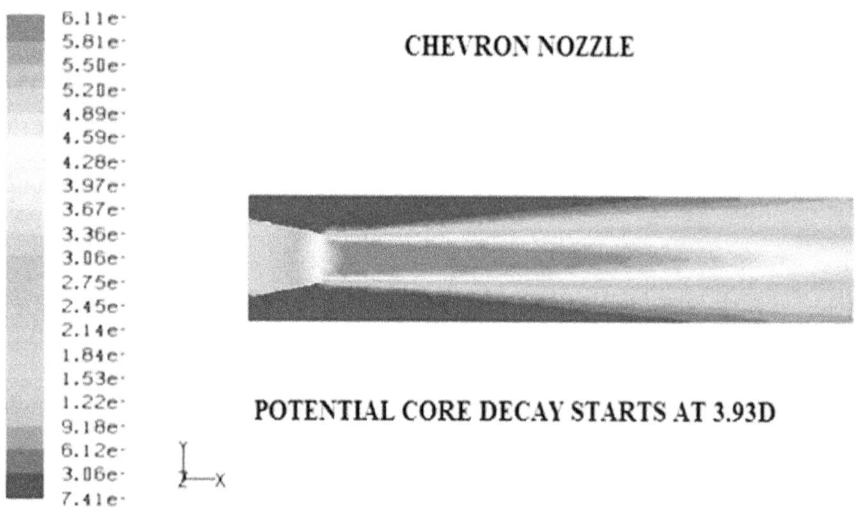

Figure 10. potential core length for chevron nozzle with wedge

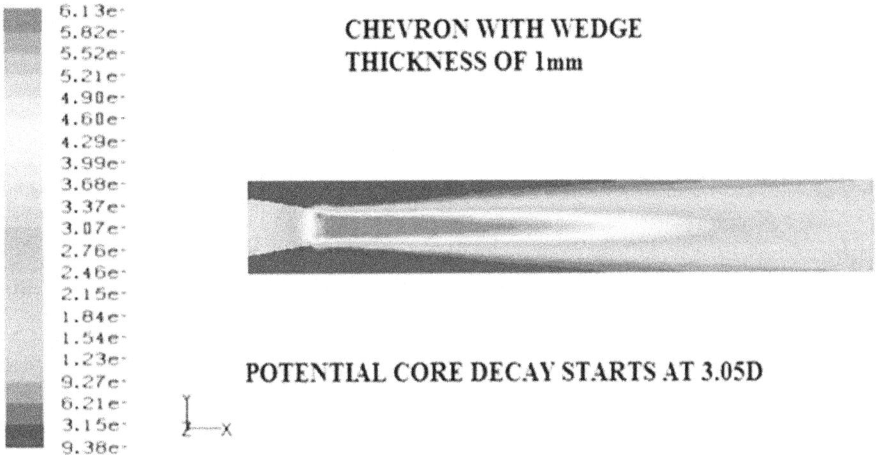

CONCLUSION

The proposed chapter provides the importance of the nozzle, classifications of them, problems associated with the operation of nozzle. In presents days, the aviation sector's main issue of jet noise created at the exit of the nozzle is addressed with its jet mixing. Two different set of 3D convergent nozzle were designed by CATIA and analyzed by using ANSYS Fluent. For jet noise, the noise created at the exit of the nozzle was measured in decibel and the flow mixing was measured in terms of potential core length. From the results it is abundantly clear that the wedge made in the chevron cut improves the jet mixing with the cold ambient air. The set of small vortices are created by the small wedge present inside each triangular saw-tooth cut. The energy and stability of the small vortices are relatively higher than big vortices created by conventional chevron. This leads to the enhanced mixing. As a consequence, the sound of the mixing is reduced in dB. The sound frequency of the chevron nozzle was 205dB. The modified chevron with wedge reduced the sound noise level by 4dB. The core length reduction achieved by wedge was 0.88D.

REFERENCES

Bharathi, Balamanikandan, Paul, Sundar, & Vikash. (2015). Effect of Chevron on Transonic Nozzle. *International Journal of Engineering Research & Technology*, 4(4), 1413–1417.

Daniel, C., & Petre, C. (2013). Acoustic characteristics of the flow over different shapes of nozzle chevrons. *ICAS Bulletin Cojan*, *5*(3), 53–59. doi:10.13111/2066-8201.2013.5.3.6

Dippold, V. (2008). CFD Analyses and Jet-Noise Predictions of Chevron Nozzles with Vortex Stabilization. Glenn Research Center.

Grigore, C., Virgil, S., & Daniel-Eugeniu, C. (2014). Acoustic control study of turbofan nozzles with triangular chevrons. *INCAS Bulletin*, *6*(1), 13–20. doi:10.13111/2066-8201.2014.6.1.2

Jenifer, N., & Preethi, S. (2016). Numerical Analysis of Chevron Nozzle with Various Configurations for Noise Reduction. *International Journal of Innovative Research in Science, Engineering and Technology*, *5*(5).

John, D. A. (2017). *Modern Compressible Flow*. Mc Graw Hill.

Kaleeswaran, P., & Shanmughasundaram, P. (2016). Experimental and Statistical analysis on the noise reduction using chevron nozzle in supersonic free jet, U.P.B. *Scientific Bulletin, Series D*, 78(3), 21–30.

Sasi Kumar, M., Abirami, K., Sandhiya, K., Vijay, G., & Vishnu Varthan, S. (2018). Noise reduction and thrust enhancement in various modified chevron nozzle. *International Journal of Development Research, 8*(1), 18540-18544.

Sivasankar, G. A., & Balamurugan, S. (2017). Virtual Analysis of Chevron Nozzle Lobe Design Modification. *Asian Journal of Research in Social Sciences and Humanities*, *7* (1), 79–86. doi:10.5958/2249-7315.2017.00007.7

Suvagiya, M., & Sharma, S. D. (2018). The effect of geometric parameters of chevron nozzle on generation of streamwise vortices in high subsonic jets. *20th Annual CFD Symposium*, 9-10.

Chapter 10
Shock Reflections and Intersections

Paramaguru V.
ACS College of Engineering, India

Suresh C.
ACS College of Engineering, India

Sankaran A.
Nehru Institute of Engineering and Technology, India

Bernard A. P. Francis
Deepwater Technology, Singapore

ABSTRACT

This chapter is dedicated to the shock wave reflections and intersections. Each topic offers a pictorial representation of the physical and shock polar plane for a better understanding of the shock wave reflections and intersections. This chapter contains an introduction to the shock-shock interference under the various real-life examples of the intersection of different and same family shock waves in a solid boundary, wave reflections from the free boundary, Mach reflections (lambda shock wave), intersection of shock from different or opposite families (Type I Interference), intersection of intense shocks of opposite families forms normal shock (Type II Interference), intersection of strong and weak oblique shocks of different families (Type III Interference), intersection of normal shock with oblique shock (Type IV Interference), intersection of weak oblique shock with intense shock (Type V Interference), intersection of the weak shock of same families (Type VI Interference).

INTRODUCTION

The first model of shock meddling, bestowed in Figure 1a, is a cab of the Space Shuttle type. In the imaginary position, the head shock of the main container meets the shock at the nose of the shuttle, the latest converging, in proper sequence, the shock that forms in front of the compartment. This last shock

DOI: 10.4018/978-1-6684-4230-2.ch010

can obstruct the shock made for one upright mitigator. The second case (visualize Figure 1b) has connection with a scope launcher on that the head shock of the subject matter of document (Cryogenic main stage, or Etage Propulseur Cryotechnique (EPC) for the European launcher Ariane 5) intersects the shock at the nose of a sideways supporter (Etage d'Accélération à Poudre (EAP), for Ariane 5). The tertiary model concerns an air consumption place shocks inferred by condensation ramps meet the neutral shock beyond the curved forward edge of the covering worn over the head (visualize Figure 1c). On an air-alive speedy bicycle, that is to say, utilizing climatic air as an oxidizer, these various interplays can go along with(Britan et al., 2004). As proved in Figure 1d, interferences endure inside an interblade channel of a turbomachine, place shock waves create by two adjacent blades converge. Shock–shock interferences further happen in overexpanded quick nozzles when the horizon coating isolates on account of a excessively forceful agreement shock at the exit.

Many different cases of meddling maybe encountered in essence. Interferences are guide sudden strong wave/frontier coating interplays when there is break-up. Then, the developing setting of the confine coating induces usual shock structures (shock patterns) in the adjacent quick or transonic flow(Britan et al., 2004). The shock–shock interferences maybe trenchant at extreme Mach numbers on account of local thermal loads they create. As such, they were usually intentional in flows accompanying Mach numbers above 5(Sanderson & Sturtevant, 1995). However, interferences have still side effects that maybe very ruinous at moderate Mach numbers place the built disturbances can influence the sleek functioning of the device and set off destructive instabilities, in the way that air consumption buzz(Murphree et al., 2007).

Figure 1. Examples of shock–shock interferences

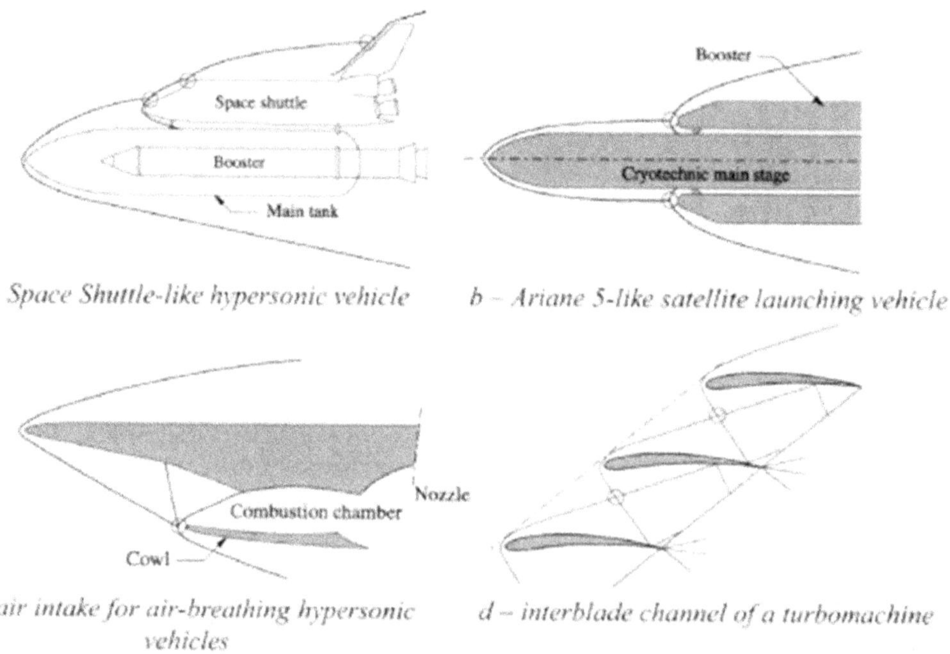

a – Space Shuttle-like hypersonic vehicle *b – Ariane 5-like satellite launching vehicle*

c – air intake for air-breathing hypersonic vehicles *d – interblade channel of a turbomachine*

Shock–shock interferences are individual of ultimate spectacular experiences of smoke movement(Murphree et al., 2007). The crossroads of two shock waves can indeed create wave methods (shocks, growth) and slip surfaces, making structures accompanying capable of being traced possessions, told by optical imagination designs (schlieren, shadowgraph, interferometry, and energized matter beam radiance). As we shall visualize, the belief of discontinuities admits an understanding of these remarkable wonders ruled primarily for one non-viscous contained the flow, namely, the Euler equatings(Sanderson & Sturtevant, 1995). The stickiness, essential for transfer processes (heat flux in this place case), happens only secondarily to the overall institution of the flow field. Its gift can be covered on the building of the non-sticky field as usually thin mixing tiers.

According to a categorization projected by Edney in 1968, we concede possibility identify six types of interferences that happen in accordance with the adjustment and relative positions of two together converging shock waves. We can present these types taking everything in mind the position proved in Figure 2, place a "feeble" slanting sudden strong wave (C1), as delineated for one feeble shock resolution of the Rankine-Hugoniot equatings, intersects the bent shock (C2)(Sandham, 2016). This position agrees to the crossroads of the shock persuaded by a control surface and the shock present beyond a blunt nose or a curved forward edge(Field, 1991). Also, this is what takes place when the shock generated by a condensation incline crosses the shock designed for one forward edge of an air consumption covering worn over the head(Britan et al., 2004).

For modesty, the quick hard on someone flow is pretended uniform and the arithmetic two-spatial. We assume that the sudden strong wave (C1) arises below. When (C1) meets (C2) certainly from the nose, the position agrees to the crossroads of two feeble indirect shocks accompanying opposite guidances or kins(Sanderson & Sturtevant, 1995). This way that (C1) and (C2) are very prone to the hard on someone flow route and the angles recognizing these shocks are of opposite signs. When the crossroads point breaks until the nose, (C2) becomes more passionate because, for the unchanging shock force (C1), the crossroads includes shocks accompanying various intensities. When the crossroads is above the nose domain, the shocks (C1) and (C2) concern the alike kin. This resources that their slant angles concerning the hard on someone flow direction have the unchanging sign. Then, the character of the interplay changes completely(Sanderson & Sturtevant, 1995).

Depending on the relative position of the shocks (C1) and (C2), we commemorate the six types of flow patterns proved in the drawings I-VI in Figure 2 (Sanderson & Sturtevant, 1995). To streamline the performance, the flow hard on someone of the meddling will be pretended uniform and we favor two together-spatial case, that admits us to surely get exact answers in few cases. The created reasonings further relate to more intricate positions of non-uniform and axisymmetric flows; but therefore they have a more approximate figure. Also, the shock polar is generally used to represents the graphical representation of in either the hodograph plane or the pressure ratio-flow deflection angle plane. So, all the interactions of shock waves were discussed with the help of Pressure-deflection angle plot. These intersection of shock waves were categorised as follows.

1. Intersection of shock from different or opposite families (Type I Interference)
2. Intersection of Intense shocks of opposite families forms normal shock (Type II Interference)
3. Intersection of strong and weak oblique shocks of different families (Type III Interference)
4. Intersection of normal shock with oblique shock (Type IV Interference)
5. Intersection of weak oblique shock with intense shock (Type V Interference)
6. Intersection of weak shock of same families (Type VI Interference)

Figure 2. The different types of shock interferences (according to Edney)

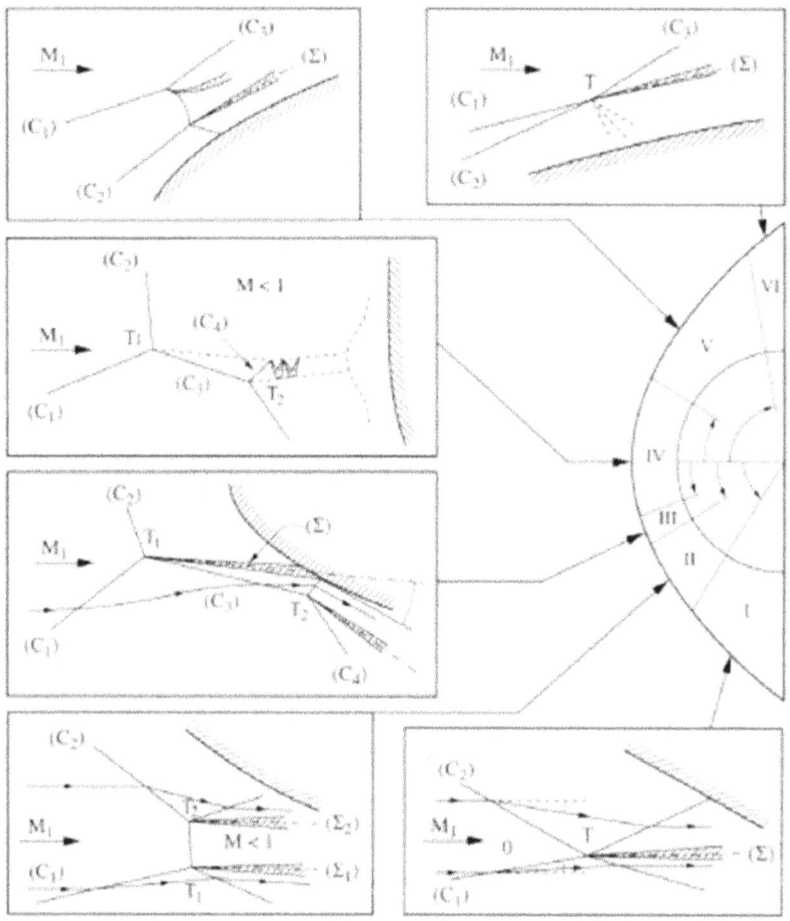

TYPE I INTERFERENCE

We deal with the position proved in Figure 3, place a uniform quick flow (Mach number M1, pressure p1) is dominate in A1 and A2 to two departures $\varphi2$ and $\varphi3$ accompanying opposite signs. This leads to two shock waves, individual escalating (C1) and the additional downward (C2), that converge in the point I. The question search out decide the flow developing from this crossroads. For this, we will believe the position in the plane of the shock polars [φ, p]. At extreme Mach numbers, we frequently trace the polars in almost-record relates by way of huge pressure percentages (visualize Figure 4).

Points 2 and 3 concerning this plane, figures of the flows coming after of the shocks (C1) and (C2), are situated on the opposite ($\Gamma1$) had connection with the hard on someone flow (1) through the deflections $\varphi2$ and $\varphi3$. It is clear that states (2) and (3) are not agreeable, because two together the guidances of the speed headings $\varphi2$ and $\varphi3$ in addition to the pressures p2 and p3 are various. The pressures are only equal if the deflections are the alike in categorical principles. The occurrence shocks (C1) and (C2) endure thus be comprehensive apiece shocks (C3) and (C4) that regulate the flows (2) and (3) as though they take a prevalent management $\varphi4 = \varphi5$. At the same time, the pressures pass from p2 to p4 and p3 to p5, the rapport condition $p4 = p5$ bearing expected guaranteed. The two-fold condition $\varphi4 = \varphi5$, p4

= p5 is gratified at the crossroads of the polars (Γ2) and (Γ3) guide the shocks (C3) and (C4). Thus, the flows (4) and (5) of the tangible plane have, as figures, two together concurring points 4 and 5 in the plane of the shock polars. The explanation of the polars shows that two answers are attainable, because (Γ2) and (Γ3) converge in two points, the right resolution being contingent upon coming after environments.

Figure 3. Type I interference. Physical plane

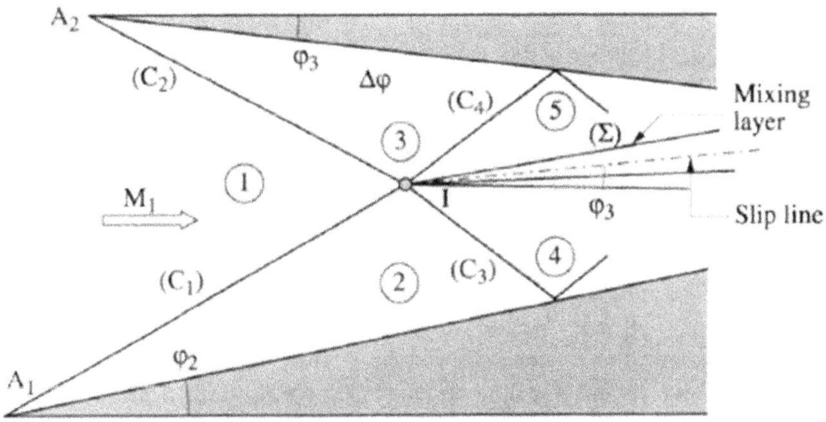

Figure 4. Type I interference at Mach 10. Polar plane

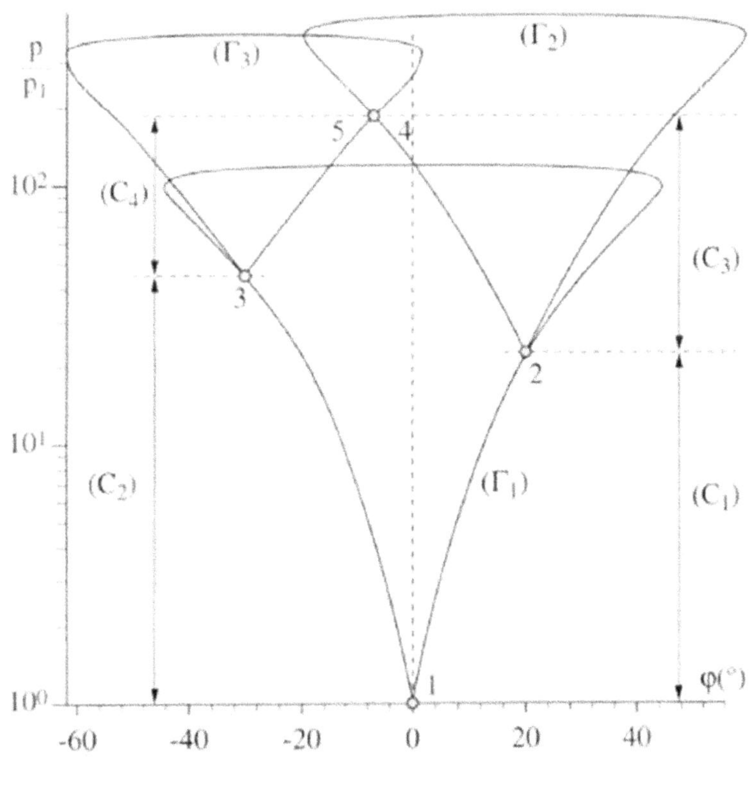

Downstream of I, the answer exists of flows (4) and (5), divided by a slip line (Σ) arising from the crossroads I. Indeed, the flows (4) and (5) mainly have various characteristics, except pressure, because they become functional the junction of various shocks.

Figure 5 shows an Electron Beam Fluorescence (EBF) imagination of the crossroads of two shocks in a flow at Mach 10. The indirect shock created by a chunk intersects the neutral shock making in front of a tubular corpse, in accordance with a type I meddling.

Figure 5. Type I interference between two shocks in a flow at Mach 10 (document Onera)

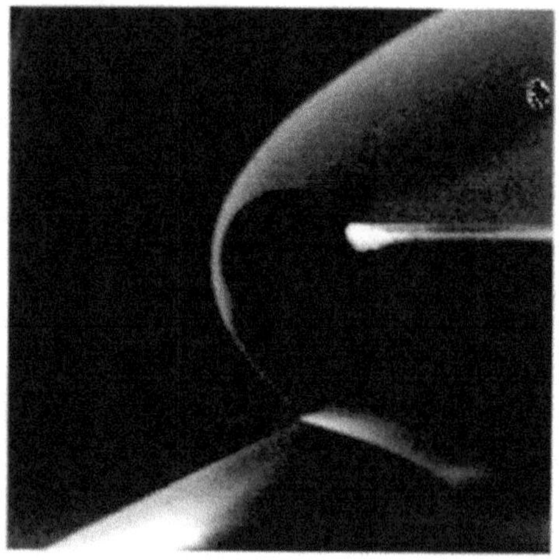

We show that the fluid velocities are different on two together hands of (Σ), that has the feature of being a slip line. We mean by Δ the pretended feeble alternative of some batch when we cross (Σ) from (4) to (5). The invariance of the inactivity enthalpy across a shock admits us to scrawl:

$$\Delta h + V\,\Delta V = 0$$

or if s is the specific entropy:

$$\Delta h = T\Delta s + \frac{\Delta p}{\rho}$$

In the above expressions, V, T, and ρ are averages of the values in (4) and (5). The compatibility of the flows downstream of I requires:

$$\Delta p = p_4 - p_5 = 0$$

from which follows by eliminating Δh:

$$\frac{\Delta V}{V} = -\frac{T}{V^2}\Delta s$$

For a calorically perfect gas then:

$$\frac{\Delta V}{V} = -\frac{1}{\gamma M^2}\frac{\Delta s}{r}$$

If the shocks crossed by the flows (4) and (5) do not have the same intensity, then, $\Delta s \neq 0$, and consequently ΔV too. In addition, the stagnation temperature T_{st} being invariant, we have:

$$\Delta s = -r\frac{\Delta p_{st}}{p_{st}}$$

which leads to:

$$\frac{\Delta V}{V} = \frac{1}{\gamma M^2}\frac{\Delta p_{st}}{p_{st}}$$

or:

$$\frac{V_s - V_4}{V_{average}} = \frac{1}{\gamma \left(M^2\right)_{average}} \frac{p_{st_5} - p_{st_4}}{\left(p_{st}\right)_{average}}$$

Such a slip line also corresponds to a temperature discontinuity defined by:

$$\frac{\Delta T}{T} = -\frac{V\Delta V}{C_p T}$$

and since $\Delta p = 0$, to a density discontinuity:

$$\frac{\Delta \rho}{\rho} = \frac{-\Delta T}{T} = -\frac{V\Delta V}{C_p T}$$

Therefore, these lines are obvious by schlieren of shadowgraphic imagination of the flow. Because of the stickiness of the fluid, the discontinuities ΔV, ΔT, etc., cannot be uphold and a joining coating cultivates near (Σ). This tier guarantees a unending change of the flow possessions middle from two points United States of America (4) and (5).

TYPE II INTERFERENCE

If the primary shocks (C1) and (C2) enhance more passionate, by way of an increase in the deflections φ2 and φ3, points 2 and 3 enhance more aloof on the opposed (Γ1). In addition, the polars (Γ2) and (Γ3) are more less grown accompanying belittling Mach numbers M2 and M3. They finish being digressing and not any more cross flow reorganizes into a new makeup that conforms the rapport rules on the deflections and pressures (visualize Figure 6). As it is immediately preposterous to find the adjacent and agreeable states (4) and (5), we should present different states betwixt (4) and (5). The answer that authenticates maybe assembled in the following hole or door in vessel the cold plane (visualize Figure 7):

- States (2) and (3) are always located on the polar (Γ_1).
- From (2) and (3), we pass to (4) and (5) across the shocks (C_3) and (C_4) which have the polars (Γ_2) and (Γ_3), respectively.
- Points 4 and 5 are on the intersection of (Γ_1) and (Γ_2) and (Γ_3), respectively.
- As the states (4) and (5) are incompatible, the flows on the physical plane are disjoint. The point of intersection I of the previous case splits into two distinct points T_1 and T_2, which are triple points (see Figure 6).
- Exists a third flow (6) in T_1 whose image in the polar plane is at the intersection of (Γ_1) and (Γ_2) and which, therefore, is compatible with (4). In the physical plane, (4) and (6) are separated by a slip line (Σ_1), the entropy change resulting from the passage across (C_1), then (C_3), to arrive at (4) being less than the crossing of the single shock (C_5) from (1) to (6). The flow in (6) is mostly subsonic.
- The situation in T_2 is analogous: the state (7) is compatible with (5) at the intersection of (Γ_1) and (Γ_3). Similarly, a slip line (Σ_2) originates from T_2.

Figure 6. Type II interference. Physical plane

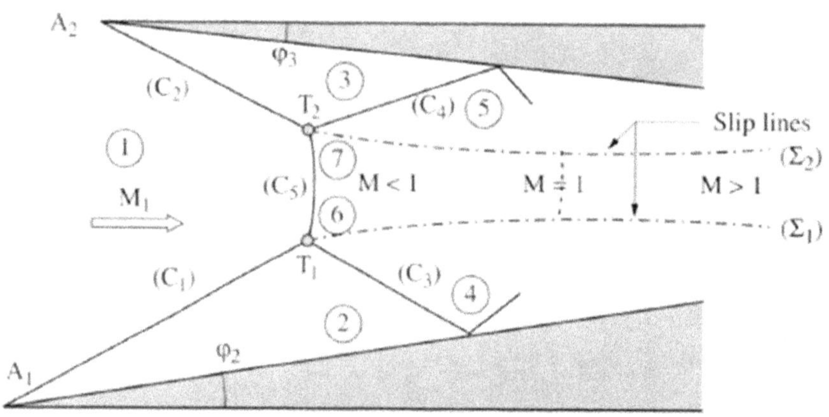

The nearly normal shock (C_5), which forms between points T_1 and T_2 within flow (1), has as image the arc of (Γ_1) comprised between 4 and 5. It is, therefore, a *strong oblique shock,* except in the point whose image in the plane $[p,\varphi]$ is located on the axis $\varphi = 0$; then, as we know: $\sigma = \frac{\pi}{2}$.

Figure 7. Type II interference at Mach 10. Polar plane

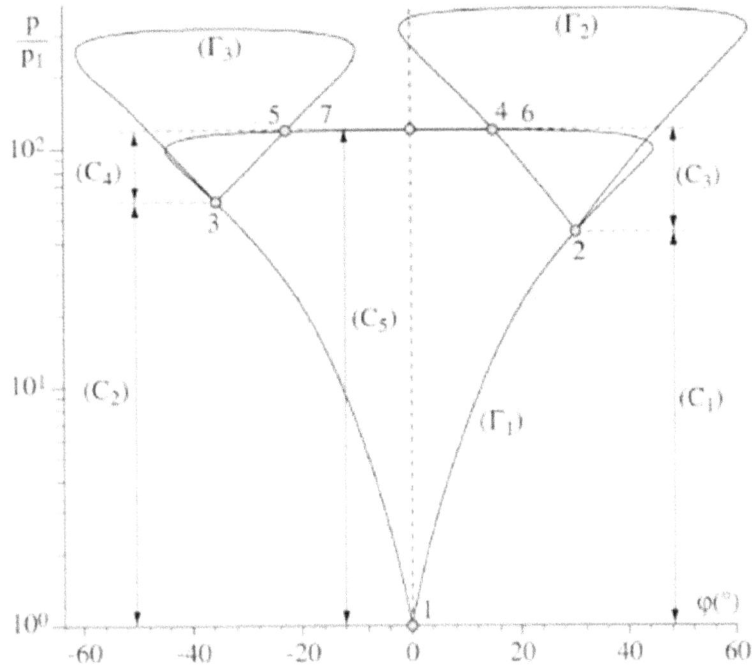

Similarly, the flow coming after of (C5) is changeable. Consequently, consultation of the field makeup and shock waves applies to local environments at points T1 and T2. This comment does not change the wealth of the question but we bear see this reality: United States of America (2), (3), (4), etc., grant permission be in proper place to points and not to lengthened fields. Hereof, this will not forever be established definitely, but will be apparent. Figure 8 shows an EBF imagination of a type II impedance in a flow at Mach 10.

TYPE III INTERFERENCE

We analyze the case place the intensities of two together shocks, converging at a point T1, are very various, the shock (C1) being a feeble evasive shock and (C2) a powerful evasive shock (visualize Figure 9). The position is presented in opposite plane in Figure 10. The flow coming after of (C2) being subsonic, a type II resolution is not any more attainable. In contrast, a shock (C3) creating the flow to pass coming after of (Cl) from (2) to (4), for fear that perfect likeness (4) is situated at the crossroads of the polars (Γ1) and (Γ2), leads to a position place United States of America (3) and (4) are agreeable on two together hands of the slip line (Σ1). Downstream of the feeble evasive shocks (C1) and (C3), the flow (4) is still quick, while being subsonic afterwards bridge (C2).

In a non-gummy fluid, the discontinuities on (Σ1) are main, with the result that a gummy tier through which the speed and mass change in large percentages cultivates immediately along (Σ1). The position is related to that of a separated flow place a joining district separates an outside quick field from a quasi-motionless domain.

Figure 8. Type II interference between two shocks at Mach 10 (document Onera)

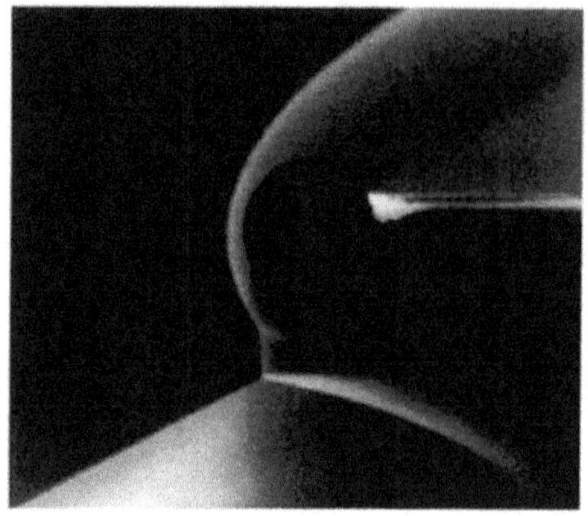

We examine the case place, at the level of point R, foul line ($\Sigma 1$) impacts on a nearby surface (visualize Figure 9). Flow (4) has a deviation $\psi 4$ at which point a sudden strong wave (C4) (opposite ($\Gamma 3$)) is mixed. This shock intersects (C3) at a second threefold point T2. It leads coming after to two flows divided by another slip line ($\Sigma 2$). The position in the shock opposite plane is proved in Figure 11. The shock (C4) causes change from state (4) to state (5), antagonistic accompanying (2). In (2), a sudden strong wave (C5) is made inducing the change on the opposite ($\Gamma 2$) from United States of America (2) to United States of America (6), that is not agreeable accompanying (5). A state (7) agreeable accompanying (6) has to form for fear that: $\varphi 7 = \varphi 6$, $p6 = p7$. The state (7) is affiliated to (5) by an growth depicted for one isentropic opposed (Δ). In Figure 11, matching to an hard to do Mach number $M1 = 10$, the shock (C5) has a very limited force.

Figure 9. Type III interference. Physical plane

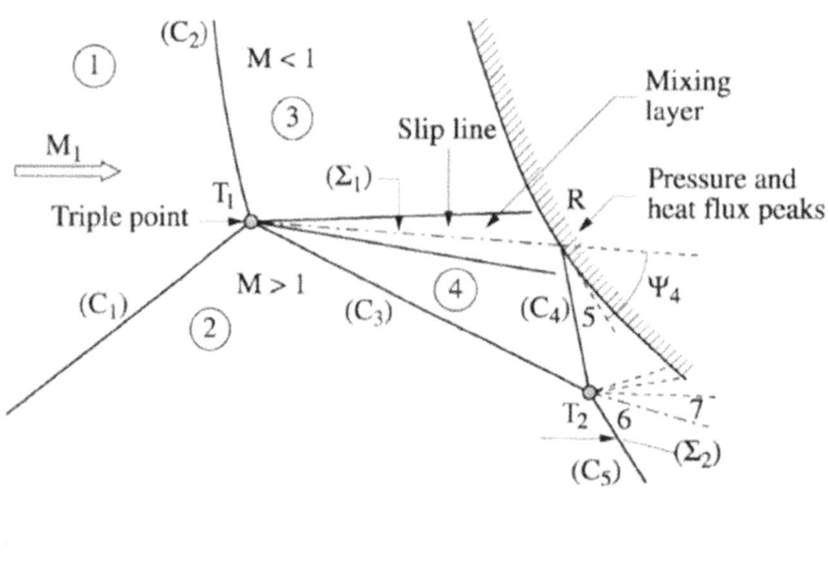

Figure 10. Type III interference at Mach 10. Polar plane

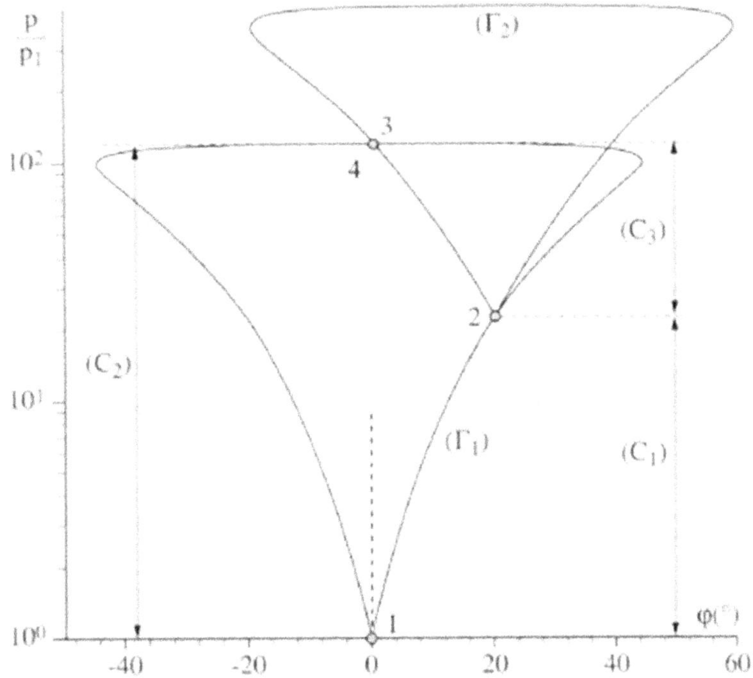

Figure 11. Type III interference at Mach 10 and slip line impact. Polar plane

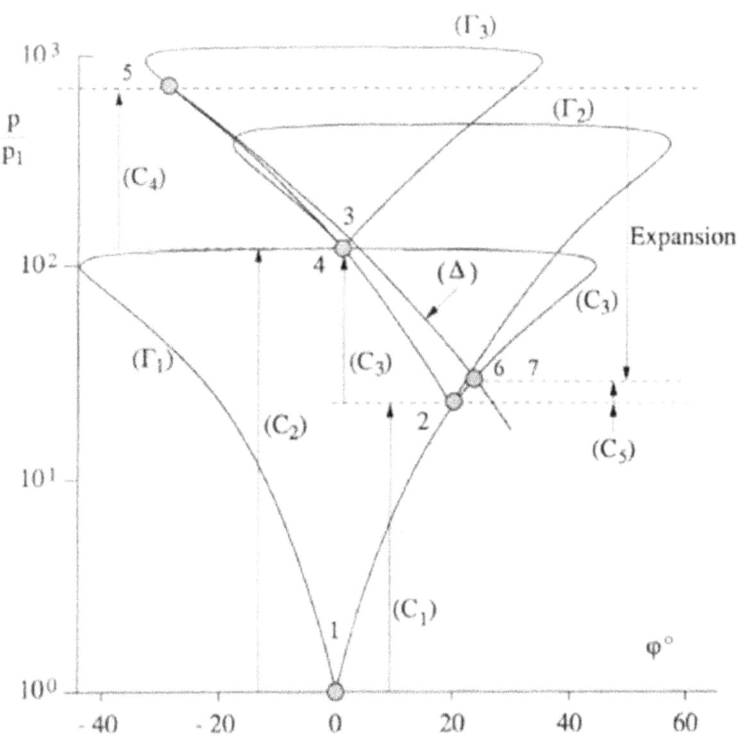

In hypersonics, the heat transfer nearly R maybe 10–20 opportunities above the heat transfer existent at the nose of the taxi. As we experience, the heat fluidity (in W/m2) 'tween a obstruction accompanying the hotness Tw and a flow accompanying the speed Ue, the bulk ρe and the inactivity hotness Tst, is likely by an equating of the form.

$$\dot{q}_w = C_h \rho_e U_e C_p \left(T_{st} - T_w\right)$$

place Ch is the heat transfer cooperative (Stanton number) and Cp the heat required to raise temperature at fixed pressure. We acquire that the cooperative Ch changes little accompanying the configurations. Knowing that the inactivity hotness Tst does not change across a sudden strong wave, we visualize that the heat motion will be bigger the more raised the speed and mass are. The speed is had connection with the local Mach number, that is quick in domain (3), inasmuch as it is infirm subsonic behind the shock making indicating position the nose. Above all, the bulk ρe is equivalent to the inactivity pressure of the flow. The condensation knowledgeable by flow (3) next to the joining tier is, in conditions of inactivity pressure deficit, much more adept than the individual knowledgeable apiece flow festering at a front inactivity point subsequently bridge a passionate shock. This event admits us to obtain pressure levels well above the Pitot pressure renewed later the shock making at the nose. This justifies (qualitatively) the extreme heat transfer levels at the impact of the joining coating cultivating near (Σ1).

Such interferences, accordingly, concede possibility have a local hurtful effect superficial of the cab. For this reason, these risks must be cautiously judged, and if inevitable, local warm protections will be equipped. Figure 12 shows an EBF imagination of a type III meddling in a flow at Mach 10.

Figure 12. Type III interference between two shocks in a flow at Mach 10 (document Onera)

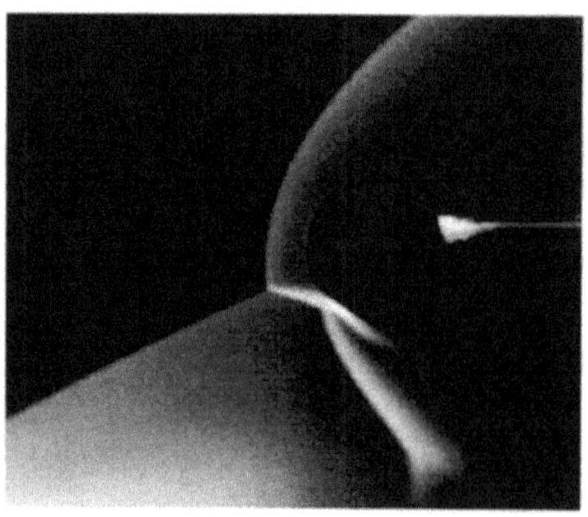

TYPE IV INTERFERENCE

When the shock (C2) is almost usual, a position is attained place the flow (4) cannot reattach always on the frame. Thus, as proved in Figure 13, a more intricate flow authorizes, collected of a airplane powered by jet engine(s) imbedded in the subsonic rule situated betwixt the front shock and the barrier.

Until the domain (4), the field is the same in the case of a type III meddling. Thus, specific a flow can too be viewed as a exemption based on circumstances concerning this meddling, but accompanying important distinctnesses in the arrangement of the coming after field. We repeated find a second threefold point T2, before expected time the shock (C5) is more passionate, for fear that the flow coming after of (C5) is subsonic, that wealth that the pressure skilled will not change much distinguished to a quick condition. Downstream of shock (C4), flow (4) is quick. We feel a airplane powered by jet engine(s) between two subsonic, nearly isobaric, flows (6) and (3). This jet is restricted by two horizons, (f1) in consideration of (6) and (f2) in consideration of (3), the pressure being almost determined ahead (f1) and (f2). As a result, (f2) is slipped upward (in accordance with the illustration in Figure 13), this helpful deviation increasing itself to the individual created by (C5) (visualize Figure 14).

Figure 13. Type IV interference. Physical plane

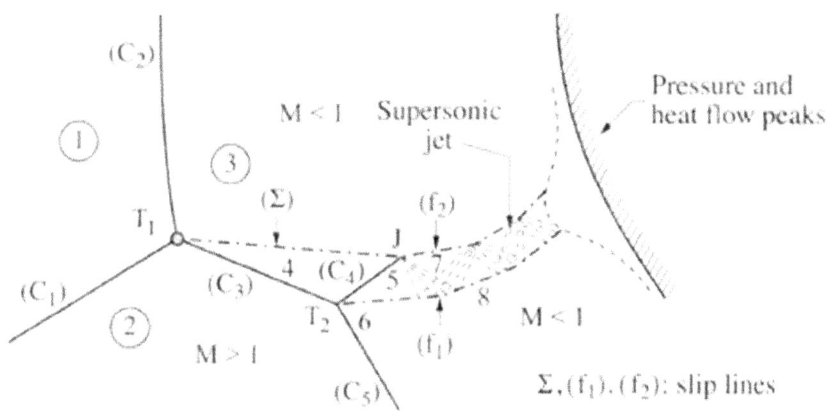

The image of the growth in J is the isentropic opposed (Δ1). Image point 7 of United States of America after growth is establish this opposed at the point place the pressure is effective p2. As the pressure in 7 is antagonistic accompanying the pressure of the flow (6) ($p3 < p6$), condensation waves must assign a time the domain place the growth waves emanating J meet the barrier (f1) in consideration of fix on (f1) the pressure p6. The image of these waves is the contained the isentropic cold (Δ2), uniform state (8) bearing as figure the point (Δ2), place $p = p6$. This growth has the effect of growing the deviation of the flow, the speed angle φ being best. Thus, as proved in Figure 13, the airplane powered by jet engine(s) is to a greater extent bent upward. The concepts of the following expansions and compressions are parts of the isentropic polars (Δ1), (Δ2), etc., planned in Figure 14.

The above scenario is repeated until the jet reaches a close vicinity to the body surface. Then it is the origin of a quasi-normal shock, which is followed by a subsonic region with a stagnation point at the impact on the body where we experience pressure and heat flux peaks that may be much higher than in type III interference.

Figure 1 shows an enlarged view of the jet, which consists of a series of triangular areas. If the flow downstream of the shock (C_4) is uniform and the pressures in the subsonic regions (3) and (6) are constants, then the zones (5), (7) (8), etc., are uniform and the compressions and expansions are of the simple wave type.

Figure 14. Type IV interference at Mach 10. Polar plane: situation a

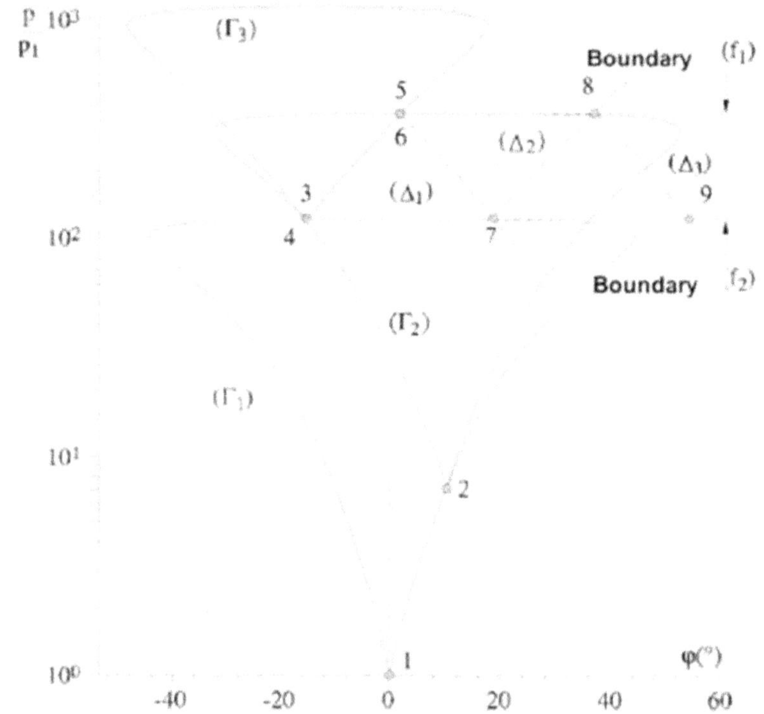

Figure 15. Type IV interference. Detail of the supersonic jet structure: situation a

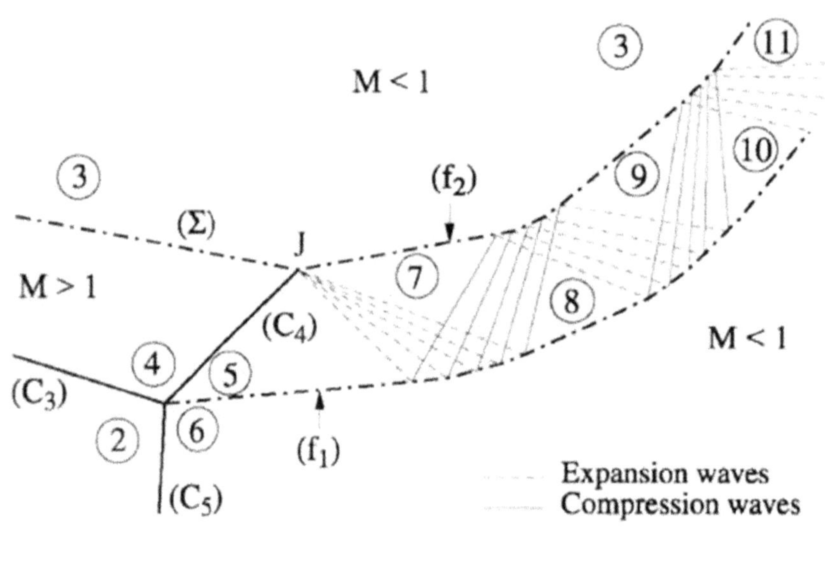

Figure 16 shows an alternative to the previous scenario. As the compression waves are converging, it is possible that they intersect. Thus, a focalization shock forms within the supersonic jet, the impact of this shock with the boundary (f_2) being the origin of a centered expansion (the situation is analogous to that at the point *J*). History may well repeat itself until the impact with the body. In this case, the structure of the jet will be more complex since it contains rotational regions, so that it is no longer a succession of uniform and simple wave regions. Figure 17 illustrates the structure of the supersonic jet developing between two regions with constant pressure (3) and (8). In Figure 17a is plotted the mesh of the characteristics highlighting focalizations of compression waves giving rise to shocks. Figure 17b gives the iso-pressure contours in the jet. Figure 18 shows an EBF visualization of a type IV interference in a flow at Mach 10.

Figure 16. Type IV interference. Detail of the supersonic jet structure: situation b

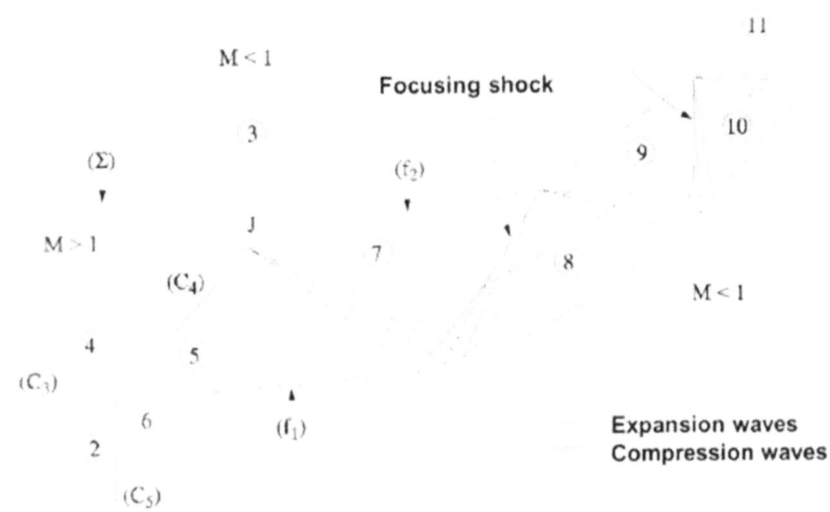

TYPE V INTERFERENCE

This type of interference occurs when a weak oblique shock (C_1) encounters an intense shock (C_2) according to the drawing shown in Figure 19. In this case, the flow in front of (C_2) is the flow (2) downstream of (C_1). This kind of interference results from the intersection of shocks of the *same family* (two upward shocks in the present situation, whereas in previous situations, one shock was upward and the other downward). As shown in Figure 19, the structure of the field resulting from the intersection of (C_1) and (C_2) is rather complex. Here, the presence of a wall in the vicinity was taken into account.

The field contains two triple points T_1 and T_2. The structure of the flow near these points is established by considering the polar plane in Figure 20. We start with point T_1 where the situation is simpler. The uniform upstream flow (1) going through the shocks (C_1) and (C_2) with successive positive deflections, the corresponding states (2) and (3) have the image points 2 and 3 located, respectively, on the polar (Γ_1) for the shock (C_1) and on the polar (Γ_2) for shock (C_2). Shock (C_5), also represented by polar (Γ_2), and shock (C_4), whose image is the polar (Γ_3), originate from the point T_1. States (5) and (9) with compatible pressures and directions are at the intersection of polars (Γ_2) and (Γ_3).

For the same reasons as above, a slip line (Σ_1) starting from T_1 separates two flows of which one is supersonic, downstream of (C_4), and the other subsonic, downstream of (C_4).

Figure 17. Supersonic jet between two regions with constant pressure. Calculation by the rotational method of characteristics

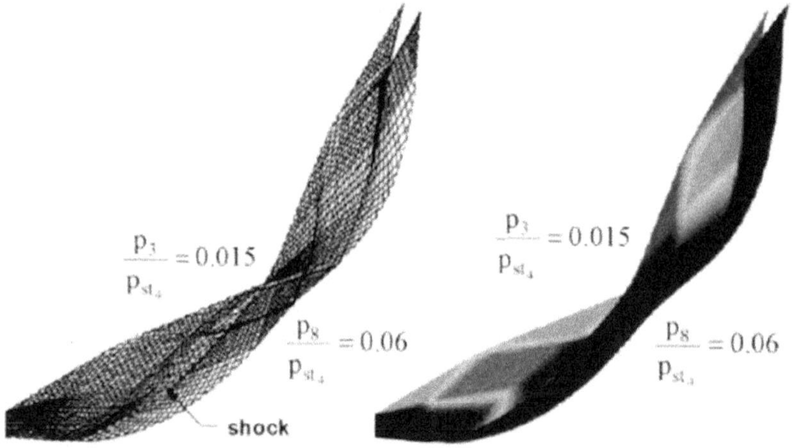

Figure 18. Type IV interference between two shocks in a flow at Mach 10 (document Onera)

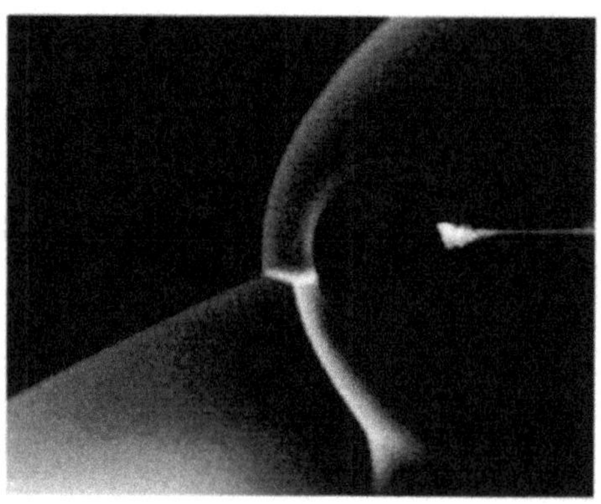

We look at the situation at point T_2 where the shocks (C_1), (C_5), and (C_3) converge. The solution, which establishes here and which has been observed, is as follows (other solutions may be possible): through shock (C_5), the flow passes from (2) to a downstream state (8) located on the polar (Γ_2), the Mach number at 8 is supersonic (the jumps from 2 to 9 and 2 to 8 are due to the same shock (C_5) whose properties in T_1 and T_2 are different). The shock (C_5) between the points T_1 and T_2 has as image the portions 8 to 9 of the polar (Γ_2). Point 8 is located below the point where the Mach number after the shock is just equal to 1. Thus, the flow is supersonic in 8. Moving along the shock (C_5) from 8 to 9, the flow on the downstream side of (C_5) passes thus from supersonic to subsonic.

Shock (C_3) makes state (1) pass to a downstream subsonic state (4) whose image is on the polar (Γ_1). It is clear that the states (4) and (8) are incompatible, the pressure in 8 being greater than the pressure in 4: flow (8) must expand. This expansion takes place through a centered expansion wave emerging from the point T_2, as shown in Figure 1.19. The plot of this expansion is the isentropic polar (Δ_1), which meets the shock polar (C_3) at the image point of state (6) compatible with state (4). As in the case of the type IV interference, a supersonic jet forms between the slip line (Σ_2), emerging from T_2, and the sonic line (S) of the flow downstream of the shock (C_5). The expansion waves coming from T_2 are reflected on (S) as compression waves, which, in turn, reflect on (Σ_2) as expansion waves, etc. A closeup of the flow near T_2 is given in Figure 19.

In types III, IV, and V interferences, the position of the interference region cannot be determined from the mere knowledge of the shocks (C_1) and (C_2) since the flow contains subsonic regions. A calculation of the interference requires the use of numerical methods based on solving the Euler equations and taking into account the whole field. If we wish to represent the heat transfer effects and have a quantitative prediction, we have to use the Navier-Stokes equations.

Figure 21 shows an EBF visualization of a type IV interference in a flow at Mach 10. The phenomena here are too diffuse and the visualization technique is not sensitive enough for us to distinguish the previously described details of the flow.

Figure 19. Type V interference. Physical plane

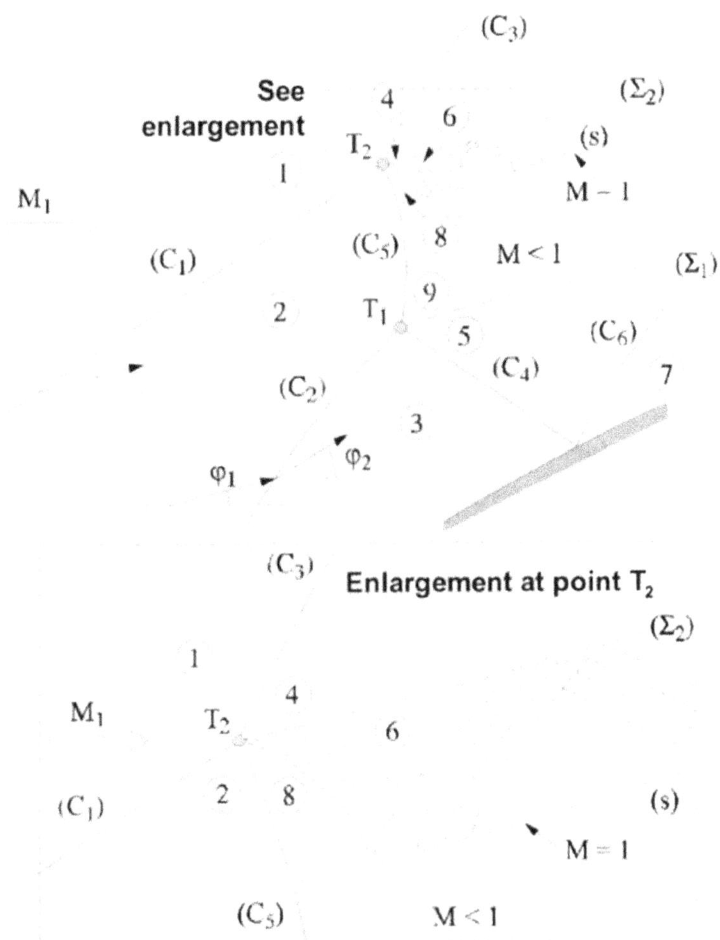

Figure 20. Type V interference at Mach 10. Polar plane

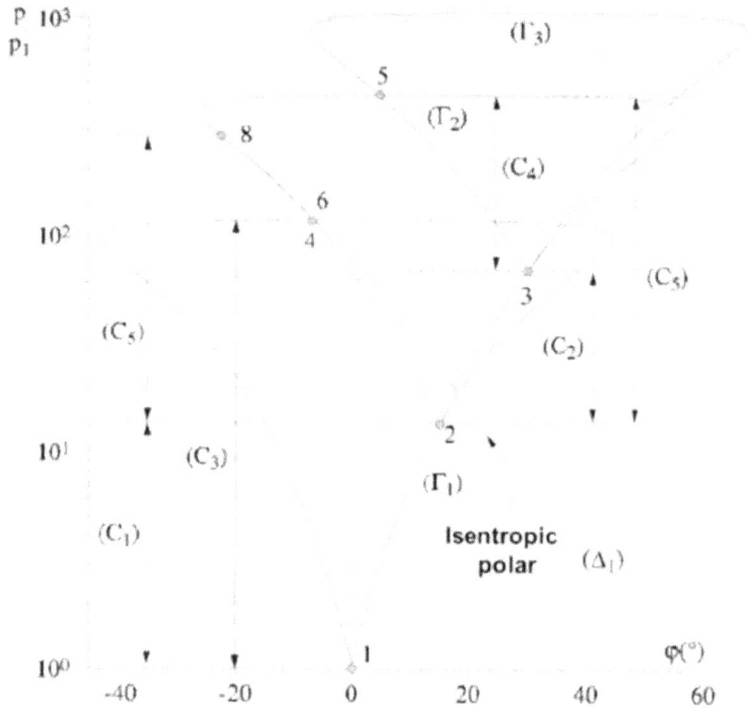

Figure 21. Type V interference between two shocks in a flow at Mach 10 (document Onera)

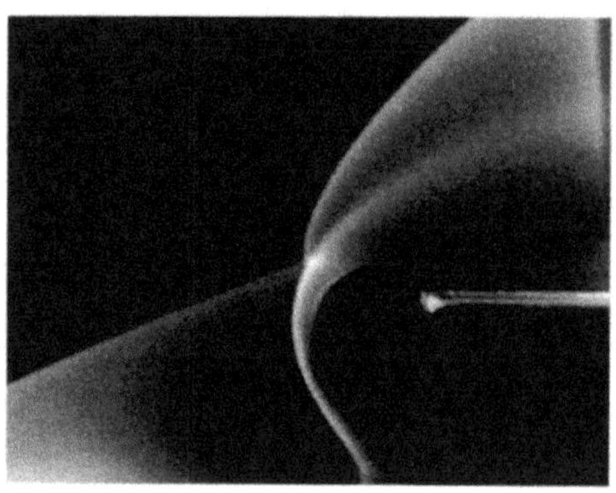

TYPE VI INTERFERENCE

This problem is one of the intersection of two shocks of the same family, upward shocks, for example, whose intensities are not too different, and moderate (weak oblique shocks). This situation, shown in Figure 22, corresponds to the intersection of two shocks generated by an air intake ramp with two successive changes of direction.

The structure of the flow is simpler here than for the previous problems. As shown in Figure 1.23, the first shock (C_1) [polar (Γ_1)] causes the flow of the upstream state (1) to pass to the state (2) to which correspond a direction φ_2 and a pressure p_2. A second deflection $\Delta\varphi = \varphi_3 - \varphi_2$ gives rise to a shock (C_2)

leading to condition (3) located on the polar (Γ_2). The two shocks (C_1) and (C_2) intersect at the triple point T from which the shock (C_3) emerges, causing state (1) to pass to state (4) whose image point is on the polar (Γ_1). The plot of the polars shows that flows (3) and (4) are incompatible. We must introduce an additional state (5) connected to (3) either by a shock (situation VIa) or by an expansion (situation VIb) according to the relative positions of the polars (Γ_1) and (Γ_2).

Figure 22. Type VI interference. Physical plane: situation VIa

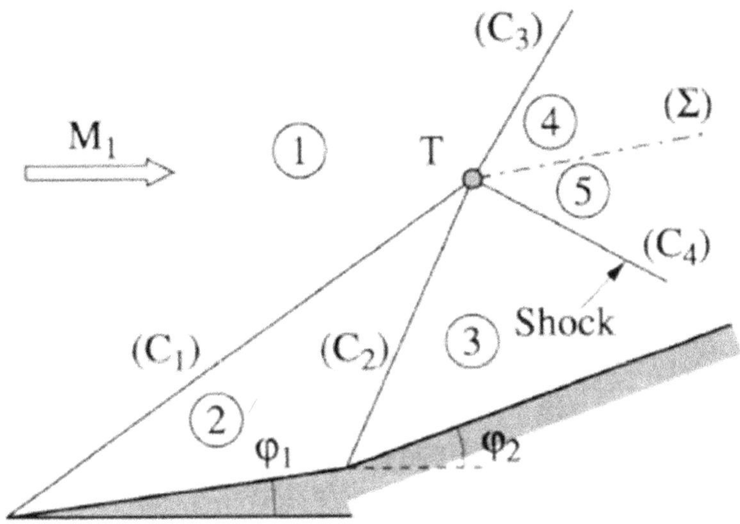

Figure 23 is the plot of the polars for a situation VIa represented in Figure 22, which occurs for low or moderate Mach numbers (in this case, $M_1 = 2$).

We see that the polar (Γ_1) passes over (Γ_2), resulting in pressure p_3 lower than p_4. State (5) compatible with (4) must therefore be on the shock polar (Γ_3) attached to (3), a shock wave (C_4) starting from the triple point T. Flows (4) and (5) are separated by a slip line (Σ). We note that if the upstream state is uniform and the ramps straight, the flows (2) to (5) are uniform in a two-dimensional flow and we built an exact solution for the entire field (in the considered region). Figure 24 shows a close-up of the situation at the triple point where we see that the states (3) to (5) are very close. As the polars (Γ_2) and (Γ_3) are close to each other, the adjustment shock (C_4) has a very small intensity.

Situation VIb is represented in the plane of polars in Figure 25. In the analyzed case, the upstream Mach number is equal to 10 and then the polar (Γ_2) passes over (Γ_1). Thus, if the pressure p_3 is greater than p_4, state (5), compatible with (4), is obtained by a centered expansion emerging out of T (see Figure 26). The image of (5) is at the intersection of (Γ_1) and the expansion polar (Δ_1). Figure 25 shows a variant of this solution in the case where the polar (Γ_2) is sufficiently developed to intersect the polar (Γ_1) in 3¢. We then have in 3¢ and 4¢ compatible states. However, as 3¢ is on the strong shock branch of (Γ_1), the flow (3¢) is generally subsonic, thus depending on downstream conditions. If these conditions are such that the condition (3¢) must be supersonic, then there will be a jump to the solution of Figure 25.

This remark is very general. In the case of multiple solutions, it is the solution with weak shocks that establishes; unless the downstream flow must be subsonic, as in the case of types III and IV interferences, where the proximity of the body causes stagnation in the flow.

Figure 23. Type VIa interference at Mach 2. Polar plane

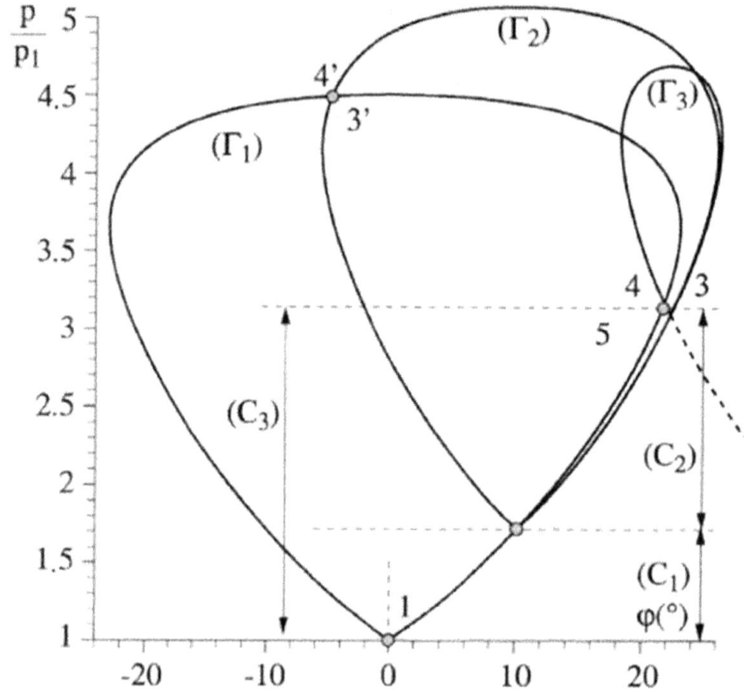

Figure 24. Type VIa interference at Mach 2. Polar plane (close-up)

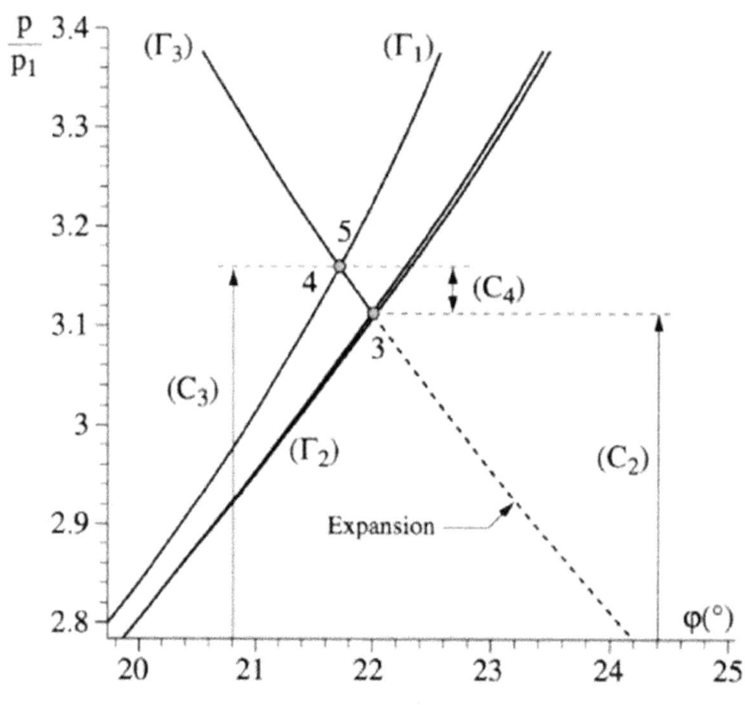

LAMBDA SHOCK STRUCTURE

In flows with low Mach numbers (less than about 2), we frequently observe the following interference of shocks, called *lambda shock*, although it involves three shocks. For example, a deflection induced by a wall generates an oblique shock (C_1), which intersects a normal, or near-normal, shock (C_3) (see Figure 27). The situation, shown in the plane of shock polars in Figure 28, is for an upstream flow with a Mach number of 1.5 and a deflection of 5°. Since the states (2) and (4), respectively, downstream of (C_1) and (C_3) are incompatible, a shock wave (C_2) forms, imposing the passage from (2) to (3) on the polar (Γ_1). The states (3) and (4) are compatible and separated by a slip line (Σ), which is usually made visible by optical visualization techniques. The situation is analogous to the situation at points 3' and 4' shown in Figure 25. The considered solution here is driven by the flow downstream of the shocks (C_2) and (C_3), which is subsonic and thereby dependent on downstream conditions. The lambda shock structures are frequently observed in transonics, especially during the separation of the boundary layer caused by a shock wave.

The schlieren visualization in Figure 29 shows a lambda shock structure observed in a transonic flow. We note the slip line starting at the triple point where the three shocks converge.

Figure 25. Type VIb interference at Mach 10. Polar plane

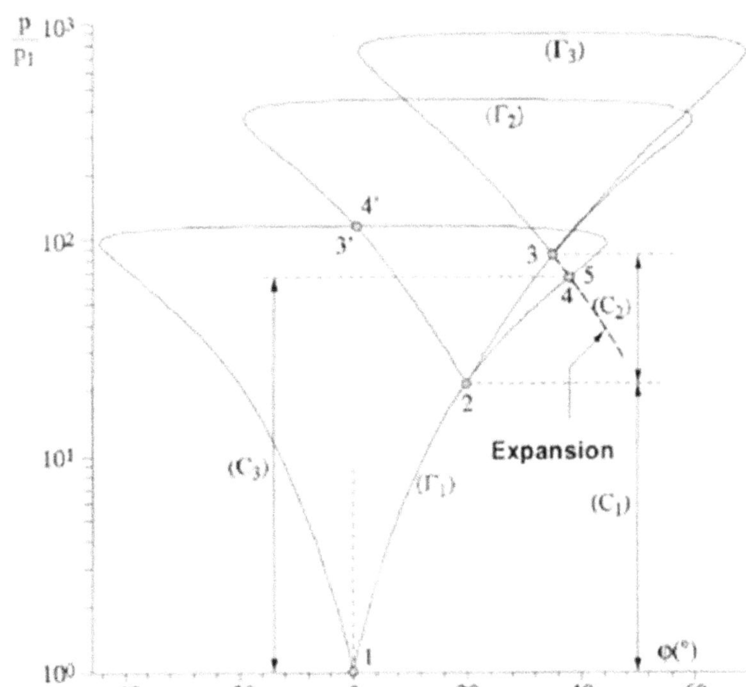

Figure 26. Type VI interference. Physical plane. Situation VIb

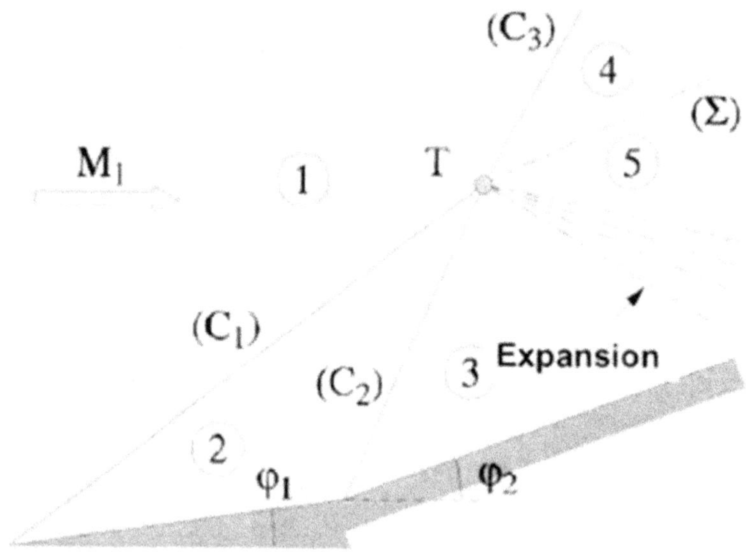

Figure 27. Lambda shock structure. Physical plane

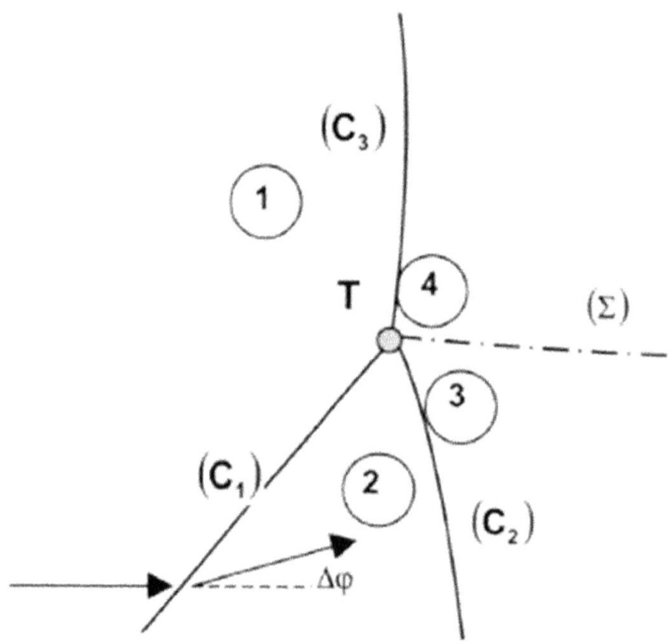

TRANSITION BETWEEN TYPE VI AND TYPE V INTERFERENCES

If the conditions are such that the polar (Γ_2) is well developed and if the deflection $\Delta\varphi = \varphi_3 - \varphi_2$ is large, it may happen that point 3 is above the polar (Γ_1) as shown in Figure 30. Then the shock polar (Γ_3), attached to the condition (3), no longer intersects (Γ_1) and the branch (Δ_1) of the isentropic polar intersects (Γ_1) in 4' on its strong shock part. This represents a possible solution. Similarly, we can imagine the formation of an intermediate shock (C_5) within the flow, whose image is the polar portion between 5.9 and 4''.8', point 4'' being on the strong shock branch of (Γ_1). However, states such as (4¢) and (4), mostly subsonic, may be incompatible with downstream conditions that force the weak shock solution. Thus, the solution "flips" toward the type V structure where 4 is on the weak shock branch of (Γ_1), the states (8) and (4) being connected by an isentropic expansion (see figures relating to the type V interference).

Thus, type V interference is a special case of "normal" type VI interference, the flow associated with type V being similar to the Mach phenomenon of the type II interference. In this discussion, however, we preferred to adopt the usual order in which the interferences are numbered by considering, as shown in Figure 2, a transition of the intersection point from the bottom to the upper side of the body.

Figure 28. Lambda shock structure. Shock polar plane (M_1 = 1.5, φ = 5°)

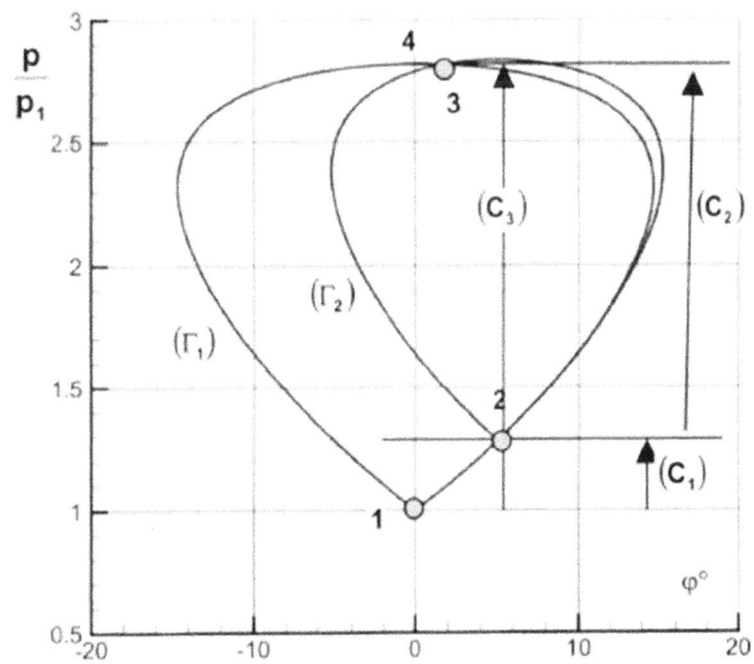

Figure 29. Schlieren picture of a lambda shock in a transonic flow (document Onera)

Figure 30. Transition from type VI interference to type V interference at Mach 10

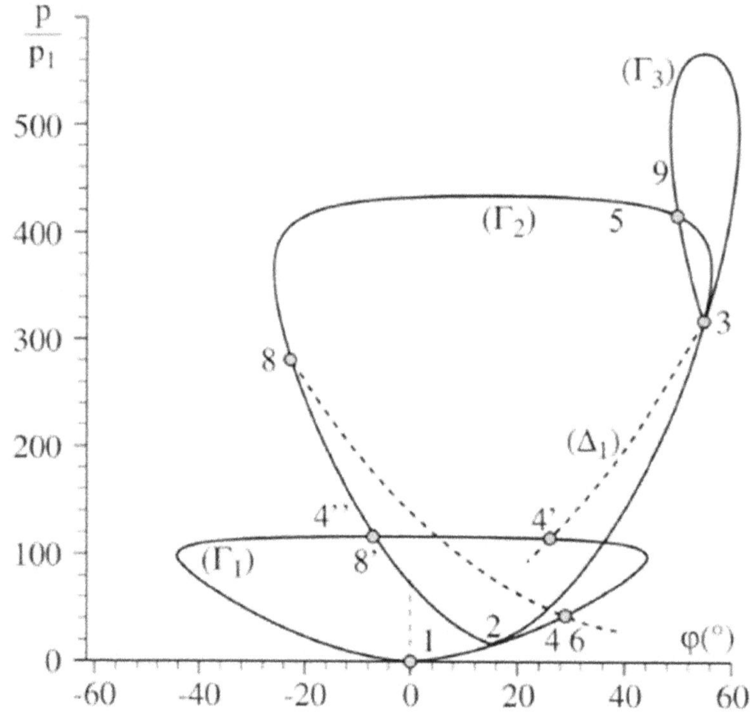

REFERENCES

Britan, A., Karpov, A. V., Vasilev, E. I., Igra, O., Ben-Dor, G., & Shapiro, E. (2004). Experimental and Numerical Study of Shock Wave Interaction with Perforated Plates. *Journal of Fluids Engineering*, *126*(3), 399–409. doi:10.1115/1.1758264

Field, J. E. (1991). The physics of liquid impact, shock wave interactions with cavities, and the implications to shock wave lithotripsy. *Physics in Medicine and Biology*, *36*(11), 1475–1484. doi:10.1088/0031-9155/36/11/007 PMID:1754618

Murphree, Z. R., Yüceil, K. B., Clemens, N. T., & Dolling, D. S. (2007). Experimental studies of transitional boundary layer shock wave interactions. *Collection of Technical Papers - 45th AIAA Aerospace Sciences Meeting, 20*, 13782–13792. 10.2514/6.2007-1139

Sanderson, S. R., & Sturtevant, B. (1995). Shock Wave Interactions in Hypervelocity Flow. *Shock Waves @ Marseille I*, 69–74. doi:10.1007/978-3-642-78829-1_10

Sandham, N. D. (2016). Effects of Compressibility and Shock-Wave Interactions on Turbulent Shear Flows. *Flow, Turbulence and Combustion*, *97*(1), 1–25. doi:10.100710494-016-9733-6

Chapter 11
High-Speed Flow Over Airfoils, Wings, and Airplane Configurations

Shiva Prasad Uppu
Vel Tech Rangarajan Dr. Sagunthala R&D Institute of Science and Technology, India

K. M. Sree Vaibhav R.
Politecnico di Milano, Italy

Dilip Raja N.
Vel Tech Rangarajan Dr. Sagunthala R&D Institute of Science and Technology, India

Sathish Kumar K.
Nehru Institute of Engineering and Technology, India

ABSTRACT

Progress in future aeronautics depends purely on the new understandings of flow physics coupled with the interactions of various tools and disciplines. Emerging numerical computing tools and experimental aptitudes play a key role in the technological progress of aeronautical studies. This chapter presents an insight on the air foils, three-dimensional geometries attached to an airplane with an emphasis on computational tools. A countless number of small and large steps have taken place over many other disciplines. Design evolution has resulted in many geometrical changes in air foils, wings, fuselages, and stabilizers come in a whole range of shapes and sizes, both in the aerospace industry and in nature – really, nothing is standard. The application the airfoil operates and dictates its shape and size. Finite wing and infinite wing shapes are still sprouting today, driving the new challenging flight conditions. More efficient flights will drive the new and intelligent wing designs to obtain better load factor and reduced drag.

DOI: 10.4018/978-1-6684-4230-2.ch011

INTRODUCTION

To make an aeroplane, you need to know what the plane will use. Aeroplane mission requirements are developed in various methods, based on the kind of aeroplane and, in some cases, the user's needs. In terms of the current aviation trend, consumers are pushing for faster flights to cut flight duration. It is even reliant on the aircraft's mission. Once the design stages are accepted, the plane must go through many analyses to evaluate the performance of the wing or airfoil(Beckwith et al., 1969; Russo, 2011; Von Karman, 2003). Many aircraft never make it through the preliminary design stage. In truth, the vast majority do not. What follows after the preliminary design process is heavily influenced by the preliminary design outcomes and the actual or projected market interest. The wing is subjected to the subsonic and supersonic wind in transonic flight, making it the most difficult for an aeroplane to fly. At 1/4 per cent of the distance from the wing's leading edge, the subsonic airfoil's aerodynamic centre is located. The control and stability of a supersonic aircraft are dramatically altered when its aerodynamic centre returns to 50% of the chord of the wings (Federal Aviation Administration, 2009; Taylor & Maccoll, 1933). Transonic flight may occur when a subsonic aircraft with a Mach number of 0.75 travels too fast and enters transonic flow. One of the key benefits of the arrangement is that it delays the creation of shockwaves to a higher Mach number when utilised with high-speed aircraft. Sections of the general airfoil are made out of uneven circular arcs, and double-wedge airfoil are summarised in this chapter. This chapter also discusses various aspects of supersonic wings, such as the shock-wave expansion approximation, finite-span wings, and computational approaches.

AIRCRAFT CONFIGURE URATIONS CHART

As shown in Figure 1, human-made aircraft are categorised as heavier or lighter than air depending on the available or employed power. In general, aircraft classifications are based on a variety of characteristics. Here is one attempt to help the reader comprehend.

Aircraft Evolutions

As shown in Figure 2, the evolution of heavier-than-air vehicles began centuries ago in the name of vaimanika sastra; here it is Bejanet et., all (Bejan et al., 2014). They observed that the evolutionary tendency for fliers in nature (e.g., birds, bats, insects) has been toward greater and larger size throughout "Big History" (history from the Big Bang to the present). High-speed flying is the next frontier in the aviation industry as it promises for substantial time savings, especially on long-distance international trips. In actuality, supersonic aircraft travel at much higher speeds than typical transonic aircraft. This is, as expected, tremendously enticing for business and combat applications. However, flying above the clouds is a different level. Military aircraft are now the only ones capable of breaching the sound barrier. Except for one thing: a few decades ago, when the Concorde and Tupolev Tu-144 made supersonic flight practicable transport for people.

Figure 1. Types of aircraft's classifications

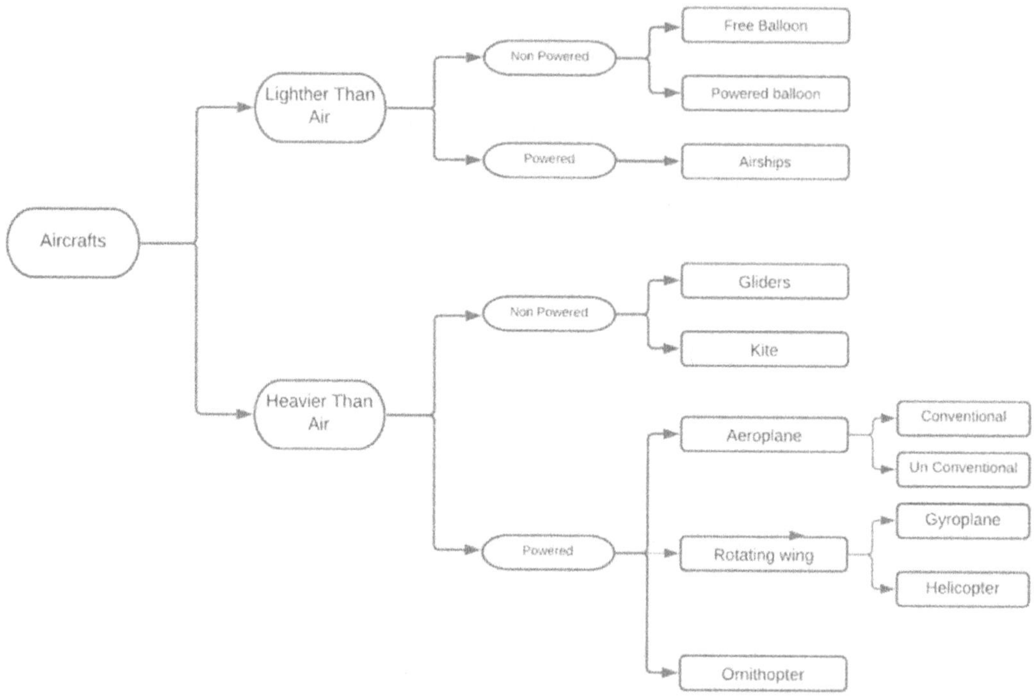

Figure 2. a) Evolution of Aircraft over the years (Bejan et al., 2014); b) Top Airplane Speed from wright to now ()

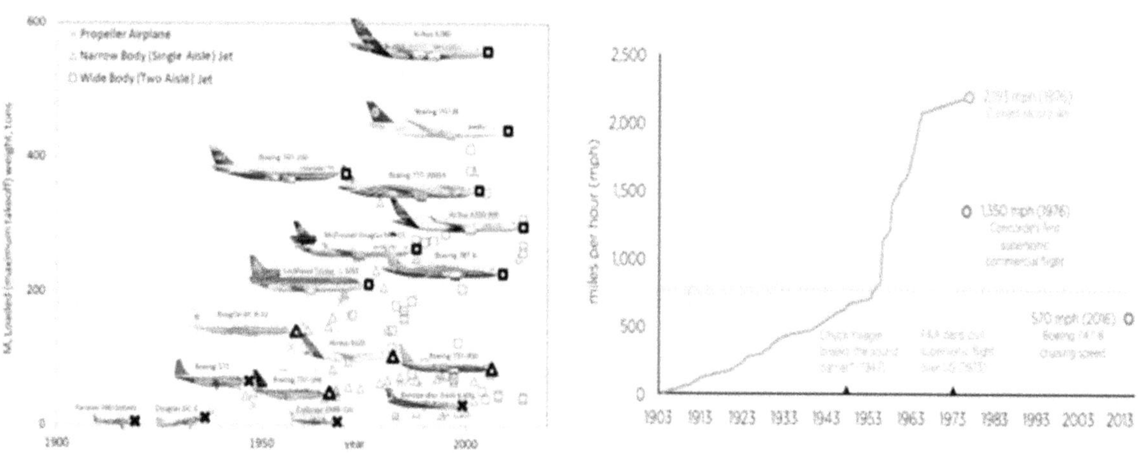

Infinite Wing Nomenclature

The leading edge is the front of the airfoil, as depicted in Figure 3. The airfoil's top and bottom surfaces meet at the leading and trailing edges. The chord of an airfoil is the distance between the leading and trailing edges. The thickness of an airfoil is a significant design parameter always expressed as a percentage of the chord. By expressing the thickness as a percentage, it is possible to make an airfoil design independent of the chord length, allowing for the definition of a single airfoil profile (form) for each wing of a given chord length. Camber is the measure of the imbalance between the top and lower surfaces. Camber is often added to an airfoil to increase its maximum lift coefficient, lowering the aeroplane's stall speed. The camber line is a line that is equidistant between the upper and lower surfaces at all points along the chord. Airfoils with more camber provide more lift than those with less camber, while those with no camber have perfectly symmetrical upper and bottom surfaces.

Figure 3. Aerofoil geometrical details

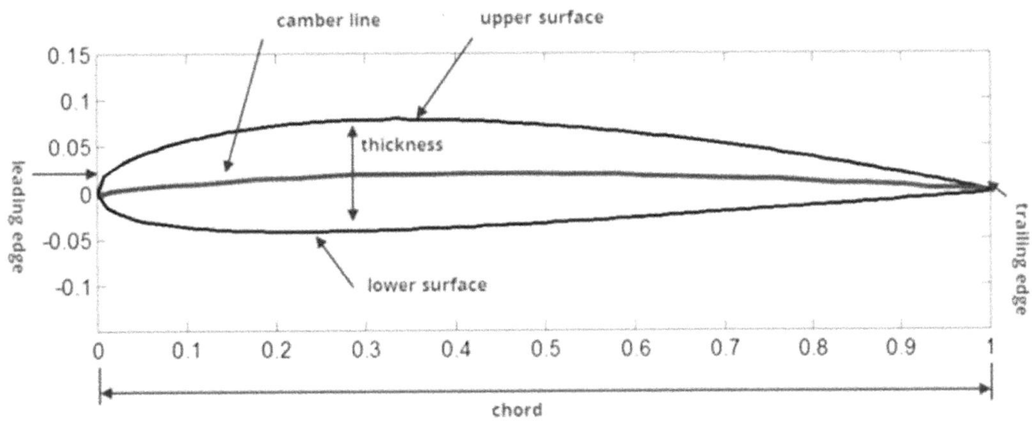

Infinite Wings in Compressible Flow

A subfield of aeronautics research in high-speed aerodynamics. Because the compressibility effects of air cannot be ignored in this flight regime, it is referred to as compressible aerodynamics (Taylor, 1974; Wirz, 1975). When a flight is slower than the speed of sound, it is classified as subsonic; when it is closer to the speed of sound, it is classified as transonic; when it is faster than the speed of sound, it is classified as supersonic; and when it is significantly faster than the speed of sound, it is classified as hypersonic. In each of the flying regimes, different flow phenomena can be found.

The ratio of the speed of a moving object (v) to the speed of sound (c) in a fluid is known as the Mach number (M).

$$Md = \frac{v}{c}$$

- Sub-sonic Regime — Mach numbers below 0.75 M
- Tran-sonic Regime range — 0.75 M to 1.2 M
- Super-sonic Regime — 1.2 M to 4 M
- Hyper-sonic Regime — greater than 4 M

The Mach number is a nondimensional number used in aerodynamics. The shock wave will occur when the Mach number is greater than 1. A supersonic object's shock wave is a cone of overlapping spherical wavefronts. Wavefront's form travels radially outward at c and expands in diameter until they reach radius a. The source, which is moving at the same rate as v, is ahead of the destination. These two displacements, which make up the leg and hypotenuse of a right triangle, can be utilised to calculate the Mach angle at the shock cone's vertex S shown in Figure 4.

$$\sin \mu = \frac{c}{v}$$

Figure 4. Sound waves propagating at different Mach numbers

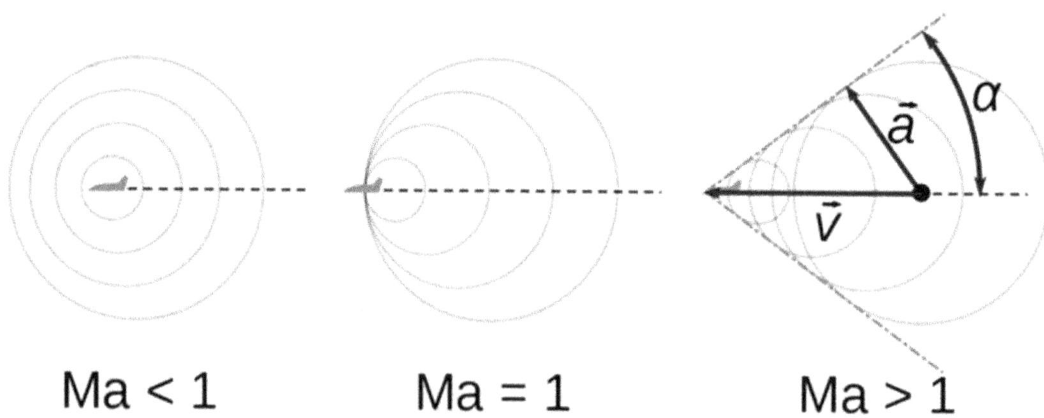

When a source's speed equals that of sound, wavefronts cannot escape. Because of the vast amplitude "sound barrier," prolonged flying at this speed is difficult and unsafe, as shown in Figure 5.

Example: The phrase "sound barrier" or "sonic barrier" originated during WWII. Many things went wrong when Figure hter pilots made high-speed dives. Aerodynamic drag rose faster than typical as speed climbing, and lift and manoeuvrability fell rapidly. When these effects initially occurred, pilots felt it was impossible to fly faster than sound. But they were wrong.

A pressure change conveyed at the speed of sound "warns" the air ahead of a subsonic flight. This warming causes the perspective to shift aside before the aircraft arrives, allowing it to pass effortlessly. A speed travelling at the speed of sound cannot detect pressure changes in the air ahead of it. Instead, the air particles build up in front of the plane, reducing flow velocity and increasing pressure and density.

As the aircraft flies faster than sound, the compressed air in front of it expands in pressure and density and spreads out. The air particles remain wholly undisturbed for a short time before being exposed to drastic fluctuations in temperature, pressure, density, and velocity. As seen in Figure ure5, a shock wave separates the compressed air from the undisturbed air. So, this occurs when a supersonic airstream is cambered and then decelerated to subsonic speed after reaching maximum camber. There is a shock wave between supersonic and subsonic. When shock waves happen on an aeroplane, airflow separation, other problems may occur. Above the Mach number, shock waves, buffeting, and airflow separation happen. At or near a jet plane's critical Mach number, it's most effective. When the airflow over any

Figure 6. Normal shockwave an experimental visualisation (Rodi et al., 2002).

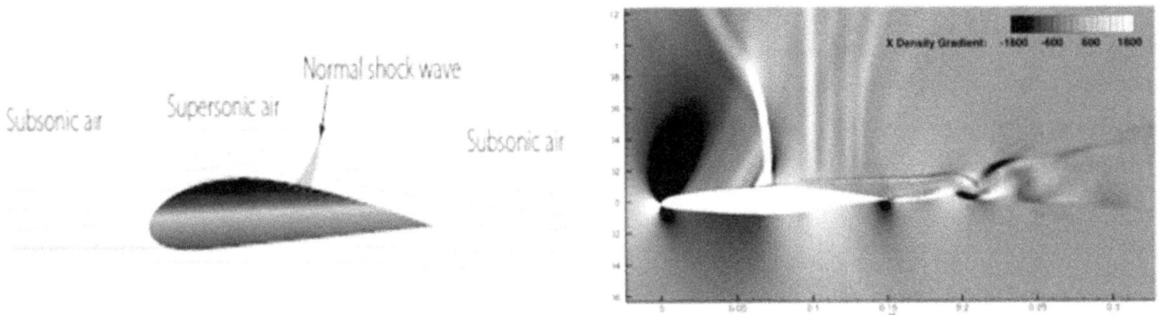

Figure 7. Supersonic aerofoil with oblique shock waves and expansion waves.

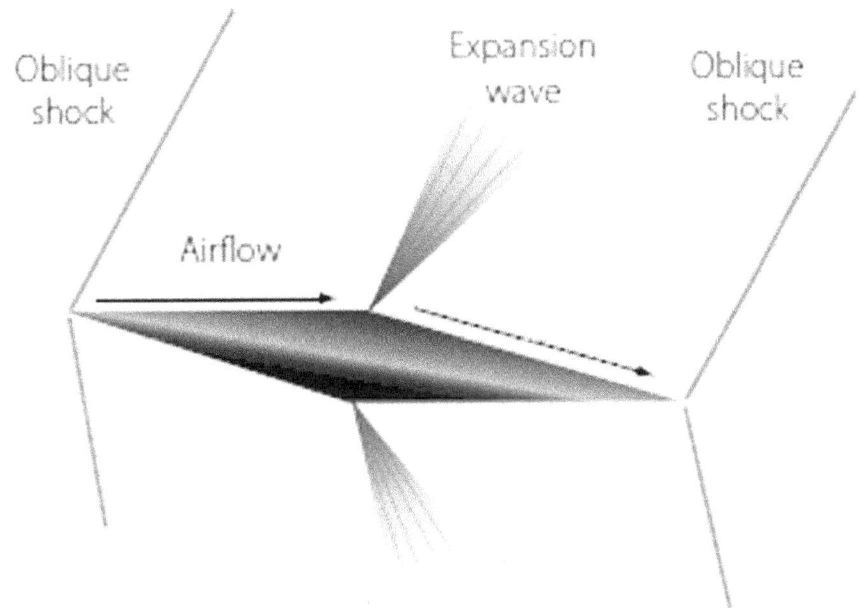

Expansion Wave

When travelling at supersonic speeds, air acts as a compressible fluid, as mentioned in section infinite wings in compressible flow. As a result, when given the chance, supersonic air seeks to expand outward. The air expands and adopts a new path as the wing surface shifts away from the flow direction when travelling over a wing in supersonic air. When the flow direction changes, an expansion wave occurs, as seen in Figure 8—the velocity behind the expansion wave increases while the static pressure and density decrease. An expansion wave is not the same as a shock wave.

Finite Wing

Span (S) uses a single known variable to begin the wing design process. All of the other fifteen wing characteristics should be created once all of the design requirements have been taken into consideration. The wing must produce enough lift while creating as little drag and pitching moment as possible. These design goals must be accomplished in concert (AeroToolbox, 2022; Chandler, 2019; Federal Aviation Administration, 2009).

THE NEED FOR SPEED

Since the birth of flight, there have been many ways to make air travel safer, more cost-effective, and faster(Beckwith et al., 1969; Houghton & Carpenter, 2003). While the first two of these goals are improved with each new generation of aircraft, the third seems to have been forgotten lately. The Concorde, the last (and virtually only) commercial supersonic aircraft, was retired from service in 2003(Talay, 1975). Because it could fly at speeds more than twice as fast as any other commercial aircraft could (or does), the Concorde slashed travel times between major cities by half. In the years after its demise, the fastest commercial jets could only fly at Mach 0.85, a speed that was possible for commercial airliners as recently as the 1950s(Talay, 1975). The reader may ask why air travel has remained stagnant in recent decades and why modern aircraft can't seem to break over Mach 0.85, a perceived speed barrier.

Overcoming the barrier of high-speed flying will be the focus of this section, covering the causes and solutions to these issues. One such example is X-1 test flights that continued throughout the fall of 1947, taking it higher and higher than Mach.85, the speed at which statistics on the subsonic flight had all but vanished. However, the X-1 test crew had a growing belief that their aircraft could successfully complete the historic flight. Even when huge aircraft took to the sky, their size created new challenges. The transonic region quickly increased in drag. Slow acceleration wasted fuel and caused control concerns throughout this flying phase. At Langley, Richard T. Whitcomb became interested in transonic drag. Whitcomb's studies led him to conclude that the fuselage-to-wing link of an aeroplane contained the solution. Whitcomb proposed the *Area rule* as a solution to the drag problem based on Adolph Busemann's transonic airflow characteristics.

The Area Rule

During WWII, engineers constructing Figure hters discovered that something peculiar happened when aircraft reached transonic and supersonic speeds. Shock waves would develop around the aircraft, sapping its energy. The "Whitcomb area rule" was developed to restrict the regions where shockwaves might build up in an aeroplane's cross-section. How engineers solved the challenge, generating solutions still utilised in current flying. At Mach 1, the area rule design cut peak drag enough so that real planes could fly supersonically by the middle of the 1950s. It was a huge success. Figure 8a & 8b compares the M-dependent changes in the drag coefficient for an area-ruled and a non-area-ruled aircraft. The area rule can often cut peak drag by two at Mach 1.

Figure 8. a) YF-102A without area rule; b) F -102A with area rule (Houghton & Carpenter, 2003)

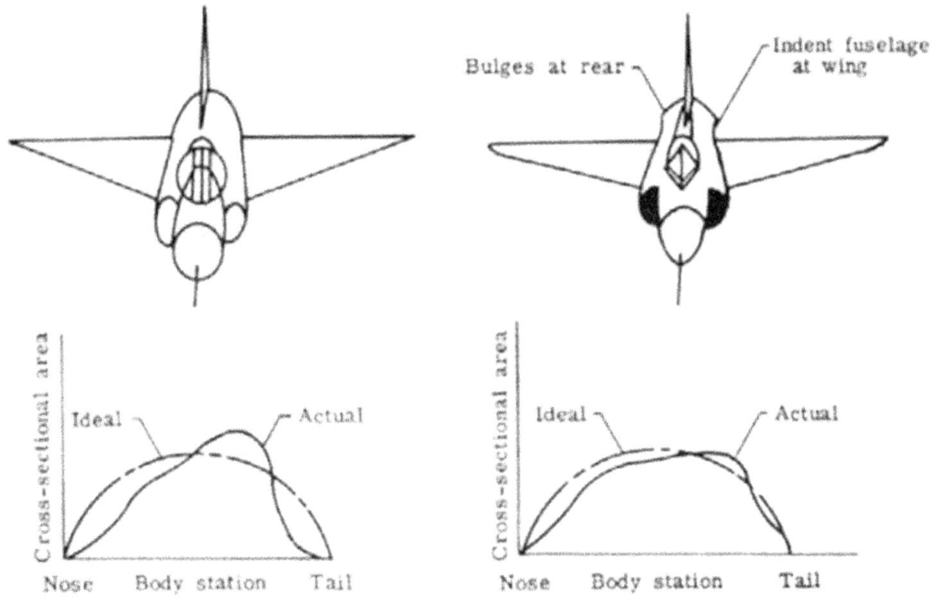

What does this have to do with where the engine is and the anti-shock body's unique drag-reducing properties?

Here's the area rule. The area rule says that an aircraft's wave drag is determined by how much cross-sectional area it has over its length. The wave drag on a plane can change if you cut the aircraft across its width at every point and look at the result. Any change in the cross-sectional area of one part of the plane would cause the whole plane to have more drag. This is because transonic airflow isn't compressible like lower-speed airflow. To smooth the change cross-sectional area by pointing at the plane's nose and tail, the whole plane's cross-sectional area must be considered.

Figure 9. Comparison aircrafts with area rule and wet area distribution

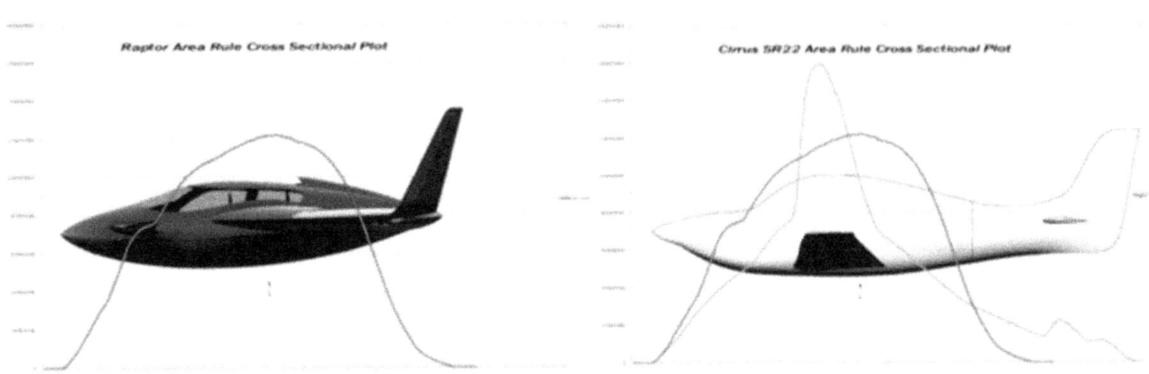

This Figure 9 clearly illustrates how close the Raptor's design comes to complying with the area criterion. As a result, the plane's pressure rises and falls gradually rather than fluctuating as the air flows around it. As a result, there is less turbulence and drag is reduced even more. When the cabin or wings first appear on a tractor, the cross-sectional area changes dramatically, and then it changes again when the tractor is finished. The tail part then expands and contracts. Conventional planes have a lot of airflow turbulence because of these sudden changes in cross-sectional area, which causes a lot of friction. This issue is nearly entirely resolved with the new design. Compare the Raptor's area Rule distribution to the SR22s. As you can see, the design is flawed. The cockpit's widest part runs parallel to the wing's length. So the Raptor's maximum cross-sectional area is 33% larger (160 vs. 120 in the chart). While the Raptor's cabin is larger than the SR22's, its maximum cross-sectional area is less, letting it to move through the air more efficiently. This makes the design much faster and more efficient.

How does area rule work?

According to Whitcomb, the plane's drag was directly proportional to the amount of discontinuity in the cross-sectional area along its length. This is where he came to the idea that an added wing adds more volume, and that reducing the volume of the fuselage and indenting it would result in an even distribution of surface areas and hence reduce waves. The same distribution of cross-sectional area may be seen in the middle two shapes. This regulation was first implemented in the Convair F-102A Delta Dagger conversion. There was too much wave drag on the F-102 and it couldn't attain supersonic speeds. Whitcomb's area rule dictated that the fuselage area be lowered (indented) at the waist of the aircraft, resulting in the F-102A designation. Mach 1.22 was eventually achieved by the F-102A when significant reductions in transonic wave drag were implemented.

By indenting the fuselage (reducing its volume) as close to the waist as practicable, it is possible to maintain a consistent and continuous cross-sectional area. As a result of this, Whitcomb's area rule reduces but does not eliminate wing-induced wave drag. The area-rule design, such as the tail-mounted engines and streamlined components, are still in use today. There are numerous examples of this rule in effect, including the Boeing 747 plane's design. It was necessary to use the area rule during near-transonic flights in order to maintain a streamlined cross-sectional surface area. It is less significant in modern supersonic designs because of greater engine power, which favours manufacturing efficiency over cheap cost.

In order to understand better about the area let us take an example on area rule

Example: Consider the table below in order to evaluate the lift to drag ratios of two aircraft.

Table 1.

Details	Boeing B-47	Avro Vulcan	Units
Max. takeoff weight	202,000	204,000	MTOW(lbs.)
Reference wing Area	1428	3965	S(ft2)
Wetted Surface Area	7070	9600(approx.)	Swet(ft2)
Wing Span	116	111	b(ft.)
Aspect ratio	9.42	3.1	AR(=b2/s)
Max. Wing Loading	141.5	51.5	W/S(lb./ft2)
Max. Span loading	1741	1834	W/b(lbs./ft)
Max.Lift/drag Ratio	16.1	16.8	L/Dmax

Swept Wing

As the aircraft's intended to travel beyond the speed of sound, there was another hurdle of shock cone/ Mach cone over the aircraft, which has invaded with results that the design of the wing travelling greater than the speed of sound must have wing sweep and the sweep angle should be less than mach angle. previous to this mach cone theory disclose it was a fiction for aerospace engineers to go at more than the speed of sound. The swept-wing decreases the speed normal to the leading edge by around 30% at a sweep angle of 45°. Lower U_N speeds will result in a minor pressure difference between the top and bottom surfaces of the swept wing and hence less lift. The pressure distribution between the top and bottom wing surfaces is dictated by the leading-edge velocity depicted in Figure 10. However, the wingtip discontinuity and other issues, such as interference between the wing root and the plane's fuselage, are not considered (Sauerwein, 1967; Sikroria et al., 2018).

Transonic and supersonic aircraft often make use of wing sweep (Dunbar, 1996; Russo, 2011). When high-speed air compressibility causes shock waves to occur on a wing's surface, the sweep reduces the frequency of these waves shown in Figure 10. The total drag of the aeroplane is reduced as it approaches Mach 1 by delaying the production of shock waves.

How to set sweep angle?

Oblique shockwaves are distinguished from conventional shockwaves. Normal shock waves grow parallel to the triggering surface (like a wing's top surface), while oblique shock waves arise at an angle to the upstream flow direction. This occurs when a supersonic flow meets a corner and compresses. Oblique shocks originate on pointed wedges, like an aircraft's nose. A normal shock's downstream velocity must be subsonic, whereas an oblique shock's Mach number could be supersonic (weak shock wave) or subsonic (strong shock wave) (strong shock wave). A subsonic wing profile may be produced by carefully constructing the aircraft form to utilise shockwaves to slow the air down to subsonic speeds depicted in Figure 11.

The oblique shock's angle changes with the speed of the aircraft. Speeds of 1.3 and 2 provide an angle of 45 degrees, respectively. Even though the cruising speed of a supersonic plane is well-known, the wing may be swept to expose it to subsonic airflow, causing the aircraft to act as if it were flying at subsonic speeds. A shockwave's subsonic flow cone contains a subsonic leading edge. When a wing is a supersonic wing, it has a supersonic leading edge.

Figure 10. Flow variation over the rectangular and swept wing

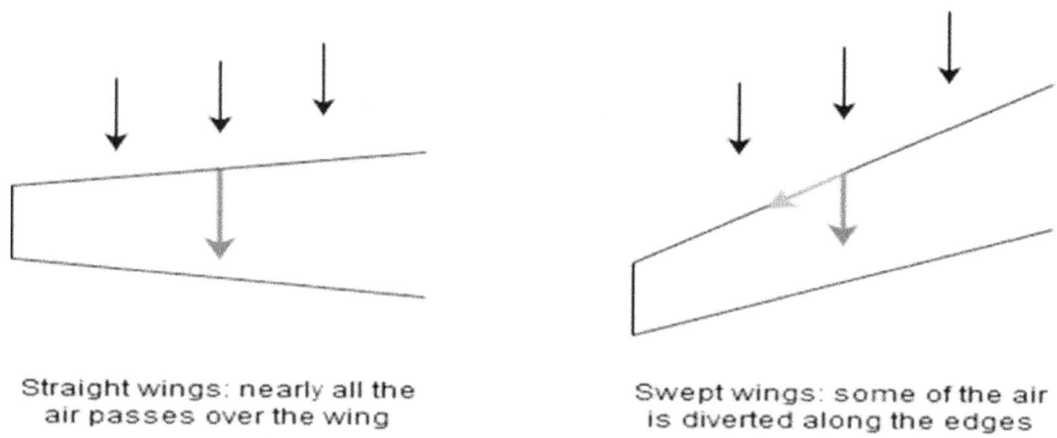

Figure 11. Subsonic and Supersonic Leading Edge (Rakich & Cleary, 1970)

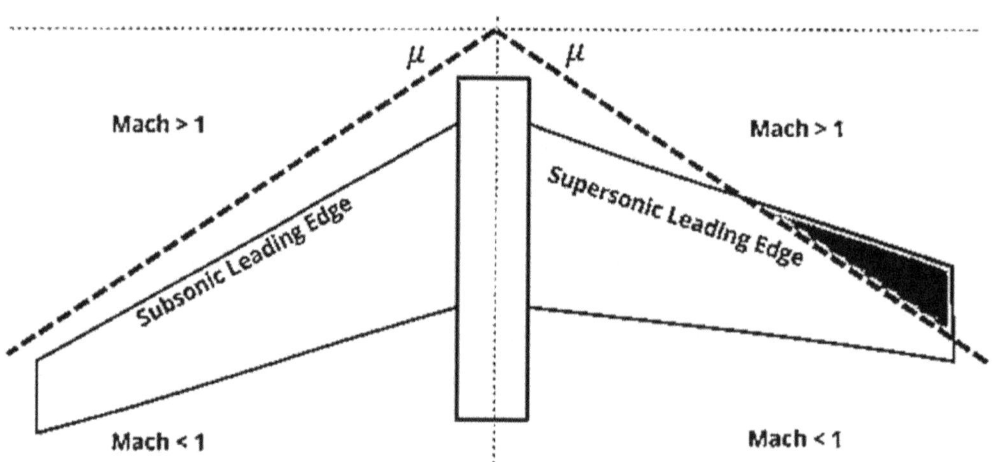

In the supersonic zone, a supersonic leading edge causes a considerable normal shock on the leading edge, increasing drag significantly throughout the wing and limiting normal wing function. Even if the freestream Mach number is more than 1, the top and lower wing surfaces interact as in subsonic flow with a subsonic leading edge. The airflow across the wing is accelerated to subsonic speeds due to the aircraft's design, which creates moderate oblique shocks. When compared to a normal shock on the wing leading edge, this reduces the drag rise.

Consider the trade-offs and design constraints while choosing the sweep angle. At supersonic speeds, a sweeping wing with a fully subsonic leading edge will perform well, but not at subsonic speeds (Aircraft Systems, 2022; Rakich & Cleary, 1970). A swept-wing produces less lift than an unswept wing, which increases stall speed and reduces manoeuvrability. Sweep angles are frequently less than 40° at velocities greater than Mach 0.8. A plane that can travel at Mach 1.5 or higher normally has a sweep angle of up to 60 degrees. There are benefits and drawbacks to wing designs with higher sweep.

- Sweeping the wing of any plane flying at speeds above Mach 0.5–Mach 0.6 is essential to avoid huge increases in drag caused by shockwaves being created in places where the flow is accelerated to speeds above Mach 1.
- For a supersonic plane to keep flying, it must slow the air that passes over its wing down to subsonic speeds. This way, the attachment can produce enough lift to keep the plane in the air without having a significant spike drag spike caused by normal shocks on the wing's leading edge.
- Sweeping a wing could make the subsonic flight less safe, so be careful. If you sweep your wings, they get shorter, making them have more drag when flying at a lower speed. Less adequate airflow travels over the wing, which means less lift is generated at the same angle of attack and a faster stall speed (Taylor & Maccoll, 1933).

Here, we'll look at how to manage this flow over the wing in light of what we learned previously in this chapter about the shock wave that occurs when an item travels quickly. Flow separation and reattachment are important in many engineering fields, including civil, mechanical, aeronautical, chemical, and environmental. To avoid it, the designer must understand the hazards of flow separation. At cruising speed, a well-designed aircraft should have no flow separation zones. It ensures optimal efficiency during cruising since flow separation increases pressure drag. An increase in AOA occurs as the plane slows from cruise speed, causing separation regions to emerge and grow.

How to control separation over the wing at supersonic flows?

When a shock hits a wing, the surface slope distribution on the upper surface is vital to the section form. Many things happen at high Mach numbers, and the trailing edge has a significant impact (Houghton & Carpenter, 2003; Rakich, 1969; Rakich & Cleary, 1970). Often, the separation starts in a specific area of the span because of high local loads. A planform can be changed to reduce the peak loads, for example, by extending the leading edge. Shock-induced separation is often caused by flow field interference between two surfaces next to each other. The leading edge of a wing's shock might break up the boundary layer on a fuselage, which would change the wing's flow through a vortex. There are many different shock wave patterns in Figure 12. This is because the wing that sweeps back has a lot of parts.

However, front, rear, or outboard shock waves are categorised, not equally on both wings. Ideal-yawed wing flow dominates near an outboard shock. Therefore, separation is predicted initially near the outboard shock. Unswept wings with narrow aspect ratios and relatively high lift coefficients need three-dimensional front and rear shocks. Themselves aren't Root-flow affects these two inner shocks, transferring some of the action to the outboard shock at the contact point. Modifying the root geometry, which involves body-shaping, is one of the most effective ways to increase airflow on the wings (Aircraft Systems, 2022; Owczarek, 1964; Rakich & Cleary, 1970). In Figure 13, a shock-induced critical separation is depicted by a curve AB.

Figure 12. A typical pattern of shock waves on a swept-back wing (Chang, 2014)

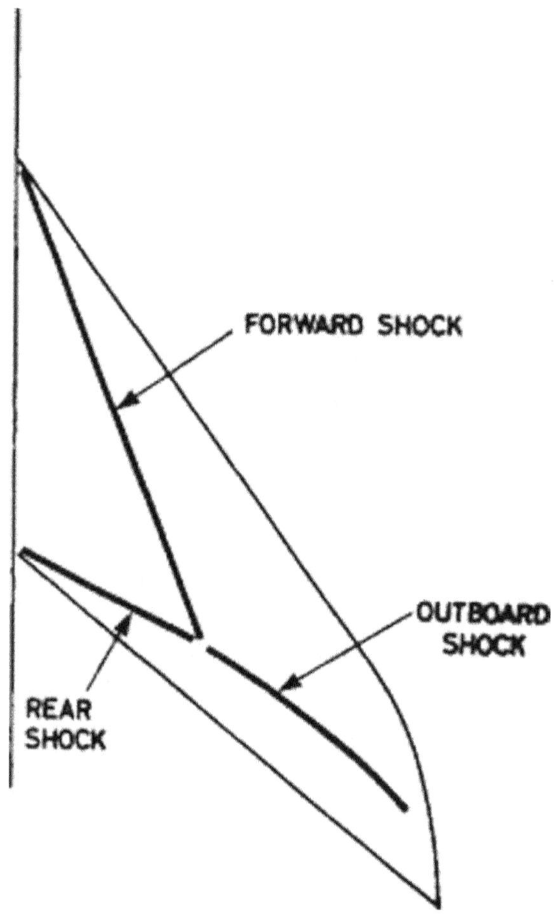

The shock begins at position A, which indicates 0% lift, and continues to point B, which indicates 10% lift. In cases when the critical separation is gradual, the left side of B is stretched to show the beginning consequences of separation. To denote divided flow, we use the BB¢ line. Shock separation can be avoided by elevating and relocating the AB boundary to the right. (Chang, 2014) Shock-induced separation is decreased by sweepback and narrow aspect ratios. Stream tube constriction is minimised and supersonic surface expansion is avoided at local sonic speeds because the aspect ratio is reduced. Wing sweepback delays separation and lessens shock waves at lower Mach numbers (Chang, 2014; Talay, 1975).

There is a wide range of variability in airflow, which has been extensively studied theoretically and experimentally in wind tunnels depicted in Figure 14. At low speeds, the pressures created by an aeroplane's passage through the air are negligible in comparison to the pressure of the surrounding atmosphere. The tension between these two forces defines the aerodynamic flight corridor. Dynamic pressure grows with velocity squared. It is 25 times greater at Mach 6 than at Mach 1.2. If the speed is high enough, dynamic pressure increases allow for extended flights at high altitudes with low air pressure. Changes in airflow's elastic and viscous characteristics affect pressure force as it travels around an aircraft (Chang, 2014). As previously noted, extreme flow variations occur at high speeds.

Figure 13. Onset boundary and subdivision of regime for separation effects (Talay, 1975)

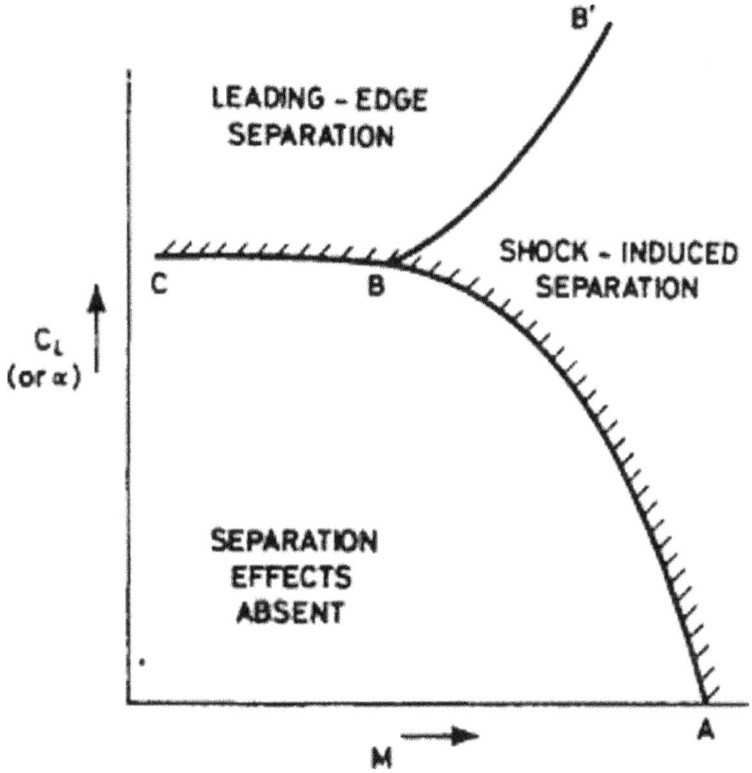

Figure 14. Free-flight models of the X-15 being fired into a wind Tunnel vividly detail the shock-wave patterns for airflow at Mach 3.5 (left) and Mach 6 (right).(Talay, 1975)

Due to air's elasticity (compression and expansion), flying at high flights has a significant impact on airflow. At the slowest speeds, subsonic, the effects are not evident. Pressure builds up ahead of the aircraft's various components as airflow speeds rise, resulting in distinct pressure waves, or shock waves, that may be seen from the air's surface. Transonic airflow is where shock waves originate on an aircraft, even if they are localised. Its stability is jeopardised by the turbulent mixed-flow between subsonic and supersonic velocities. As speed increases, little subsonic areas evaporate, and the flow becomes supersonic. The air pressure has plummeted. When an object creates a cone or a wedge in the air, shock waves are now trailing behind it. This is in stark contrast to the thickness of a shock wave, which is often less than.001-inch thick. These impacts are far-reaching, even causing sonic booms on Earth. The parameters of these shock waves may be calculated using aerodynamic theory.

When the aeroplane reaches supersonic speeds, the surfaces and shock waves form acute angles. These interactions between shock waves and flow field defy supersonic flow models. The shock waves lower the aerodynamic forces, reducing the wing and tail surface lifting effectiveness. More importantly, the friction of air moving down a surface raises the temperature of the air by several degrees. Aerodynamics now incorporates thermodynamics due to hypersonic airflow. Owing to its ambiguity, it is usually assumed that speeds over Mach 5 are covered. This location has several shock waves and interference effects, as shown in Figure 28. The difficulty for the aerodynamicist arises from the impact of discontinuous flow at each shock wave. Each new geometric form requires rethinking. Optimising the wing, tail, and fuselage proportions may increase an aircraft's efficiency. But when airflow enters a new flight phase, that unique design may cause unwanted interference. A layout appropriate for subsonic to hypersonic speeds requires several sacrifices.

A CASE STUDY ON DOUBLE WEDGE AEROFOIL

Here in order to extend the user understating related to shock waves a numerical case study with oblique shock waves on the double wedge model is presented in this section. Modelling is done using CAD software. The model's physical dimensions vary, but for 0 degrees, 5 degrees, 10 degrees, and 15 degrees AOA, the half wedge angle is fixed at 5 and 10 degrees. Specifications for the model: Domain dimensions are 22 m x 10 m. (For 5deg half-angle model) Four aspects Defining Conditions: Wedge surface = Wall, Domain Walls: Far Field Pressure The second model specification is as follows: The domain is 22 metres by 10 metres in size (For ten half-angle angle model) The following conditions define the perimeter: Wedge surface = Wall, Domain Walls: Far Field Pressure FLUENT was used to import mesh model files. The K-epsilon and the viscous models (Butler, 1960; Chushkin, 1968; Holt, 2012; Roache, 1976) are two-equation models used to analyse turbulence. The Epsilon model is an excellent choice if the flow features separation, recirculation, or scales smaller than the cell size. Two additional transport equations were included in the model to account for turbulent aspects of the flow and historical impacts such as convection and diffusion of turbulent energy. Figure 15a shows a half-angle aerofoil. 15b. At a 10 deg angle, half-angle aerofoil.

Figure 15. a) 5deg half wedge angle 9; b) 10deg Half wedge angle (Prasad et al., 2017)

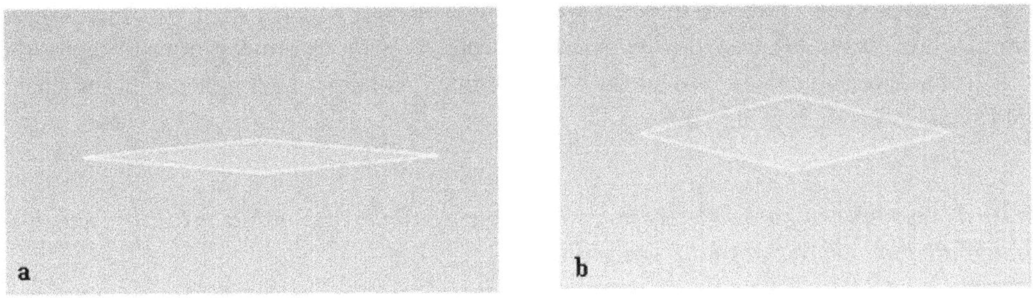

Computational Analysis

A Mach number range of 1.2-1.8 at zero- and five-degrees AOA is analysed. A density-based solver with a K-epsilon viscous model is employed in this problem. Once the Standard Equations have converged, iterations are carried out until they do. The Mach number was changed, and charts were studied. AOA and the half-wedge angle were modified. A similar study was performed at other Mach numbers for further adjustments in the angle of attack. No shock was seen at the leading edge because the half-wedge angle matches the AOA. Specifically, the top half of the plane is parallel to the free-flowing direction of travel. As a result, no flow is prone to self-reinforcing. The AOA was extended to 10 and 15 degrees. At a value greater than half the wedge angle, At the leading edge of the top, an expansion wave may be visible as seen in Figure 16 to 20 and Figure 16, indicate that when the Mach number is the same but the AOA is different, the expansion wave on top and the shock wave at the lower surface becomes stronger.

Using velocity profiles and oblique shock relationships, engineers concluded that plots necessary to comprehend how the attack angle varies. Variables such as dynamic pressure, Mach number, temperature, and surface turbulence are compared. The C_l, C_d, Lift, and Drag Vs Mach numbers were manually calculated and shown using post-processing tools. Wedge angle change CAD The model was created in a modelling application and read fluently at a 10-degree half wedge angle (deflection angle). Several Mach values were employed in the investigation at zero AOA, ranging from 1.2 to Contours and plots presented in Figure 16.

Case Study Summary

Airfoil with a 0 degree AOA wedge. Calculations performed by hand produce precise outcomes. When the shock angle is more than 90 degrees, the de-attached shock is shown in Figure 23. Increasing the Mach number reduces pressure coefficient. As intake Mach increases, so does peak Mach and turbulent kinetic energy. Because the wedge gets smaller, this is the case. At higher speeds, the effect is more noticeable. This is the same as the angle of attack (AOA). AOA>half wedge angle causes an expansion wave to form at the top of the leading edge. Top and bottom leading edge expansion waves are more powerful when AOA is more than half a wedge angle (AOA). Angle of the AOA half-wedge Although the pressure coefficient decreases as AOA rises, the reverse is true when AOA climbs. Mach number, dynamic pressure and turbulent kinetic energy all exhibit the same trend. Lift and drag coefficients As M and AOA increased, Cl and Cd (Coefficient of Lift and Drag) decreased. Lift and drag are boosted by M and AOA. When the angle of the Wedge is changed Using the same MPH as before Shock and expansion wave forces are increased by a half-wedge angle increase at zero AOA. While Cp rises initially in Figure 23, it begins to fall back down towards the end. As with Dynamic pressure graphs, the same is true. As the wedge angle rises, so does the Mach number. As the wedge angle decreases, so does this energy (Taylor, 1974; Rodi et al., 2002).

Figure 16. a) Velocity contours on 5 deg half wedge angle at 0 degAoA at Mach 1.2; b) Detached shock wave (magnified image) on 5deg half wedge angle

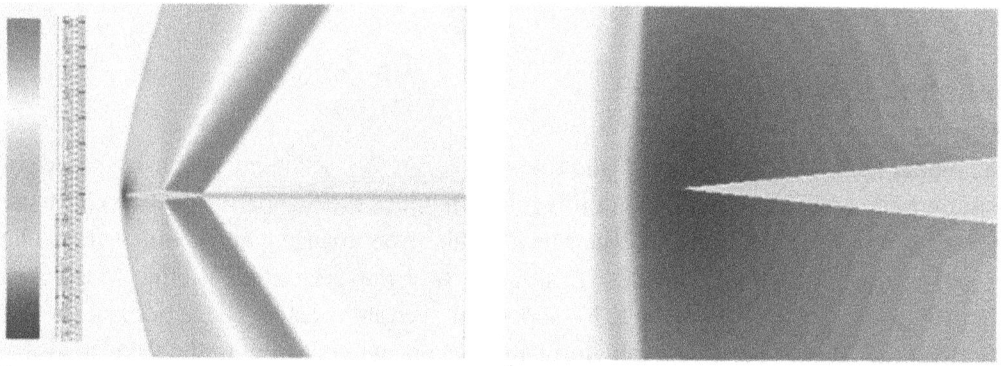

a b

Figure 17. a) Mach 1.6 velocity contour over 5deg half wedge angle; b) Mach 1.6 attached shock (magnified image)

a b

Figure 18. a) Velocity contour at 0 degAoA respectively with 5 deg half wedge angle at Mach 2 freestream velocity; b) Velocity contours at five degAoA with five deghalf wedge angle at Mach 2 freestream velocity

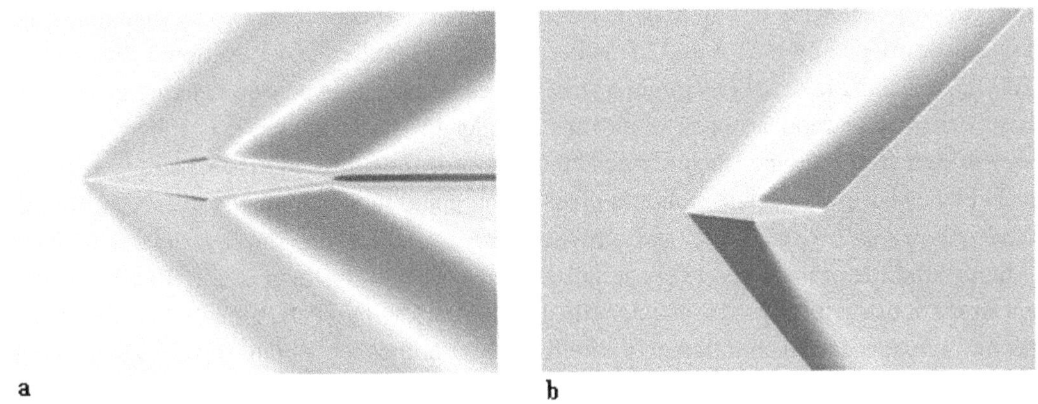

Figure 19. a) Expansion wave observed at the top leading edge when AoA exceeds the half wedge angle; b) velocity contours at Mach 3, AoAat 10 degAoA

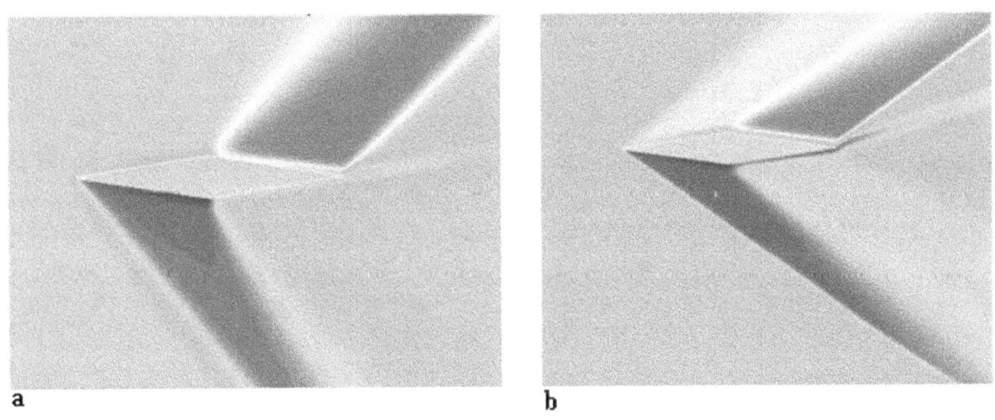

Figure 20. Velocity contours at Mach 3, Angle of attack at 15 deg.

FUTURE AIRCRAFTS

The cantilevered swept-wing design may be replaced in the next decade or two by the "transonic truss-braced wing" developed by NASA and firms like Boeing (Chandler, 2019; National Research Council (US), 2002). Research organisations prefer a strut-braced wing. These new designs are quieter and more eco-friendly than prior models, saving more fuel. Under the descriptions, it is stated how much fuel efficiency is increased and when technological maturity may be achieved in ideal circumstances. These days, little planes have wings with standard cross-bracing and no apparent V form. New, thin Carbon-reinforced polymers(CFRP)aerofoils and structural bracing made possible by slender CFRP trusses allow wings with little or no sweep to fly at these speeds(Aircraft Systems, 2022; Sikroria et al., 2018). Unswept wings are structurally more efficient than swept-back wings(Peterson, 2020). Besides that, new concepts for high-speed transportation may significantly alter the nature of air travel. New transportation technologies may be available within the next decade (Sadraey & Sadraey, 2012).

Figure 21. Coefficient of lift at different AoA concerning Mach Number b. Coefficient of drag at different AoA concerning Mach Number

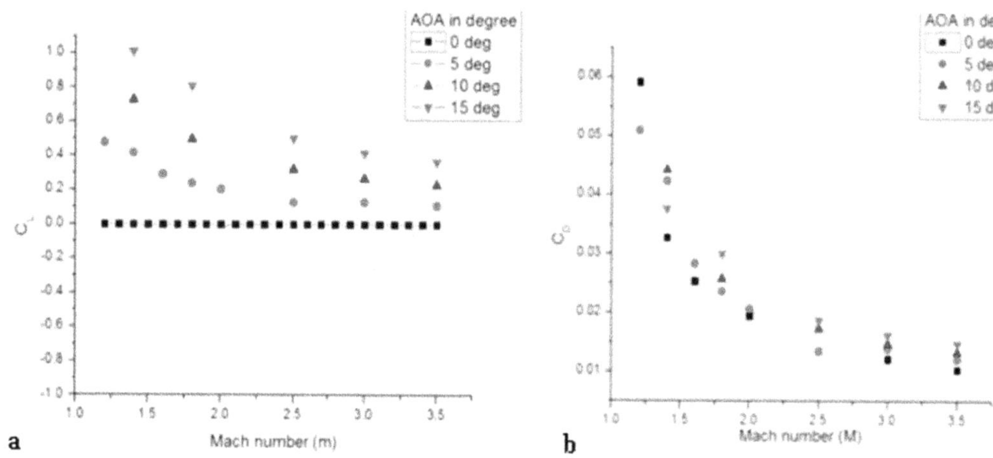

Figure 22. a) Velocity contours at Mach 3.5 with 5 deg half wedge angle; b) Velocity contours at Mach 3.5 with 10 deg wedge angle

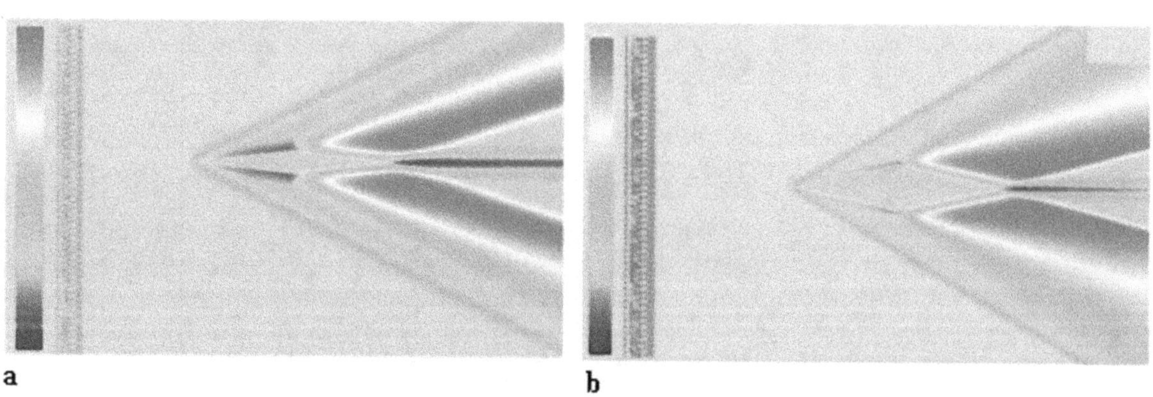

Figure 23. a) Coefficient of pressure vs Length(m); b) Mach Number over the surface vs Length of the airfoil.

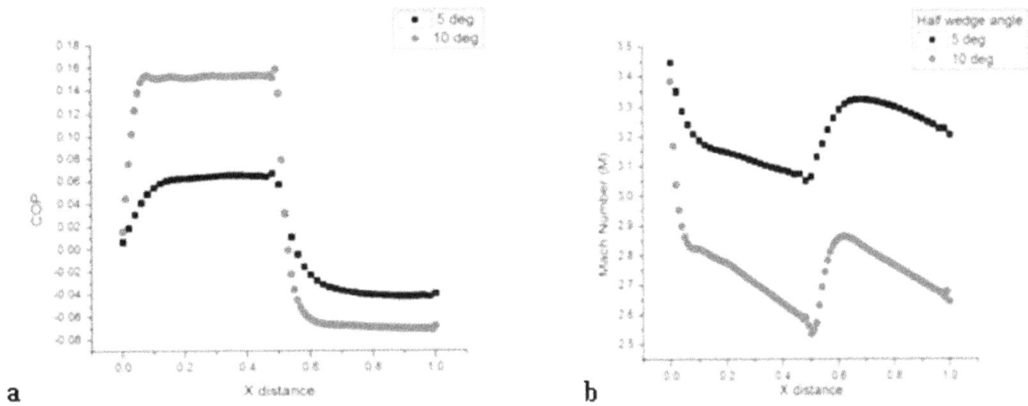

Figure 24. Future large "airplane" development opportunities (Ros, 2018)

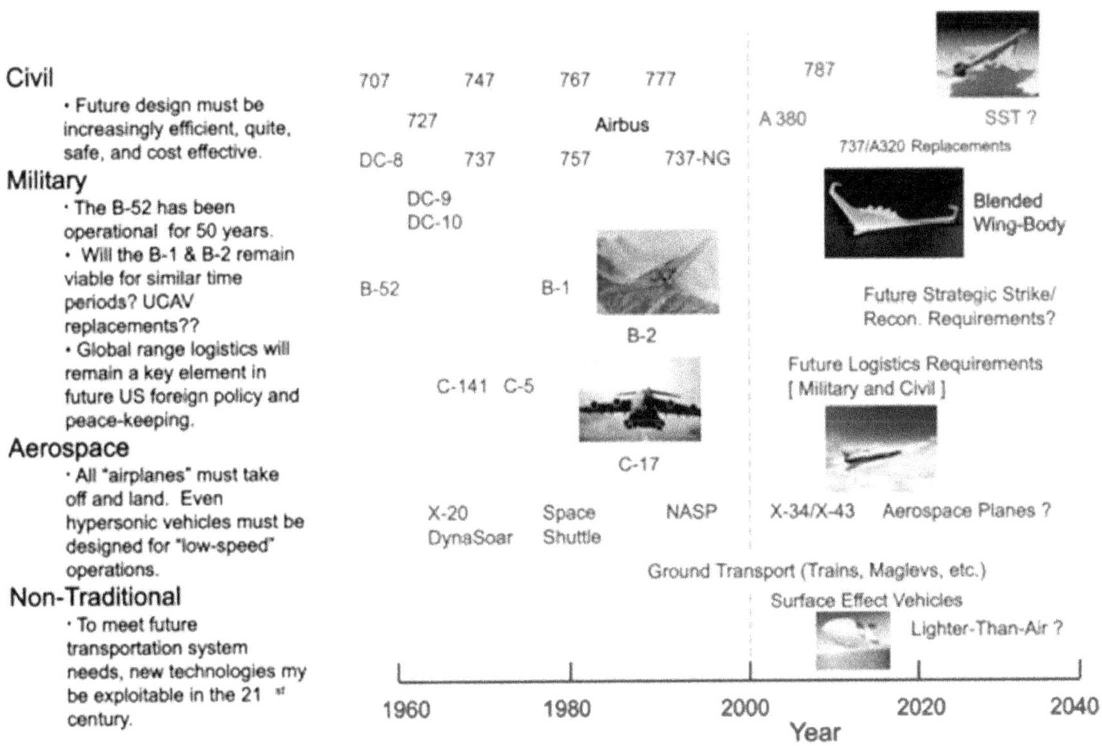

As it is indicated in the above Figure 24 the future aircraft are in need of travel of commercial planes as presented in the section of the aeroplane. The commercial aircraft industry looks to have slipped backwards in one area. Even the wealthiest people today have to settle for subsonic speeds when flying across the Atlantic. Flex jet, a fractional jet operator, has already placed a \$2.4 billion contract for 20 of the AS2's Mach 1.5 jets, despite the fact that it is still in development. If you're looking for the ultimate in speed, however, you'll have to look elsewhere. The German Aerospace Research Institute (DLR) is leading the development of a spacecraft that would travel at 25 times the speed of sound to the furthest reaches of space. So, for example, London to Australia may be reached in about 90 minutes.

REFERENCES

AeroToolbox. (2022, May 18). *Introduction to aircraft airfoil aerodynamics.* Retrieved from https://aerotoolbox.com/intro-airfoil-aerodynamics/

Aircraft Systems. (2022, May 27). *High speed flight – aerodynamics of Flight.* Retrieved from https://www.aircraftsystemstech.com/2017/05/high-speed-flight.html

Bai, S. (1959). *Introduction to the theory of compressible flow*. van Nostrand.

Beckwith, T. G., Buck, N. L., & Marangoni, R. D. (1969). *Mechanical measurements* (Vol. 5). Addison-Wesley.

Bejan, Charles, & Lorente. (2014). The evolution of airplanes. *Journal of Applied Physics, 116*(4).

Butler, D. S. (1960). The numerical solution of hyperbolic systems of partial differential equations in three independent variables. *Proceedings of the Royal Society of London. Series A, Mathematical and Physical Sciences, 255*(1281), 232–252. doi:10.1098/rspa.1960.0065

Chandler, D. (2019). *MIT and NASA engineers demonstrate a new kind of airplane wing*. Academic Press.

Chang, P. K. (2014). *Separation of flow*. Elsevier.

Chushkin, P. I. (1968). Numerical method of characteristics for three-dimensional supersonic flows. *Progress in Aerospace Sciences, 9*, 41–122. doi:10.1016/0376-0421(68)90004-3

Dourado, E., & Kotrous, M. (2016, July 20). *Airplane Speeds Have Stagnated for 40 Years.* Mercatus Center, George Mason University. Retrieved from mercatus.org/publications/technology-and-innovation/airplane-speeds-have-stagnated-40-years

Dunbar, B. (1996, September). *High-speed research (HSR) – smoothing the flow over Supersonic Wings.* NASA. Retrieved from https://www.nasa.gov/centers/langley/news/factsheets/HSR-Wings.html

Elert, G. (n.d.). *Shock waves*. The Physics Hypertextbook. Retrieved from https://physics.info/shock/

Federal Aviation Administration & United States Federal Aviation Administration. (2009). *Pilot's handbook of aeronautical knowledge*. Skyhorse Publishing Inc.

Holt, M. (2012). *Numerical methods in fluid dynamics*. Springer Science & Business Media.

Houghton, E. L., & Carpenter, P. W. (2003). *Aerodynamics for engineering students*. Elsevier.

National Research Council (US) Transportation Research Board, Committee for a Study of Public-Sector Requirements for a Small Aircraft Transportation System, & National Research Council. (2002). *Future flight: A review of the small aircraft transportation system concept.* Author.

Owczarek, J. A. (1964). *Fundamentals of Gas Dynamics* (1st ed.). International Textbook Company.

Peterson, B. (2020, December 14). *These New Planes Could Change the Way You Travel.* Conde Nast Traveler. Retrieved from https://www.cntraveler.com/story/new-planes-that-will-change-the-way-you-travel

Prasad, U. S., Ajay, V. S., Rajat, R. H., & Samanyu, S. (2017, May). Aerodynamic Analysis Over Double Wedge Airfoil. *IOP Conference Series. Materials Science and Engineering*, *197*(1), 012076. doi:10.1088/1757-899X/197/1/012076

Rakich, J. V. (1969). *A method of characteristics for steady three-dimensional supersonic flow with application to inclined bodies of revolution.* National Aeronautics and Space Administration.

Rakich, J. V., & Cleary, J. W. (1970). Theoretical and experimental study of supersonic steady flow around inclined bodies of revolution. *AIAA Journal*, *8*(3), 511–518. doi:10.2514/3.5698

Roache, P. J. (1976). *Computational fluid dynamics.* Computational Fluid Dynamics.

Ros, M. (2018, February 16). *What will passenger planes look like in 2068?* CNN Travel. Retrieved from https://edition.cnn.com/travel/article/passenger-planes-future-look/index.html

Russo, G. P. (2011). *Aerodynamic measurements: from physical principles to turnkey instrumentation.* Academic Press.

Sadraey, M. H. (2012). Wing Design. In M. H. Sadraey (Ed.), *Aircraft Design: A Systems Engineering Approach* (pp. 161–264). John Wiley & Sons. doi:10.1002/9781118352700.ch5

Sauerwein, H. (1967). Numerical calculations of multidimensional and unsteady flows by the method of characteristics. *Journal of Computational Physics*, *1*(3), 406–432. doi:10.1016/0021-9991(67)90048-4

Sikroria, T., Soria, J., & Ooi, A. (2018, December). A novel method for determination of convective velocity of coherent structures in high speed flows. In *Proc. 21st Australasian Fluid Mechanics Conf* (pp. 1-4). Academic Press.

Talay, T. (1975). *Introduction to the aerodynamics of flight* [NASA SP-367]. Langley Research Center. Retrieved from https://practicalaero.com/wp-content/uploads/2010/04/NASA-SP-367.pdf

Taylor, G. I., & Maccoll, J. W. (1933). The air pressure on a cone moving at high speeds.—II. *Proceedings of the Royal Society of London. Series A, Containing Papers of a Mathematical and Physical Character*, *139*(838), 298–311. doi:10.1098/rspa.1933.0018

Taylor, T. D. (1974). *Numerical Methods for Predicting Subsonic, Transonic and Supersonic Flow.* Advisory Group for Aerospace Research and Development Paris (France).

Von Karman, T. (2003). Compressibility effects in aerodynamics. *Journal of Spacecraft and Rockets*, *40*(6), 992–1011. doi:10.2514/2.7046

Wirz, H. J. (1975). *Progress in Numerical Fluid Dynamics.* Springer-Verlag. doi:10.1007/3-540-07408-2

Yao, Y. F., & Sandham, N. D. (2002). DNS of turbulent flow over a bump with shock/boundary-layer interactions. In W. Rodi & N. Fueyo (Eds.), *Engineering Turbulence Modelling and Experiments 5* (pp. 677–687). Elsevier. doi:10.1016/B978-008044114-6/50065-X

Zucker, R. D., & Biblarz, O. (2019). *Fundamentals of gas dynamics*. John Wiley & Sons.

Zucrow, M. J., & Hoffman, J. D. (1976). *Gas Dynamics* (Vols. 1-2). Academic Press.

Section 4
Hypersonic Aerodynamics

Chapter 12
Concise Study of Hypersonics and Its Flow Characteristics

Naren Shankar R.
Vel Tech Rangarajan Dr. Sagunthala R&D Institute of Science and Technology, India

Irish Angelin S.
Vel Tech Rangarajan Dr. Sagunthala R&D Institute of Science and Technology, India

Habib Gurbuz
https://orcid.org/0000-0001-5157-6227
Suleyman Demirel University, Turkey

ABSTRACT

Hypersonic vehicles attain a speed five times more high-speed than the speed of sound, which is beyond Mach 5 and nearly equal to 6174km/h. Hypersonic flow has certain characteristics, essentially a reedy shock layer, formation of entropy layer, viscous interactions, high temperature, and low-density flow. Only if these characteristics are subjected in the flow can the flow can be called hypersonic. While designing a hypersonic flight vehicle, the characteristics of the hypersonic atmosphere should be understood clearly. This comes as a major design constraint. A brief overview of the hypersonic vehicle design and high-speed flow characteristics will be presented in this chapter. A few characteristics of hypersonic flow and the ideal design conditions will also be reviewed.

INTRODUCTION

Later the discovery of an airplane, engineers and designers produced a new plane intended for a numerous usages. Airplanes have become an elemental part and parcel of everyday life expectancy. Aircraft had been used up often to deliver the shipment and also as a means of transport. Sooner or later, the speed of an aircraft vehicle has been increased. Aircraft or Airplanes are categorized primarily grounded totally upon by what method they can rapidly fly. These groups are termed the speed regimes of an aircraft. A hypersonic vehicle attains a speed five times more high-speed than the speed of sound, which is beyond

DOI: 10.4018/978-1-6684-4230-2.ch012

Mach 5 which is nearly equal to 6174km/h. Unlike other flows, in hypersonic flow merely considering the Mach number is no more the major parameter in lieu of studying the other flow characteristics for hypersonic flight; in fact, it is frequently considered as a secondary parameter of importance for studying the flow characteristics. The necessity of Mach number with some similar parameter may show a discrepancy with the shape of the body. Temperature which is being as another important property parameter for the dependence of hypersonic flow should also be considered as a serious factor in studying hypersonic flows.

Hypersonic flight vehicle design comes across by taking into account the characteristics of the outer layer of the atmosphere when considering the vehicle's design constraint. Certain flights are invented for rapid deceleration of the outer atmosphere throughout reentry. As a result, these aircraft have long flight duration at a higher angle of attack, hence a blunt nose configuration is commonly chosen for these hypersonic vehicles(Girka, 2015). On the other side, a sharp nose and small angle of attack vehicles are designed when focusing the mission on short flight duration with high deceleration closer to the earth's surface(Cheng, 1963). Blunt shapes are also designed aiming to provide favorable hypersonic aero-thermodynamic properties for low heat generation and hypersonic aerodynamic properties aimed at maneuverability and stability. A sharp-edged slender body which is used for cruisers or accelerators can minimize the drag thereby providing good lift for its air-breathing engines. Such a body is designed to handle increased heating loads.

Hypersonic flight has unique characteristics and flow behavior, several of which can be visible in each hypersonic flight. The compressed layer of air formed between the body surface and the freestream is called as shock layer. There will be a reduction in the gap between the shock formed and body surface as the Mach number increases. In hypersonic flow, this region becomes very thin and is called a thin shock layer. As the distance between the shock formed and the body surface is very small, the shock layer interacts with the boundary layer which creates the flow field analysis more complex.

Shockwave is a localized pressure wave that increases continuously which is also termed as an pressure gradient. This shock wave in turn increases the drag thereby decreasing the performance efficiency of the vehicle. Dissipation of energy, variations in velocity, pressure, temperature, and the flow turning are some of the factors allied with shock waves. On subjected to the instantaneous flow property (like pressure and temperature) change they take along the medium of propagation of shock waves are categorized as strong or weak. Due to the sudden variations in flow property, shockwaves are regarded as discontinuities with extremely localized irreversibility. The interaction taking place in the middle of the shock wave plus boundary layer leads to a degradation of the performance of the vehicle. On hypersonic vehicles, the interaction of shock waves with the boundary layer strongly affects the aerothermodynamic loads (surface heat transfer, pressure, and skin friction) (Kianvashrad & Knight, 2017). In case the shock formed found to be strong enough in order to detach the boundary layer then it leads to a complex phenomenon that is responsible for an increase found in the frictional loss.

In case the boundary layer is neglected in a flow, a complicated discontinuity in surface pressure will be created when a wave produced by shock intersects over a solid object which possibly leads to an unbalanced reflection of the shock. But in actuality, the fluid velocity drops to zero by the side of the walls to a laminar boundary layer or a turbulent boundary layer, and therefore interaction takes place in between the shock wave and the boundary layer. This is called shockwave boundary layer interaction (SWBLI).

The surface of the body and the slope impact majorly towards the interaction produced between the shockwave and boundary layer near the sharp edges, at the ramp, flare, or in front of any projected surfaces like fins. Though the SWBLI starts to initiate at the transonic regime and extends up to the hypersonic speed, the shock formation becomes stronger with higher magnitude and intensity when it reaches the hypersonic regime.

One special added principal feature of the bow shock waves is that every single streamline pattern passing from end to end of the shock encounters a distinct entropy increase which leads to the formation of vorticity. This entropy layer makes the situation challenging to study the boundary layer. This interface is named vorticity interaction. Boundary layer therefore is created nearby the body surface of the vehicle due to no-slip condition. No-slip condition is seen in a viscous flow over the solid surface where the relative velocity between the surface and the adjoining fluid elements will be zero. This boundary layer transfers the outer inviscid flow. This relation stuck between the in viscid flow as well as the viscous boundary layer is termed as viscous interaction. In hypersonic vehicles, this viscous interaction phenomenon performs a vital role in affecting the aerodynamic efficiency of the vehicle as well as affecting the heating behavior(Gülhan et al., 2021).

Three main theories are used for designing a hypersonic vehicle. One is Newtonian Theory is adopted to study about the pressure distribution above the body surface of the vehicle and to calculate the forces and moments at hypersonic conditions. This theory is used as one of the methods for designing the shape of a hypersonic vehicle. Hypersonic small disturbance theory (Brazeal et al., 2020) and thin shock theory (Roe, 2005) are the other theories used are which is further used to analyze the flow over the vehicle and the shock formed .

Taking into account all the above stated features, the object of this chapter is to give a brief description of the hypersonic vehicle design and the physical characteristics of hypersonic flow lay emphasis on the fluid dynamic property that is responsible for the Mach number increase. An enormous research work has put into effect to understand the flow behavior, shock waves, their interactions with the boundary layer, and the nature of shock wave boundary layer interactions which is one of the major inevitable problems faced in hypersonic flow. In recent years hypersonic has become an important research area.

TYPES OF HYPERSONIC VEHICLES

Hypersonic vehicles layout is an exceedingly hard procedure regarding numerous areas such as aerodynamics, aerothermodynamics, control systems, avionics, navigation systems, propulsion units, and their structural design. These disciplines are highly interlinked and frequently have an impact on one another due to the fact that they pursue opposing goals. As a result, synergistic interactions among vehicle subsystems and functions are expected to result in an optimal multidisciplinary vehicle layout with significant overall performance and economical improvements.

The design of hypersonic aircraft is regarded by strict requirements such as providing a sharp leading edge to increase the lift-to-drag ratios while undergoing extreme ambient gas temperatures that can reach 10,000 K. Designers believe that the rate of convective heat transfer between the hot gas and vehicle surface increases thereby reducing the size of these leading edges, causing the most common technical materials such as aluminum or titanium to deform, melt or even vaporize. Basic research into a collection of materials used called as Ultra-High Temperature Ceramics has sparked as a result of this knowledge (UHTCs). This material shows a unique combination of refractory and oxidation resistance properties that allow it to withstand the high thermal conditions encountered during hypersonic flight conditions.

Constructing a vehicle that can withstand hypersonic speeds and the high temperatures generated is not an easy deed. There are three main types of hypersonic vehicle design that has been designed so far. They are waveriders, winged bodies and reentry blunt bodies.

Figure 1.

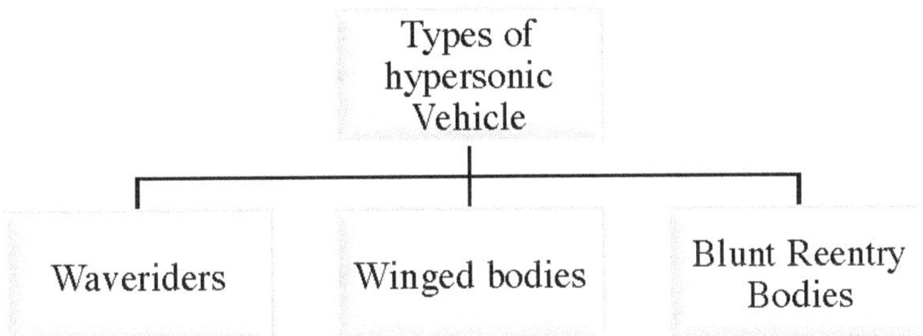

Waveriders

Waveriders is a type of a hypersonic vehicle that ought to have a wedge shaped fuselage which in particularly aimed to produce lift by means of piercing over the shock wave that are created by means of its own vehicle blows through the air at a high speed. This is especially important at high altitudes, when the air density is kept low, making it difficult for vehicles with conventional wing designs to generate lift. Hypersonic vehicle doesn't require large wings like any other commercial aircrafts. Hence, the vehicle is designed with an extra streamlined structure. This decreases the drag and lets in it to hold its speed over a far distance.

The waverider make use of the shockwave to generate a high pressure region along the windward side of the aircraft. In a winged aircraft, when travelling at hypersonic speed underwent many problems which increased the drag. The drag is linked with the wings and stabilizers. The high pressure formed below the wing surface interface highly with the low pressure region formed on the top surface of the wing. This causes a spillage of air from the lower surface to the top surface of the wing. This leakage consents for the high pressure flow with the drift to communicate with the lower pressure flow and produces a higher wave drag. The vehicle's lower surface is designed in such a way that the flow stream surface will interact with the shock wave formed and this intersection creates the leading edge of the vehicle. The vehicle's upper surface is typically designed considering it is a free-stream surface. This vehicle is useful when the mission necessitates a high L/D ratio, or lift to drag ratio, in order to achieve a longer range at higher Mach numbers.

By any means, significantly improving L/D would decrease the technological challenges of a hypersonic aircraft's long-range flight. Theoretically, "waverider" designs will be able to boost a hypersonic vehicle's L/D range to a higher value, say six. These employ a wedge shape that resembles the shock wave pattern of air flow all around the vehicle at a predetermined speed and altitude, enclosing a portion of the shock wave beneath the vehicles to provide additional lift.

Figure 2. Forebody of waverider

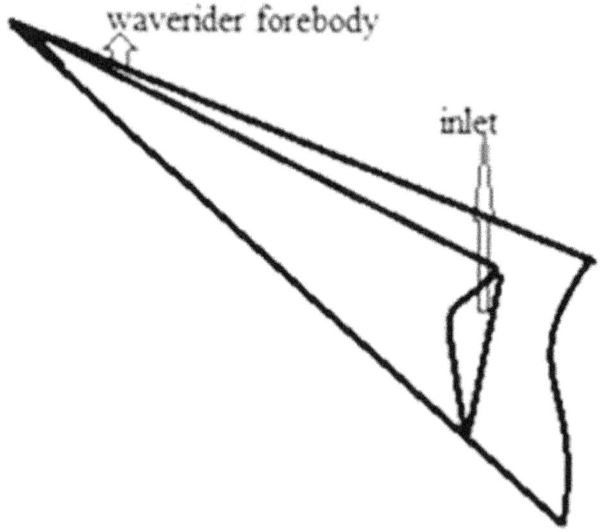

Winged Bodies

Adjusting the aerofoils at hypersonic speeds isn't a small mission; due to the fact the shock wave could have complicated interactions with the air flowing over the vehicle's surfaces, ensuing in a nonlinear behavior.

A fixed-wing aircraft is an aircraft that are capable to fly using their wings which are responsible for creating the lift. This lift is produced by means of the vehicle's forward motion airspeed as well as its wing shape. NASA's Space Shuttle and North American X-15 are the two main designs which were verified and tested using the fixed wing body. A most familiar hypersonic wing design is the Space Shuttle formally known as Space Transport System (STS). The Space/shuttle is made up of three primary components: a winged orbiter which is considered as the shuttle the aforementioned, two reusable solid rocket boosters, and one non-reusable external fuel tank. The orbiter resembles an aeroplane in shape. Wings on the orbiter provide lift. It has a double-delta wing layout to achieve the most efficient flight throughout hypersonic flight while also delivering a good lift-to-drag ratio upon landing.

The most familiar winged hypersonic vehicle is one known as a Space Shuttle. Space Shuttle exists as a fixed wing aircraft that can produce the lift as well as endorsed for the controlled descent in distinction from Space at Mach 25. Due to the high velocities that were attained by the shuttle the vehicle experienced tremendous amount of heat transfer owed to the dissociation of the air flow everywhere by creating a plasma surroundings.

The North American X-15 which is a type of hypersonic rocket powered vehicle also with a first manned vehicle though it had its first attempt to fly in 1960. The vehicle touched an extreme Mach number of 6.72. It made use of stubby wings for controls and stability. This type of winged bodies offered a problematic situation when attaining hyper velocities. Because of the attachment of wings and stabilizers the vehicle experienced a high drag. The high pressure interacts greatly with the low pressure region because of the spilling from the bottom of the wing to the top which in turn decreases the performance of the vehicle(Mason, 2018).

Reentry Blunt Body

Clearly the difficulty of reentry to Earth's atmosphere became a tremendous challenge for the spaceflight researchers before time, as they took into consideration how exceptional to overcome the heat created with the aid of using resistance as a spacecraft slowed withinside the atmosphere.

Nevertheless, not all reentry vehicles had been spacecraft. Some vehicles had no preference however to fight through the atmosphere. Vehicles along with atomic warheads launched higher than the ballistic missiles. They could take a flight up in a cannonball arc directly above the maximum of the atmosphere after which originated screaming again through it at possibly 20 times the speed of the sound, heating up extremely. The heat-sink's problems turned out to be even further clear when it was measured for a space vehicle.

Controlled atmospheric reentry is extraordinarily mature and properly understood nowadays it stays real that any earth orbital project for the payload should be recovered or interplanetary challenge centered for the planet with an atmosphere should cope with the difficulty of the way to rise up in addition to the way to get down. Atmospheric entry vehicles ought to offer managed dissipation of both kinetic energy and potential energy related to the reentry vehicle's velocity and altitude at the entry boundary. By managing entry it is assumed that each dynamic and thermal load is maintained in the desirable restrictions throughout the entry. This requires a cautiously planned flight trajectory and regular precision steerage to gain a favored result. Controller of vehicle in reaction to steerage system indicates control of lift and drag during the flight.

Reentry Bodies are vehicles with low ballistic coefficients on reentry into the environment so that you can bring them to subsonic speeds for landing. Reentry bodies were used for the reason that humans were first launched into space at the Mercury Program. They make use of a massive detached shockwave just like the winged vehicles so that you can deplete the heat load that takes place on reentry. There aren't any control mechanisms to be used on reentry but so maneuvering is non-existent. They are broadly used on launch vehicles nevertheless nowadays with the Orion Capsule at the Space Launch System and the Dragon Capsule from SpaceX.

HYPERSONIC FLIGHT VEHICLES BASED ON DESIGN CONSTRAINTS

Based on the design limits enforced from by the mission specifications, hypersonic flight vehicles are classified in four different types (NPTEL-Aerospace, n.d.).

Reentry Vehicles: Rocket propulsion systems are generally used to launch these types of vehicles. Control surfaces take control during re-entry into the atmosphere. Blunt nosed with a high angle of attack are the most required feature to design these types of vehicles. US space shuttles, Russian vehicle Buran, Japan vehicle Hope are some examples.

Vehicles for Cruise and Acceleration: These vehicles are often constructed for high heating loads and have an ablative cooling system. The key structures of these flights are streamlined narrow designs with a low angle of attack. Ramjet engine or scramjet engine which is a type of air breathing propulsion engine is mostly preferred for designing such vehicles. Sanger is one of the hypersonic vehicle, which is of the type of two stage to orbit (TSTO) or two stage rocket, has air breathing propulsion at its first stage and the second stage is propelled with the help of a rocket. Therefore the first stage of Sanger comes under a Cruise and Acceleration Vehicle (CAV) type planned for a separation at Mach 7.

Figure 3. Types of hypersonic vehicles based on design constraints

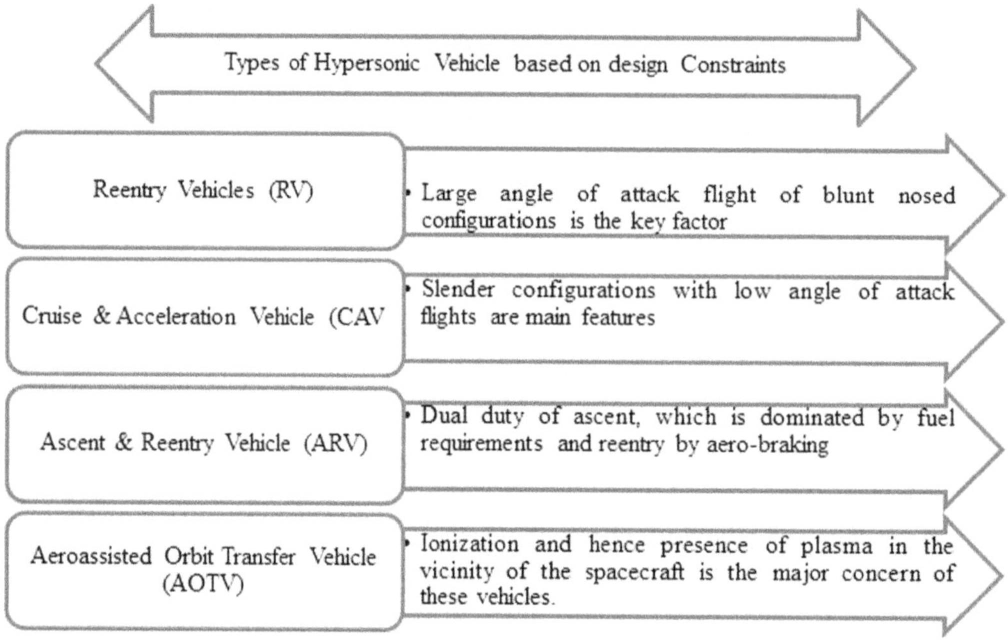

Ascent and Reentry Vehicles: Because of its twin duty of ascent, which is controlled by fuel supplies, and reentry, which is controlled by aero-braking, this type of spacecraft is guaranteed to have contradicting design requirements. In most cases, rocket or even air breathing propulsion engines are used to build these trips. This type of vehicle includes the US National Aerospace Plane, the Japanese Space Plane, and the British Aerospace HOTOL for Horizontal Takeoff and Landing.

Aero-Assisted Orbit Transfer Vehicle: Aero-Assisted Orbit Transfer Vehicle, AOTV make use of the atmosphere in the interest of provide brake at the time of returning from a geosynchronous orbit or a planetary mission. The minimum altitude for air brakes is generally 255,000 feet at the equator. The time for braking is usually 480 seconds from 400,000 feet to 255,000 feet and reversing about 8 minutes. The problematic of this type was designing a control system which could handle density difference. This is another type of hypersonic vehicles. Ionizations and there after a presence of plasma in the environs of the spacecraft is the most important concern while designing these types of vehicles.

Figure 4 makes available the trajectories for velocity-altitude designed aimed at various flight at hypersonic speeds structures that explain the velocity ranges. When interpreting the curves, keep in mind that the flight speeds U stated in kft/s are statistically corresponding to the flight Mach number M at 35 km height above sea level in the stratosphere, where the speed of sound is almost equal to 1 kft/s. Figure showed is mentioned all the way through to remain the text in piecing together with various theories. For the time being, it will enough to note that the current state-of-the-art in hypersonic technology prevents hypersonic systems from attempting to match high Mach numbers at low altitudes in continuous-level flight, where the heat wall prevents further growth. High enthalpies, high Mach numbers, and higher Reynolds numbers are all part of this forbidden region, which is carefully avoided by all current aircraft paths of man-made objects.

Figure 4. Path of different hypersonic flight system (Urzay, 2020)

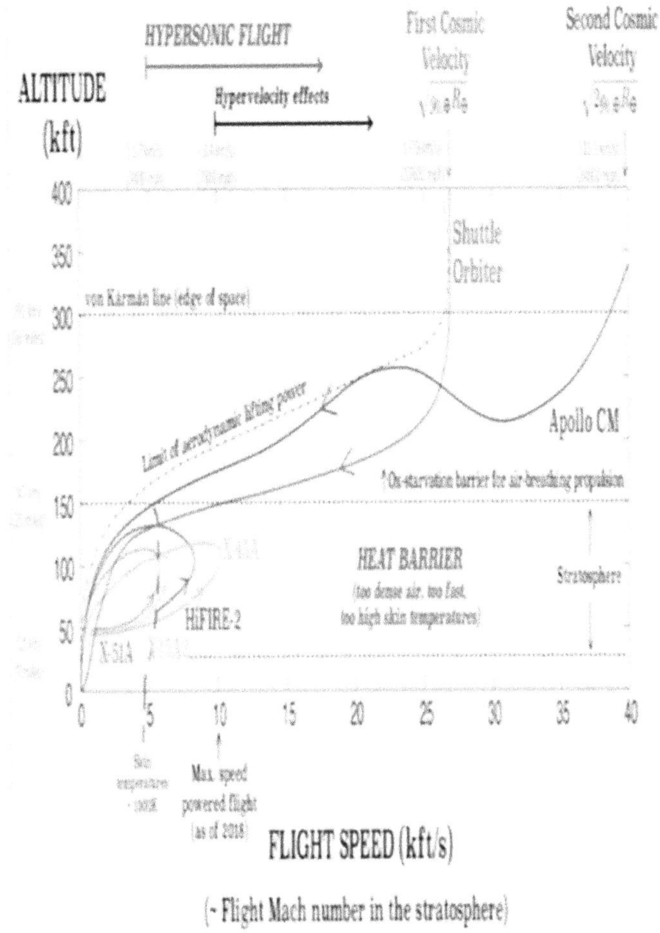

In hypersonic flows the kinetic energy is transformed into thermal energy which takes place all over the place shocks in the downstream. In that region, the flow is compressed and decelerated, as seen in the diagram. When compared to the freestream kinetic energy and post-shock static enthalpy that occur at supersonic speeds, the freestream static enthalpy and post-shock kinetic energy are insignificant at hypersonic speeds. The freestream's stagnation enthalpy is usually of kinetic origin, with the equation $h_{0\infty} \sim V^{2}_{\infty}/2$. As like the post shock shown in Figure, most part of the freestream kinetic energy develops and converted into enthalpy that is local in the areas such that the gas is intensely reducing its acceleration in a negative way. The gas flowing downstream of the bow shock originating from the vehicle's nose is extremely hot, and as the Mach number approaches 20 and above, it may release temperature to the fuselage.

The above observations regarding the conversion of kinetic energy to thermal energy can also be applicable qualitatively when considering the boundary layers that are hypersonic, even though the energy mechanism conversion is basically different. The law of "adiabatic wall temperature" describes how the influence of friction along the wall, which is induced by essential passivity with the non-slip boundary condition, can cause an increase in temperature up to some desired levels that can be quite considerable (Urzay, 2020). This can be clearly understood from the Figure 5.

Boundary layers that are hypersonic in nature are likely to remain thick and the temperature seems to be high when compared to their low-speed flows. Due to the increasing temperatures, reactions will happen chemically in boundary layer which causes the ionization and separation of the air. For space related vehicles, a tool was used to estimate the aerothermodynamic heating on the vehicles surface which is used as a thermal protection system known as boundary layer transition tool (Berry et al., 2006).

Figure 5. Flows that are hypersonic forebody that is blunt and flat.

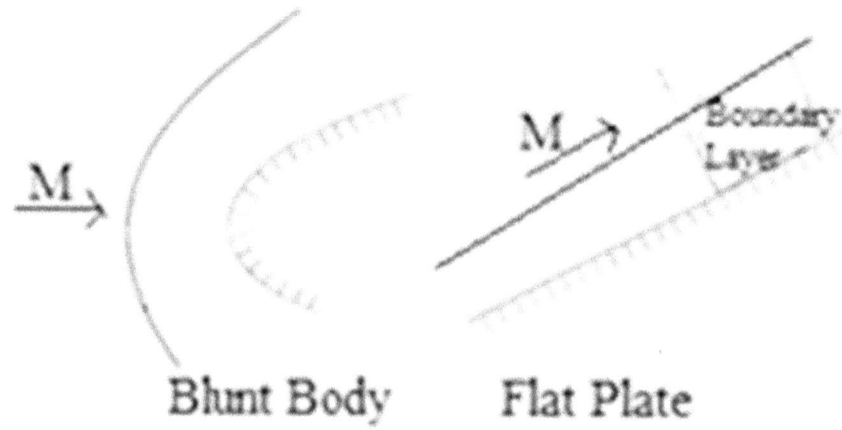

HYPERSONIC PHYSICS

Atoms or molecules often experience quantum transitions as a result of radiative energy transfer. In hypersonic flow, gases with high enthalpy are subject to this physical phenomenon (Shang & Yan, 2020). Shock waves are created when vehicles travel faster than the speed of sound, causing temperature, air pressure, and density to bounce quickly. At hypersonic speeds, however, the jumps are frequently so large that they have significant thermodynamic implications.

The stagnation enthalpy, which represents the total thermal energy approaching air molecules and atoms if they slowed to a halt using the method of a steady, adiabatic process, is a useful metric for describing the influence of a shock wave on a flow. Stagnation enthalpy increases with a vehicle's Mach number and is conserved throughout a shock assuming a stable, adiabatic flow state. It depicts the fact that the momentum of airflow upstream of a shock wave generates thermal energy within the shock layer, the compressed air layer between the shock wave and the vehicle surface. As soon as the stagnation enthalpy exceeds about 5 MJ/kg, which corresponds to about Mach 10 for an a high speed vehicle hovering at 30 km altitude, the amount of energy set down within the shock layer is of such high quality that the internal energy modes of the air molecules—electronic, vibrational, and so on—equilibrate with the newly active surroundings at distinct time scales, which can be similar to the time balances of the flow. Consequently, the air molecules will no longer be defined by a single temperature.

HYPERSONIC CHARACTERISTICS

Mach number greater than 5 is usually considered as a hypersonic flow. However it is not always to be so when focusing on the hypersonic aerodynamics. Understanding the aerodynamics of hypersonic flow is unlike as of a supersonic flow. As how the aerodynamic characteristics of speed regime of high subsonic range (say M=0.8 to 0.99) and low supersonic range (say M=1 to 1.2) is difficult to study, even in hypersonic range, the flow does not "instantly change its characteristics. Hypersonic flow, on the other hand, is defined as the regime in which a number of physical flow phenomena become more prominent as the Mach number increases. One or more of these events may become substantial at Mach 3 in some cases, while they may not be convincing until Mach 7 or above in others.

The atmosphere thickens as a well-known NASA space shuttle reenters, and the aircraft's control surfaces prove to be workable. At a higher angle of attack, the winged orbiter re-enters the atmosphere (say about 30 degrees). Space shuttle's body is protected by the heat resistant slating installed underneath such a high angle of attack. This slating directs most of the aerodynamic heat to the underneath of the vehicle.

The characteristics of flow which distinguish supersonic flow from a hypersonic flow field are listed below.

1. The greater temperatures predominant in the gases post-shock,
2. A small dwelling periods of the gaseous molecules all over flight system,
3. An increased proportions of heating of the fuselage aerodynamically,
4. Linearity does not appear in the inviscid streamlined body boundary.

The re-entry of spacecraft into planetary atmospheres, as depicted in Figure 6 (Urzay, 2020), is a notable application of hypersonics. A spacecraft attempting to settle on Earth from LEO or an interplanetary trajectory must completely eliminate the kinetic energy of its speed relative to the Earth's surface. Deboosting operations using retro-rockets to lower the orbital altitude and set the spaceship on a trajectory away from the Earth's surface will convert a small fraction of this energy into potential energy in space. Ionization, and thus plasma in the vicinity of the spacecraft, is the primary concern of these vehicles.

Figure 6. Atmospheric reentry

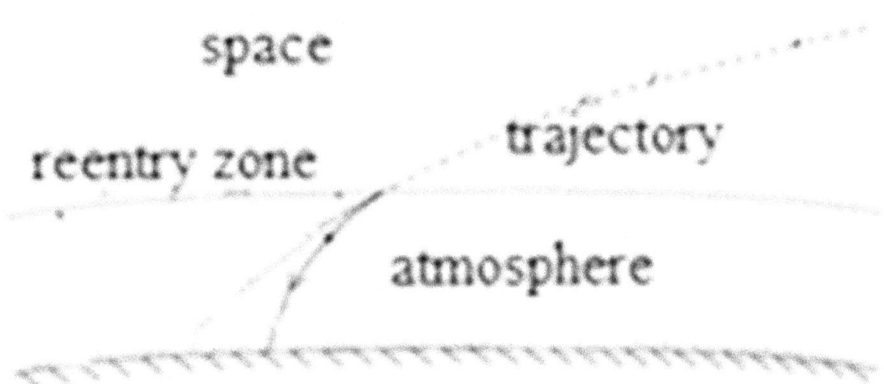

HYPERSONIC AERODYNAMICS FLOW PREDICTION TECHNIQUES

Hypersonic aerodynamics may vary under different conditions like viscous flow, inviscid flow and high-temperature flow.

Figure 7. Variations of hypersonic aerodynamics based on flow

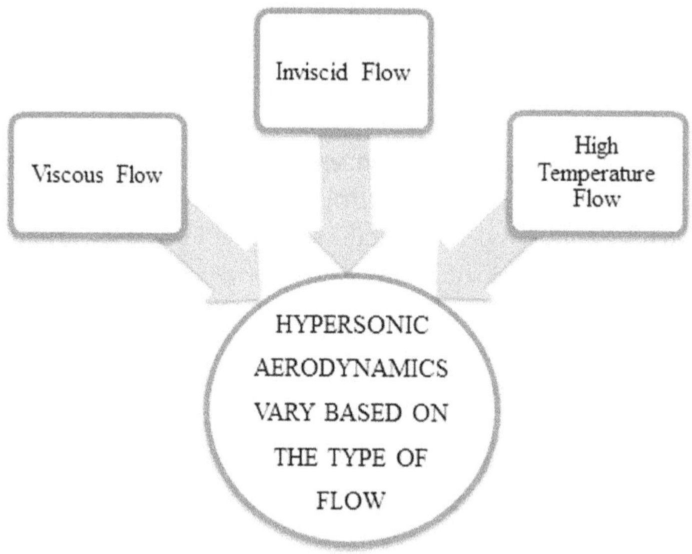

Hypersonic flows are inherently nonlinear. For the evaluation of flow variables, unlike the supersonic case, tending the Mach number to infinity cannot lead to any linear theory for small deviations in the hypersonic region. For predicting the flow properties in hypersonic flow mainly for pressure predictions, a local inclination method is used where it is not necessary to solve for the entire flow field.

Figure 8. Flow prediction techniques

Flow Prediction Techniques				
Newtonian Technique	Modified Newtonian Technique	Tangent Wedge Method	Tangent Core Method	Shock & Expansion Method

When the freestream Mach number increases the angle between the vehicle's surface and the shock wave formed tends to decrease. This variation can be seen from Figure 9.

Figure 9. Mach number versus angle

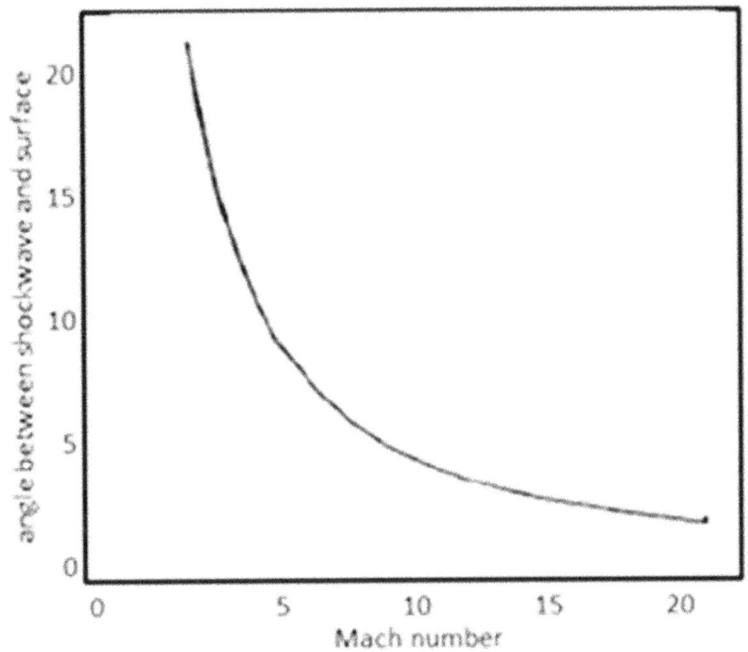

FORCES

Drag, the resistance a fluid causes to whatever is passing through it, is a terrifying enemy for hypersonic motor designers. The drag on a flying object increases in proportion to the square of its velocity, making it particularly crippling at hypersonic speeds. A glider flying at Mach 5 is subjected to twenty-five times the drag pressure of one flying at Mach 1, while one flying at Mach 20 faces four hundred times the drag of one flying at Mach 1.

The energy drain from a vehicle is even more acute since it pushes the molecules of air ahead and aside: it will increase as the velocity cube increases. As a result, a glider flying at Mach 5 will waste energy 125 times faster than one flying at Mach 1; one flying at Mach 20 will lose energy 8,000 times faster. The kinetic energy that flows from the glider to the surrounding air is also converted to thermal energy and shock waves, which is problematic. Some of that energy is sent back to the vehicle in the form of heat: leading edges of boost-go with flow guns flying at Mach 10 or higher can reach temperatures of over 2,000 K for extended periods of time. One of the most difficult problems facing engineers is protecting a vehicle from this extreme heat.

A hypersonic vehicle, like any other flight, needs to produce lift—a lift force perpendicular to its direction of motion—to stay aloft and turn. (A flight turns through by banking or making a horizontal factor of the lift force in any other way.) The lift, on the other hand, is also proportional to the square of the velocity. Furthermore, the aerodynamic approaches that generate lift inevitably cause drag. The lift-to-drag ratio, L/D, is a vital indicator of a glider's performance because it compares the lift force, L, to the drag force, D.

The L/D values that can be achieved with hypersonic motors are far lower than for regular aircraft. The ratio for a subsonic vehicle could be 15 or higher. Despite years of research and development, U.S.

hypersonic guns tested in the last decade appear to have L/D values substantially lower than three. Low lift and severe drag limit the speed and range of hypersonic flight. This also limits agility and raises surface temperatures.

As if that weren't enough, at hypersonic speeds, the physics and chemistry of air travelling beyond an object become noticeably remarkable. The surrounding air dissociates when heated to hundreds of degrees, converting molecular oxygen to free atoms that might ionize and wash away the vehicle's surface. As the missile continues to burn, the heat creates a bright infrared signal that satellites will be able to detect.

Another important aspect of hypersonic flight is that changes in the ratio of specific heats can have a significant impact on the pressure distribution over an object or its control surfaces. This is because, which is also the isentropic exponent, affects the flow's rate of expansion or compression at the same time. This can manifest itself in the setting of control surface angles, or possibly in the trim angle of attack for an object, which may be larger than that predicted for the perfect gas value of two to four degrees. During the first Shuttle reentry, this difference was readily visible. The pressure distribution along the vehicle's windward side was calculated for both a non-reactive, ideal gas boundary layer and a chemically reacting boundary layer.

The shock wave tends to move closer to the surface as the Mach number increases, reducing the angle between the shockwave and the surface.

At high altitudes, hypersonic flow can be achieved at comparatively low ambient pressures. This suggests that the Reynolds numbers are low and the thickness of the boundary layer is rather substantial.

NEWTONIAN THEORY

Newton viewed fluid movement as a homogeneous rectilinear stream of components, much like a haze of pellets produced by a powerful blast. Newton theorized that when particles collide with a surface that is inclined at an angle to the stream, the particles' standard momentum would be transferred to the body surface while their tangential momentum would be preserved. As a result, after colliding with the surface, the elements would immediately move alongside it(Shang & Yan, 2020).

The nearest instance for the Newton's model in the area of fluid mechanics is the think shock layers formed around the hypersonic bodies. The pressure distribution over the body surface of any hypersonic vehicle is typically approximated or estimated using Newtonian theory (John D Anderson, 1989).

Despite the fact that the information of the flow around a surface in hypersonic flow is difficult to analyse due to the difficulty of the phenomena involved, the pressure distribution around a surface located in a hypersonic flow can be predicted fairly accurately using an approximate method described below. The Newtonian model is named after the flow model that is essentially identical to the one that was mistakenly advocated by Newton for the computation of forces on bodies in incompressible flow.

To measure the Newtonian model some calculations are given below,

The mass flow happening on the body surface is:

$$\rho_{¥} (A \sin\theta) V_{¥} \qquad [1]$$

The change in the normal velocity component is

$V_{¥}\sin\theta$ [2]

The time rate of change of momentum = mass flow x change in normal component of V [3]

$= \rho_{¥}(A\sin\theta)V_{¥} \times V_{¥}\sin\theta$ [4]

$= \rho_{¥}V_{¥}^{2}A\sin^{2}\theta$ [5]

Therefore, the surface forces is given by

$N=\rho_{¥}V_{¥}^{2}A\sin^{2}\theta$ [6]

The normal force per unit area is given

$=\rho_{¥}V_{¥}^{2}A\sin^{2}\theta/A$ [7]

$N/A=\rho_{¥}V_{¥}^{2}\sin^{2}\theta,$ [8]

Therefore,

$p-p_{¥}=N/A=\rho_{¥}V_{¥}^{2}\sin^{2}\theta$ [9]

Where

$C_{p} = p-p_{¥}/q_{¥}=2\sin^{2}\theta$ - [10]

The equation [10] is called as a Newton's sine-squared law.

To put it another way, the pressure coefficient is proportional to the sine of the square of the angle formed by the tangent to the body surface and the freestream flow direction(Katyal et al., 2013).

If the calculation is expressed in terms of the angle in between a normal to the body surface and to the freestream directionϕ, then the, $C_{p}=2\cos^{2}\phi$

As the angle varies the pressure coefficient also vary, say when $\theta = \pi/2$ or 90°andϕ=0°, then at the point of stagnation, Cp tends to 2.We know that, for incompressible flow pressure coefficient becomes equal to 1 that is Cp = 1 at the stagnation point. Without doubt, the stagnation coefficient of pressure Cp rises without any interruption starting from 1.0 at freestream Mach number $M_{¥} = 0$ until it reaches to 1.28 and for freestream Mach Number $M_{¥} = 1.0$ to 1.86 and at free stream Mach Number $M_{¥}$ tends to $_{¥}$ when $\gamma = 1.4$ As the coefficient of pressure reaches 2 that is, Cp®2, freestream Mach Number reaches ¥ which can be seen clearly from the 1-D momentum equation.

$p_{¥}+\rho_{¥}V_{¥}^{2}=p_{2}+\rho_{2}V_{2}^{2}$ [11]

where

$\rho_{¥}V_{¥}^{2}>>\rho_{2}V_{2}^{2}$ [12]

then,

$$p_2\text{-}p_{\text{¥}}=\rho_{\text{¥}}V_{\text{¥}}^{2} \qquad [13]$$

or

$$C_p=p-p_{\text{¥}}/q_{\text{¥}}=2 \qquad [14]$$

For a huge but fixed Mach number, the coefficient of pressure Cp at the stagnation point is very much less than two. For blunt bodies, the coefficient of pressure at the downstream of the stagnation point is given by,

$$C_p = C_{p,max}\sin^2\theta \qquad [15]$$

The equation [15] is considered as a simple form of Modified Newtonian law.
In addition, the Newtonian theory is more accurate at

(1) Larger free stream Mach Number M¥ and inclination angle
(2) 3-Dimensional bodies

Let us now see about the drag and lift of wings of a vehicle at hypersonic level speed by considering the Newtonian final output for a plate that is flat at some angle of attack. Variations of lift coefficient C_L and drag coefficient C_D versus angle of attack α.:

Case (i) At Subsonic condition:

Coefficient of lift is directly proportional to the angle of attack C_Lµα (before stall)and the coefficient of drag is directly proportional to the square of the angle of attack, $C_D=C_{D0}+kC^2$ that is C_Dµα^2

Case (ii) At Supersonic condition:

The relation between the lift coefficient and the angle of attack is given by,

$$C_L = \frac{4\propto}{\left(M_\infty^{\;2}-1\right)^{1/2}} \qquad [16]$$

And the coefficient of drag is given by,

$$C_D = \frac{4\propto}{\left(M_\infty^{\;2}-1\right)^{1/2}} \qquad [17]$$

Case (iii) At Hypersonic condition; according to the Newtonian theory, on the lower surface the coefficient of pressure equation is given by $C_{p,l}=2\sin^2\alpha$. According to Newtonian physics, the upper surface of the flat plate has no direct influence on the free stream flow elements. The upper surface of the flow is considered to be in the shadow. As a result, only the free stream pressure operates on the upper surface of the flat plate, which is compatible with the fundamental concept of Newtonian flow, and because the coefficient of pressure at the upper surface Cp,u = 0, therefore

$$C_n = 1/c\int_0^c (C_{p,l} - C_{p,u})dx \quad [18]$$

$$= (1/c)\ 2\ \sin^2\alpha c \quad [19]$$

$$=2\sin^2\alpha \quad [20]$$

$$C_l = C_n\cos\alpha = 2\sin^2\alpha\cos\alpha \quad [21]$$

$$C_d = C_n\sin\alpha = 2\sin^3\alpha \quad [22]$$

Lift-Drag ratio (L/D) = $\cot\alpha$. [23]

Hypersonic flows are generally governed by a non-linear velocity potential equation and hypersonic flow cannot be defined by a linear theory same like a transonic flow. Hence, both of these flow types are fundamentally nonlinear systems, even at low angle of attack α.

Let's start with a flow across a flat surface that's angled at a given angle to a hypersonic flow stream. For a moment let us consider merely the flow above the upstream face of the flat surface. Figure 10 demonstrates a flow over the flat body inclined at some angle. The upper surface of the flat plate, in term of Newtonian theory, has on no account had a direct influence over the free stream flow particles. The upper surface of the flow is considered to be in the shadow. Only the free stream pressure acts on the upper surface of the flat plate from now on, as dictated by the fundamental concept of Newtonian flow, and as we have coefficient of pressure at the upper surface Cp,u = 0, therefore

Figure 10. Hypersonic flows over a plane surface

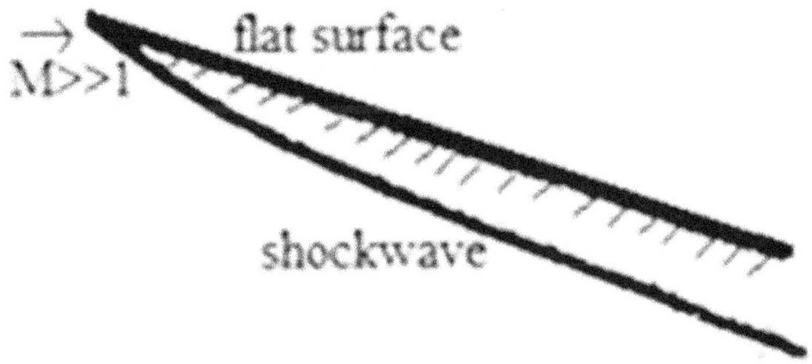

In a view of the fact that the shock waves is to be found so adjacent by to the surface in hypersonic flow, the flow for all intentions and purposes will continue to be unpretentious by the flat surface up to the flow touches the surface, that is, till it "hits" the surface, next to that point it will straight away turn out to be parallel flow to the surface. Henceforth, the flow over the upstream of an aircraft surface at hypersonic speeds bears a resemblance to that shown in the Figure 11.

Figure 11. Hypersonic flows on a flat surface using a Newtonian model

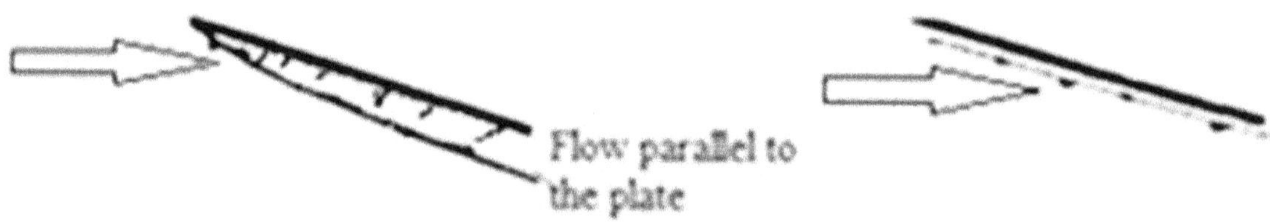

MODIFIED NEWTONIAN THEORY

At this instant, let us consider a hypersonic flow over an arbitrary shape over a symmetrical body as shown in the Figure 12.

Figure 12. Form of body being considered.

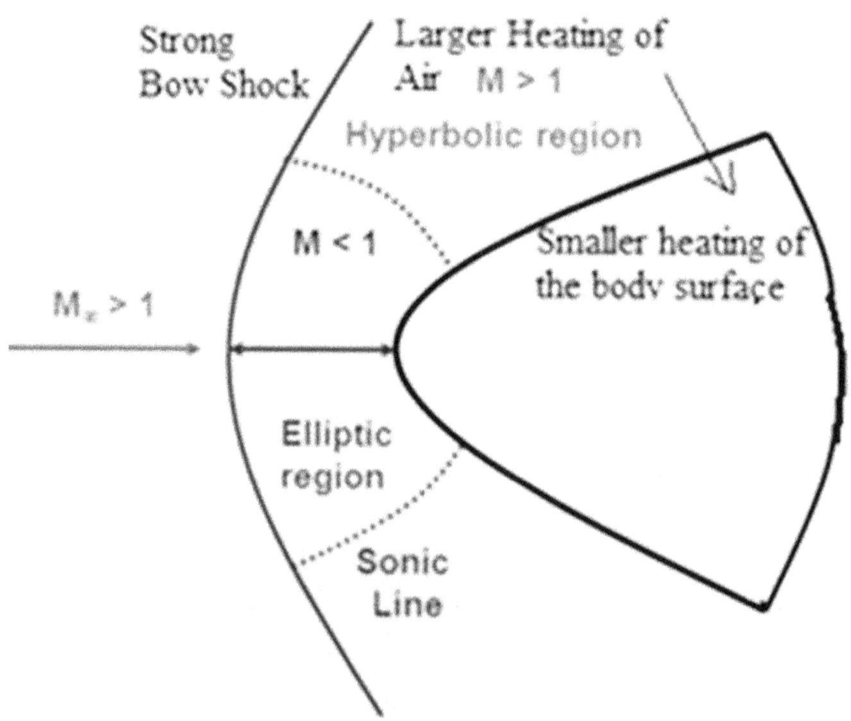

The pressure at any point on the surface is given by:

$$p_S - P_\infty = \rho_\infty V_\infty^2 sin^2\theta \tag{24}$$

On the stagnation point wherever θ = 90° and wherever the sin θ = 1, the pressure, p_S, is assumed by:

$$p_S - P_\infty = \rho_\infty V_\infty^2 \tag{25}$$

Henceforth:

$$\frac{p_S - P_\infty}{\frac{1}{2}\rho_\infty V_\infty^2} = 2 \tag{26}$$

Therefore, the coefficient of pressure at the stagnation point is specified by:

$$C_{pS} = 2 \tag{27}$$

As of such relatives it surveys that the pressure distribution round the external surface can be mentioned as:

$$\frac{C_p}{C_{pS}} = sin^2\theta \tag{28}$$

It can also be written as:

$$\frac{p - p_\infty}{p_s - p_\infty} = sin^2\theta \tag{29}$$

Newtonian theory, on the other hand, isn't applicable at the point of stagnation. However, the shock wave in this area is, in effect, a regular shock wave. As a result, normal shock relations may be used to locate the pressure on the surface at the stagnation point, and the Newtonian relation can then be utilized to govern pressure distribution throughout the body.

The earlier mentioned equation can also be written as:

$$\frac{C_p}{C_{pSN}} = sin^2\theta \tag{30}$$

In which C_{pSN} denotes the coefficient of pressure at the stagnation point as per assumed in the relations for normal shock. This is, fundamentally, the Newtonian equation after modification.

Let's recall that the normal shock relations give:

$$\frac{p_S}{P_\infty} = \frac{\left[\frac{\gamma+1}{2}M_\infty^2\right]^{\frac{\gamma}{\gamma-1}}}{\left[\frac{2\gamma}{\gamma+1}M_\infty^2 - \frac{\gamma-1}{\gamma+1}\right]^{\frac{1}{\gamma-1}}} \quad [31]$$

It is also noted that

$$C_p = \frac{P-P_\infty}{\frac{1}{2}\rho_\infty V_\infty^2} = \frac{\frac{P}{P_\infty}-1}{\frac{\gamma M_\infty^2}{2}} \quad [32]$$

Combining the above equations then gives:

$$C_{pSN} = \left[\frac{\left[\frac{\gamma+1}{2}M_\infty^2\right]^{\frac{\gamma}{\gamma-1}}}{\left[\frac{2\gamma}{\gamma+1}M_\infty^2 - \frac{\gamma-1}{\gamma+1}\right]^{\frac{\gamma}{\gamma-1}}} - 1\right] / \frac{\gamma M_\infty^2}{2} \quad [33]$$

If M_∞ is extremely large the above mentioned equation tends to: Miscellaneous

$$C_{pSN} = \left[\frac{\left[\frac{\gamma+1}{2}\right]^{\frac{\gamma}{\gamma-1}}}{\left[\frac{2\gamma}{\gamma+1}\right]^{\frac{\gamma}{\gamma-1}}\left[\frac{\gamma}{2}\right]}\right] \quad [34]$$

When $\gamma = 1.4$ the above equation contributes the restrictive significance of C_{pSN} for greater values of $M \rightarrow \infty$ as 1.839. Considering a perfect gas and also a greater freestream Mach number M_∞, the modified Newtonian theory reaches the C_p value that is nearly equal to $1.839\sin^2\theta$. As the Mach number is very high, the temperature after the typical shock wave in the stagnation point area grows so high that high-temperature gas special effects become more prominent, causing the value of to be distributed as C_{pSN}.

FEATURES OF HIGH-TEMPERATURE FLOWS

1. The thermodynamic properties (e, h, p, T, p, s, etc.) are entirely dissimilar from other supersonic conditions.
2. The transport properties say like μ and k vary entirely. In addition, the other transport mechanism of diffusion turn to be more significant, with the related diffusion coefficients, Di,j.
3. Great heat transfer rates are typically a main feature of several high-temperature applications.
4. The ratio of specific heats, $\gamma = C_p / C_v$, is just a variable constants. Indeed, intended to the simulations of high-temperature flows, γ can lose the significance it devours in place of the standard constant γ flows.
5. Instead of having theoretical forms and solutions, high-temperature gas flow investigations require some kind of numerical analysis.
6. If the temperature is high enough to commence ionization, the gas partially transforms into ionizing plasma, which has a low electrical conductivity. As a result, if the flow stream is within the presence of an external electric or magnetic field, electromagnetic body forces act on the fluid constituents at that time. Magneto hydrodynamics is a term used to describe the compass of a location (MHD).
7. If the air temperature is great at a sufficient amount, there will be non-adiabatic after effects owing to the radiation of the gas.

The composition of the atmosphere after a normal shock wave in front of a hypersonic aircraft will be very different from the composition of the atmosphere before the shock wave. Diatomic molecules are tremblingly stimulated and dissociated to certain level, resulting in a partially ionized atom and the rest of the molecule. As a result, the notion that molecules that survive collisions thermally behave like perfect gases is no longer true. The vehicle boundary layer has enough viscous dissipation to alter the flow chemistry, which might result in a chemically reactive boundary layer. Since the specific heat is no longer the same due to vibrational excitations and chemical reaction, the ratio of γ = cp/cv is also no longer constant and is temperature dependent. For air it starts at around 800K.

ELEMENTS OF A HYPERSONIC FLOW

A thin shock layer is one of the main characteristics of hypersonic flow. A practical impact of a thin shock layer is that it causes a critical interaction between the inviscid flow behind the shock and the viscous boundary layer over the vehicle's surface on a regular basis. Therefore hypersonic main considerations related to boundary layer are mentioned below(Fletcher, 2004):

1. A hypersonic vehicle in general fly at high altitudes wherever density ρ, and Reynolds's number Re is low as a result the boundary layers appear to be thick.
2. Boundary layer thickness on a streamlined body is nearly proportional to the freestream Mach Number M_∞ therefore; the high Mach numbers as well contribute to the boundary layer getting thicker.

Figure 13. Elements of hypersonic flow

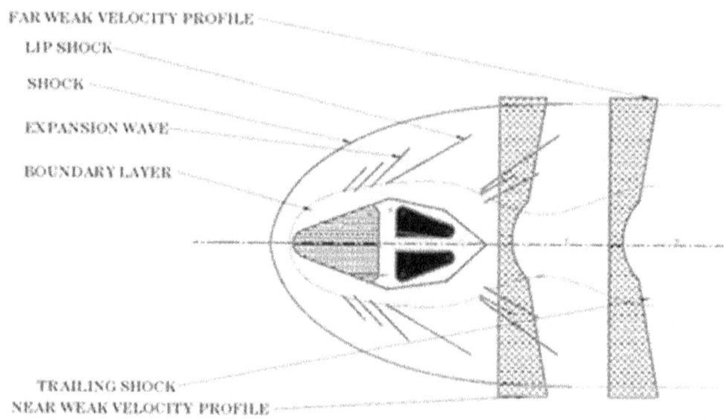

In many cases, the boundary layer thickness is in the same size such as the shock layer thickness when travelling at hypersonic speeds. Beneath such conditions, the shock layer remains to be fully viscous, and the shape of the shockwave and the surface pressure distribution have an effect on such cases of viscous effects. This is referred to as a viscous interaction phenomenon, in which the viscous flow prominently disrupts the exterior inviscid flow and vice versa.

Figure 14. Various flow phenomena

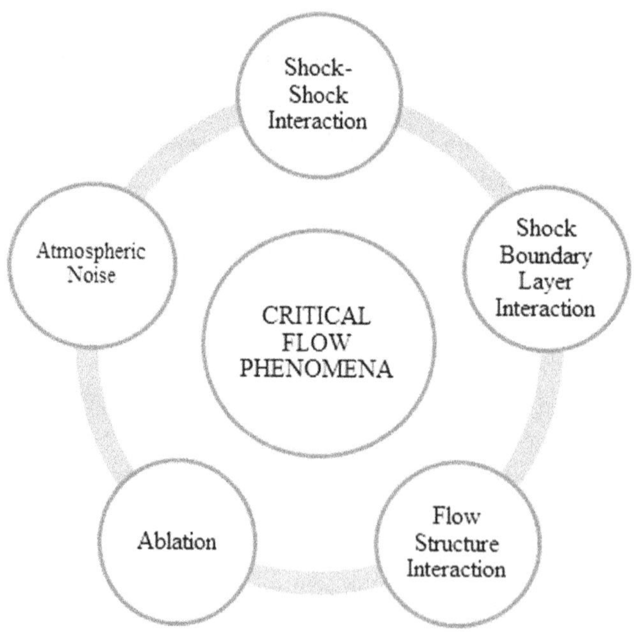

Furthermore due to the influencing of the aerodynamic forces, such as high pressure, aerodynamic heating at the leading edge also increases.

The other principal feature of hypersonic flow is high temperature in the shock layer (Viviani et al., 2012), in addition to the large aerodynamic heating of the body.

At high temperature:

1. Molecule dissociation takes place.
2. An enormous quantity of heat transfer to the surface of a high speed vehicle takes place where, the thermal radiation = σT4, therefore on high temperature, radiation is a governing element.

Hypersonic flows are distinguished from other appropriate supersonic flows by the features of thin shock layer viscous interaction, as well as high temperature, chemical responding, and radiation impacts(Katyal et al., 2013).

Even though the complete shock layer all over a hypersonic vehicle is decided by excessive pressure and temperature, a huge part of the viscous drag and heating the aircraft vehicle studies may be trailed again to a thin area close to the surface called the boundary layer. As it travels through the thin layer, which may just be a few millimetres thick, the air can drop from thousands of meters per second to zero relative to the vehicle. Normally, the flow within the boundary layer that surrounds the vehicle's nose is laminar, which means it is structured into streamlines that run virtually equivalent to the body surface. The boundary layer may have frequently evolved to a chaotic, turbulent state by the time the floe reaches the vehicle's rear side. Understanding how and where that shift occurs has long been a concern in hypersonic aerodynamics.

AERODYNAMIC HEATING IN HYPERSONIC VEHICLE

Aerodynamic heating is a key issue in hypersonic flow research due to its strong relevance to the safety of ultra-high-speed flight. (John D Anderson, 1989)Accurate knowledge about where and why, and with what strength, strong heating Under real flight conditions, laminar-to-turbulence transition is one of the most important sources bringing about uncertain aerodynamic heating that might adversely impact a vehicle's thermal protection system (TPS) peaks occur at the wall is of crucial importance to the performance and safety of hypersonic vehicles. It is generally acknowledged that enhanced heat transfer is related to the evolution of flow disturbances and onset of turbulence, which in turn modifies the skin friction.

In order to protect spacecraft entering Earth's atmosphere from aerodynamic heating, they require thermal protection systems. Thermal protection systems (TPS) are external materials applied to vehicles to keep the external structure of a hypersonic vehicle at an acceptable temperature, particularly during the reentry phase. Material selection for a TPS is based on its resistance to high temperatures and light weight. Through a combination of thermal insulation and surface materials that are resistant to extreme temperatures, the TPS system used by a hypersonic vehicle inhibits heat transmission to the inner surface of the vehicle.

When designing a hypersonic vehicle, say a reenter vehicle that enters the Earth's atmosphere, one of the main challenges is precisely defining the convective heat flux which is also called as aerodynamic heating that is generated at the surface of the vehicle. Based on the numerical integration of the Navier-Stokes equations, a detailed analysis of flow and heat transfer parameters of a reentry vehicle under hypersonic flight conditions may be carried out in dense atmospheres where continuity of the gas medium is assumed.

Challenges of Hypersonic Vehicle

To design and construct a hypersonic vehicle, there are many challenges.

There are few technical challenges faced by engineers in designing hypersonic vehicles are listed below (David Hunn, 2019) .

- **Heating** – Hypersonic speeds create high levels of heat caused by friction and air resistance, which must be controlled by using proper heat shielding systems and thermal protection. These extreme conditions also require equipment that is safe and secure.
- **Material Selection** – In order to manage extreme heat and speed we need to devise and deploy advanced materials and composites that are resistant to extreme conditions which is one of the challenges in hypersonic vehicle when coming into material selection.
- **Operation Maneuverability** – It must be capable of overcoming a variety of defenses since hypersonic systems must operate in contested environments. Managing hypersonic flight at high speeds requires extensive calculations and planning.
- **Communication System** – During hypersonic flight, basic operations, such as communications, become increasingly important. Communication systems and sensor systems that can operate in these high-speed environments are vital to keep personnel connected to operators and decision-makers.
- **Propulsion System** - At hypersonic speeds, keeping an aircraft or another air vehicle in flight is extremely energy-intensive(Wall, 1964). Jet engines aren't typically capable of handling this. Launching a rocket requires an enormous amount of fuel. The use of ramjets and other propulsion systems can achieve this, but the propulsion system must be accelerated to a certain velocity before the vehicle is operated.

When flying at Mach 5 at a 20 km altitude, an aircraft's surface temperature can exceed 2200 K. With increasing Mach values, the temperature of air travelling over an aeroplane rises. Knowing where and when those high temperatures occur is critical to an aircraft's performance and safety.

To transform mechanical energy into thermal energy in a hypersonic flow, three techniques are commonly used. The first approach is viscous kinetic energy dissipation caused by shear strains close to a surface. The second is analogous to viscous dissipation, but it is caused by normal stresses acting on compressed air. The work done by pressure changes acting on compressed air is the third way.

The two viscous energy transformation mechanisms in the preceding technique are unidirectional, converting mechanical energy to thermal energy, and so dissipate kinetic energy to heat energy at all times. As a result, the high-shear strains associated with the start of turbulence continue to heat the surface. Approaches to handle the heat are drafted in Figure 15.

Three different approaches are used to reduce the heat of a hypersonic flight. They are:

- Hot Structure
- Insulated Structure
- Internally Cooled Structure

Figure 15. Methods to Reduce Heat

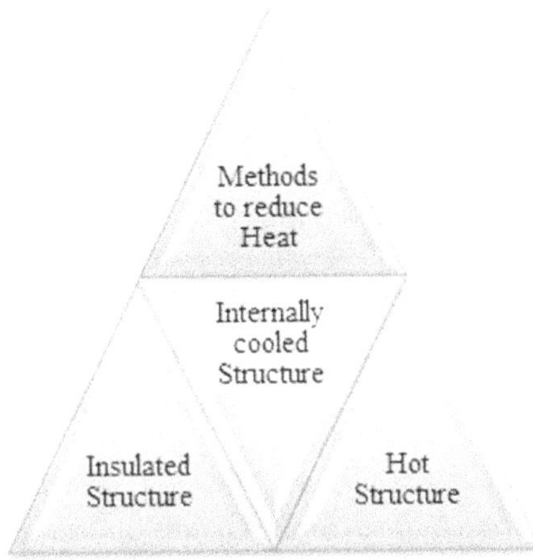

Hot Structure: It is also called as Radiation Cooled Structure. Based on the radiation and conduction of the thermal energy from the structure to keep up a sense of balance with a thermal inputs, reflexive and shape are maintained stable.

Insulated Structure: It is other called as Thermal Protection System (TPS). The most important structural components are protected from the direct effect of the hot environment by means of a shield. The thermal Protection System can be ablative, passive, or semi active.

Internally Cooled Structure: The structural method utilizes a coolant which will circulate over a structure and it's moreover retrieved or jettisoned once used.

Among the above three types the radiation cooled structure is very much suitable for hypersonic vehicles as it is much acceptable for using in the hypersonic design structure because of its simplicity and the allowable range of aerodynamic characteristics, reliability and as well as its minimum weight and less space consumption.

The atomic flux to the surface will contribute to some substantial additional heating over the mechanism of surface-catalyzed recombination, where the energy is released as heat on the vehicle's surface, depending on the materials that make up the vehicle's thermal protection system.

When considering the material selection, it is always important to recall the three laws of high temperature chemistry, chemical catalyst reactions because of its various reaction rate. Some materials oxidize readily when exposed to a high temperature in the air.

CONCLUSION

Few of the hypersonic flow characteristics features have been discussed in this chapter. Basic hypersonic physics, its aerodynamics and its aerothermodynamics were briefed. A hypersonic vehicle slugs through the atmosphere soon after its flight into orbit for a reentry vehicle say like space shuttle. But a hypersonic will continue to remain within the Earth's atmosphere all the way through most of its climb. However the Shuttle uses the thrust from its engines. Generally hypersonic propulsion within the atmosphere can be of turbojets, ramjet, scramjet, ducted rockets. A hypersonic vehicle would make use of the aerodynamic lifting forces which are generated on its wings and fuselage. A design constrains for the above mentioned features were also summarized followed by a brief description of materials and how to handle heat in a hypersonic vehicle.

REFERENCES

Anderson, J. D. (1989). Hypersonic and High Temperature Gas Dynamics. The McGraw-Hill Company.

Berry, S. A., Horvath, T. J., Greene, F. A., Kinder, G. R., & Wang, K. C. (2006). Overview of boundary layer transition research in support of orbiter return to flight. *AIAA Paper*, 2006–2918.

Brazeal, J., Wuetcher, A., & Wang, X. (2020). Revisit of Hypersonic Small-Disturbance Theory for Perfect Gases. *AIAA 2020-0102. AIAA Scitech 2020 Forum.*

Cheng, H. K. (1963). Recent Advances In Hypersonic Flow Research. *AIAA Journal, 1*(2), 295–310.

Douglas, E. (1964). Wall A Study Of Hypersonic Aircraft. *Fourth National Airport Conference.*

Fletcher, D. G. (2014) *Fundamentals Of Hypersonic Flow-Aerothermodynamics* RtoAvt Lecture Series On "Critical Technologies For Hypersonic Vehicle Development", Held At The Von Kármán Institute, Rhode-St-Genèse, Belgium.

Gülhan, A., Willems, S., & Neeb, D. (2021, November 12). Shock Interaction Induced Heat Flux Augmentation in Hypersonic Flows. *Experiments in Fluids, 62*(12), 242. doi:10.100700348-021-03336-y

Hunn, D. (n.d.). *The Road to Hypersonics -Key Challenges, Advantages and Disadvantages*. Lockheed Martin Missiles and Fire Control.

Katyal, A., & Srivastava, S. (2013). Hypersonic Aerodynamics of Aerospace Vehicle Design: Basic Approach and Study. *Advances in Aerospace Science and Applications, 3*(3), 209-214.

Knight, D., & Kianvashrad, N. (2021). Simulation of Hypersonic Shock Wave Laminar Boundary Layer Interactions. *11th Ankara International Aerospace Conference.*

Mason, W. H. (2018). *Hypersonic Aerodynamic*. Configuration Aerodynamics.

NPTEL-Aerospace. (n.d.). *Module 1: Hypersonic Atmosphere Lecture 1: Characteristics of Hypersonic Atmosphere.* Author.

Rajendrasing, R., Girka, Y., Zaharenko, V., Mallikarjuna, V., & Boopathiraja, A. (2015). *Aerodynamics and Thermal Features of Reentry Spacecraft with Blunt Body Capsule.* Aerodynamics, Dynamics, Ballistics and Flight Control of Aircraft. Aviation and Space Equipment and Technology No. 2 (119) UDC 532.58.

Roe, P. (2005). Thin Shock Layer Theory Revisited. *4th AIAA Theoretical Fluid Mechanics Meeting.* 10.2514/6.2005-5194

Shang, J. J., & Yan, H. (2020). High-Enthalpy Hypersonic Flows. *Advances in Aerodynamics*. doi:10.1186/s42774-020-00041-y

Urzay, J. (2020). The Physical Characteristics Of Hypersonic Flows Center For Turbulence Research. Stanford University.

Viviani, A., Pezzella, G., & Golia, G. (2012). Aerothermodynamic Field Past a Reentry Capsule for Sample Return Missions. *28th International Congress of the Aeronautical Sciences*.

APPENDIX

Acronyms

AOTV - Aeroassisted Orbit Transfer Vehicle
ARV - Ascent & Reentry Vehicle
CAV - Cruise & Acceleration Vehicle
D – Drag
L – Lift
M – Mach number
RV - Reentry Vehicles
STS - Space Transport System
SWBLI - shockwave boundary layer interaction
TSTO - Two Stage To Orbit
UHTC - Ultra-High Temperature Ceramics

Symbols

ϕ - Freestream direction
A – Area
C_{pSN} - coefficient of pressure at the stagnation point
C_D - drag coefficient
C_L - lift coefficient
Cp - coefficient of pressure
Cp,l - coefficient of pressure at the lower surface
Cp,u - coefficient of pressure at the upper surface
$h_{0\infty}$ - freestream's stagnation enthalpy
K = Proportionality constant used in the modified Newtonian theory
M_∞ – freestream Mach number
N - Surface force
p_S - Stagnation pressure
q_∞- dynamic pressure
V_∞- freestream velocity
γ – Specific heat ratio
θ - Flight path inclination
α – Angle between the flow and aircraft's surface element.
ρ_∞ – freestream density

Chapter 13
Study of Drag Reduction on a Hypersonic Vehicle Using Aerospike

Naren Shankar R.
Vel Tech Rangarajan Dr. Sagunthala R&D Institute of Science and Technology, India

Irish Angelin S.
Vel Tech Rangarajan Dr. Sagunthala R&D Institute of Science and Technology, India

Vitalii Pertsevyi
Dnipro National University of Railway Transport, Ukraine

ABSTRACT

Humans longing to fly higher and quicker have prompted the improvement of hypersonic vehicles. Typically, hypersonic streams are described by high temperature fields and a thin layer of shock close to the object wall or the body surface. To ease the reduction of thermal loads, a blunt nose is forced in a hypersonic vehicle which is more imperative. In any case, increase in the wave drag is one of the quick outcomes of a constrained bluntness. Consequently, investigation in the hypersonic field is constantly fixated on the wave drag decrease. The flow features around the blunt body get changed because of the attachment of spike in front of the vehicle. This chapter aims to give a detailed review of a hypersonic vehicle that involves an aerospike design in front of the blunt body, which tends to reduce the drag at the forebody. Views of various researchers are investigated, and efforts are taken to summarize the reported results on how the drag has been reduced using aerospike technique.

INTRODUCTION

The speed of an aircraft is usually denoted in Mach Number, M. Based on the speed regime the aircraft can be classified into different types (i) $0 < M < 0.8$ subsonic, (ii) $0.8 < M < 1.2$ transonic, (iii) $1.2 < M < 4$ supersonic, (iv) M ^{3} 5 hypersonic. If $M = 1$ it is called as sonic condition. In hypersonic flows, the

DOI: 10.4018/978-1-6684-4230-2.ch013

flow velocity is much greater than the sound velocity, which is the velocity of propagation of small disturbances. Hypersonic vehicles in general fly at higher altitudes where the density and Reynolds numbers are considered to be low and where the boundary layer formation is found to be thick. Reducing the drag and aerodynamic heating are the two main factors in critically designing a hypersonic vehicle. Reducing the drag in the hypersonic vehicle is necessary as it helps in saving the fuel, extending the range, simplifying the propulsion unit and increasing the payload to take-off the gross weight ratio. A hypersonic vehicle creates a bow shock in front of the blunt body which leads to a higher surface pressure load and aerodynamic drag. This surface pressure can be reduced if the conical shock wave is created rather than a bow shock wave. This can be achieved by using a spike in front of the blunt body. This chapter aims to review various papers related on drag reduction techniques by using aerospike passive method. The objects moving at high speed experience forces that will in general tend to slow objects down. This confrontation withinside the considered volume is named such as a drag that is a solitary significant concern throughout the time of planning in designing a high speed vehicle. Hypersonic vehicles are those that travel at Mach above 5. Usually, missiles, rockets, reentry vehicles travel at hypersonic speed. It is often planned to reduce surface heat flux by designing hypersonic space vehicles with blunt noses. By any means, this detached shock increases the wave drag, making space flights more expensive. Thus, the spaceship configuration should consider the best balance between wave drag and surface heat flux. Subsequently, in the future, there will be more development and improvement of the hypersonic drag reduction technique as it becomes a significant research area within the hypersonic field. A number of strategies have been put forth in this area.

The streamlined aerodynamic layout of hypersonic vehicles must be completed considering the intricacies associated with the hypersonic flow. Typically, hypersonic flows are described by high temperature field and a thin layer of shock close to the object wall or the body surface. Such shock and viscous dissemination came about raised temperature close to the space vehicle surface which led to inordinate warming of vehicles flying at speeds above Mach 5, which are considered as hypersonic speed regimes.

The nose area of a space vehicle which is considered as the most serious place of space vehicle that undergoes most noteworthy heat flux, blunt nosed plan of the equivalent design has been recognized as a powerful strategy to drive the shock layer away from the body, to lessen the surface heat flux. (John D Anderson, 1989). In the meantime, the aerodynamic heating appearing in the spot of stagnation changes conversely which is equal to the blunt nose radius root, possessing a huge blunt nose area is worthwhile all things considered. Besides this design measure improves the drag reduction. For space missions increase in drag makes higher fuel utilization and lesser payload limit. In this manner, the expense of space missions increments radically. Along these lines, the decrease of wave drag has stayed as a significant field of exploration in the field of hypersonics. Wave drag is triggered by the development of shock waves around the vehicle in supersonic flight and hypersonic flight. Also, to reduce fuel usage and to expand the payload limit of the space vehicle, decreased heat load or lower drag is required for a hypersonic flight or mission.

To ease the reduction of thermal loads, a blunt nose is forced in a hypersonic vehicle which is more imperative. In any case, improving the wave drag is a quick outcome of constrained blunt body. Consequently, investigation in hypersonic field is constantly fixated on the wave drag reduction. In order for hypersonic vehicles to succeed, reducing drag is a vital component. Extreme aerodynamic heating and aerodynamic forces are imposed at the blunt-nosed surface, owed to the drastic bow shock to be found at the forefront of the nose when a vehicle travels at hypersonic speed.

To manage the aerodynamic heating and diminish the drag force imposed on the body by the stronger shock, a few strategies were proposed, for example, the connection of air spike, counter-jet flow, jet spike counterflow method and blunt body breathing nose.

WHAT A SPIKE DOES

A spike connected at the blunt nose and protrusive in front of the body alters the bow shock to oblique shock, in this way producing a noteworthy decrease of drag and aerodynamic heating. The perception of designing a mechanical aerodynamic featured spike remains a successful technique to lessen the pressure drag appertaining to the blunt body happening in hypersonic flow.

Spike is an elementary thin contraption that points at a noticeable rate from the nose segment of the space vehicle. Due to this insertion of the spike, the flow in at the frontal region is varied significantly. The bow shock formed in forward-facing out from the blunt body is altered to a conical shock spike as a consequence a low pressure recirculating region is created over the surface of the body. The structure of this altered the shock and the recirculating region together forms a streamlined body profile like that one of a slim frame in front of a blunt nosed body. Subsequently, the developed low pressure area adjoin to the body significantly decreases the dynamic pressure taking place over the body, consequently wave drag as well. Nevertheless, the separated and recirculated flow further reattaches preceding the wall of a blunt body somewhere downstream of the nose area. The reattachment about the flow on the surface of the blunt-body exists prominently which is being influenced by way of the freestream Mach number, length of the spike, spike design, and Reynolds numbers. At the reattachment point, the flow rotates through a reattachment shock to allow an even flow applied over the aft body surface.

The spike which is being located in the forebody of the vehicle modifies the flowfield around the blunt body. Bow shock which is usually formed on ahead of a blunt body is changed into a pointed shock and reattachment shock. This reattached flow field eliminates the strong bow shock created in front of the forebody and prominently reduced the shock wave drag. The reattached of the shock and the afore-mentioned conical shock can be driven in addition far from the focal line because of the increase in the spike length, altogether bringing down the strength and converting the aerodynamic features(SUN et al., 2019). Considering all these issues, this chapter aims to review the ways to reduce the drag at hypersonic forebody vehicles by using an aerospike which is one of the passive techniques.

DEVELOPMENT OF AEROSPIKE

An American Submarine Launched Ballistic Missile named Trident C4 or UGM-96 Trident I which was built by Lockheed Martin Space System developed in the year 1979 has this concept. It was expected to make sure an improved range of nearly 550 km. The aerospike used in Trident is made up of a flat circular plate that is attached to an extendable boom that is extended in a while after the missile leaves the submarine and penetrates the water surface subsequently being launched from the submarine. With the aerospike, the nose shape could be blunted, allowing a much larger internal volume to accommodate payload and propulsion without causing any increase in drag. As a result of adding the third propulsion stage to the Trident I C-4, it was able to increase range by about 15% in excess of the Poseidon C-3 the aforementioned supplanted. In order to apposite the third-stage motor into the existing submarine launch tubes, the vehicles of the reentry phase had to be mounted around the motor

In the middle of 1970's KB Mashinostroyeniya developed an aerospike for a 9M39 which is a surface to air missile of Russian-made man-portable infrared homing surface-to-air missile defense system named as 9K38 Igla MANPADS reduced the wave drag. In comparison with Igla-1, the improvements included much better resistance to flares and jamming, a high-sensitivity seeker, the ability to engage fighters in an all-aspect manner under favorable conditions which has all hemispherical engagement capability and improved performance in altitude.

To generate air-spikes, a mechanism was developed with an open-ended mechanism. A pulsed laser or electric arc torch can be used to project concentrated energy forward commencing the frame, which results in a low-density area of warm air. A shock wave is notable for its reduced drag, as opposed to structural aerospikes, because the air density behind it is lower.

During the 23rd Aerospace Sciences Meeting in 1995, an aerospike-protected missile dome was successfully tested to Mach 6. This test provided quantitative data on air pressure and temperature increases to estimate the possibility of utilizing aerospike structure on a hypersonic missiles. At hypersonic speeds, aerospikes can reduce aerodynamic drag dramatically if they are sized appropriately for the body diameter.

Figure 1. Formation of shock on a blunt body at high speed without aerospike.

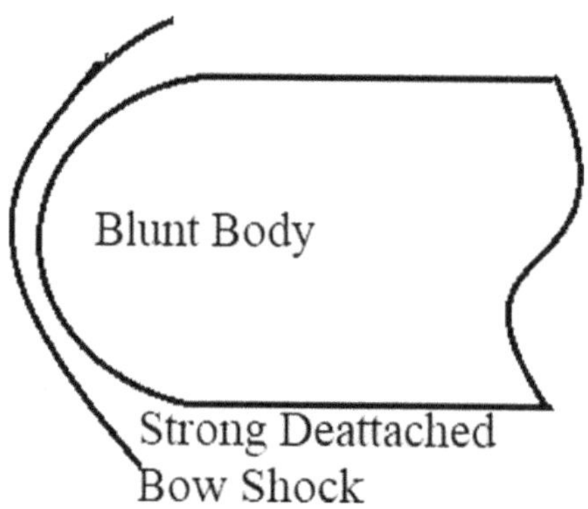

NUMERICAL APPROACHES REDUCING THE DRAG USING AEROSPIKE OVER A BLUNT BODY

(Senthilkumar et al., 2021) involved in a numerical approach study over a spike configuration in a blunt nose by considering various angle of attack and their effects on reducing drag. The author did a comparative study between aerospike and aerodisk. The works indicated that a design of aerodisk shape is an exceptional with that of an aerospike configuration. Drag reduction may be possible with an aerodisk with a suitable length, diameter, and nose structure. The design processes were carried out using CATIA. Numerical results are attained by ANSYS Fluent CFD software. At a Mach number of 2, flow observations are made on two blunt bodies with a flat-faced aerodisk and hemispherical aerodisk or conical. Different flow features are observed among the flat-faced aerodisk and hemispherical aerodisk or conical. A study of lift and drag coefficients with aerospikes and aerodisk is conducted. The spike attached at the forebody of a hemispherical blunt nosed body, when attached, amends the flow field

Figure 2. Blunt body with different configurations of aerospike at high speed

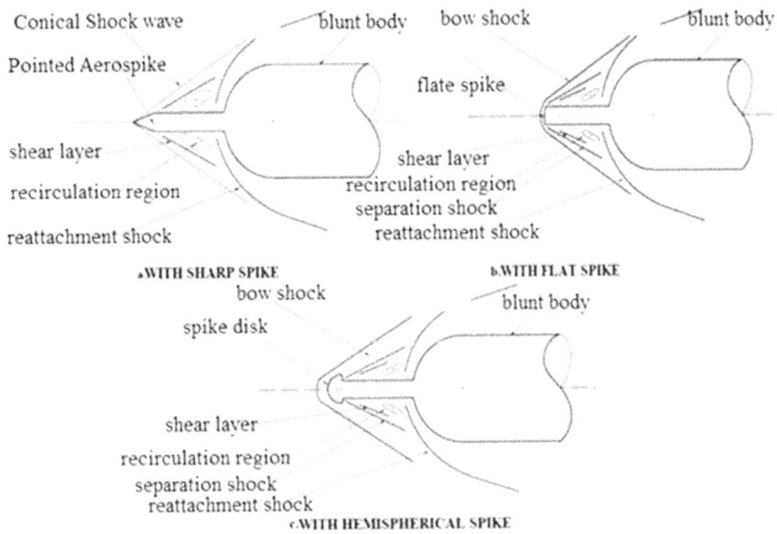

in a significant manner and therefore reduces the drag by ensuring a separation zone at the point of stagnation. Related to the back-disk geometry, the flow field directly behind the aerodisk is complex. This is unlike the conical spike. Choosing the optimal length of the spikes with the proper geometrical configuration of the nose will allow you to capitalize on the forward-facing spike to reduce drag more efficiently. It is cross examined that as soon as the flow turns distorted because of the growth in an angle of attack, there is an increase in the lift coefficient. The drag decrease is likewise elevated from a normal blunt nosed to aerodisk.

(Ahmed & Qin, 2010) using three-dimensional steady and unsteady flow fields, the authors examined the rationality of the axisymmetric assumption induced by long and short spikes in hypersonic freestream. As long as the flow field is steady-state and there is no angle of attack, the assumption is reasonable except for the separation zone. As for the asymmetric contours for an unsteady flow field by long spike, every axial cross section showed an extreme degree of asymmetry. Nevertheless, a small variation in drag coefficient can be detected in the separated region within the shear layer. An asymmetry of flow features is particularly pronounced in cases of unstable flow. A spiked hypersonic vehicle in the air at zero incidences with a forebody scaled to Mach 6 flying at a height above sea level of 60kilometers is tested with two models. In both models, the forebodies are hemispheres while the spikes are thin circular cylinders with a 10° semi-apex angle. This first model will yield a stable flow pattern because its spike length equals 1.5 the forebody diameter. Analysis was performed on a structured multi block grid.

An aerodisk model of a hypersonic hemispherical structure bedecked with an adaptable length of spike and an aerodisc in a hemispherical shape with an adjustable size was also developed and evaluated numerically in laminar hypersonic freestream conditions. Based on the effective body's shape, the authors developed a mechanism explaining the drag decline and the root reasons for an uncertainty in the flow. A study has been presented conducted on a drag reduction and its relationship to the spike's specific design. A certain aerodisk size is optimal for the models studied in this paper, and this size is inversely correlated with the spike length. Flow pressure does not vary significantly across the entirety of the separated area. Despite the fact that the flow decelerates at the reattachment point, the increase in pressure is due to compression waves as the shear layer reaches the surface of the body. Because the body interferes with the pressure gradient, the boundary layer and shear layer are likely to separate and form when the spike is present. The shape of the effective body and the state of flow stability change as the total pressure in the dividing streamline increases. As a result of the aerodisk's performance, lessening the drag was primarily determined by two actions: first, lifting the partition of the insulating ring and (generally) pushing its division point outward after the fixed spike hub, producing a slimmer compelling body; and second, lowering the pressure field generally speaking downstream of the aerodisk by the activity of its development fan. Models with a spike longer than the model breadth showed a steady stream for the models examined here. While models with more limited spikes can benefit from a bigger aerodisk, the stream field can be balanced out by a larger aerodisk. Both the length of the spikes as well as the size of the aerodisk contributes to drag reduction. There be existent a size of aerodisk with a particular base drag for a point of interest pike length. It was considered contrary to the length of a spike that this ideal size should exist.

(Peng, 2015) proved that the spike used for the reduction of drag is not exaggerated by way of the angle of attack on free-stream streams, whereas the drag decrease rate has almost no change in hypersonic flows. Generally, the typical cylindrical or pointed spiked bodies are seldom examined in present research; instead, different forehead configurations are used for flow control. The reattachment of separated flows from a spiked blunt body using peak heating was described in detail by Mehta et al. studied about the peak heating in a for reattachment of separated flow over a blunt body with spike attached. In addition to attaching a disk in flat shape at the front body of the spike, the author proposed that it would improve the anti-drag effect as well as alleviate the drag caused by aerodynamic heating. Finite volume discretizations and multistage Runge-Kutta time stepping methods are used to solve a Navier-Stoke equation which is of an axisymmetric compressible type. A Mach number at freestream condition impacts the length of a separable region on the spike. It has been shown by the computed results that the freestream Mach number furthermore disturbs the greatest heat flux formed on the nose of the blunt-body. Mach number and wall heat flux take place on the surface of a blunt body as the freestream Mach number increases. With no-slip conditions and isothermal conditions applied to the surface of the wall, it is assumed the wall is impermeable. The relative difference in the drag coefficient is about 1.5%. With an increase in Mach numbers, the length of the flow separation decreases. All the Mach numbers for freestream observe a similar pressure peak and heat flux along the blunt-nose spot. Through the conical-shock-reattachment interaction, the shear layer formed over the spike zone is affected by every reattachment shock wave.

(Raman, 2015) studied that the pressure coefficient plot reveals that shock reattachment on the forebody is delayed as the aero disk diameter increases. It is also found that the aerodisk diameter increases as the radius of the shock is increased, also influencing the pressure coefficients on the forebody and lowering the drag coefficient. Thus the aerodisk diameter increases as the radius of the shock increases, indicating the importance of an aerodisk configuration on shock reattachment. Research is currently being conducted on additional configurations. A factor that affects the forebody drag is the diameter pertaining to the aero disk. The report put forward that the analysis work carried on aero disk diameter compared to forebody diameter of the vehicle should be deliberated to be further significant.

In a simulation, (TU, 2014) investigated the effects of spikes with disk caps on the flow field, and demonstrated that decreasing drag was effective even at small angles of attack. The researchers have recognized that reattached flow can generate severe aerodynamic heating on the surface despite the benefits of simplicity compared with other drag reduction schemes. A slight difference in static temperature distribution can be explained by the length dimension of the spike ahead and the diameter in the lead of the aerodisk.

(Yadav et al., 2014) in their paper analyzed previous research on aerospikes in retorting hypersonic regime flows in addition to proposed the design by way of using two types of hemispherical projections placed near to the head and intermediate respectively. Furthermore, this would help mitigate the reheating taking place at the surface of the blunt body as well as increase the drag reduction rate. A hemispherical cylindrical shape of a conventional blunt body reentry model, fixed axisymmetrically by means of an aerodisk shaped type aerospike next to the nose radius is examined with a flow solver monetarily accessible control volume stationed as an axisymmetric flow at Mach 10.1 numerically. The Navier Stokes Equations is used aimed at solving a compressible turbulent flow in hypersonic condition, axisymmetric flow in two-dimensional case. The findings indicated that the spiked body blunt nose with two aerospikes in sequences of 30 mm and 20 mm in that order and the length diameter ratio is 1.5 exhibited a decline happening around the uttermost reattachment high temperature flux and greatly reduced aerodynamic drag and making it a potential candidate for hypersonic blunt noses with high lift efficiency. Based on the results of the study, shorter aerospikes offer only small reductions in an aerodynamic heating's, aerodynamic drag and highest reattachment heat flux by ratios of 1, 1.25 and 1.5. Comparing two aerospikes with a particular aerospike, it was found that the main body would be less heated by the aerodynamics of the dual aerospike arrangement. The fundamental outcomes recommend that an appropriate aerodisk shaped aerospike plan using two types of aerospikes arranged in sequences consumes an ability of diminishing the aerodynamic heating attributed to a reemergence vehicle much lower than completely violent circumstances. Hence, assuming that aerospikes remains to be utilized in place of hypersonic vehicles, and it must be using various aerodisk shaped aerospikes in successions.

(Fan & Huang, 2022) embraced the approximately coupled strategy to concentrate on its drag efficiency and heat decrease, in access to the response impact of wall high temperature ascend by considering aeroheating. The simulations were done by the CFD and ABAQUS. The aeroheating at the forebody of the vehicle was found to be decreased when an aerodisk spike model was used whereas the model without aerospike had a high aeroheating on its surface. The spiked prototype using a plan aerodisk took the finest staging on drag besides the heat decrease surrounded by every type of the ideal model. Also, if the spike length be expanded, it will decrease the drag and high temperature at the front of the body surface. By means of incrementing the length, the transformation paces of drag, pressure, and high temperature diminished continuously. Expanding the aerospike breadth additionally diminished the heat of the blunt forebody, even though the productivity of forebody drag decrease previously expanded and afterward diminished. Taking everything into account, the hotness and drag decrease should be thought about thoroughly for the ideal plan of the spike. An insertion calculation is utilized to accomplish the information move of the coupling factors. The fundamental rule is planning the lattice hubs on the coupling surface from the 3D actual space to the 2D control plane, and the information move is accomplished on the control plane. The coupling strategy followed by the creator. A coupling methodology is a technique in which the run times of different codes are coupled together. The ABAQUS uses a technique called as co-simulation which can couple with any of the third-party analysis program like CFD. This method can be employed to resolve complex flow and structure interaction as displayed in Figure 3. With the advancement of estimation, the changing speed of the coupled determines the outcome which diminishes bit by bit. The minimum drag was obtained by the flat type spiked design; moreover, the planar shaped aerodisk additionally obsessive the smallest amount of temperature and had the preeminent non-ablative characteristics.

Figure 3. Coupling methodology

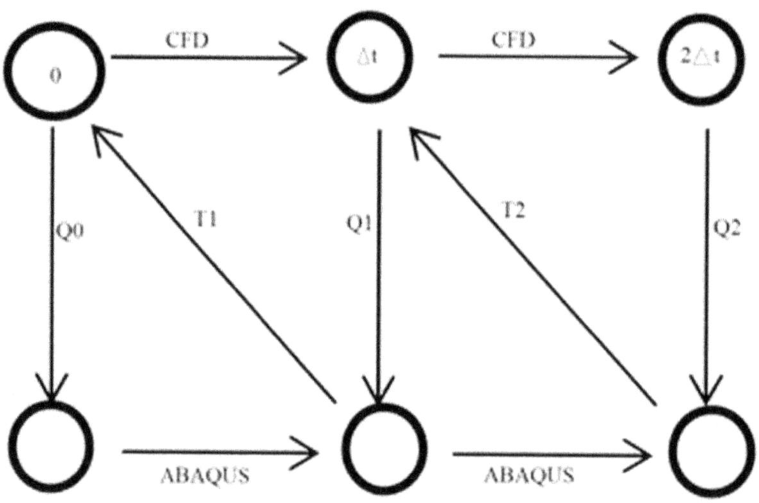

(T & CM, 2017) completed a mathematical examination on different models with spike and without spike, and observed the drag taking place at the spiked blunt body is a lot of lesser than the structure without using spike. Aerodisks achieve their effectiveness in lowering drag by increasing both the spike length and the total diameter of the aerodisk. So the geometric shape of the tip will be a major determinant of drag decrease. Unlike spike-on cases, spike-off cases have a much higher total drag because of the solid bow shock. In addition to this, without spike structure shows a temperature dispersion is much greater than the design with spike. Furthermore, a hole was proposed on the aerodisk, and based on the results; we postulated that the maximum hole under the same length of aerospike would result in the maximum base drag. Drag reduction can also be influenced by the body's compelling force and distribution's effect. More importantly, as the measures of spike and whole increment are increased, the effectiveness of drag decrease would diminish. Based on the results obtained, we have found that the drag coefficient can be affected by the stream designs which include solid bow shock waves, extension fans, shear layers, stream partitions, distribution districts, pressure waves, and so on. The bow shock originates at the aerospike and affects the entire hemispherical body. Depending on the boundary settings and choppiness model of the fundamental body surface, the division point along the aerospike will affect the strain and temperature distributions along that surface. For aerodisks and blunt primary bodies, pressure drag is the most dominant; however, thick drag would play a minor part in the overall drag of spiked obtuse bodies. If the spike length and whole width are the same, then the greater the total width, the lower the total drag.

(MohanT & Annie Ghevarghese, 2015) concluded that hemispherical aerodisk model was less efficient when compared to the spiked configuration in reducing the drag where as a blunted aerodisk was found to produce a maximum efficiency in reducing both heat transfer and drag over the surface. As the angle of attack increases a self-aligned aerodisk should be preferred which will align itself in the direction of the freestream flow for reducing the drag.

(Mehta, 2010), (Mehta, 2000) carried out a numerical study at two different Mach Numbers, say M=6 and M=6.8 separately in different research papers using spike attacked at the nose of the blunt body. Analyses were carried out at Mach 6, having constant length to diameter and at zero angle of attack to investigate the effects of different shapes of the spike on the drag reduction. The Hemispherical shaped spike gave a better results compared with other shapes of the spike. The results have been validated with the experimental results. In the other study, the author carried out a numerical analysis at Mach 6.8 varying the length of the spike and the peak pressure and the conical shock has been studied.

REVIEW OF EXPERIMENTAL DEMONSTRATIONS USING AEROSPIKE MODELS

(Menezes et al., 2002) carried out a sequence of examinations to assess the achievability of utilizing various aerospike/aerodisc configurations as a device to decrease the retractable drag for a number of various cone angles flying at hypersonic speeds. An inside insertable accelerometer balance has remained exceptionally planned, created as well as utilized aimed at estimating the streamlined drag over the blunt body cone made with aerospikes shaped types say, aerodiscs. Drag estimations are accomplished at various approaches designed on the blunt nosed body cone in both cases: with aerospike structure and without aerospike structure (Menezes et al., 2003) . The outcomes demonstrate nearby 55% drag decrease for the blunt body attached with plate slanted spike congregations at zero degree.

(Kalimuthu et al., 2010) studied force measurements, flow visualization, and pressure estimations over a hemispherical spike fixed in the blunt body at Mach Number 6 using a hypersonic wind tunnel. Speculative estimates of drag lift, pitching moment, and strain are discussed when spike boundaries are used. A hemispherical nose body and two front-oriented hemispherical spikes create a distribution area around the stagnation point of the blunt body, thereby creating an optimal flow profile and reducing drag. In a 1:5 L/D ratio, the coefficient of drag decreases by 62%; in a 2:0 ratio, it decreases by 78%. A second thought must be given to the expanded pitching moment. It was noticed that there were two different spike lengths of the prototype in the Schlieren photos. Around 50 degrees, the greatest surface tension is caused by the pressure over the blunt body. In terms of dynamic pressure, the deliberate pressure is non-dimensional. This is because of the pitching moment coefficient, which displays a straight variation depending on the approach, due to the unpolished blunt body design without the spikes. In a half-of-the-globe spike setup, the CM variety is considered to be nonlinear. As L=D expands, so does the pitching-moment coefficient. The vehicle will therefore be controlled by an additional control force.

(D'Humières & Stollery, 2010) used a hypersonic wind tunnel to study the laminar flow on a spiked cone ending in a spherical cap. The schlieren picture shows that there is no flow instability within the tested peak length range, and the force measurement shows a extreme decline of 77% of the blunt body resistance. This model is based on a 1:100 scale Apollo plan to reenter the aircraft, but has a sharp edge in between the sphere-shaped cap and the tapered rear body. The model is installed on a force balance that uses strain gauges to measure resistance. Nevertheless, no major longitudinal flow instability was found, and minor lateral vibration of the spikes was noticed. Compared with the hemispherical body, the shorter ear length presents a resistance advantage. The resistance generated by this spiked object can be similar to the pressure resistance created by the cone structure, which is modified by a constant. In the absence of spikes, the calculated shock separation distance is to a certain extent good.

(Kalimuthu et al., 2008) studied about the heat transfer at high speed over a forward facing spiked shaped body surface by zero angle of incidence. Various aerodisk models were considered for studying the velocity gradient, stagnation point, sonic line, heat flux at stagnation point, shock standoff distance. When compared to flat face aerodisk, hemispherical aerodisk experienced a greater wall heat flux.

Flow visualization were carried to capture the shock and to analyze the spike shape to optimize the drag reduction and heat flux.

(Srinath & Reddy, 2010) investigated using a hypersonic shock tunnel a model with a large blunt cone angle which is flying at hypersonic Mach number. For the study the author considered two different spike configurations. One is a conical tip spike model and the other is an adaptive aerospike fixed with disk shapes for which the diameter was decreased in the opposite flow direction. The aerodynamic forces are estimated utilizing an accelerometer-based force balance framework and the rate of heat transfer are estimated using a sensor of thin film made up of platinum. At angles of attack beyond 2° it is set up that the telescopic aerospike has better performance in drag reduction whereas the performance of the plain aerospike is superior designed for angles of attack which are closer towards zero degrees. The drag decrease for the adaptive spike design is because of this getting out of the interaction of shock over the body surface and at the leeward side there is a low pressure zone. In any case, on account of the plain spike the stream separation is not that enormous and in this way the drag increments. It is observed that this increment in the heat transfer rate is bigger for the adaptive aerospike than for the plain aerospike. At approaches, the shock cooperation creates some distance from the nose on leeward side and the hotness transition rate diminishes while the shock communication moves towards the nose of the dull cone body on the windward side and the hotness motion rate achieves a greatest level at a 6° approach.

(Holden, 1966) emphasized a study on the factors that have an effect on the heat-transfer rates established within the reattachment region and also the overall aerodynamic heating of a spiked frame. The whole heat transfer to the spiked bodies varied noticeably with Mach number, profile of the vehicle, length measurement of the spike. Total heat-transfer rates wide-ranging from a most of double to a minimum of common fraction the fundamental cone value. Spike cones with larger length, the efficiency of spike stimulated separated flow in decreasing total heating redoubled by means of growing body bluntness in addition to increasing the freestream Mach number. The amount of the reattachment heat-transfer rate in a very face reattachment region was established to be directly proportionate to the reattachment angle.

(Panaras & Drikakis, 2009) argued that for certain structures or configurations, the flow intensely are influenced by the initial considered flow field conditions, hysteresis effect and the unsteady flow conditions. As the flow fluctuates the development of shock and shock interactions are not much sensitive to the initial conditions whereas for steady flow, the progress of the flow be strongly dependent on the initial conditions presumed.

Wind-tunnel measurements were conducted by (Guenther & Peter Reding, 1977) to study the flow separation over an area of spike-induced and in a region of unstable pressure environment. Continuously, towards the sides of the spike base, exceptionally large pressure fluctuations were measured at transonic speeds due to the non-axisymmetric and vertical nature of separated flows at least 1 degree above zero degrees. In supersonic speeds and at angles of attack greater than 7 degrees, high pressure fluctuations were detected on the nose cap windward side. All along the spike and nose cap but the region aft of the disk, there are distinct differences in the spectra between windward and leeward sides. As the pressure fluctuations from the reattachment are influenced by the oblique shock coming from the concavity, the reattachment shock will be greatly enhanced. Shock-shock interactions caused the maximum pressure fluctuations measured.

COMBINATIONAL STUDY OF AEROSPIKE METHOD AND COUNTER FLOW METHOD

Having a jet counterflow jet in a hypersonic vehicle can change the bow shockwave generated around the blunt body, which is likely to reduce the heat loads and drag. In an attempt to increase propulsion efficiency, numerous measures have been taken to decrease the drag of hypersonic vehicles(Kim, 2020). By injecting gas towards the freestream, counter-flow jets reduce drag. The counter-flow jet is one method of reducing drag, in which a jet of gas interacts with the freestream and enables the jet to reduce both drag and heat at the same time as shown in Figure 4. When counter flowing jets and aerospikes work together, the efficiency of both will be improved (Ou et al., 2018). It will be possible to resolve the spike ablation problem by combining them with jet. The opposing jet, on the other hand, has the capability to realize its own alignment concept together with angle of attack to vanquish its defect in flow.

Figure 4. Counter flow jet physics

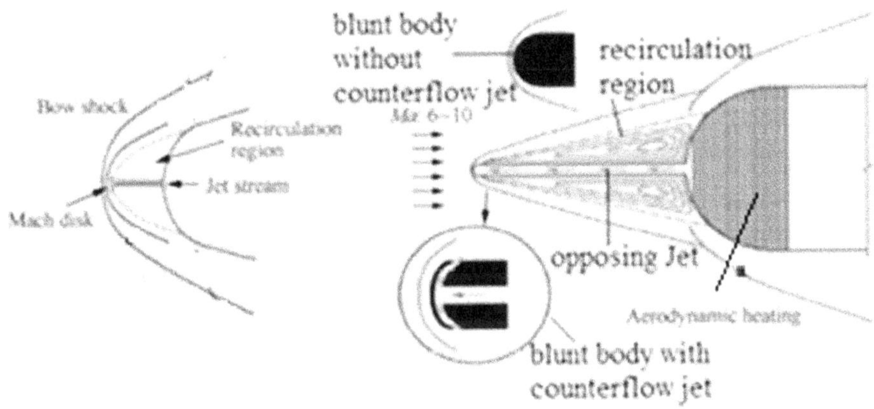

According to the numerical results,(Liu & Jiang, 2013) attempted to discharge the coolants into the spike angle in both axial and lateral ways, respectively. These experiments produced significantly larger reduction of drag effects as well as a vast reduction in aerodynamic heating at the spike tip and blunt surface, respectively. In addition, the on line proportionate spike and also the opposing jet concepts produced similar results as resolved by (J. Huang et al., 2019) optimized the aerospike length diameter proportion and also the jet pressure ratio, which became important factors in axial opposing jet and spike theory, according to a large number of simulations, whereas Ou et al. estimated numerically the effects of nozzle width, ratio of length-to-diameter, and the ratio of jet pressure taking place over the flow field surface, aerodynamic drag, and the heat properties over the blunt cone-structure combined with opposing jet or counterflow and the spike. In the later paper the author considered a RANS equation in two-dimensional case combined using the SST k-x turbulence model has used and existed to study the resistance and heat flow reduction mechanisms in a collective thermal protection system with opposing jet and spikes. With the intention of advancing the accurateness of numerical analysis, the grid Reynolds number ought to be established to a suitable value. When L/D increases, the standards of wall heat flux, along with the static pressure as well the drag coefficient declines significantly. The summary of the reverse jet forms a recirculation zone upstream of the nose head. When the L/D increases, a recirculation zone initiates at the downstream of the air disc. Both recirculation zones significantly alleviate the high temperatures experienced by the gas pan. For the single aerospike arrangement, the single-equation namely, S-A model plus the medium grid has the maximum simulation precision. The SST k-x model collected with the medium grid is the best solution for simulating the reverse jet. With the purpose of improving the accurateness of numerical analysis, the grid Reynolds number must be set to a suitable value, and the appropriate value in this paper is about 5.0. After installing a counter-flow system with a 4mm nozzle, the Stanton number St is significantly reduced, and the decrease in St increases with the increase in pressure ratio.

In hypersonic flow, (Barzegar Gerdroodbary, 2014) suggest injecting gases from a spike root. As a result of the cool jet, the aerodisk temperature and average heat load behind the nose part cone were reduced. Carbon Dioxide and Helium which are considered as coolants are injected from the nose cone tip to cool the recirculation area. Gases are assumed to have an ideal structure. The computational domain is axisymmetric. The results suggest that the counter flowing jets reduce the heat load formed on the nose cone by altering the flow field. The Helium flow stream is set up to make sure that it has a moderately extra powerful cooling execution. Understood CFD code is utilized which utilizes moderate, cell-focused, control volume detailing. The traditional SST model that is Shear-Stress Transport tempestuous stream model is taken on.

In both (Morimoto et al., 2014) and (Zheng, 2014) experiments, numerical simulations were conducted, but they gave contradictory results. Zheng commented that in lieu of an effective thermal protection by the side of reduced energy input, a strong opposing jet and an extended nozzle should be produced, while Naoki discussed the simultaneous decrease in the aerodynamic heating and drag caused by strong jets. Subsequently numerical codes, simulation assumptions, and operating conditions may possibly affect the results; variant inferences may not certainly deny the other results. As a result, with the intention of avoiding the contrasting results generated by numerical methods, high-accuracy numerical programs should be used for the simulation analysis of such complex flow fields in the days yet to come, including Implicit Large Eddy Simulation (ILES), Large-Eddy Simulation (LES), Detached Eddy Simulation (DES), and Direct Numerical Simulation (DNS).

(Eghlima & Mansour, 2017) scrutinized in detail regarding the flow field about a hemispherical nose cylinder under a free flow with a Mach number of 6 using a new combination of sharp peaks and counter-current jets. Air is injected at the speed of sound through a nozzle over a hemispherical model at the nose part of the vehicle. In the numerical analysis study which has been carried out, the axisymmetric Reynolds average Navier-Stokes equation is resolved by using k-ω which is of SST turbulence model form. The obtained results are verified and validated using the experimental results of spiked bodies without spray conditions. From there on the outcomes of countercurrent jets with different lengths of spikes and different pressures are given. The results show that the drag coefficient is significantly reduced by about 86-90% related with the actual model (L/D ratio equal to 1.5 and 2) of the spherical cylindrical model without jets and spikes. In addition, the study also shows that when the spike length increases, the resistance decreases even more. With an increase in the distance of spikes say, L/D ratio from 1.5 to 7.5, the shock formed nearby the nose radius of spikes takes away energy from the main sphere, causing the shock that is formed near the main sphere to become fragile and more capable of reducing drag. The L/D scaling is shown to be inappropriate for structure when the M/L is increased to large values.

(W. Huang et al., 2015) analyzed numerical results and found that they were similar to the experimental data on the level, and the grid scale only detained its influences on the numerical results to some extent.

In addition to reducing the jet pressure ratios and increasing the aerospike angles, increasing aerospike length-to-diameter ratios increases the drag reduction coefficient, while decreasing jet pressure ratios reduces the maximum drag reduction coefficient. The pressure ratio of 0.4 leads to an extreme reduction in drag coefficient which is of 65.02 percent. The top of the peak pressure is roughly between 40degree and 55degree, and the location is essentially the same regardless of variations in the jet pressure ratio. Static pressure profiles in walls don't appear to be influenced much by grid scale. In this case, turbulence modeling may not be accurate enough to induce the prevalent disagreement at the intersection between the opposing jet and the airstream flow.

(Han & Jiang, 2018) directed an exploratory review using a hypersonic wind tunnel considering 6 as the testing Mach number. The author showed that over a blunt body every interaction stuck between shock - shock is diminished by around 70% lower than a 4 degree angle of attack because of the jet infusion and the high pressure at the reattachment point. The conclusion of flow visualization and the pressure measurements is that by spike the bow shock formed in the front of the blunt body is converted to a conical shock, which can decrease the drag formed due to shock drag. In addition, the thermal aerodynamic around the vehicle is also reduced. During the flight, when the pitching angle is applied, the lateral jets which are injected help to move the conical shock out of the way originating at the surface of the blunt body.

(Jiang et al., 2009) used second order dissipative schemes for convective term analyses and second order central differencing scheme foe viscous terms. In front of the blunt body structure a combination of spike and jet injected sideward near to the spike nose were constructed. The pressure gradient created by means of the reattachment caused the disturbances to develop, that in turn influenced the conical shock and also the reattachment zone. As a result, an active stability or equilibrium will be established between the conical shock formed, the reattached shock as well as the flow all over the place that is dominant in decreasing the aerodynamic drag.

(John et al., 2021) explored to study on an active and passive methods by means of counterflow jet in addition to spike in front of the spike at Mach 8. Finite volume centered compressible flow solver assuming two dimensional axisymmetric natures. As the pressure ratio of the counterflow injection is increased to some particular value the drag decreases. Beyond that pressure ratio value the drag turn out to be inconsequential. In cooperation to the stagnation point infusion of counter flow and front oriented actual spike remain to be deliberated to relieve the overabundance drag as well as the heat transfer be subjected by means of a blunt body addressing the nose fir cone part of a hypersonic vehicle. Ultimately, in light of recognized drag decreases, the current review proposes comparability instances amongst the two techniques. The problem is made known that a sharp spike of value one for the L/D ratio gives practically a similar drag decrease as per the counterflow infusion jet with 8.25 as a pressure ratio. The situation is seen that by means of the underlying expansion in infusion pressure distribution, the reattachment point moves downwards in addition the stagnation zone pressure falls significantly. As the infusion pressure builds, the energy of the infused stream additionally increments and assists the infused liquid with infiltrating additional obsessed by a shock layer. This also can likewise be noticed that along with an expanded infusion pressure proportion, the extent of the infused flow additionally grows. Along with the enlarging of the stream cone point, the pivotal energy of the fly decays and also the outlet pressure force increments marginally. With the diminished pivotal energy of the infused fly, the impending supersonic fly can bind and power the shear layer to append on the front facing locale of the dull body much successfully. The review included examination of two unique kinds of spike, obtuse and pointy, each one by means of L/D=0.5 as well as L/D=1. The examination recommends in particular that as the downstream shifts further to the stream reattachment point, as on interpretation of L/D=1 compared to L/D=0.5, the lesser would be the stagnation area pressure for 0.5 ratio that will decrease the drag. In addition, when the length measurement of the spike is expanded, the point of tendency of the detachment region diminishes give rise to the increment in the size of the distribution zone which thusly would diminish the strain and thickness behind the shock inside the distribution locale. This shows that despite the fact that the spike can give same in general drag as counterflow infusion; the last option is dependably a superior choice thinking about the thermal protection and structural integrity.

SUMMARY

This chapter summarizes studies on drag reduction techniques. A brief discussion of a spike and counterflow jet combination is also provided in this chapter. Spike design the simplest technique makes it easier. Researchers have observed that flat disc aero-spikes of critical length can reduce drag to the greatest extent. Several studies have shown that drag reduction techniques using aerospikes could also reduce heat flux to a greater extent; however, not all investigations have adequately demonstrated that reduction of heat flux using aerospikes could be achieved. In addition, the laminar nature of incoming flow fields should be ensured in order for aero-spike to operate optimally. The experimental studies clearly explained about the shock and shock interactions.

REFERENCES

Abhijith, Krishnan, Mohan, Gokul, & Ghevarghese. (2015). Effect of Blunted Aero-Disc on Aero-Spiked Blunt Cone-Nose at High Speeds. *International Research Journal of Engineering and Technology, 2*(9).

Ahmed, M. Y. M., & Qin, N. (2010, January–February). Drag Reduction Using Aerodisks for Hypersonic Hemispherical Bodies. *Journal of Spacecraft and Rockets*, *47*(1), 62–80. doi:10.2514/1.46655

Anderson, J. D., Jr. (1989). Hypersonic and High Temperature Gas Dynamics. The McGraw-Hill Company.

Barzegar Gerdroodbary, M. (2014, September). Numerical analysis on cooling performance of counter flowing jet over aerodisked blunt body. *Shock Waves*, *24*(5), 537–543. doi:10.100700193-014-0517-4

d'Humières, G., & Stollery, J. L. (2010, February). Drag reduction on a spiked body at hypersonic speed. *Aeronautical Journal*, *114*(1152), 113–119. doi:10.1017/S0001924000003584

Eghlima, Z., & Mansour, K. (2017). Drag reduction for the combination of spike and counterflow jet on blunt body at high Mach number flow. *Acta Astronautica*, *133*, 103–110.

Fan, B., & Huang, J. (2021). Coupled Fluid-Thermal Investigation on Drag and Heat Reduction of a Hypersonic Spiked Blunt Body with an Aerodisk. *Aerospace 2022*, *9*, 19.

Han, H. G., & Jiang, Z. (2018). Hypersonic Flow Field Reconfiguration and Drag Reduction of Blunt Body with Spikes and Sideward Jets. *International Journal of Aerospace Engineering*, *2018*, 7432961. doi:10.1155/2018/7432961

Huang, Liu, & Xia. (n.d.). Drag reduction mechanism induced by a combinational opposing jet and spike concept in supersonic flows. *Acta Astronautica*.

Jiang, Z., Liu, Y., Han, G., & Zhao, W. (2009). Experimental demonstration of a new concept of drag reduction and thermal protection for hypersonic vehicles. *Lixue Xuebao*, *25*(3), 417–419. doi:10.100710409-009-0252-8

Jie, Yaoa, & Shan. (2018). Numerical investigation on drag and heat reduction mechanism of combined spike and rear opposing jet configuration. *Acta Astronautica*.

John, B., Bhargava, D., Punia, S., & Rastogi, P. (2021). Drag and Heat Flux Reduction using Counterflow Jet and Spike - Analysis of their Equivalence for a Blunt Cone Geometry at Mach 8. *Journal of Applied Fluid Mechanics*, *14*(2), 375–388.

Kalimuthu & Mehta. (2008). Experimental investigation on spiked body in hypersonic flow. *The Aeronautical Journal*.

Kalimuthu, R., Mehta, R. C., & Rathakrishnan, E. (2010, January–February). Drag Reduction for Spike Attached to Blunt-Nosed Body at Mach 6. *Journal of Spacecraft and Rockets*, *47*(1), 219–222. doi:10.2514/1.46023

Kim & Kim. (2020). Drag Reduction Effect by Counter-flow Jet on Conventional Rocket Configuration in Supersonic/Hypersonic Flow. *Journal of Aerospace System Engineering, 14*(4), 18-24.

Liu, Y., & Jiang, Z. (2013, March). Conceptual Study on Non-ablative TPS for Hypersonic Vehicles. *AIAA Journal*, *51*(3).

Mehta, R.C. (2000). Numerical heat transfer study over spiked-blunt body at Mach 6.8. *J Spacecraft & Rockets, 37*(5), 700 – 701.

Mehta, R. C. (2010, April). Numerical simulation of the flow field over conical, disc and flat spiked body at Mach 6. *Aeronautical Journal*, *114*(1154), 225–236. doi:10.1017/S0001924000003675

Menezes, . (2003, October). Experimental Investigations of Hypersonic Flow over Highly Blunted Cones with Aerospikes. *AIAA Journal*, *41*(10).

Menezes, Saravanan, Jagadeesh, & Reddy. (2002). Aerodynamic Drag Reduction Using Aerospikes For Large Angle Blunt Cone Flying At Hypersonic Mach Number. *22nd AIAA Aerodynamic Measurement Technology and Ground Testing Conference*.

Morimoto, N., Aso, S., & Tani, Y. (2014). Reduction Of Aerodynamic Heating And Drag With Opposing Jet Through Extended Nozzle In High Enthalpy Flow. *29th Congress of the International Council of the Aeronautical Society*.

Ou, M., Yan, L., Huang, W., Li, S., & Li, L. (2018). Detailed parametric investigations on drag and heat flux reduction induced by a combinational spike and opposing jet concept in hypersonic flows. *International Journal of Heat and Mass Transfer*, *126*, 10–31. Advance online publication. doi:10.1016/j.ijheatmasstransfer.2018.05.013

Peng. (2015). Research on aerodynamic characteristic of the hypersonic projectile with drag-reduction spike. *Kexue Jishu Yu Gongcheng*, *15*(9), 142–147.

Rolf. (1977). Fluctuating Pressure Environment of a Drag Reduction Spike. *J. Spacecraft, 14*(12).

Senthilkumar, Mudholkar, & Sanjay. (n.d.). Comparative Study on Aerodynamic Drag Reduction of a Blunt Nose Body using Aerospike and Aerodisk – Numerical Approach. *IOP Conference Series: Materials Science and Engineering.*

Srinath & Reddy. (2010). Experimental Investigation of the Effect of Aerospike Geometry on Drag Reduction and Heat Transfer Rates at Hypersonic. *International Journal of Hypersonics*, *2*.

Sun, Huang, Ou, Zhang, & Li. (2019). A survey on numerical simulations of drag and heat reduction mechanism in supersonic/hypersonic flows. *Chinese Journal of Aeronautics*.

Tu, W., & Jin, H. D. (2014). Numerical simulation study on missile antidrag spike drag reduction. *Computational Simulation*, *31*(4), 87–91.

Yadavn, R., Velidi, G., & Guven, U. (2014). Aerothermodynamics of generic re-entry vehicle with a series of aerospikes at nose. *Acta Astronautica*, *96*, 1–10. doi:10.1016/j.actaastro.2013.11.015

Zheng, Y. Y. (2014). Supersonic flow field investigation of the effect of a mechanical spike with and without counterflow jet *J. Heat and Mass Transfer*, *10*(1), 75–83.

Section 5
Applications of Aerodynamics

Chapter 14
Aerodynamic Effectiveness of Bio-Mimic Shapes at Different Reynolds Numbers

G. Gowtham Gajapathy Gajapathy
Vel Tech Rangarajan Dr. Sagunthala R&D Institute of Science and Technology, India

Vishal Naranje
https://orcid.org/0000-0003-1076-6803
Amity University, Dubai, UAE

A. H. M. Hussein
Helwan University, Egypt

Sundharasan R
Jaya Polytechnic College, India

ABSTRACT

Nature helps a lot to create and recreate new concepts to enhance our existence. Using nature-inspired means to address new engineering problems gives a better solution that is very quick, easy, and environmentally friendly. The goal of this chapter is to explore the overall performance increase by bioinspired design elements. Simulations of computational fluid dynamics are utilized to compare existing designs to experimental data in various flow regimes.

INTRODUCTION

Reliable solutions are obtained from nature to complex engineering problems. Especially in the field of aeronautics, nature helped a lot to achieve this much within a short period when compared to the time cycle of evolution. Right from the first experiment of flight to the current miniature model of MAV are inspired by Mother Nature. Remarkable advances have been made in the Mimication of birds or insects' flapping-wing mechanisms and investigation of Bio mimic-wings based aerofoils for efficient

DOI: 10.4018/978-1-6684-4230-2.ch014

lift-to-drag-ratio. Bio-Mimic is a broad term that unites biological objects which can be stimulated to research on non-biological science. Nature-inspired or bionic or Biomimicry and bio-inspiration are the trendy words that will be used in this chapter. The various bioinspired devices were discussed in detail (Goldfield, 2018)

As far as aerodynamics is considered, the reduction of drag, the increase of lift, or increasing the lift to drag ratio is the ultimate goal of achieving any kind of perfect aerodynamic shape. To achieve the above aerodynamic characteristics a wide knowledge of Reynolds numbers at various fluid flow regimes is necessary.

ROLE OF REYNOLDS NUMBER IN AERODYNAMICS PERFORMANCE

The Reynolds number (Re) is a dimensionless number that is used to classify fluid systems in which the effect of viscosity is crucial in determining fluid velocities or flow patterns. In various fluid flow scenarios, the Reynolds number aids in the prediction of flow patterns. At low Reynolds numbers, laminar flow dominates, whereas, at high Reynolds numbers, turbulent flow dominates. Turbulence is caused by changes in the fluid's speed and direction, within a fluid that is subjected to relative internal movement due to varying fluid velocities, the Reynolds number is the ratio of inertial forces to viscous forces. which can sometimes intersect or even move in the opposite direction of the flow's main direction. Based on several recommendations, it is assumed that a Reynolds number less than or equal to 2100 indicates laminar flow, and a Reynolds number greater than 2100 indicates turbulent flow. The Reynolds number would be used in a variety of situations, from fluid flow in a pipe to airflow over an aircraft wing. It's used to anticipate the transition from laminar to turbulent flow, as well as the scaling of similar but different-sized flow conditions, such as between a wind tunnel model and the full-scale version of any shape. In the last two decades, there has been a greater emphasis on how to build aerodynamic shapes at lower speed ranges. Due to the relative cost of other methods, small, low-speed vehicles are currently being utilized to survey huge areas.

A key trait across all vehicles is the small geometric scales on which the spinning wings, or rotors, must work, resulting in very low Reynolds number (Re) flow around the blades. Because of the low atmospheric density and temperature on Mars, low-Reynolds-number aerodynamics play a critical role in aerial vehicle design for the Mars expedition. Low-Reynolds-number flows are characterized by the relative importance of viscous and inertial forces inside the fluid. As a result, boundary-layer physics includes flow separation, re-attachment zones, and the quantity of laminar/turbulent flow on the shape change, so while adopting geometrically scaled models from full-scale to small-scale regimes, caution must be exercised because they may not deliver the highest potential aerodynamic performance.

CLASSIFICATION OF BIO-MIMIC MODELS

Bio-Inspired surfaces, shapes, or models can be classified by the origin of inspiration such as Birds, Insects, land animals, or sea creatures. In another case, Bio-Mimic can be categorized by the principle of operation such as adopting the movements of animals, birds, and insects, in which the kinematics moments are traced and recreated exactly following the shape mimic. Jellyfish, and squid inspired to create pulsed-jet-powered underwater vehicles. Protuberances or Humpback or serrated geometrics along the leading edges, of whale flippers, can improve aerodynamics, and hydrodynamics performance is relatively novel and inspiring one especially in the case of low Reynolds no.

SHAPE-BASED BIO-MIMIC

Serrations or wavy trailing edges on aerofoils, wings, propeller blades, or even nozzles have been studied to obtain positive aerodynamic characteristics for feature flight. this kind of serrations improves aerodynamics stall also increases the maximum lift in low Reynolds number. The serrated wing generates streamwise-aligned vortex pairs which delay flow separation along the wing surface under given conditions. There is an unlimited number of possible combinations in terms of wavelength, and amplitude to create different geometrical configurations with favorable aerodynamic characteristics within the given flow environment. The aerodynamics of sinusoidal and smooth leading-edge wings. with different leading-edge configurations (smooth and sinusoidal), and different trailing-edge shapes were constructed. Wind tunnel tests were conducted to compare these wings for a wide range of angles of attack and different low Reynolds numbers. A smooth leading-edge wing advances the stall point while a sinusoidal leading-edge wing delays the stall for the same Reynolds no, and the maximum lift coefficient significantly enhances for sinusoidal leading-edge wings. it was identified that using serrated/sinusoidal leading-edge causes a higher lift coefficient compared to those without serrations on all Reynolds numbers. It has been concluded that a sinusoidal leading-edge wing, helps to delay the stall and produce maximum lift coefficient. Figure 1 shows the whale humpback flippers and the pectoral flipper with leading edge tubercles.

Figure 1. Whales humpback flippers and the pectoral flipper with leading-edge tubercles
Source: Nature-inspired solutions to bluff body aerodynamic problems: A review, 2021

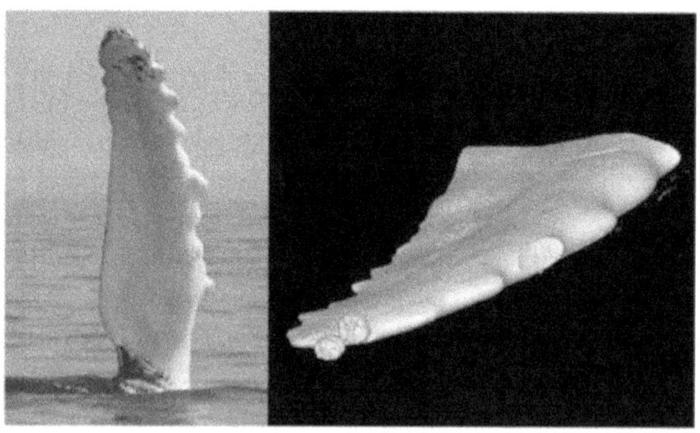

The peculiar leading structure of the whale's flippers drew researchers' notice. It works by causing flow excitation, which delays the stall for high angles of attack. Initially, a simulation was conducted on a wing with and without tubercles and discovered that adding tubercles boosts lift by 4.8 percent, lowers generated drag by 10.9 percent, and improves the lift to drag ratio by 17.6 percent. The stall delay was then reported by using over-idealized humpback whale flippers in a wind tunnel experiment. They discovered a 40% delay installing when compared to a smooth wing, as well as reduced drag and enhanced lift. The flippers' higher lift at a large angle of attack was the cause of the stall delay. Flippers maintain minimal drag at high lift during post-stall. During feeding and moving, this technique is beneficial to whales.

An aerodynamic model for whale flippers that explains why they stall later than expected. To see the consequences of bumps, they altered the anatomy of a smooth wing according to whale flippers. The amplitude of the bumps affects the control qualities of flippers; as the amplitude grows, the lift curve flattens and remains indifferent with wavelength. A water tunnel experiment discloses yielded similar findings. Tubercles based on the modification of airfoil NACA0021 by simulating flippers discovered that the tone noise of the airfoil is reduced due to the streamwise production of the vortices modified by tubercles. It's thought to affect the boundary layer's stability. In further studies, these tubercles perform better on NACA 65-021 than on NACA 0021. Lift coefficient and stall are both improved with higher amplitudes. Furthermore, it performs even better during post-stall by delaying the stall angle, raising the maximum lift coefficient, and other performance factors up to a certain wavelength. The flow behind the smooth and tubercle's wing was also reported to interact with the oncoming flow from tubercles lateral to crest and downstream vortices. Due to tangential velocities, the pair of vortices from the inward-facing side change direction towards the trailing edge of the wing. The tubercle peak also has the effect of speeding up the process. After merging with the vortex pair, flow further downstream. The separation of the atoms is halted as a result of these effects. The stall line was shifted backward due to the passage of downstream tubercles. The impact of these tubercles is once again highlighted. As a result, such protrusions on whale flippers are not unique to tubercle-free whales. Alternate forms are also proposed that offer similar benefits in terms of agility and maneuverability. As shown in Figure 2. which details the various forms of wavy patterned foils for comparison with whale tubercles conducted for a comparative study. With numerous possibilities, such as boundary layer momentum, angle of attack, vortex lift, and effective angle, the flow mechanics behind these improvements are still unknown. Many experiments have examined a competing hypothesis in this area, which will be valuable for future research. It was thought that the force for swimming comes from the flukes of cetaceans, whereas the lift and torque for maneuvering come from the flippers. However, a recent estimate showed an extra function of the whale flippers for the first time. Flapping provides lift and boosts propulsive thrust with its dynamic stroke. Massive forward-oriented forces can be generated by these strokes, which can improve lung feeding performance. Most of the living species are having corrugations in its biological part and the functions of them in their respective way was explained (Buckholz, 1986). The numerical investigation was carried out for two different corrugated aerofoils in an unsteady condition (Ho and New, 2017).

The bioinspired corrugated aerofoils were discussed in detail about its performance (Ho and New, 2013). The performance of the corrugated dragonfly airfoil was compared with the plain airfoil at low Reynolds numbers (Tamai et al., 2007). Micro air vehicles with the wing design of dragon fly was incorporated and the numerical and aerodynamic performance analysis were carried out for fluid flow (Gaurav and Jain, 2014). At lower Reynolds number of 8000, the design of simplified dragon airfoil was tested and its aerodynamic characteristics were calculated (Levy and Seifert, 2009). The flow separation was controlled by different modifications on the tip of the airfoil and the surface features of corrugated airfoil was measured in the flow control (New et al., 2014). The normal plain and dragonfly airfoil aerodynamic characteristics were compared (Okamoto, Yasuda, and Azuma, 1996).

Figure 2. Wavy wing section mocked from tubercles of whale
Source: Effect of Airfoil-Preserved Undulations on Wing Performance and Wingtip Vortex, 2020

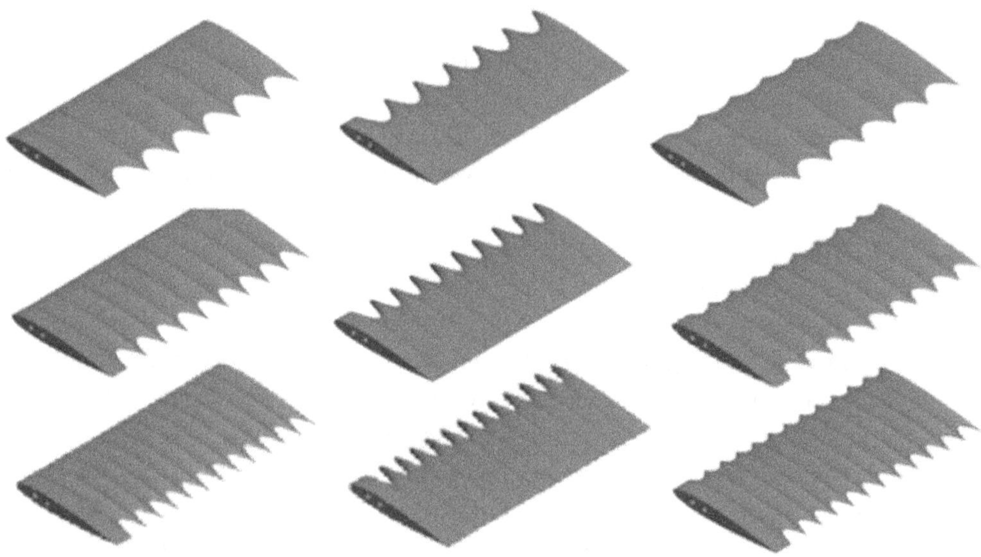

AERODYNAMICS OF RIBLETS

The intricate shape of sharkskin micro-grooved surfaces that reduce drag in turbulent situations was first discovered which opened the path of exploring drag reduction techniques by surface mimicking, the thickness of the viscous sublayer is the cause of this drag reduction. If the thickness of the contact surface is greater than the roughness of the contact surface, the rough surfaces will be buried in it, converting friction to viscous resistance. Sharkskin has micro-grooved structures made up of riblets that evolved. These grooved surfaces are thought to be the cause of lower viscous drag and turbulence intensity. The longitudinal grooves in sharkskin riblets reduced drag by up to 8%. They maintained the same order of groove dimensions as the turbulent wall streaks and bursts. A steep peak groove with valley curvature is also the best rib form. Figure 3 depicts the skin of a white shark. The effect of riblet characteristics on the hydrodynamic has been investigated. The viscous sub boundary layer caused by the shark's longitudinal rib has been studied theoretically, and the origin of the velocity profile from the riblets was predicted by the theoretical computation. The 'protruding height' notion states that it resides below the tips of riblets, with a distance of 10-20% of rib spacing. Through oil tunnel studies, the 5-8 percent claimed drag reduction was enhanced to 10% when compared to smooth surfaces. Sharkskin riblet has been modified to fit technical difficulties by drawing inspiration from nature. The experimental results on converging-diverging riblets in the turbulent boundary layer with zero pressure were observed. The thickness of the turbulent boundary layer is greatly affected by large-scale periodicity in the spanwise direction. As a result, the thinned boundary layer is caused by an increase in local mean velocity and drop-in turbulence intensity. Particle image velocimetry was used to explore the effect of streamwise riblets in turbulent boundary layers.

Figure 3. Textured skin of a white shark
Source: Nature-inspired solutions to bluff body aerodynamic problems: A review, 2021

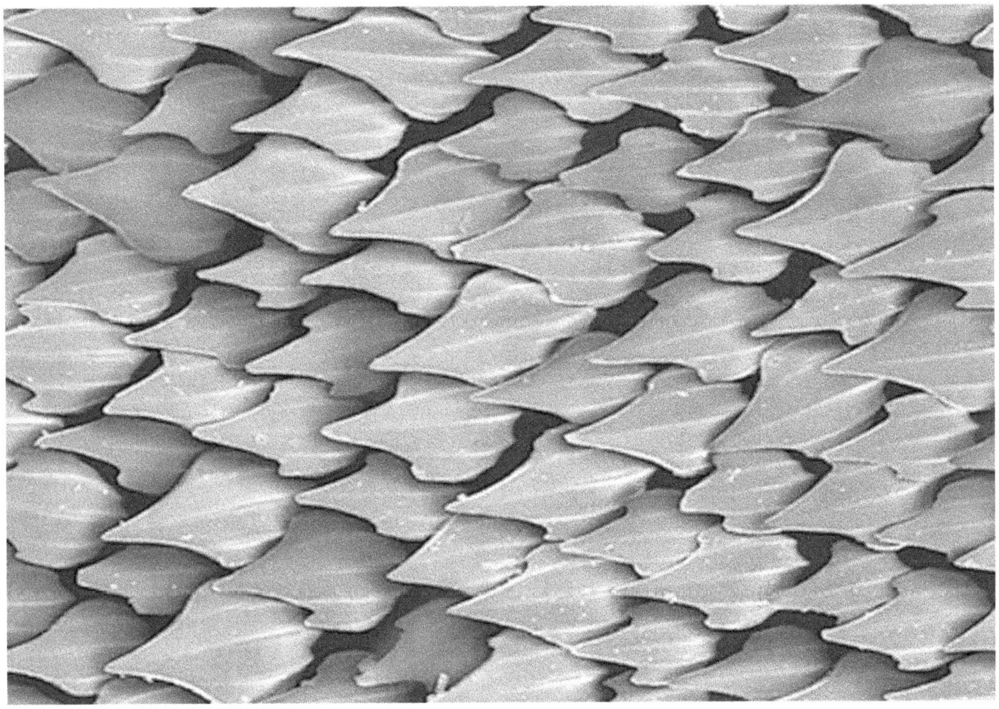

A decrease in friction velocity and Reynolds stress within the turbulent boundary layer, resulting in a reduction in drag. Increases in streamwise riblet surface diminish hairpin vortices, according to the link between hairpin vortices and momentum distribution. In contrast to the smooth surfaces, these are rough. Similar research has been done in this area, making sharkskin riblet an acknowledged strategy for reducing drag.

However, the impact of riblet cannot be fully appreciated until all of the elements that influence shark mobility are established. The sharkskin can bristle its scales while in motion. An increase in momentum was seen near the slip area that arises above the scales in the experiment. As a result, the rise in velocity can be attributed to separation control of the boundary layer. This is a different issue that needs to be looked into. A unique perspective on the angle of attack across sharkskin scales experimented. where it comes to lowering drag and turbulence intensity, the angle of attack is a huge role. The scales fluctuate depending on the swimming circumstances. They further claimed that scales' complex morphology acts as a super-hydrophobic surface with a contact angle of more than 150 degrees. Due to viscosity effects, this causes boundary slippage at the fluid-solid interface, which can lessen the velocity gradient as well as resistance. Another important consideration is the function of nanochain mucus, which stretches within the boundary layer, resulting in a more stable and consistent flow. It was also noted the diversity of shark riblets throughout the body, noting that no two sharks had the same second rib on the whole surface. a new type of riblet inspired by denticles that improved the wing's aerodynamics significantly. At a low angle of attack, this inspired device shown in Figure 4 can improve the drag to lift ratio by 323 percent, outperforming conventional vortex generators. This was due to a change in flow pressure caused by separation bubbles in the wake of denticles, which improved suction. The streamwise vortices were responsible for the loss of momentum caused by skin friction. The significance of this study is that it has introduced the lift and lift to drag ratio metrics, which are important in understanding the role of denticles, in addition to the drag reduction process is addressed.

Figure 4. Surface structure inspired by shark denticle
Source: Nature-inspired solutions to bluff body aerodynamic problems: A review, 2021

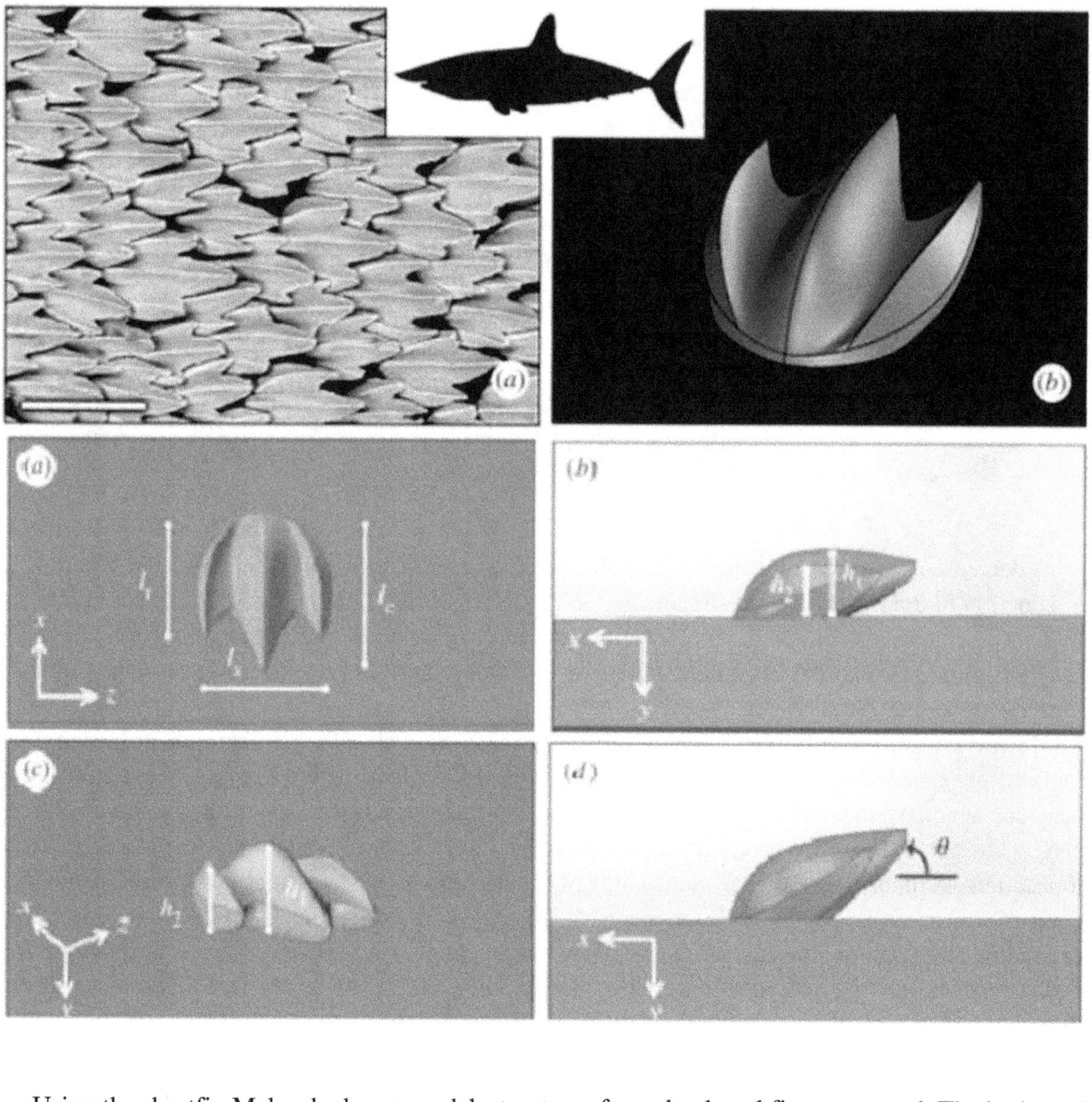

Using the shortfin Mako shark as a model, structures from the dorsal fin are created. The body and the dorsal fin have slightly different structures, and the replicated created structure was rounded, semi-rounded, and lengthy, as illustrated in Figure 5. that rounded and lengthy denticles reduce drag the most. Long and rounded at a 9.5-degree angle of attack with 5 m/s speed has the lowest drag coefficient of 0.011.

Figure 5. Shark dorsal fin denticles encouraged structures. (a) Rounded inclined and (b) Rounded horizontal
Source: Nature-inspired solutions to bluff body aerodynamic problems: A review, 2021

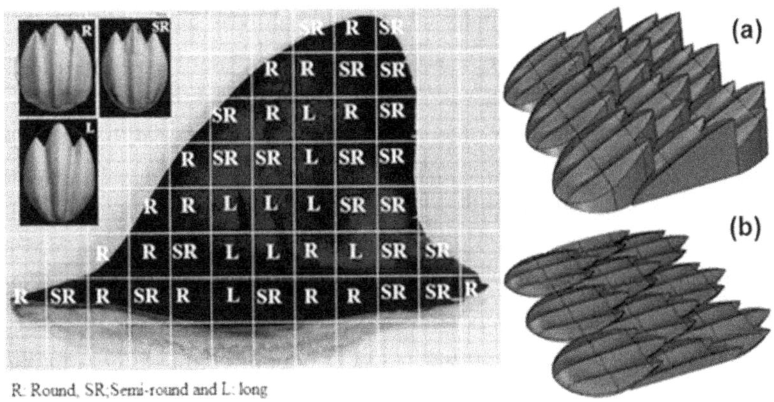

Direct numerical simulations have been done by adding sinusoidal surface roughness to superhydrophobic surfaces. It has been discovered an asymmetric secondary flow inspired by microgrooves that fluctuate in a streamwise orientation. The transverse shear strain on top of the sinusoidal microgroove is discovered to be similar to a Stokes spatial layer (SSL). A simulation was conducted and experimental tests were also conducted to determine the influence of the micro-grooved surface on the blade of an air engine. As a result, it has been discovered that a micro-grooved blade reduces drag better than an untextured blade. They also improved the texture placement on the blade surface. Inspired by riblets, a slew of application-oriented studies has been carried out Similarly, the protrusions on the surface of sailfish have been studied for their ability to reduce friction drag. In another experiment, it was discovered that the drag caused by sailfish riblets did not improve significantly. The reported reduction in skin friction drag was only 1%. The shark riblets, on the other hand, have better aerodynamic performance. The aerodynamic impacts of the fishes' surface structure have been examined in this section. The physical knowledge of structures like Riblets and shark denticles led to the invention of new types of vortex generators that reduce drag significantly. The basic physical understanding of forms, wavy shape, tail and fin geometries, and the surface structure of fishes constitute a fascinating field of inquiry. These are some of the ideas that help us better comprehend fluid flow.

GRAY'S PARADOX

The overall drag is made up of pressure and friction drag, both of which are affected by the surface's geometry and nature. Friction contributes a considerable amount of drag in streamlined bodies, and in bluff bodies, friction contributes a big portion of drag. Body shape and appendages, as well as the active and passive-interface, are used to manage flow around the fish. Only the body and protrusions will be examined, with the interface and dynamic flow manipulation left out. Because eddying motion transfers the outer momentum to the inner wall region, which delays separation, laminar flow is more prone to separation than turbulent flow.

Some species use roughness at the maximal girth location to induce turbulence. Scombroid is a fast-moving fish family that developed so that the laminar boundary layer changes to a turbulent layer near the maximal girth or body width, delaying flow separation. At Reynolds numbers more than 2×10^6, the flow is considered transitional or turbulent. This relationship between body shape and size is linked to active swimming in the ocean's epipelagic zone. Because they can't handle the ocean waves below 15-30 cm in length, nektonic fishes are between 15 to 30 cm in length reported in a similar argument. There are already differences in shape and size of the fishes if they live in the same environment. One argument seems that the pressure changes depending on how deep a fish can swim, however, this is based on a purely evolutionary method that needs to be addressed. Gray's paradox is the theory that the force required to overcome water drag is greater than the fish's available muscle power. the dolphin's flippers, dorsal fin, and flukes, as well as streamlined designs projected to reduce drag by delaying flow separation. The fineness ratio (FR) is a proportion of maximum length to a maximum thickness that is used to describe the extent of a body's streamlines. The drag will be minimal at maximal volume for an ideal FR. For dolphins and whales, the FR ranges from 3.3 to 8. This supports the idea of, mathematically connected bodies of revolution of the hull and established the FR minimal resistance. A decrease in hair density indicates a reduction in drag. A link between drag and maximum girth is also coined in view of drag estimation in dolphins. Dolphins' maximum thickness is between 34 and 45 percent of their total length. By preserving the pressure gradient and laminar boundary, this shape is similar to any hydrodynamic foil as a result, flow separation occurs later in the body, near the dorsal fin. When the experiment was conducted on live dolphins swimming in a bioluminescent (a living organism emitting light) sea, no significant separation was seen. Skin friction drag, on the other hand, is unaffected by body shape. As a result, fish must act to maintain laminar flow for the most part to avoid excessive flow separation. This can only be accomplished by using streamlined shapes. Moreover, at a high Reynolds number, for the ideal shape of a fish that shows attached flow with a laminar boundary layer. Dolphins have been observed to have no separation and no-wake vortex as a result of this. Furthermore, the volumetric friction coefficient (C_v) is determined by the Reynolds number rather than the geometry of the body.

This is also applicable for laminar attached flow and the minimal drag on a rigid body revolution. The least achievable value ranges from 0.0015 C_v to 0.002 C_v, which can only be achieved by an animal with a slim shape. Similarly, a three-dimensional simulation of a model fish discovered that when the swimming movement increased, the drag increased as well. There is a critical velocity at which force and drag are identical, indicating that drag and swimming velocity are linearly related. Gray's paradox motivated researchers to study the swimming behaviours of fish during the second and third quarters of the twentieth century. An experiment was conducted on a little fish in a water tunnel because it was believed that all previous studies had problems. The findings contradicted commonly held beliefs about laminar and turbulent flow. Furthermore, the quanta's peculiar motion violated the fundamental laws of motion. Related to this gap, two hypotheses were developed, notably the vortex peg and scale the force, according to which propulsive forces are due to the vortices of two-thirds of the fish body, as well as after fins and tail surfaces. These forces are generated by centrifugal forces and angular momentum within the vortex core. The drag calculation of the streamlined shape does not work for undulating fish shapes because of this variance. The fish's great swimming speed is due to a progressive wavy-type motion. On a 222-centimeter rubber sheet, a water tunnel experiment is conducted and discovered that when $(c/U_0) = 1$, each trough has an isolated vortex, but when $(c/U_0) > 1$, there is no vortex, where c is the fish's wave velocity and U_0 is the uniform water velocity.

Swimming reduces wall shear stress by shifting the turbulent boundary layer into the laminar boundary layer. The concept of waving motion was no longer a mystery that wavy motion increased movement in fish and flyers. Choosing a big aspect ratio wing to replace the bird's wing and fishtail as a model revealed a general tendency in which the wing used the wave's transverse velocity for optimal mobility. As a result, the wing's incidence increased, increasing thrust at the crest and trough.

Figure 6. Approximate dimension of sailfish
Source: https://www.pinterest.com/pin/atlantic-sailfish--723672233866097560/

Figure 6 shows swordfish and sailfish, which have the fastest speeds in the ocean at roughly 110 km/h. Sailfish drag coefficients are 0.0075 and swordfish drag coefficients are.0091, according to hydrodynamic research on sailfish and swordfish inside a wind tunnel. These coefficients are far lower than those seen in tuna and tiny fish like dogfish. The absence of flow separation from the entire body, even without the bill, was revealed by velocity measurements.

During gliding and cruising, sailfish fold their first dorsal, first anal, and pelvic fins. But swordfish is not followed in such a case. Examined the usage of the long bill in sailfish and swordfish for the first time and discovered that it is employed to feed on prey by entering into the school of prey without showing off an attack. Form drag is comparable to that of swordfish, killer whales, and manta rays. So, if the fastest species under the sea do not differ in terms of friction drag, what additional factors contribute to their speed? Adequate information about the form drag that elaborates the effect of shape on aerodynamics is not available, more research is needed.

AERODYNAMIC CHARACTERIZATION OF BIO-INSPIRED BOXFISH

The use of nature-inspired body shapes in engineering has a long history, with applications in both aeronautical and ground vehicles. Even if a large drag reduction was obtained with such an unattractive appearance, even the shape does not look appealing. This would result in a fuel consumption reduction of around 12.64 percent. Similarly, the body form of a sailfish has been considered as a possible replacement for an airplane's fuselage. Sailfish are well-known for being the fastest animal on the water. The nose of the fuselage was redesigned as a result of this idea. The features of the sailfish proved to be more aerodynamically efficient than the traditional fuselage. This resulted in a 10% reduction in drag force, demonstrating the importance of an elliptical cross-section over a circular cross-section. This has been addressed since the potential for this design to be used on the front of tractors and buses is enormous.

Sailfish droplets at the torso fin used to rise from pockets of air and water scales were embedded into a P1 hypercar similarly. Inside the duct that connects to the engine, this texture was put. It improved the car's efficiency by 17 percent by increasing the volume of air delivered to the engine. It was a required chore because the electric Hypercar's 903 horsepower necessitates a lot of air. Similarly, the boxfish's odd shape inspired Mercedes-Benz to create a completely new automobile design. Figure 7 depicts the shape as a substantial adoption, even though precise details are not available in the open literature. This design sparked debate, with some claiming that it would be difficult for an automobile to turn in a straight line due to morphological reasons. However, boxfish have increased agility as a result of this destabilizing moment.

Figure 7. Car model inspired from box fish
Source: Bio-inspired design: aerodynamics of boxfish, 2014

In view of course stabilization of boxfish caused by vortex lift created from the edges will bind a car to a track like a train, and aberration from this straight path demands significant energy. As a result, the advantages can the exceptional maneuverability of a boxfish be transferred to a car traveling at a considerably faster speed adopting a total shape design is an area that requires further exploration. Adapting several of the boxfish's characteristics, however without specifying the characteristics themselves. Since the goal is to approximate the shape of a boxfish, which resulted in a drag coefficient of 0.28. In figure 8, the mesh model of box fish is shown.

Figure 8. Generated mesh for the boxfish model
Source: Bio-inspired design: aerodynamics of boxfish, 2014

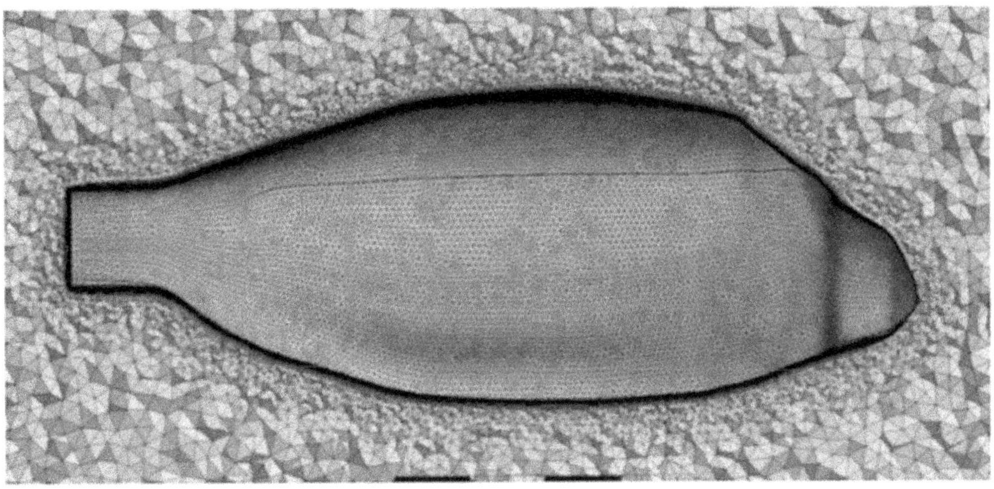

Vehicle design is far more involved than simply looking for aerodynamic advantages. As a result, it's critical to seek key qualities that may be duplicated in other components of the vehicle. The use of non-smooth surfaces on airplane fuselages, which did not entice the researcher to apply the same concept to ground vehicle design. For the first time, experiments attempted to apply it to a vehicle structure. The size and dimensions of the cart with a non-smooth surface. With such non-smooth surfaces, only the engine cover lid and car body cap were evaluated. Controlling the boundary layer, they claimed, resulted in a 10.31% reduction in drag. The burst and loss associated with turbulent kinetic energy were reduced as a result of this control. Figure 9. Gives the surface characteristics.

Figure 9. Surface streamline characteristics (a) front view; (b) rear view.

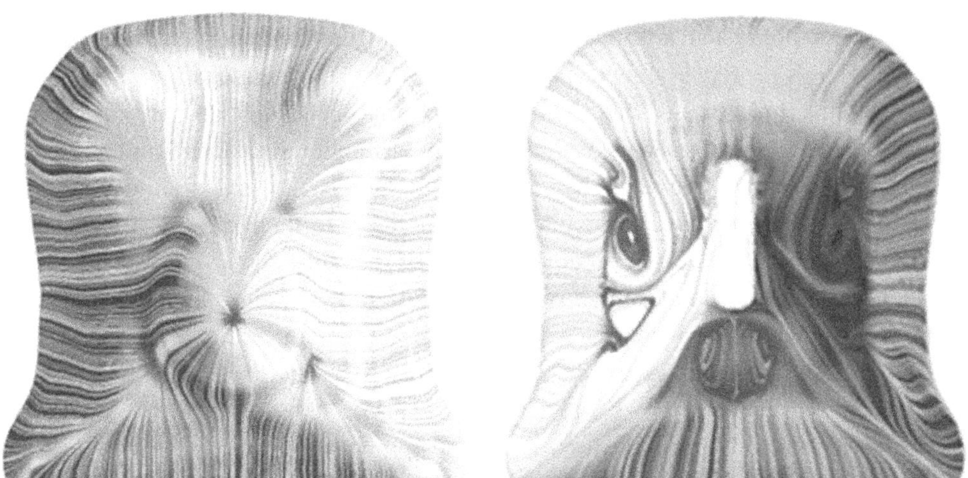

Figure 10. Velocity pressure model (a) velocity vector for the simplified model; (b) static pressure characteristics of the simplified model.
Source: Bio-inspired design: aerodynamics of boxfish, 2014.

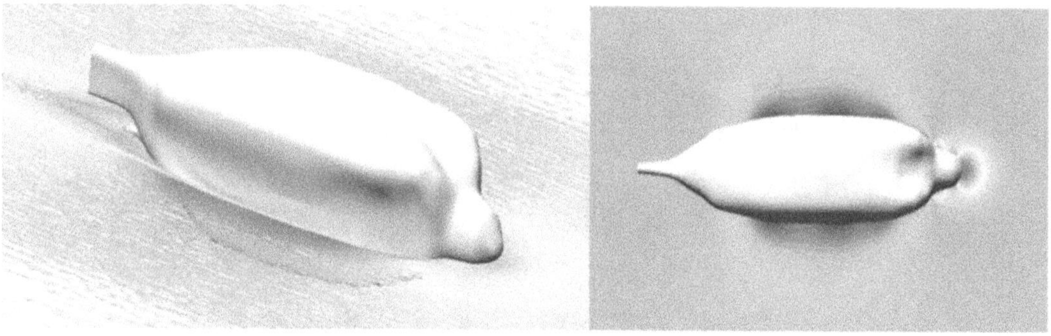

Circumstantial forces led to the evolution of a bluff body form in the water, in contrast to the expected streamlined shape for superior aerodynamics. A boxfish (Ostracion Meleagris) with a similar form to a bluff body. The boxfish has a unique design that allows it to operate in a small space with near-zero turning radiuses and a 180° shift in direction. The significance of the pectoral and caudal fins is also becoming more essential. Swimming is rapid and has a long-endurance, according to quantitative research. The speed of the boxfish fluctuates depending on how much body length is traversed in one second and is dynamically stable. Their parasitic drag is similar to that of a well-streamlined body. Figure 10 gives the velocity vector of the box fish model.

The formation of strong longitudinal vortices coming out of the keels of fish varied with the angle of attack which showed the development of strong longitudinal vortices coming out of the keels of fish varied with the angle of attack. The pressure was low on the carapace surface, which resulted in a concentric vortex. The ventral keels of all boxfishes are predicted to make self-correcting motion looked into the differences in the aerodynamics of real fish and a model fish. The stability of live and prototype boxfish is higher, also the strength of vortex circulation in live fishes was larger than in the model due to the pectoral fin interaction with kneel-induced flow. The ability of boxfish to use their fins to change the underlying self-correcting system is critical for stability and maneuverability. The most important feature is that the keel system's passive system has established such a stable equilibrium. The created vortices automatically stabilize the boxfish; otherwise, flow separation would occur, resulting in significant drag, but such repercussions are managed at no expense. Ostracion Meleagris and Lactophrys triqueter boxfishes have a speed of six body lengths per second, which is a combination of uncommon design elements. Vortex intensity increases as the pitch angle increases, although it was higher in live boxfish. The contact of the pectoral fin with the keel flow could be the cause of the higher vortex. Despite this, past research has produced a conundrum: how can a boxfish be forwardly stable while spinning quickly is because that drag reduction performance in boxfish is lower than the normal fish morphology through experiment and modeling. The drag coefficient falls with speed, with a minimum of 0.26 being reported. They've also discovered unstable times that are accountable for the best maneuverability. It's because the active fins have various purposes. Experimental and computational research of the aerodynamics of boxfish shape in terms of drag coefficient to develop a way for applying bioinspired design to vehicles. It is discovered that the unique shape of boxfish has a drag coefficient of 0.10, which is significantly lower than any other model on the market. It contributes to the recovery of pressure and reduces the drag coefficient due to diffusion from all sides. The drag coefficient convergence characteristics of the simplified boxfish model for the standard k-turbulence model. All tests were carried out at inflow velocities of 20 to 100 km/h, which correspond to the Reynolds number shown in Figure 11. As results finalized from current research, boxfish may not be the best choice for automobile manufacturers due to the self-correcting stabilization of what is known as 'vortex lift,' which keeps boxfish in a straight path, hence thorough research is required. To take a turn with such a self-correcting mechanism, a car would need a lot of energy. However, because not all of the activities of such fishes are understood, and the results are conflicting, this argument appears to have a lot of room for more investigation. Lower Re numbers showed that k- underpredicted CD whereas higher Re numbers showed that k- overpredicted CD. However, the accuracy of the wind tunnel employed for lower Re numbers has been questioned.

Figure 11. Variation of drag with Reynolds number

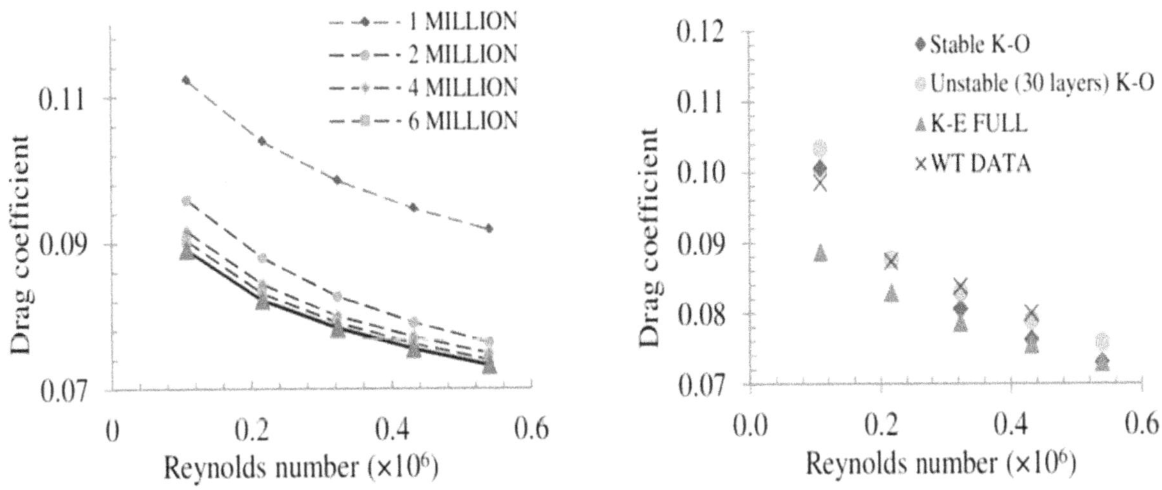

This section summarised the significant progress achieved in comprehending natural shapes such as streamlined and blunt bodies. The topic of streamlined shape emphasizes the evolution of flow behaviour knowledge. Natural flow management strategies are depicted in three different body shapes: rigid, flexible, and wavy. However, the streamlined shape isn't a unique feature of nature, as the blunt-body is said to have a low coefficient of drag. As a result, the debate over shape obscures the veracity of beliefs about the secrets of true energy harvesting processes. Now, through the design of boxfish, nature has struck humanity's ability to comprehend fundamental flow problems once more. Moreover, it is the comprehension of underlying principles that we have formulated, not just a change in structure. Because streamline and bluff body are two natural geometries, how the flow behaves in a general sense has yet to be determined. The lowering of the drag coefficient is interesting for such a form since water resistance is substantially higher than air resistance.

AERODYNAMICS OF BIRD FEATHERS

The wings produce the majority of the lift and drag in a bird's flight. The Helmholtz hypothesis states that the wing is what causes the generated drag at the tip. Birds' basic anatomy and morphology, on the other hand, vary greatly, from insects to gigantic soaring birds. Some birds fly at low Reynolds numbers, while others fly at extremely high Reynolds numbers Birds flying over land differ from those flying above the sea in several ways. The architecture of bird feathers is critical for understanding why such unique arrangements are directly linked to improved flying performance. A bird feather is made up of a shaft and vanes, with the vane being resistive to aerodynamic forces due to the barbs' inclination. There is a moment due to variances in vanes and shaft, yet, it cancels out due to many feathers. The mechanical behaviour of the wings is influenced by the morphology and functioning mechanism of feathers. Primary feathers, particularly those at the tip, are more resistant to stresses than the interior ones. The tip feathers of a bird's wing reduce drag by allowing air to pass through them, and tip reversal upstroke is used to reduce drag.

The Reynolds number remains around 105 due to the relatively small wing size and speed. The Reynolds number in large birds can reach the transition zone, reducing drag and boosting lift coefficient. The Reynolds numbers of most of the large soaring birds range from 75000 to 106, with 75000 being the critical Reynolds number for a smooth bird-like airfoil. The shape of the bird feather has a significant impact on aerodynamic performance. The overall lift and drag coefficient are affected by their size and shape. During the second wingbeat of take-off, substantial acceleration has been seen, and feathers contribute significantly to acceleration when compared to other regions of the bird's body. Figure 12. Shows various types of feather arrangements in the wingtip. The major feather tip gets vertically curled with slots between them during the trip. The span of a feather is increased by this form. The flat wing, whose shape is constant from root to tip, does not suppress vorticity as effectively as a non-planar surface does in the case of a bird feather

Figure 12. Wing tip shapes (a) rounded (b) pointed (c) concave (d) convex
Source: Nature-inspired solutions to bluff body aerodynamic problems: A review, 2021

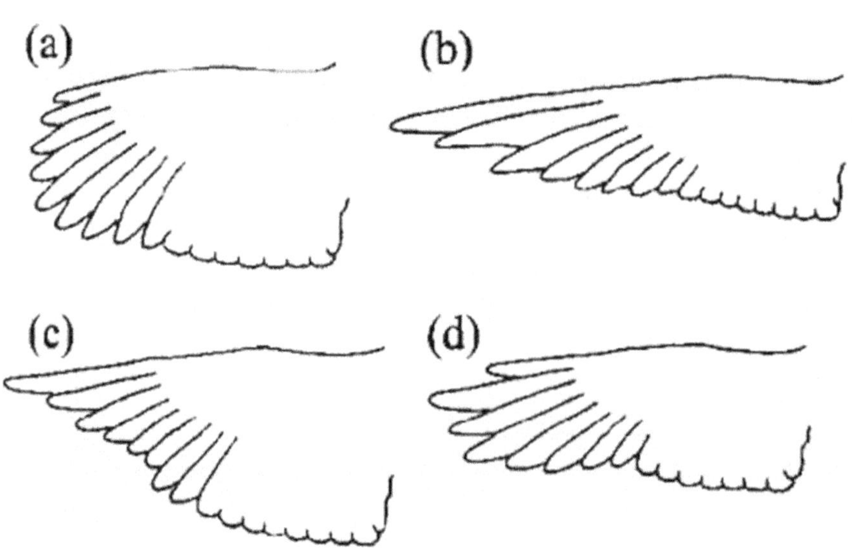

In-flight hawk, the half wing is 1.06 m long from root to tip, and 0.92 m long from the head-on side. The bird feather immediately minimizes the produced drag by following this form, which disperses the vorticity vertically. During landing and take-off, induced drag accounts for a large amount of the drag. The feather's span, area, and size are all altered by a vertically slit feather tip. All of these parameters have an impact on the induced drag, either directly or indirectly. In general, separating the feathers vertically at the tip improves wing efficiency, but maximum performance is gained by spreading the feather over a high dihedral angle.

The major criteria for reducing induced drag are higher aspect ratio, elliptical loading, and low coefficient of lift, all of which are supported by the classical theory of aerodynamics. However, due to design implementations, these criteria have limits. The weight constraint keeps track of the increase in aspect ratio. If only the planar wing is considered, then decisions must be taken. To get out of this situation, a non-planar wing is assumed that may lower induced drag while also stating that the induced drag could not be less than C_d minimum. planar tandem wings could not reduce drag much. An analytical solution for non-planar wings. Similarly, it was proposed that advanced methods for reducing induced drag may be developed based on the lifting line theory, with the major shift being the development of non-planar surfaces that disperse drag in a vertical direction, such as wingtip forms. Arguments and methods for recovering energy from the wingtips, the nonplanar wing, have been developed using this approach. The main lesson from the preceding discussion is that due to their peculiar configuration during flight, bird wings are naturally non-planar. Because the surface of feathers is distinctive for aerodynamics, they can lower both induced drag and profile drag. A flacon is tested in a wind tunnel and found that the highest lift to drag ratio was achieved at a speed of 12.5 m/s. With increased speed, the bird's lift coefficient, wing area, and span decrease.

The aerodynamic properties of a bird's wing at Reynolds numbers of 104 and compared them to insects wing at Reynolds numbers of 1 to 5 and real airplane wings and discovered that bird wings have a large minimum drag coefficient of.03-0.13, a low maximum lift coefficient of 0.8-1.2, and a low lift to drag ratio (L/D) of 3-17 due to low Reynolds number operating. The bird has a low efficiency compared to conventional airfoils due to its high profile drag. Another experiment on the other hand makes it plain that the flow structure across the bird's wing is consistent with aerodynamic theory. A gliding bird in a wind tunnel evaluated the wing aerodynamics of two different species. The wingspan of both the Falcon and the Vulture is reduced, allowing them to glide freely. The induced drag increases as the wingspan increase, however, because the bird's wing has an aerofoil section, it has a minimum profile drag coefficient around one, whereas a conventional wing has a value of 0. The same method was used to study a Harris hawk floating freely in a wind tunnel at speeds ranging from 1.1 to 16.2 m/s. The maximum lift to drag ratio was 10.9. The relationship between airspeed, glide angle, and wingspan is obvious since as speed or glide angle increases, the bird's wingspan decreases. Similar research has been carried out to determine the wingspan and its relationship to drag and lift.

An asymmetric wing has three significant effects: the entire span can be modified, which influences drag and lift; it will cause an asymmetrical yawing and rolling moment; and finally, the turning performance would be harmed. The issue of additional lift in insect flight in a paper published in Nature. It means that insects can't fly, according to acronymic theory, because flapping produces more lift than stable flight, although the explanation for such high lift isn't known. flow visualization shows a leading-edge vortex created on the top of a wing. This vortex is caused by a stall in the dynamics, and it is unaffected by rotational lift. However, it was later discovered that this unusual lift production system is not exclusive to insects, as swifts also exhibit a leading edge vortex. This leading-edge vortex starts at a large angle of attack in normal conditions. This vortex must be close to the wing to be used. Insects beat their wings and swifts sweep their wings backward to attain this purpose.

This vortex must be close to the wing to be used. Insects beat their wings and swifts sweep their wings backward to attain this purpose. Another change in the form and span of the bird's wings is due to the birds' active management of their shape and span. Aerodynamic effects are produced by this natural way tested on a swift in gliding flight and regulate the wingspan as well as the sweep angle during glide based on the same notion. They discovered that stretched wings perform well in turns and sluggish gliding, while swept wings perform well in quick gliding and turning. Swept wings are not good at providing lift, but they can withstand heavy loads. The force coefficients are controlled by the bird's wing forms, speed, and angle of attack. Furthermore, due to the overlapping of vanes and expanded shafts, the primary feathers of swift are extremely rough, with a roughness height of 1-2 percent of the chord length at the upper surface, when compared to a sailplane surface that is 10,000 times larger, this roughness height is significant. The main difference between a traditional sailplane and a swift surface is that sailplanes use a smooth surface to reduce drag and enhance the laminar boundary layer area, whereas the swift uses roughness to do so. In a low-turbulence wind tunnel, they measured the laminar flow over a rapid wing. During gliding motion, the laminar surface over the quick wing is roughly 69 percent of the total, which boosts their flying distance and duration. It can reduce drag due to the extended run of the flow without separation.

AERODYNAMICS OF THE CORRUGATED AIRFOIL AT A LOW REYNOLDS NUMBER

In recent years, micro air vehicles (MAVs) with a wingspan of 15 cm or less and a flight speed of roughly 10 m/s have gained a lot of attention. The aerodynamic performance of an airfoil is measured using the chord Reynolds number Re, which is based on the chord length and flight velocity of the airfoil. Traditional macroscale aircraft have chord Reynolds numbers in the range of 10^6–10^8, whereas MAVs have chord Reynolds values in the range of 10^4–10^5. Due to the large difference in chord Reynolds numbers, aerodynamic design techniques relevant to classic macroscale aircraft may not be applicable to MAVs. Several insects, such as locusts, dragonflies, and damselflies, have non-smooth or simple cambered surfaces on their wings. The wings have well-defined corrugated patterns in their cross-sections. Such a corrugated form was discovered to be critical to the ultralight wings' stability in handling the spanwise bending stresses and mechanical wear that occur during flapping. Several studies on corrugated dragonfly wings in steady flow or gliding flight have led to an extraordinary conclusion: in the low Reynolds number regime, a corrugated dragonfly wing could have comparable or even better aerodynamic performances (i.e., higher lift and bigger lift-to-drag ratio) than conventional streamlined airfoils. Despite differing explanations for the fundamental process for increased aerodynamic performance, most researchers agree that corrugated dragonfly airfoils or wings perform well in low Reynolds number regimes, implying that such corrugated airfoils or wings could be used in micro air vehicles. With this in mind, we undertook the current study to see if we could use the corrugation feature of dragonfly wings to improve the aerodynamic efficiency of MAV designs using unconventional bioinspired corrugated airfoils. The pressure contours of corrugated profile and NACA 0015 are shown in Figure 13 while the velocity contours are in Figure 14.

Figure 13. Pressure of corrugated profile, pressure of NACA 0015
Source: Aerodynamic characterization of bio-mimicked pleated dragonfly Aerofoil,2021

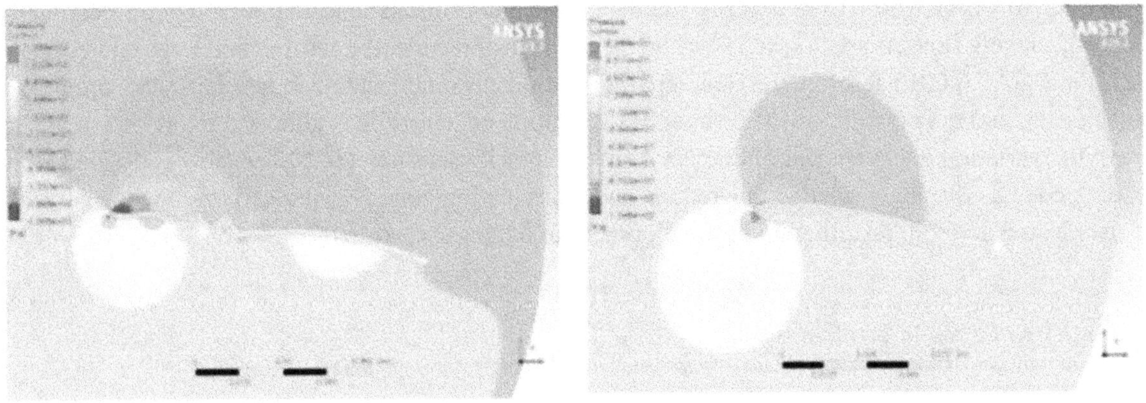

Figure 14. Velocity of corrugated profile, velocity of NACA 0015Source: Aerodynamic characterization of bio-mimicked pleated dragonfly Aerofoil,2021

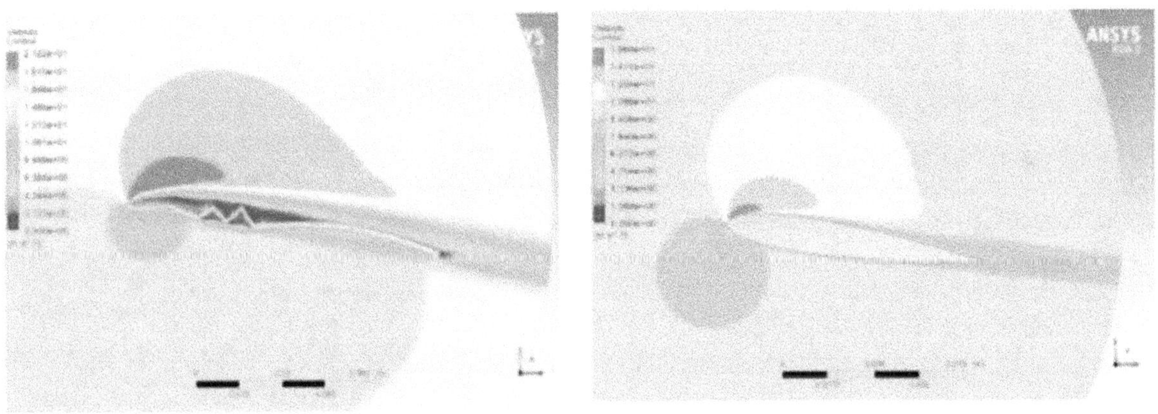

The experimental and numerical simulation comprehends the performance parameters of the bio-mimicked corrugated dragonfly wing cross-section and the traditional NACA0012 aerofoil at two distinct Reynolds values of 75000 and 115000. According to CFD simulations, the pleated corrugated shape performs significantly better than the smooth aerofoil at all angles of attack. When compared to the typical aerofoil model explored here, the corrugated aerofoil lift-to-drag ratio is higher, ranging from 4 to 8 degrees, and it has a high maximum gliding ratio.

At 40% of the forewing chord, the pleated aerofoil design adopted, bio-mimicked from 'Pantala flavescens,' provides much superior and compatible outcomes. Finally, we discovered that the CFD analysis solution provides more insight into the aerodynamic behavior of the corrugated wing structure and that the natural design specification plays a critical role in reducing the formation of wake due to flow breakdown (i.e., adverse pressure gradient), which causes a recirculation zone inside the V-shaped grooves, and the flow reattaches to the edge of the next corrugation. Finally, it was discovered that the dragonfly wing's pleated structure provides improved aerodynamic performance at low Reynolds numbers, i.e., a high L/D ratio, and aids in the reduction of shear drag, which is important for lightweight and structurally stable MAVs and UAVs.

As the AoA is increased to 12 degrees, the stagnation point shifted beneath the crest of the corrugated aerofoil, causing flow circulation and swirl on the top face of the pleated aerofoil and large eddies on the traditional NACA 0015 aerofoil, the flow separation at the upstream and boundary layer also progresses much faster, and the pressure on the camber of the trailing edge profile is almost zero, as shown in Figure15 the variation of coefficient of drag vs Angle of attack was shown in.b. The flow is divided at the leading edge as the angle of attack increases, and a circulation zone is generated on the upper surface of the pleated aerofoil, resulting in the development of eddies and vortices.

Figure 15. a) CL vs AoA b) CD vs AoA
Source: Aerodynamic characterization of bio-mimicked pleated dragonfly Aerofoil,2021

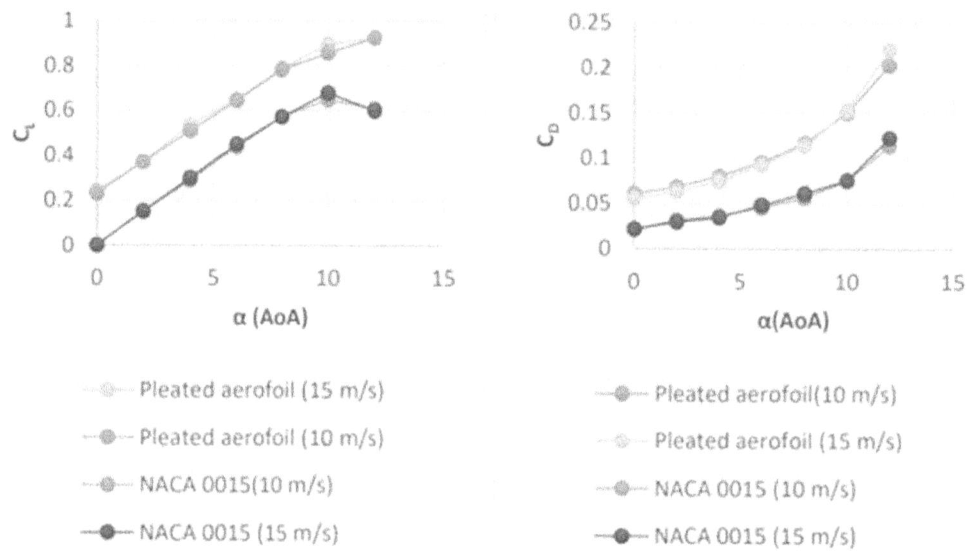

AERODYNAMIC NOISE REDUCTION

In Japan, the bullet train was recently researched to see if it may help with the noise problem. There were three unique reasons for the noise, none of which were related. The first was caused by high-speed vibrations between the railway frame and the ground. The second was an aerodynamic noise caused by the train's blunt body and pantographs that connected it to catenary wires, and the third was a sonic boom that occurred when it entered the tunnels. The creation of Karman vortex sheets, which are alternate and opposite eddies flowing in the back of the object, was the second aerodynamic noise. The serration of owl feathers was explored to overcome this problem and discovered to operate as a vortex generator. They employed a V-shaped triangular cross-section arranged on both sides of the turbulence line after modification to suppress the Karman vortex and adjust the airflow in a parallel manner. The tunnel's third issue was more serious. When the train enters the tunnel, air waves travel at sonic speeds through the tunnel end, resulting in a sonic boom. The shape of a kingfisher was studied in order to tackle this challenge.

The functioning principle of the kingfisher and train was discovered to be the same through a large-scale model and computational investigation since they both shift the pressure environment from less dense to denser. When implemented, the bill form has a unique morphology with a consistent circular cross-section, which lowered air resistance by 30% and power consumption by 13%. Figure 16 depicts the bullet train's ultimate shape. Figure 17 gives the velocity contour of kingfisher beak.

Figure 16. Kingfisher and the respective inspired train
Source: https://peterfalkingham.com/2019/05/15/kingfishers-their-beaks-and-diving-physical-modeling-and-cfd/

Figure 17. Velocity contours of kingfisher beak
Source: https://peterfalkingham.com/2019/05/15/kingfishers-their-beaks-and-diving-physical-modeling-and-cfd/

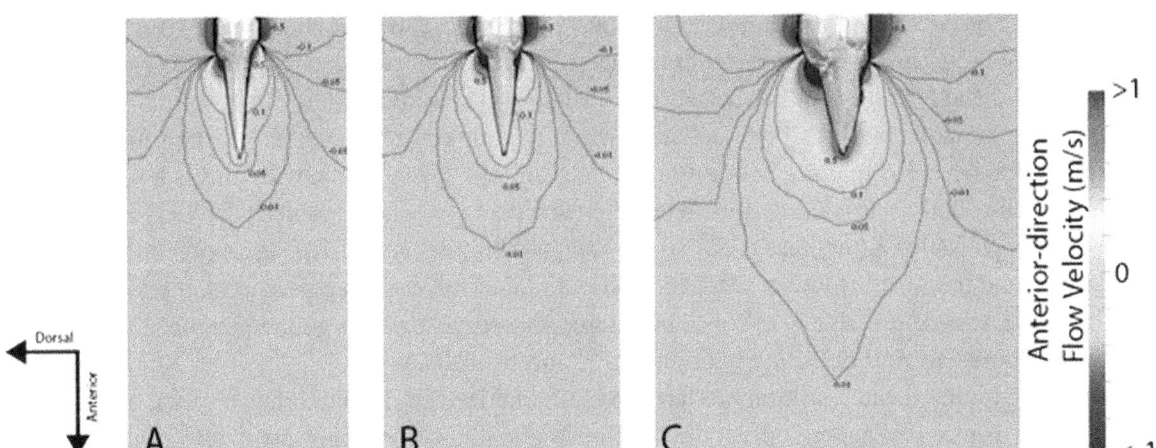

COMPARISON OF VARIOUS SHAPES

There are many models inspired by biomechanism or species that are effectively in use in many sectors. A few examples are listed in table 1 below with the evaluation method and their respective aerodynamic performance.

Table 1. Species and bio-mimic inspired models

Species / Mechanism	Evaluation Method	Bio Mimic Inspired Model	Aerodynamic Performance
Tubercles of Whale	Experimental/CFD simulation	Wind turbine Blades	C_L and Stall velocity increases
Riblets	Experimental	superhydrophobic surfaces	Drag reduces substantially
Box Fish	CFD/Experimental	Bionic Car	10.31% reduction in drag
Kingfisher Beak	CFD/Experimental	Shinkansen Bullet Train Nose	Noise level reduced to
Bird Wing Tip feathers	CFD	Vortex generators	Noise and Induced drag are reduced
Dragonfly wings	CFD simulation	MAV wings	High L/D ratio in low Reynolds No.

CONCLUSION

Three different species of fish, birds, and terrestrial animals in the above nature-inspired optimization of aerodynamic forces. The crucial issue is that these species have a highly developed and sophisticated flow control mechanism at low Reynolds numbers, with some active strategies used as and when needed, therefore not all of the methods can be mimicked from nature. The focus of the fish review was on the aerodynamic impacts of body shape and the various methodologies used by researchers, as well as the key results reached. Furthermore, the dispute over huge fish streamlining morphologies was compared with a discussion of boxfish, which has a lower drag coefficient than well-shaped fish and models. The argument between these two body types isn't over yet, therefore more research can be done on it. Passive devices such as fins, tails, and flippers have also been considered and demonstrated to be capable of reducing drag, increasing lift, and delaying stall. Furthermore, the architecture of fish, particularly sharks with micro grooves, gives greater drag reduction than a flat surface. This topic has sparked a lot of discussions, and a lot of work has been done to study it. The aerodynamic effects of the un-flapping aerodynamics around the bird's body, tail, and wing, wingtips were studied using the same methods. The study of avian aerodynamics is largely limited to aircraft applications, with only a few techniques used in the bluff body, particularly in road vehicle flow management. Secondary feathers and primary feather wingtips' self-pop-up process has a lot of potential for delaying flow separation and dispersing vorticity behind the wake. However, while considering inland animal drag prospects, it was discovered that there is less work available in the open literature. As a result, the flow around large animals like the cheetah and horse, as well as the usage of their tails, has been discussed.

Promising future research inspirations have been explored in detail after going on such a trigonal voyage to locate inspiration that can enrich the bluff body and notably vehicle flow control. The strategies proposed in this debate, as well as how they might be applied to bluff bodies, are discussed. However, it has been stressed throughout this research that nature does not support the concept of smooth surface drag reduction. Overland, fish, fliers, and animals do not have smooth surfaces. As a result, in both streamlined and bluff bodies, the non-smooth surface should be given special attention. Nature's unusual designs are challenging traditional ideas about drag and motion. Depending on the level of complexity and necessity involved in the process, nature used both aggressive and passive approaches. Overall, the ideas from three different species were combined in this review study to serve as a quick reference for a nature-inspired solution to bluff body aerodynamics.

The driving force acquired by flapping the wings mainly comes from the following three areas, First, from the effect of the tail vortexes. The Karman-Burgers theory says that the vibration of wings removes the vorticities in the boundary layer, generating reverse Karman vortex street, which carries momentum opposite to the direction of wing movement to generate thrust. Second, the suction in the leading edge. When the fluid flows through the leading edge of birds and their head with large curvature, the local flow velocity increases to form a low-pressure region, which causes the leading-edge suction to generate some driving forces. Third, the effect of inertia. When the wings are flapping, the local additional mass effect produces additional inertial forces. The output power of birds during flight is the main subject to study the principles of birds' flight. These ideas provide a theoretical basis for the research on the new aerodynamic configuration of flapping-wing aircraft. The development of bio mimic flight systems involves various disciplines such as micromechanics. Microelectronics energy power unsteady aerodynamics. At low Reynolds number. The highly turbulent inflow conditions dominants aeroacoustics mechanisms of serrated leading edges might be compressed to spanwise decorrelation effects and also effects of destructive interference. The Spectral scaling laws applied for single aerofoils results in broadband noise reduction. Though, transferring serrated leading edges to rotating machinery, results in noise radiation patterns of significantly increased complexity, impeding to allocate the observed noise reduction to the underlying physical mechanisms. The aeroacoustics transferability of leading-edge serrations. concatenating the scaling laws for stationary aerofoil and rotating-blade application low-pressure axial fans are designed, obtaining identical serrated fan blade geometries than plane leading-edge single aerofoils.

REFERENCES

Buckholz, R. H. (1986). *The functional role of wing corrugations in living systems*. Academic Press.

Gaurav, S., & Jain, K. K. (2014). Numerical Investigation of Fluid Flow and Aerodynamic Performance of a Dragonfly Wing Section for Micro Air Vehicles (MAVs) Applications. *Int J Innovat Scient Res*, *92*, 285–292.

Goldfield, E. C. (2018). *Bioinspired Devices: Emulating Nature's Assembly and Repair Process*. Harvard University Press.

Ho, W. H., & New, T. H. (2013, December). CFD Analysis of Bio-Inspired Corrugated Aerofoils. *Proceedings of Eleventh International conference of Fluid Dynamics*.

Ho, W. H., & New, T. H. (2017). Unsteady numerical investigation of two different corrugated airfoils. *Journal of Aerospace Engineering*, *231*(13), 2423–2437. doi:10.1177/0954410016682539

Levy, D. E., & Seifert, A. (2009). Simplified dragonfly airfoil aerodynamics at Reynolds numbers below 8000. *Physics of Fluids, 21*(7), 071901. doi:10.1063/1.3166867

New, T. H., Chan, Y. X., Koh, G. C., Hoang, M. C., & Shi, S. (2014). Effects of corrugated aerofoil surface features on flow-separation control. *AIAA Journal, 52*(1), 206–211. doi:10.2514/1.J052398

Okamoto, M., Yasuda, K., & Azuma, A. (1996). Aerodynamic characteristics of the wings and body of a dragonfly. *The Journal of Experimental Biology, 199*(2), 281–294. doi:10.1242/jeb.199.2.281 PMID:9317808

Tamai, M., Wang, Z., Rajagopalan, G., Hu, H., & He, G. (2007, January). Aerodynamic performance of a corrugated dragonfly airfoil compared with smooth airfoils at low Reynolds numbers. In *45th AIAA aerospace sciences meeting and exhibit* (p. 483). 10.2514/6.2007-483

Chapter 15
Aerodynamics Enhancement Using Adaptive Flow Control

Sundharasan R.
Jaya Polytechnic College, India

Gowtham G.
Vel Tech Rangarajan Dr. Sagunthala R&D Institute of Science and Technology, India

Kumaran T.
Vel Tech Rangarajan Dr. Sagunthala R&D Institute of Science and Technology, India

Elumalai K.
Vel Tech Rangarajan Dr. Sagunthala R&D Institute of Science and Technology, India

ABSTRACT

Flow separation and wind gusts affect the aerodynamic performance of low-Reynolds number flyers with a chord-based Reynolds number of 105 or below. Active flow control provides information into fluid dynamics as well as potential vehicle performance enhancements. Aircraft engines, as well as stationary flow devices, exhibit undesirable flow states that have a significant impact on noise output and aerodynamic loss. Vibrations and aerodynamic performance are likely to suffer as a result of flow and boundary layer separations. Flow management can be an effective approach for lowering noise levels, increasing efficiency and thereby lowering fuel usage. Various forms of flow can be controlled using both passive and active flow control strategies. The active alteration of aerodynamic flows utilizing small time-dependent actuators in well-chosen places is defined as adaptive flow control. Micro-adaptive flow control has a significant overall system benefit in a variety of large-scale applications, which are presented.

INTRODUCTION

Adaptive flow control is described as the selected engagement with flow instabilities on a large scale to modify flows with technical value. These instabilities are capable of extracting energy from the external flow. They can increase significantly too large amplitudes and they can be activated selectively

DOI: 10.4018/978-1-6684-4230-2.ch015

by simple actuation. A fraction of one percent of the total flow is enough for movement if the position and frequency of actuation are chosen correctly. This method allows for low-power, widely dispersed, and redundant actuation systems. The method is fundamentally more practical than prior generations of flow control systems that relied on consistent blowing to deliver high velocity flows via intricate, high-loss ducting. The amplitudes needed for stable actuation are typically an order of magnitude smaller. Active flow control has been chosen for high-payoff applications that can result in significant system performance improvements. Complex networks have helped in the identification of implementations with the highest possible advantages, which in turn enable performance gains.

In its most basic form, the flow control subsystem enables the integration of actuators, sensor systems, and response control systems. Actuators can be modulating flow valves or electrically driven devices because the actuation inputs are often time-dependent. Detector arrays and control systems are still in their infant stages, and also the current set of actions in the program is mostly an open loop. Implementing active systems also necessitates the use of design tools that account for time-dependent flows.

FLUID MECHANICS CONCEPTS

Constantly evolving the Navier–Stoke equations revolutionized continuous fluid dynamics technology and the above resulted in the idea of an effective control system via stability analysis. Important breakthroughs that revolutionized our understanding of the turbulent flow model, on the other hand, inadvertently slowed the development of control systems. Reynolds' famous inspection of conversion in pipe flows in 1883 was one such discovery. He differentiated between quiescent (laminar) and sinuous (turbulent) flow and split the real-time time velocity vector into stable and irregular elements since the latter appeared to be unpredictable. Reynolds stresses (Cho and Shyy, 2011) and the period of equations (Reynolds averaging) sparked a century of research in establishing practical forecasting tools based on this approach. It is necessary to keep in mind that turbulence is the inherent state of most flows of engineering importance. Considering the flow inside any pumps, turbines, and engines, as well as the flow over any conventional vehicle. An uneven three-dimensional turbulence action combined with significant fluid mixing distinguishes it. Mixing is now an important feature of turbulent flow because uneven motion may occur in solid, specifically those that are compliant. There is a vast variety of levels due to the irregularity of the motion, meaning that a completely predictable explanation of the flows is not attainable. It was no surprise that turbulence has been defined statistically, with pressure and velocity divided into average and information about the level. The above biodegradation developed a new system of equations that look similar to the immediate motion equations but that can be solved unless supplemental equations for new unknown factors implemented by the biodegradation (i.e., the Reynolds stresses) are computed in some way. For many decades, this would be the crux of the turbulence problem, or as it was known in the field, the completion problem. Since there have always been more unknown factors and explanations given, adding more equations to the fundamental Navier–Stokes equations ever solved this issue. It just shifted the unavoidable random conclusion to the new set of circumstances. A whole "modeling business" devoted to producing mathematical models of turbulence grew over time.

RANS algorithms have effectively established themselves as the most convenient tool known to date, but they are routinely used in industrial applications. To a large part, the accuracy of the actual inputs that all these models use influences their success. As a result, rather than projecting the behavior of a completely unique flow, they may predict the kinds of flows since there is a huge amount of information.

Traditional Reynolds averaging is seen to be harmful to turbulent shear flow control because it fails to explain the basic processes behind turbulent shear flows or propose ways for changing or regulating them. As a result of Osborne Reynolds' most fundamental finding, turbulence reduction has become fatalistic.

The use of difference equations or spectra techniques to solve the contemporaneous equations in numerical modeling of the whole flow field seems to have the possibility to be a significant advantage in AFC. However, unlike an experimental facility, the numerical simulation does not rank the parameters impacting the flow in order of importance. As a result, it produces outcomes indiscriminately, with no knowledge of the physical principles driving the flow. Direct numerical simulation (DNS) results, on the other hand, can be an If the physical characteristics governing the flow have even been vaguely recognized, this is a critical tool. Indeed, DNS results have become a key source of turbulence information not previously available from experiments. The most important innovation that led to the AFC notion was Prandtl's boundary-layer theory. It divided the fluid flows into something like a small surface of rotating fluids adjacent to a solid surface, surrounded by a massive body of inviscid unsteady fluid. The concept depicts flow around streamlined bodies accurately, as well as mechanical losses and heat transfer characteristics between the surface and the atmosphere fluid. The ability to determine the pressure gradient around a slender body using unsteady flow services is a significant practical benefit of boundary-layer theory. This resulted in the development of the separation concept, in which a viscous-retarded fluid breaks away from the surface due to a significant unfavorable pressure gradient. Experiments supported Prandtl's hypotheses as well as set the direction for their failure separations. Those that as well identified its conditions under which normally stable laminar boundary layers develop instabilities and become turbulent. Prandtl's concept laid the groundwork for the computational viscous/inviscid interaction paradigm by combining much further flow field answers to vicious ones at the boundary layer's tip. The requirements for separation naturally led to the criterion for boundary layer stability and existence. The use of turbulent predictions based on Reynolds decomposition to extend the boundary-layer theory to turbulence hampered AFC development because it claimed the randomized movement was limited by local fluid flow conditions, and therefore controls at a single place wouldn't have had lengthy implications. This had been linked to the widespread use of modeling in engineering fields, which gave rise to the famous expression "turbulence flow forgets its beginning." This belief was deeply engrained until the discovery of the massive analytical hierarchy process in turbulence shear flows. Even though the Reynolds-averaged calculations are mathematically correct, they are only useful in turbulent shear flow regions where turbulence is relatively uniform and rotationally symmetric. In an intermittently turbulent flow, such as that found in the outer half of a turbulent boundary-layer, awake, or a jet, the Reynolds-averaged equations will mix together and indiscriminately average the vertical (turbulent) fluctuations with the irrotational fluctuations lying outside the instantaneous turbulent boundaries of the flow. Along this border, the action of viscosity imparts vorticity to their spinning fluid.

By supposing that perhaps the terrific layers are continuously and that there is no island of disturbed fluids in the unsteady flow region, Corrsin and Kistler seem to have been capable of calculating the endurance T of the massive eddies there at the boundary layer's horizontal stripe. This was most likely the very first measurement of the mean range of a massive flow moving there at the boundary layer's periphery, implying that turbulence isn't as randomized as previously supposed. Few authors illustrated the constraints of a completely statistical method for turbulence using zone averaging and conditional sampling (Jahanmiri, 2012). Inhomogeneity was observed in the turbulent boundary layer's wall area Richard Wlezien (2000) and the consistency among some wide fluctuations was established (Jaganraj et al., 2015).The most important observations that began a study on patterns in turbulent were made in this double mixed surface (Shyy et al., 2008). Their Schlieren picture demonstrated that large-scale eddies dominate the turbulent mixing layer, conveying smaller, more uniform turbulent eddies inside (Lissaman PBS 1983). The spreading rate of the mixing layer is clearly linked to the production of the massive flows that devour unsteady liquid from of the nearby flows (Gad-el-Hak, 2000).

Coherence patterns in turbulent study surged and dominated practical and theoretical efforts in the 1970s and 1980s. More advanced variable-interval time-averaging (VITA) techniques replaced area averages and condition samples, which maintained and used specific temporal information associated with a well-known event. Because of their capacity to provide immediate knowledge over such a large portion of the flow instead of comprehensive-timeline, time-series- series focused on a single point, flow visualization and particle image velocimetry (PIV) techniques have risen to prominence. The scholar was able to test these thoughts using quantitative information after using modeling methods to generate new ideas. By extending Reynolds' original technique, the so-called triple provides a framework for dealing with the problem of turbulent boundary management. The tri deconstruction identifies that unstable movement can be divided between large, coherent, predictable patterns that really are anticipated and tiny, currently unpredictable structures that can only be defined using statistical methods and are thus thought for being arbitrary. Applying hydrodynamic stability theory to a turbulent flow field could be the first step in building logical flow control models, and this technique can provide the theoretical approaches needed to handle enormous swirls. Inviscid stability theory gave a good prediction of both the absolute magnitude and frequency distributions throughout the mixed layers when implemented in the induced turbulence mixed layers, but only a quantitative expectation of perturbations amplified in the path of the stream. Large coherent structures can thus be seen as the result of competing for unstable waves traveling downwards and amplification or diminishing with time, however, this perspective appears to apply better to inviscid unstable flow.

Randomized fluctuation, on the other hand, affects the development of quantitative measures. In this circumstance, they have really no choice but rather to employ some variation of the model. By use of a limited forced pulse was required to give a phase baseline for information collection using a minimal amount of hot sensors, resulting in information being supplied at only a few locations, which led to the concept of turbulence Flow manipulation and control. Even just a minor magnitude perturbation alters both normal flows and the amplitude and dispersion of turbulence Flows (Glezer and Amitay, 2002). This allowed for the control of more shear layer flow, separation controlling, as well as the isolating of several key control factors. Furthermore, limited fluctuation has been demonstrated to really be highly sensitive to the forms of interaction patterns, meaning that manipulating unstable types allows for single-molecule- molecule controlling of contaminant reactions. In recent decades, the amount of AFC research and possible applications has increased. Changes in the indirect benefits of turbulent boundary fluid flow will unquestionably have a huge technological effect as our ability to control these flows improves, affecting the measurements and patterns of wingtip stips, reflectors, engine components, surface, and submersible equipment, and other components. To put it another way, every device that deals handle liquid dynamics, whether exterior or interior, will almost certainly change in the future. In turbulence streams, flow visualization initially discovered the presence of large alignment patterns forty years ago. In recent years, the focus of fluid flow analysis had also shifted significantly, from the numerical collection of fluid flow levels of intensity as well as Reynolds stress modeling to the exploration of patterns, as well as from acknowledging the inevitable consequence of the presence of a worldwide flow velocity to manipulating and altering that uniqueness. Now the time has finally come to develop a compromise mostly on the subject and present a point of view that will hopefully influence the next generations of students and trickle down through all of the employment environment.

Figure 1. Illustrates the interrelationship of flow control goals (Courtesy: Gad-el-Hak1998)

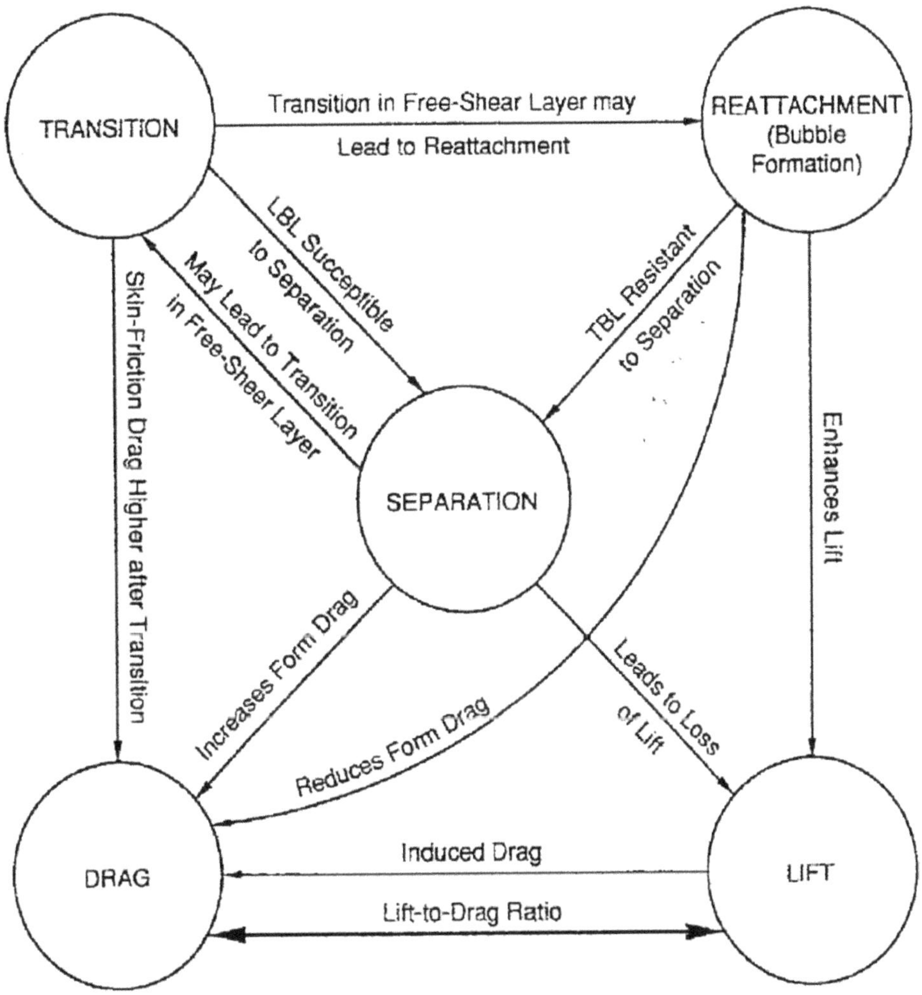

Figure 1. depicts the interrelationship of flow control goals as an essential issue. Consider a flow that develops on the exterior surface of an airplane or a submarine and is confined by a wall. Transition delay, separation postponement, lift increase, skin friction and pressure drag reduction, turbulence augmentation, heat transfer enhancement, or noise suppression can all be achieved by manipulating this type of flow. These goals do not have to be mutually exclusive. Consider the flow developing on a lifting surface, such as an airplane wing, to help concentrate the discussion. When the boundary layer gets turbulent, its resistance to separation increases, allowing for more lift at higher incidence. A laminar boundary layer, on the other hand, can have a skin-friction drag that is up to an order of magnitude lower than a turbulent one. Lower skin friction and flow-induced noise are achieved when the transition is delayed. The laminar boundary layer, on the other hand, can only withstand a very modest unfavor-

able pressure gradient before separating, resulting in a loss of lift and an increase in form drag. When the laminar boundary layer divides, a free-shear layer forms and the transition to turbulence occurs at moderate Reynolds values. Due to turbulent mixing, increased entrainment of high-speed fluid may result in the reattachment of the split zone and the creation of a laminar separation bubble. The bubble breaks down at increasing incidence, either totally separating or producing a longer bubble. In either situation, form drag increases and the slope of the lift-curve falls. The ultimate goal is to improve the performance of the airfoil by raising the lift-to-drag ratio. Induced drag, on the other hand, is created by the lift generated on a finite-span lifting surface. Furthermore, higher incidence generates more lift, but form drag also increases at these angles.

All of the preceding points to potential conflicts as a result of attempting to achieve one control aim while inadvertently affecting another. There is no such thing as an ideal technique of control that is simple, economical to build and operate, and has no trade-offs, thus the experienced engineer must make constant compromises to reach a specific design goal.

Adaptive Flow Control Methods

The Major Adaptive Flow Control techniques are summarized in the following sections, Flow actuators come in a variety of shapes and sizes, with varying energy consumption, control authority, bandwidth, and operating conditions (Joslin and Miller, 2009). Most actuators for flow control can be split into two groups based on the energy conversion mechanism: devices based on mechanical motions to impact the fluid, and devices controlling the flow field without moving parts. Synthetic-jet (zero-net mass-flux) actuators and various flaps are among the former. Due to their large bandwidth and efficient energy conversion capability with simple structures, piezoelectric materials have recently become popular for generating mechanical motions.

Boundary-Layer Aspiration

A systematic strategy to improve the performance of engine compression systems includes boundary-layer aspiration. Low-energy flow is removed from the flow stream through aspiration. This low-energy flow would otherwise limit the compressor's diffusion (and thus its work) by creating flow separation from the blades' surface (Roth et al., 1998). The feasibility of a quasi-three-dimensional design technique that allows precise prediction of boundary layer flows in response to changes in blade form, as well as the effect of removing the boundary layer fluid on the downstream behavior of the boundary layer, is critical to the approach. More aggressive blade designs (Figure 2.) have been made possible, with higher turning angles and no flow separation. Aspirating a portion of the boundary layer flow through correctly located slots on the blade surface achieves separation control.

The input could be supersonic (Figure 3), and the aspiration slot is situated near the blade suction surface, where separation is just beginning. Flow control, in this context, is a passive strategy for suppressing the growth of instabilities and energizing the boundary layer by removing the mass. Only by utilizing the synergy between blade shaping and the effects of boundary layer removal can the higher work enabled by aspiration be attained. The work of a compressor stage can be doubled (at fixed blade speed and efficiency) by eliminating as little as 1% of the flow, according to designs proven by three-dimensional studies and experiments.

Figure 2. Comparison of conventional and aspirated Blades

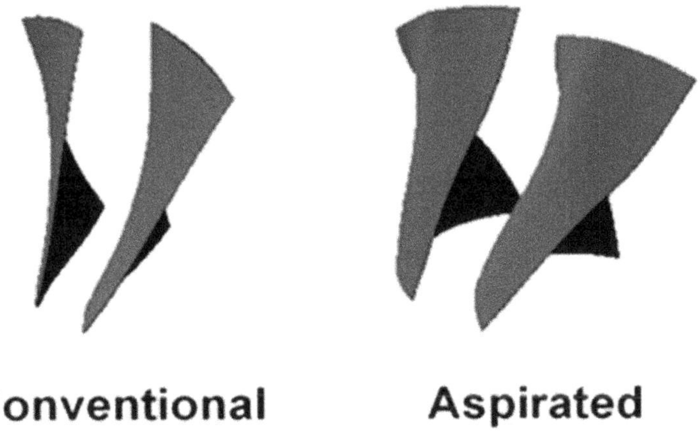

Figure 3. Computed blade flow field with aspiration

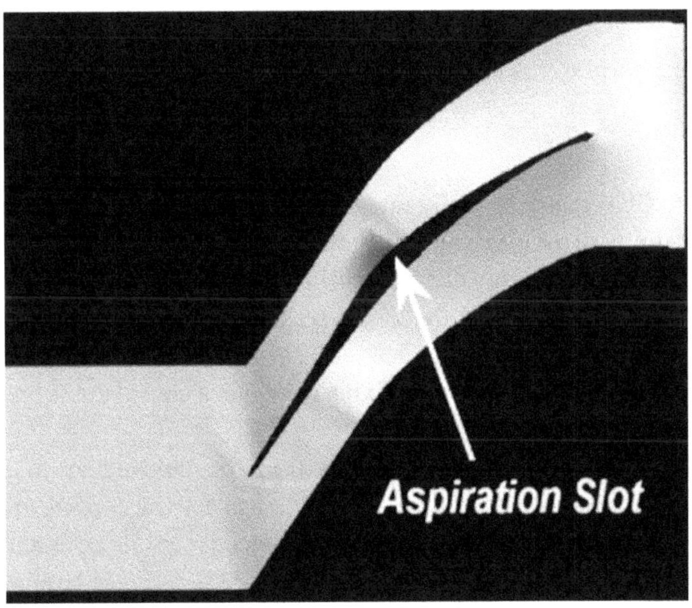

A distinct but related aspect is that removing the high entropy viscous flow might help improve compressor efficiency. Experimental results have proven the performance of one of the two phases that have been designed. At a tip speed of 750 feet per second, one stage produces a pressure ratio of 1.6, while the other produces a pressure ratio of 3.4 at a tip speed of 1,500 feet per second. The first has been constructed and tested, while the second has been developed, studied, and is currently being built. The stages have a composite wrapped tip shroud (Figure 4.) that provides structural stability for a stage with thin aspirated blades. The tip shroud also makes it easier to pull air out of the blade tips and into a suction plenum. Aspiration provides a reduction in stage count of around half while also reducing the weight of the compression system by roughly doubling the work capability of a stage. Even more dramatic decreases in stage count are achievable when the effects of aspiration and counter-rotation of succeeding stages are combined.

Figure 4. Shrouded aspirated stage design for Pressure ratio 3.5

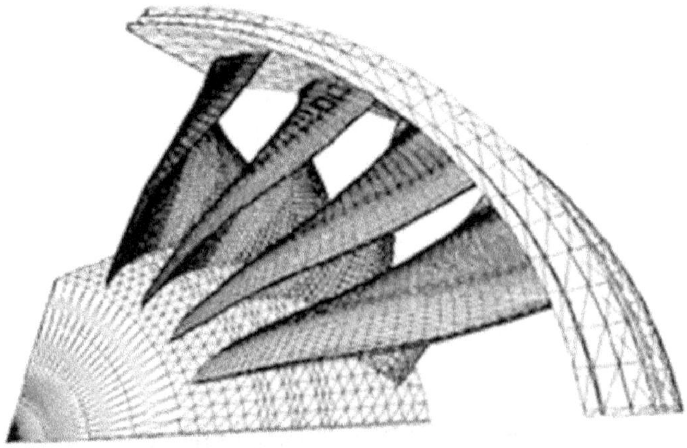

Active Core Exhaust

Active Core Exhaust (ACE) control is a fully integrated, non-intrusive pulsed injection technology that causes large-scale instability in the core exhaust plume, resulting in better mixing of the plume with the surrounding ambient air. A tiny quantity (1.5 percent) of high-pressure compressor bleed air is injected perpendicular to the core exhaust flow at the core nozzle exit to cause plume instability. Natural flow instabilities are pushed to saturation in this application by selectively activating these instabilities through injector ports at the nozzle exit. The increased mixing results in a considerable and rapid decrease in the exhaust plume temperature, as seen in Figure 5

The Core Thrust Reverser (CTR) on the C-17 aircraft can be removed due to the lower exhaust temperature. The CTR is not necessary to meet the C-17's backing needs, according to analysis and statistics, and is primarily employed to meet the load masters' human effectiveness temperature requirements during engine-running ground operations. Because ACE is a demand system, it can be employed to minimize the thermal burden on the trailing edge flaps during takeoff and descent before being turned off during the mission's cruise phase. The ACE control system will cut the cost of each new C-17 aircraft by roughly $1.2 million and the weight of each C-17 aircraft constructed or converted by about 1200 pounds.

The engine bleeds air is guided through a fluidic actuator, which injects air into ports around the nozzle exit on a regular basis (Figure 6). All tubing and injector channels are fully integrated into the nozzle shell and the present engine housing in the rebuilt nozzle. The nozzle's outward profile has not changed as a result of the redesign. At the 17th stage, bleed air from the compressor will be removed from an existing port and channeled through a fluidic actuator, which will direct the bleed air to the proper injector. This actuator is designed to give a steady load to the compressor bleed port and has no moving parts. Allowing the insertion of a plug into the nozzle provides a secondary benefit.

Figure 5. ACE plume flow mixing enhancement

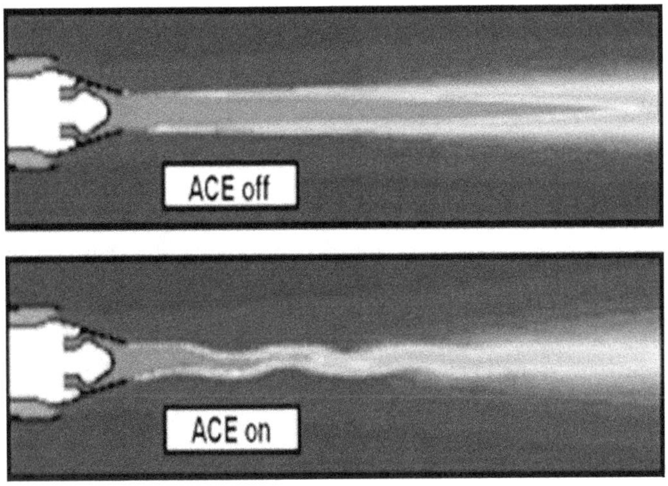

Figure 6. Nozzle system of C-17 with ACE system

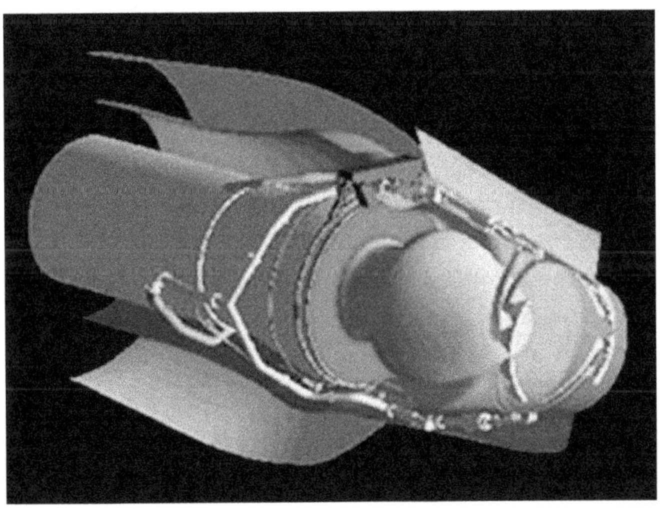

Adaptive Virtual Aero Surfaces

The Adaptive Virtual Aero surface is based on the adjustment of apparent shape using surface-mounted synthetic jet actuators to regulate the aerodynamic performance of a lifting surface Crittenden (Shlyubsky and Glezer, 2004). These actuators are distinguished by the fact that they do not require an external fluid source. There is momentum transfer into the external flow, despite the fact that there is no net mass injection into the total system. The jets can manage flow separation, adjust the effective camber of airfoils, and manipulate vortex flow phenomena both statically and dynamically. A schematic of a simple synthetic jet actuator is shown in Figure 7. An opening and cavity are used to create the jet, which is supported by an actuated diaphragm. Fluid is sucked into the cavity from all directions on the actuator down stroke; fluid is ejected from the cavity in a direction on the actuator upstroke. A two-dimensional synthetic air jet's schlieren picture stretches about 60 jet widths in the streamwise direction. The image clearly displays the formation of a vortex pair near the orifice, as well as the contour of a turbulent jet downstream.

Figure 7. Schematic of synthetic jet and schlieren visualization photo

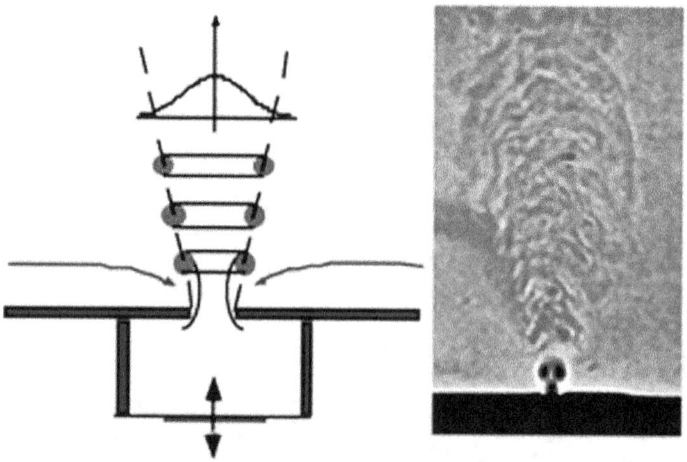

Because the synthetic jets have no net mass flux, their contact with an external flow causes closed recirculation areas to emerge. As a result, the geometry of the surface and the distribution of surface pressure appear to change. Within these recirculating zones, vorticity builds up, resulting in trapped or standing vortices (Figure 8). These vortices displace the cross flow's local streamlines, causing an 'apparent' or 'virtual' change in the surface's structure and global circulation. It's worth noting that the jet actuators' working frequency is usually set to be at least an order of magnitude greater than the flow's characteristic time scale.

Figure 8. Trapped vortex induced by synthetic jet

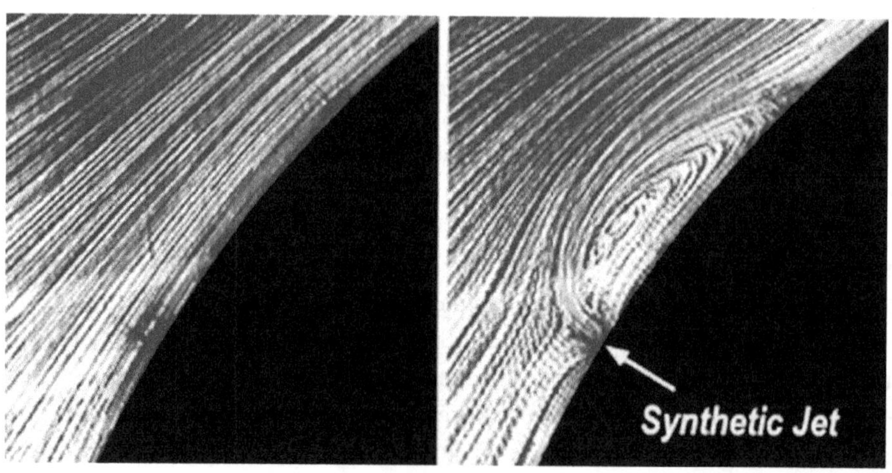

The recirculating flow zones are nearly "stationary" on the global time scale of the flow, and the resultant aerodynamic forces are unaffected by the actuators' working frequency. Different mechanisms for lift modification employing these actuators should be distinguished here. On the time scale of the flow around the airfoil, "W-type" actuation (invented by Wygnanski) is achieved by linking to an instability of the separating shear layer through pulsed blowing or zero-mass-flux actuation. This method relies on the separating shear layer's sensitivity to upstream control input and has no effect on aerodynamic performance in the absence of flow separation.

Glezer's "G-type" actuation operates on a time scale that is at least an order of magnitude shorter than the flow's typical time scale in AVIA. It is not dependent on the coupling to global instability and can be used (in theory) across the whole flight envelope. Because the placement of an airfoil's front and rear stagnation points (as well as its camber and thickness) are crucial to its aerodynamic qualities, synthetic jet actuators can change these features without the need for movable flaps. Front and rear stagnation points can be displaced by jet arrays that generate closed recirculating flow areas along the leading and trailing edges, as well as along the upper and bottom surfaces of an airfoil. The thickness of the airfoil is changing apparently (Moreau, 2007). The goal of AVIA is to create and demonstrate a technology for controlling aerodynamic flow using synthetic jet actuators. By comparing idealized linear potential solutions with fully viscous Navier-Stokes calculations for a typical vehicle, the possible aerodynamic benefits of active flow management were calculated (Figure 8.) The lift vs drag curve is given upper and lower boundaries by the potential and viscous computations, respectively. The AVIA technology is expected to give a 100 percent increase in turn rate and a 75 percent reduction in flight control weight. A new generation of high-power actuators designed for high-speed flows will be created. The effort will culminate in a planned demonstration on a semi-span model at high subsonic speeds.

The goal of AVIA is to develop and demonstrate a synthetic jet actuator-based aerodynamic flow control approach. By comparing idealized linear potential solutions with fully viscous Navier-Stokes computations for a generic vehicle, the possible aerodynamic gains gained by active flow management were calculated (Figure 9.) Upper and lower boundaries on the lift vs drag curve are calculated using potential and viscous calculations, respectively. AVIA technology is expected to boost turn rate by up to 100 percent while reducing flight control weight by up to 75 percent. High-speed flows will require a new generation of high-power actuators. The project will culminate with a planned demonstration at high subsonic speeds on a semi-span model.

RBS-Retreating Blade Stall

Greater range for worldwide self-deploy ability, higher speed and performance (maneuverability and agility) for fast agile missions, increased cargo, reduced external noise emissions, and reduced cabin noise and vibration are all anticipated requirements for future high-performance rotorcraft. Active flow control technologies can reduce unstable loads and vibration while increasing rotor maximum load capability and aerodynamic efficiency in forwarding flight. RBS (retreating blade stall) is a rotor load and flying speed limiter. The boundary layer separation along the leading edge of the rotor blade during the rapid move to a high angle of attack is the flow mechanism involved in the blade stall. In forward flight, some separation is normally present on the inboard portion of the retreating blade, but as load and speed increase, the region of separation develops rapidly. Figure 10. depicts the relatively small stall regions for a moderate load condition, and much bigger stall regions for a high load condition that is extremely close to the stalled border, based on model rotor experimental data. Under flight conditions, no effective

Figure 9. With actuation, a wide range of flows is displayed.

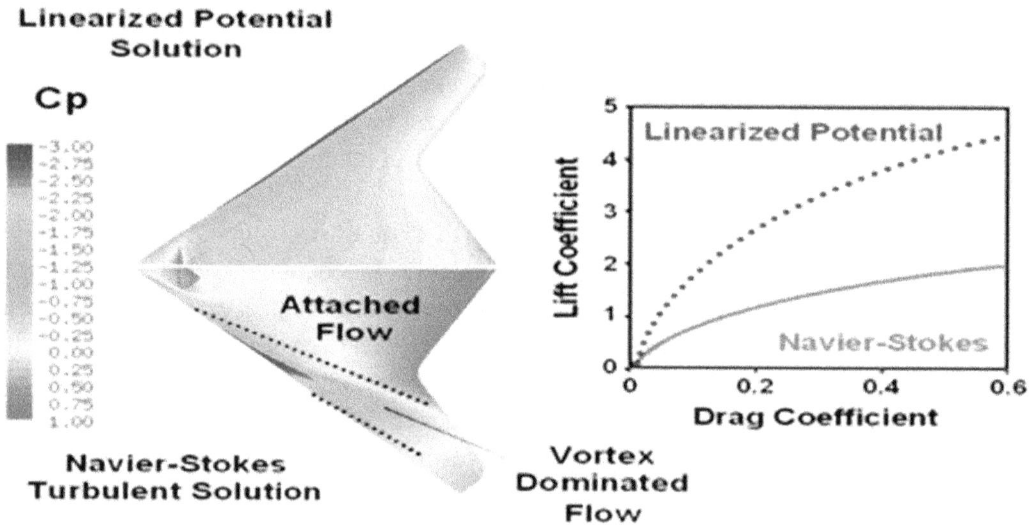

means of avoiding or considerably delaying RBS has ever been established. In an unstable, compressible, high Reynolds number flow, separation control must be successful. On the advancing side, where the relative external Mach numbers are 0.6 to 0.85, the stall management strategy used on the retreating blade side must not result in a significant drag penalty. In addition, the strategy must be implemented in a revolving blade. Mechanical high lift devices, such as leading-edge slats, have been studied and proven to be effective for delaying stall, although they have a tendency to either cause excessive drag if left in place on the advancing side or to be impractical to deploy and retract each rotor circle.

Figure 10. Retreating blade stall penalty increasing with thrust

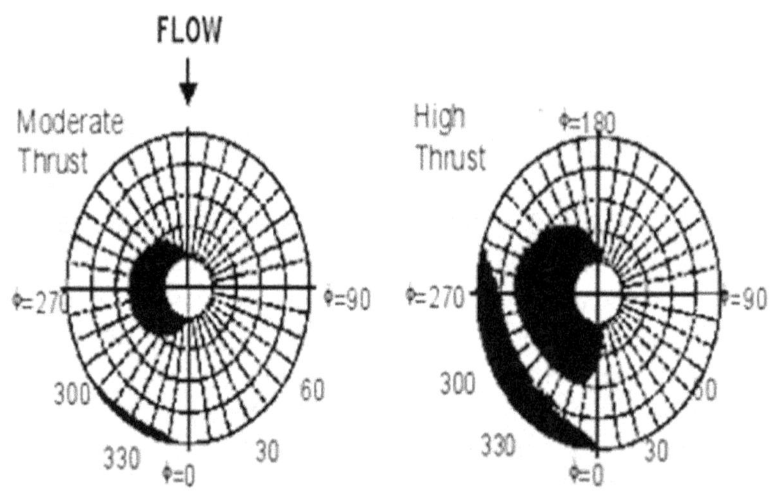

For a rotorcraft blade, "W-type" actuation has been chosen as a technique of delaying stall. A Directed Synthetic Jet is the major separation control approach (DSJ). A synthetic jet is created with an optimized exit neck for separation control (Figure 11). During the suction phase of the DSJ, low momentum fluid can be absorbed, while high momentum fluid can be released during the blowing phase. The boundary layer is energized by both phases.

Figure 11. Synthetic jets at the leading edge of the rotor

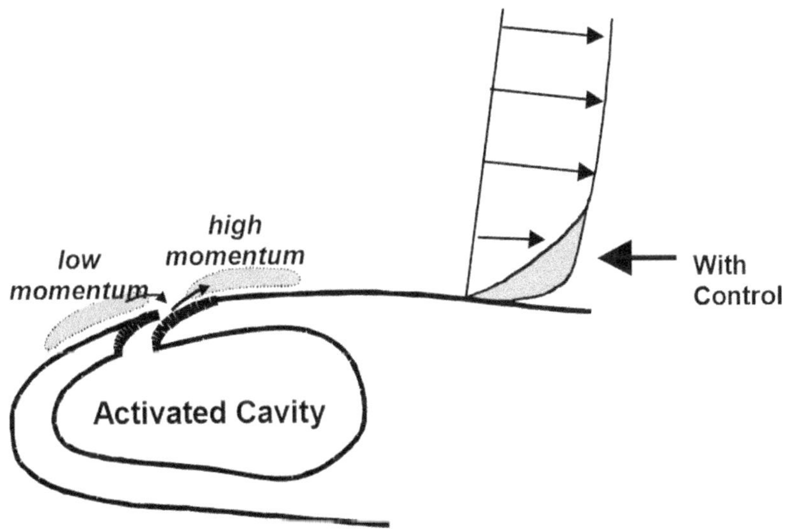

Mesoflaps for Aeroelastic Transpiration

Mesoflaps for Aeroelastic Transpiration (MAT) is a concept that allows for passive cavity recirculation to manage shock/boundary-layer interaction. When subjected to shockwave loads, the MAT system consists of a matrix of small flaps covering an enclosed cavity that are meant to undergo local aeroelastic deflection to achieve correct mass bleed or injection. In comparison to the baseline no-bleed situation, MAT improves post-shock impingement boundary-layer properties (reduced boundary layer thickness and turbulence intensity). The interaction of shock waves and turbulent boundary layers is a major factor in high-speed aircraft performance. Controlling the shock/boundary-layer interaction zone has been proposed as a possible strategy for reducing the negative consequences of intense shock waves, particularly in supersonic engine inlets. The concept of a supersonic mixed-compression intake in which boundary layer bleed is used to reduce separation at shock impingement points is shown in Figure12. Both the internal oblique shocks and the terminating normal shock interactions with the engine cowl and center body boundary layers are handled by the bleed bands. Most military aircraft with speeds greater than Mach 2 uses active bleed control, which involves piping bleed flow to an exterior surface where it is released.

The quantity of bleed required grows dramatically with Mach number, and for Mach 3 it is on the order of 10-15% of the engine mass flow. This bleed fraction is closely tied to associated penalties such as drag, weight, and overall vehicle cost.

Figure 12. Mixed compression inlet with bleed, and with MAT

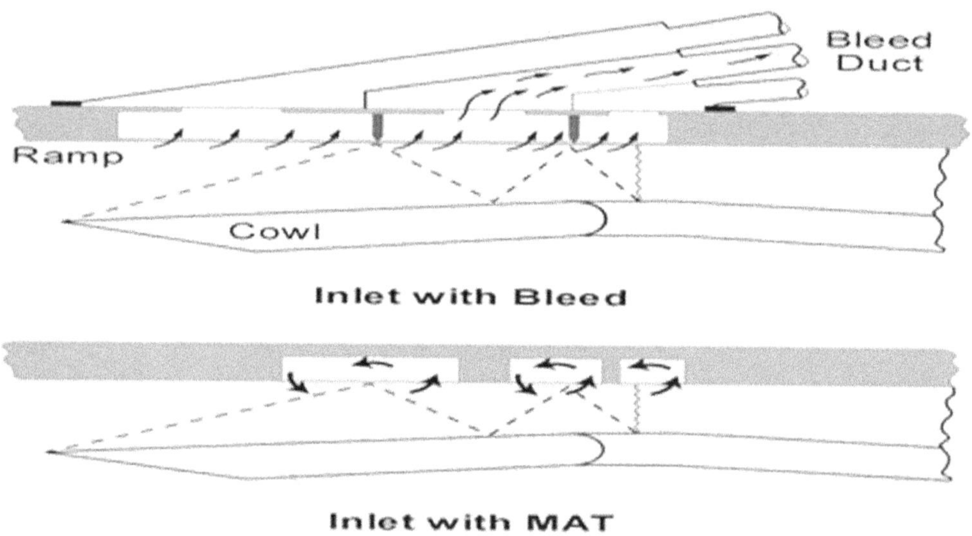

The MAT program's purpose is to efficiently prevent flow separation by shock boundary layer interaction while removing no net mass, resulting in improved downstream boundary layer characteristics. As seen in Figure 13, the MAT idea consists of a matrix of mesoflaps covering an enclosed chamber. Each flap is rigidly fastened across a tiny piece of its upstream end but can deflect aeroelastically at its downstream end due to the pressure difference between the supersonic and subsonic cavity flows. The pressures above and below the flaps are approximately identical in shock-free conditions, hence little transpiration is induced.

Figure 13. MAT with aero elastically deformed flaps

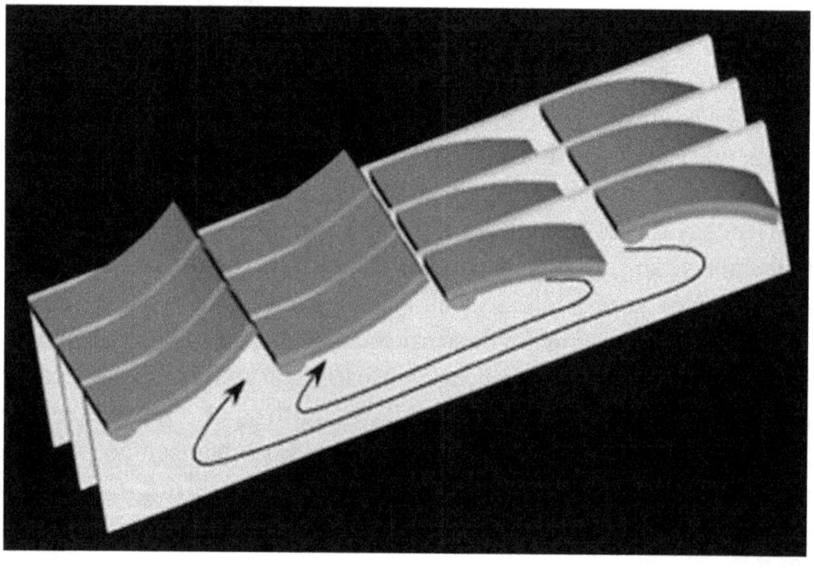

The roughening of the surface caused by typical transpiration holes or slots is avoided since the surface is nearly aerodynamically smooth in the absence of shock impingement. The MAT system is depicted in Figure 14 as if an oblique shock were to impinge on the mesoflaps between the second and third rows, causing a large streamwise pressure change. The almost constant cavity pressure will be somewhere between the pre-shocked flow's low pressure and the shocked flow's high pressure. As a result, flaps upstream of the impingement position will deflect higher, enabling angled flow injection into the boundary layer, while downstream flaps will deflect downward, allowing angled bleed from the boundary layer into the cavity.

The MAT concept keeps the simplicity of a traditional passive-transpiration system (no bleed airflow ducting or dumping), but adds three extra benefits thanks to the system's unique structural shape:

1. By utilizing angled transpiration, the geometry allows for superior bleed and injection aerodynamics (i.e., increased sonic mass coefficient).
2. It maintains angled transpiration while allowing for varied streamwise positions and sweep angles of the shock. As a result, the exact shock impingement locations as a function of flight/inlet conditions do not need to be known a priori.
3. The MAT system enables great aerodynamic efficiency in subsonic flow (i.e., skin friction comparable to that of a solid wall) because when no shocks are present, the system simply reverts to a nearly smooth flat plate.

Figure 14. Shock boundary layer interaction with solid wall and with MAT array (MAT Over Cavity)

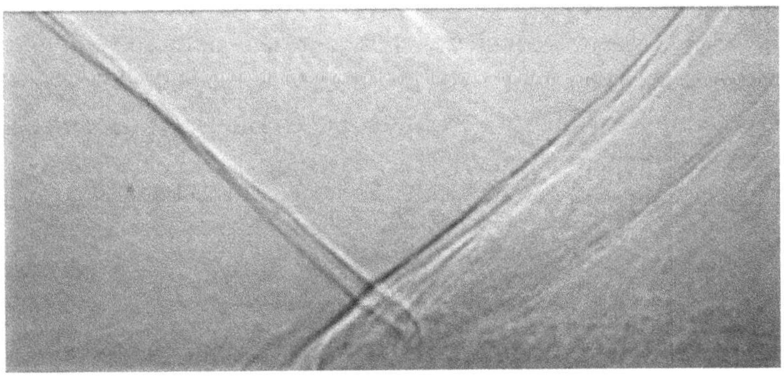

In Figure 14, the flow through the solid wall at the area of the shock impingement, the characteristic features of a classic shock boundary layer interaction are seen. The incident shock strikes the lower wall reflects, and coalesces close to the wall, forming a shock front due to the quick thickening of the boundary layer. A slip line was also seen originating at the junction of the incident and reflected shocks, demonstrating the presence of a shear layer within the boundary layer.

Wake Vortex Control

According to the phenomenological process of vortex shedding, a wide range of aerodynamic and hydrodynamic approaches for suppressing vortex shedding are categorized into three groups. The three types are: I surface protrusions, which impact separation lines and/or separated shear layers, such as helical strakes, wires, fins, studs or spheres, etc.; shrouds, which affect entrainment layers, such as perforated, gauze, axial-rod, axial-slat, etc. Splitter plates, guiding vanes, base-bleed, slits cut across the cylinder, and other entrainment layers interact. The majority of the means in the first two categories are omnidirectional, meaning they work regardless of the direction of fluid velocity. Some of the first- and third-category are unidirectional, meaning they are only effective in one velocity direction. The most widely supported active concept for reducing wake danger is to generate instabilities in trailing vortices by regularly oscillating control surfaces such that the integrated aerodynamic stresses do not change. This hastens the formation of unstable vortices, which eventually interact, pinch off, and disintegrate into harmless small-scale turbulence. A rotor blade cutting through the tip vortex shed by its predecessor is a primary source of noise and vibration on rotorcraft (blade-vortex interaction, BVI). Active solutions, such as trailing-edge flaps, aim to increase the "miss distance" between the rotor blade and the tip vortex or diffuse the vortex. AFC tries to manipulate the vortices by regulating the vortex sheet locally, as opposed to passive approaches that rely on the deflection of conventional control surfaces. AFC can be thought of as a method of managing vortices' basic features, such as their position (centroids), strength, size, and associated velocity components. The need of establishing control systems for flow over 3D bodies such as a truck or buses is unavoidable. The streamwise length of the body (which may determine the momentum thickness at separation), aspect ratio (i.e., height to width), body height from the ground, and other factors all affect vertical structures in the wake in this body type.

Actuator and Its Modeling

Passive (no-energy input) flow control can provide more versatile control measures to accomplish target performances than active flow control, which provides control techniques involving energy expenditure. Flow actuators come in a variety of shapes and sizes, with varying energy consumption, control authority, bandwidth, and operating conditions. Most actuators for flow control can be split into two groups based on the energy conversion mechanism: devices based on mechanical motions to impact the fluid, and devices controlling the flow field without moving parts. Synthetic-jet (zero-net mass-flux) actuators and various flaps are among the former. Due to their large bandwidth and efficient energy conversion capability with simple structures, piezoelectric materials have recently become popular for generating mechanical motions. Pulsed micro jets, which generate pulse mass injections, and vortex generator jets, which imitate traditional vortex generators with an angled jet injection, are two more motion-based actuators. Actuators without moving parts, on the other hand, use electric, electromagnetic, or thermal actuation mechanisms. This category includes gas discharge techniques (thermal and non-thermal plasma-based actuators) (Benard and Moreau, 2010) and combustion processes. Because these actuators do not require a mechanical component, they typically outperform mechanical actuators in terms of operation bandwidth, responsiveness, and installation. The dielectric barrier discharge (DBD) actuator, among the numerous actuation systems, requires neither mass injection components nor mechanical moving parts. The DBD actuator may be simply put on any flat surface or changed into other shapes due to its basic structure see Figure 15, which only requires electrodes and a thin dielectric insulator. By transferring momentum to the neutral flow field via non-thermal ionized particles utilizing a high-intensity electric field, the DBD actuator is known to generate a thin layer of wall jet. The capacity of this actuator to perform both steady and erratic actuations up to several kHz makes it a versatile control device for a wide range of applications. The duty cycle, or the length of time the actuator is turned on, as well as the amplitude of the applied voltage, can be varied to meet performance and power requirements, depending on the flow and flight conditions. The DBD actuation's high bandwidth allows for multi-frequency excitations to accomplish multi-frequency control goals, which is advantageous for acting on flow instabilities. Although the power module's ability to sustain gas discharge currently limits the DBD actuator's applicability to small air vehicles, recent studies demonstrate the DBD actuator's versatility as a flow actuator, particularly for low-speed flows, persuade us of its potential. The DBD actuator, for example, improves the high-angle-of-attack performance of an airfoil and a delta wing by altering wake and leading-edge vortex for chord-based Reynolds numbers in the range of O (103)–O (105). Furthermore, the DBD actuator's ease of installation and operation over traditional lift control devices encourages its use as a primary or supplemental high-lift device.

Figure 15. Flow–actuator system Schematics at Free-stream condition. A strip of the dielectric barrier discharge (DBD) actuator is located at 0.25 of Chord from the leading edge on a rectangular wing whose wing section is the SD7003 airfoil and of chord aspect ratio is h/c.

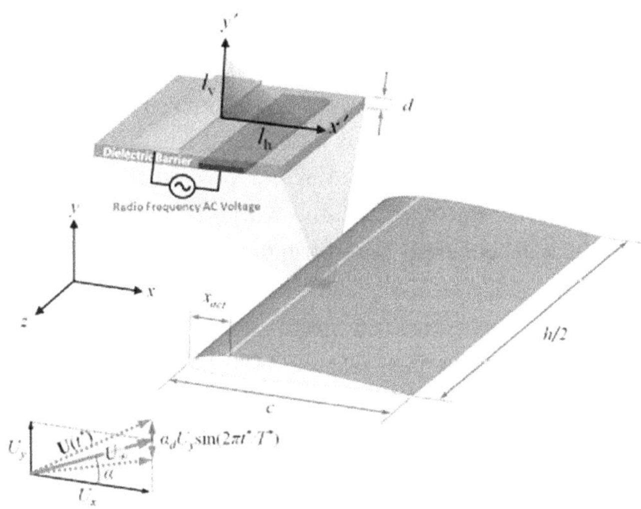

The DBD actuator has also been used to reduce cavity noise, improve jet mixing, cancel the Tollmien–Schlichting wave for a flat-plate boundary layer, and synchronize vortex shedding from circular cylinders. Although experimental studies are based on actual DBD actuator performance, difficulties probing a design space, which typically has a larger number of design parameters, and measurement challenges encourage numerical approaches, which can provide comprehensive information on unknowns in the actuation process. Despite our best efforts, we were unable to precisely capture the actuator. Although experimental studies are based on actual DBD actuator performance, difficulties probing a design space, which typically has a larger number of design parameters, and measurement challenges encourage numerical approaches, which can provide comprehensive information on unknowns in the actuation process. Despite intensive efforts to precisely describe actuator physics, a significant gap in plasma and neutral flow time scales at low Reynolds numbers renders numerical techniques wasteful and impractical for most practical tasks (Greenblatt et al., 2010). Instead of using high-fidelity discharge models, simplified reduced-order models that approximate the effective body force in the neutral fluid have been proposed, with varying degrees of simplification: a bilinear quasi-steady force field, a doublet in the potential flow, body force based on an applied electric field and prescribed charge distribution, and lumped element circuit model, to name a few. Despite their lack of comprehensive discharge physics, reduced-order models allow for the simulation of complicated flow fields at a fraction of the expense of first-principles models.

Adaptive Flow Control Based on Closed-Loop Designs

Closed-loop flow control can improve aerodynamic performance with less control effort, achieve a set target such as a desirable pitch-angle profile, and lessen the influence of disturbances when compared to open-loop control systems. Closed-loop flow control encompasses the delay/mitigation of turbulence in a channel, removal of periodic vortex shedding, decrease of flow separation, increase of mixing, and suppression of flow-induced cavity oscillations, to mention a few (Braun et al., 2009). However, the great complexity and nonlinearity of most flow dynamics problems make modelling the dynamics of a flow system and implementing an appropriate control method extremely difficult. The mathematical modelling of a flow system is a crucial topic in closed-loop flow control problems due to the complexity of the flow dynamics of interest. In general, there are two perspectives on how to operate a flow system: whether to use a dedicated flow model or a general input-output model (also called the system identification approach). Because a flow model is based on the explicit description of flow physics, it may express physical understandings of the system, which can aid control system design, such as measurement, actuation, and objectives. Furthermore, the whole flow field data accessible for the former is beneficial for flow systems with coherent flow structures as well as control systems with scattered sensors and/or actuators.

A flow control problem can be turned into an optimization problem based on the flow quantities or states of the full flow field. There are several ways, such as adjoint-based optimization and Riccati-based feedback, depending on the linearization and optimization strategies. A high-fidelity flow model that provides the whole flow data, such as direct numerical simulation (DNS), large eddy simulation (LES), and multimodel approaches, is very expensive and often unrealistic for real-time control applications, necessitating the use of reduced-order models. Furthermore, the reconstruction of high-dimensional flow field data from restricted measurements, which is crucial for practical applications, necessitates the use of a state estimator or observer. Reduced-order models with varying degrees of realism exist, ranging from vortex-based models to those based on non-dominant mode truncation, such as the proper orthogonal decomposition (POD). When the flow dynamics are dominated by coherent structures such as shedding vortices and turbulent structures, the POD-based model reduction is appealing. Stabilization of vortex shedding from a bluff body or a high-angle-of-attack lifting surface, resonance attenuation in cavity flows, and characterization of turbulent coherent structures are some of the more common uses. Despite significant efforts to construct efficient reduced-order models, concerns such as how to include a control with the POD modes and whether pivotal flow physics is conserved in the retained modes remain unresolved.

Instead of relying on specific flow models, merely the relationship between control input and measurement, as well as a collection of relevant system parameters, can be described. The general input-output model requires less information than the flow model-based approach to fulfill control goals because the reconstruction of the flow field is not required in this situation. The information in the system dynamics perspective (i.e., the relationship between actions and measured responses) is readily accessible, notwithstanding the input-output model's lack of insight into the evolution of flow structures. Furthermore, if the key dynamics of the flow system are effectively captured, this less-modeling method might lower the control system's vulnerability to modeling uncertainty, which is appealing, particularly for flow systems with complicated dynamics.

Most flow control studies use linear time-invariant models, which are represented by rational transfer functions or time-series models, the coefficients of which may be calculated using various system identification approaches. The rejection of pressure fluctuations on an airfoil, mitigation of uneven flow speed in a channel, and reduction of flow separation are some examples based on this method. Once a flow system model has been created, feedback control can be done using a variety of control laws, including classical controllers, current control designs like linear-Quadratic-Gaussian (LQG), adaptive algorithms, and combinations of the aforementioned. Adaptive control, as opposed to fixed gain control, which is usually tailored for known system parameters and statistical properties of noises, can dynamically update the controller gains in response to changes in the plant and disturbance dynamics and achieve target performance with larger parameters uncertainties. Adaptive control tries to change the system and/or controller settings in the face of disturbances, time-varying and unmolded system dynamics, and time-varying and unmolded system dynamics. Although adaptive control techniques offer the benefit of tailoring feedback gains in response to genuine plant dynamics and exogenous inputs, they do necessitate some understanding of the system to be managed. Furthermore, adaptive algorithms are effective when the number of unknown parameters is minimal and function evaluations can be completed fast.

CONCLUSION

Adaptive flow control is only now starting to be used in large-scale systems. According to system studies, the benefits of this technology are significant for a number of applications. The idea of Adaptive Flow Control for aerodynamic advantage indicates first hands-on experience with real systems, however, they will be open-loop. The interest in real closed-loop flow control beyond these initial uses of open-loop systems that suppress or saturate instabilities. We hope to see the actual payoff of this promising technology in the next generation of adaptive flow control.

Active flow control involves an interdisciplinary approach that includes fluid physics, structural mechanics, material science, acoustics, systems, multidisciplinary optimization, and stability and control, among other areas. Researchers working in their own separate fields of specialization have traditionally developed technology. Not only in colleges but also in government laboratories and industry, this trend has emerged. To cross-fertilize and integrate component technologies, a recent tendency has been to form integrated, interdisciplinary, or multidisciplinary design teams.

REFERENCES

Benard, N., & Moreau, E. (2010). Capabilities of the dielectric barrier discharge plasma actuator for multi-frequency excitations. *Journal of Physics*, 43(14), 1–14.

Braun, E. M., Lu, F. K., & Wilson, D. R. (2009). Experimental research in aerodynamic control with electric and electromagnetic fields. *Progress in Aerospace Sciences*, *45*(1-3), 30–49. doi:10.1016/j.paerosci.2008.10.003

Cho, Y. C., & Shyy, W. (2011). Adaptive flow control of low-Reynolds number aerodynamics using dielectric barrier discharge actuator. *Progress in Aerospace Sciences*, *47*(7), 495–521. doi:10.1016/j.paerosci.2011.06.005

Crittenden, T. M., Shlyubsky, D., & Glezer, A. (2004). Combustion-driven jet actuators in reversed flow configurations. *Proceedings of the second AIAA flow control conference*, 1-12. doi:10.2514/4.479892

Gad-el-Hak, M. (2000). Flow control: Passive, active, and reactive flow management. Cambridge University Press.

Glezer, A., & Amitay, M. (2002). Synthetic jets. *Annual Review of Fluid Mechanics*, *34*(1), 503–529. doi:10.1146/annurev.fluid.34.090501.094913

Greenblatt, D., Goksel, B., Rechenberg, I., Schule, C. Y., Romann, D., & Paschereit, C. O. (2008). Dielectric barrier discharge flow control at very low flight Reynolds numbers. *AIAA Journal*, *46*(6), 1528–1541. doi:10.2514/1.33388

Jaganraj,, R., Boopathy,, G., & Varun,, V., & Vigneshwar. (2015). Experimental investigation of passive flow control on bluff bodies. *International Journal of Applied Engineering Research*, *108*(2), 19793–19798.

Jahanmiri. (2012). *Active Flow Control, review report*. Academic Press.

Joslin, R. D., & Miller, D. N. (2009). Fundamentals and applications of modern flow control. *Progress in Astronautics and Aeronautics*.

Lissaman, P. B. S. (1983). Low-Reynolds-number airfoils. *Annual Review of Fluid Mechanics*, *15*(1), 223–239. doi:10.1146/annurev.fl.15.010183.001255

Moreau, E. (2007). Airflow control by non-thermal plasma actuators. *Journal of Physics. D, Applied Physics*, *40*(3), 605–636. doi:10.1088/0022-3727/40/3/S01

Roth, J. R., Sherman, D. M., & Wilkinson, S. P. (1998). Boundary layer flow control with a one atmosphere uniform glow discharge surface plasma. *AIAA Journal*, 1–28.

Shyy, W., Lian, Y., Tang, J., Viieru, D., & Liu, H. (2008). Aerodynamics of low Reynolds number flyers. Cambridge University Press.

Wlezien. (2000). Micro Adaptive Flow Control. *ICON 2000 Congress*, 2101-1 - 2101-9.

Chapter 16

Computational Investigation on the Influence of Dump Gap and Novel Flame Tube Geometries in the Hybrid Diffuser Performance

Bruce Ralphin Rose
Anna University Regional Campus, Tirunelveli, India

Bibal Benifa J. V.
Indian Institute of Information Technology, Kottayam, India

Ramzania M.
Anna University Regional Campus, Tirunelveli, India

ABSTRACT

The chapter addresses an extensive numerical investigation for the optimization of static pressure recovery in the dump diffuser of annular combustor for modern aero-gas turbine engines. The modelling and simulation processes are accomplished through the commercial ANSYS fluent module with SST k-epsilon (k-ε) turbulence model with the atmospheric conditions prevailing at 9 km altitude. The effect of velocity changes on the flow recirculation and vortex mixing phenomena is also addressed with novel dome-shaped optimization at the velocity of 25 m/s. The dome geometry can also be considered as a flame tube head, and the influence of vortex pattern and pressure distributions caused by the different dome shapes are analyzed using CFD. The dome-shaped optimization strategy addressed herein could initiate more potential studies on dump diffusers with hot and cold flow conditions to enhance the overall combustion efficiency in the near future.

DOI: 10.4018/978-1-6684-4230-2.ch016

1. INTRODUCTION

The design and analysis of dump diffusers with maximized pressure recovery is a key objective for the propulsive system engineers in the past few decades (Reddy and Ganesan. 2005). Specifically, the gas turbine combustor should completely burn the fuel because the unburnt fuel would create undesirable emissions of hydrocarbons, CO and soot. The aviation industry is striving hard to achieve the lowest possible CO_2 and NO_X emissions through novel bio-inspired airplane configurations and subsystems to reduce the climate impact (Grewe et al., 2021). The Blended Wing Body (BWB) aircraft like configurations have already proved their potential towards about 20% reduction in the fuel consumption. However, such large scale transformations are considered futuristic airplane variants which would be realized after 2030. Hence, the subsystems like winglets, power plant components, and flow control mechanisms are widely focused by the researchers towards valuable optimizations for ensuring the efficient flight at different missions (Varshney et al., 2020). The dump diffuser enhances the air pressure entering into the combustor and maintains the total pressure loss as minimum as possible (Carrotte et al., 1994). The present chapter describes the various optimization strategies to enhance the performance of the dump diffuser at different velocities and operating conditions in contrast to the conventional diffusers.

2. RELATED WORKS

The extensive study on the dump diffuser performance optimizations in the last few decades has revealed the fact that only a sparse amount of literature is available on this domain. Specifically, the diffuser geometry-based performance optimization requires a multi-disciplinary perspective as the flow physics and thermodynamic characteristics of modern aero-gas turbine engines are governed by many variables. Here, a well-structured literature review is presented to offer the key insights and comprehensive view on the previous research works on dump diffusers. The pressure recovery, dump gap ratio, shape of the dome, shape of shrouds, the annulus geometry, turbulent intensity, and pressure gradients are the few parameters considered herein to consolidate the methodological contributions in recent times.

2.1. Geometric Optimization

Sanal Kumar et al., (2009) addressed the geometrical optimization of dump diffuser using SST k-ω turbulence model and observed that the performance of the dump diffuser depends on the wall geometry. Wall geometry with $\alpha_1 = 5^o$ and $\alpha_2 = 4^o$ have delivered higher pressure recoveries than $\alpha_1 = 10^o$ and $\alpha_1 = 2^o$ respectively (i.e., 0.4% higher for the fully developed flows). Further, an increase of shroud passage height about 1.5 times of initial design has improved the static pressure distribution about 0.1%. Similarly, Leilei et al., (2014) also studied the dump diffuser optimization and suggested that the vortex generators (VG) can be used as a valuable aerodynamic tool to enhance the performance of diffusers. Here, the Reynolds stress model is used to capture the pre-diffuser separation delay that is introduced by the vortex generators (Juhasz and Smith, 1977). The velocity distribution within the dump diffuser system was visualized by the PIV system and the VG has reduced the total pressure drop about 5% than the previous values.

Mariotti et al., (2014) introduced a flow control strategy to improve the efficiency of plane diffusers which produce different flow patterns at various divergence angles. The efficiency of the diffuser increases with the addition of the number of cavities and its performance is limited by the divergence angle. The optimal cavities leads to an increase in pressure recovery for all considered configurations. Prakash et al., (2013), introduced the turbulent isothermal flow inside an axisymmetric dump diffuser and analysed the model with and without swirling effects. Here, the *k*-epsilon (k–ε) model is used and it is observed that the swirl in the inlet flow affects the pressure recovery and increases the corner recirculation (Rahim et al., 2007). Earlier, Klein. (1995) reviewed the various characteristics of combustor diffuser and stated that the air from the compressor is decelerated by the combustor diffuser with maximum rise in static pressure and minimum pressure loss. Here, the Reynolds time-averaged form of the Navier-Stroke (RANS) equations are used for the computational study with k–ε model.

2.2. Dump Gap Optimization

The early studies of Honami and Morioka (1990), concluded that a small dump gap causes inadequate pressure recovery in the pre-diffuser, thus resulting in a loss of total pressure. The findings were strongly supported through cold flow experiments in a 2D dump diffuser model with the help of Hot-wire anemometer and Laser Doppler Velocimetry (LDV). Sanal Kumar et al., (2007) studied the effect of different dump gap ratios on the flow losses that exist in the dump zone. Interestingly, it is reported that the small dump gap with high turbulence intensity produces a significant amount of flow separation and thereby increasing the pressure recovery (Adkins, 1975). Further, the optimization of dump diffuser based on the pre-diffuser geometry and dump gap must be concerned with shock boundary layer interactions because it is more sensitive in the small-scale models (Stevens et al., 1988). Later, Duncan et al., (2009) also suggested that the dump gap ratio should be as small as possible because at deeper flame tubes and larger dump gaps, the aerodynamic loss potentially increases due to increased turning. Alternatively, with fixed geometries, as the injector mass flow rate increases, the performance of the pre-diffuser drops with the high pressure loss (Aravind et al., 2019).

Reddy and Ganesan (2005) numerically optimized the pre-diffuser of a gas turbine combustion chamber and concluded that the dump gap ratio and diffuser angle are the key factors for reducing the total pressure loss with the good static pressure recovery. The best optimal value for the straight wall pre-diffuser angle is 6^0 and the area ratio is 1.6594 with the dump gap ratio of 1.08. Meanwhile, Sanal Kumar et al., (2005) also stated that the small value of dump gap ratio helps to minimize the pre-diffuser length and it results in a more uniform outlet velocity profile and prevents separation. In addition, if the head of the flame tube is located closer from the pre-diffuser exit then additional losses may be incurred.

2.3. Annular Combustor Performance

Wael et al., (2014), simulated the influence of external flows and associated effects on performance, durability, and stability of an annular can combustor. As the external flow is one of the key elements in controlling the pressure losses, the air flow distribution around the combustor liner is considered as the external flow path to study the diffuser performance (Fishenden and Stevens. 1974). Previously, Walker et al., (2009) studied an annular diffuser with large downstream blockage to enhance the characteristics of the boundary layer which influences the performance of the diffuser. It is concluded that the interaction between the diffuser and downstream geometry has resulted in the reduced total pressure loss in the range of 20% to 35%. Lewis and Backstrom. (1991) obtained symmetrical flow distribution around the dome shape through the experimental optimization of an annular dump diffuser. The optimal flow split strategy coincides with the design value and thereby 3% improvement in effectiveness and 5% overall flow loss reduction is obtained.

The early works reported on dump diffusers mainly relied on the cold flow experiments with the help of Hot-wire anemometer (HWA) and LDV. Honami and Morioka (1990) studied the flow behavior in a dump diffuser 2D model with distorted flow at the inlet and concluded that the dump gap reduction causes insufficient pressure recovery in the pre-diffuser region. Hence, the dump gap optimization has a significant effect on the overall performance and efficiency of the dump diffusers preferably at high altitudes (Wang and Luo. 2020). The numerical solutions are the best alternative to achieve unique solutions and the flow characteristics with proper grid distribution and boundary conditions. Shyy. (1985) has suggested the use of a body-fitted coordinate system for the numerical study of annular dump diffuser flows if the skewness of the meshed is controlled.

As the modern gas turbines produce high pressure ratios within a few stages of compression, the axial velocity of air entering through the dump diffuser is high enough to cause serious pressure losses. It occurs mainly due to the following reasons, viz i). skin friction and turbulence and, ii). The local rise in temperature around the flame tube head. Hence, the dump gap (the distance between the dome cowling and pre-diffuser exit plane) plays a crucial role to customize the overall thrust-to-weight (*T/W*) ratio of the engine by admitting the air through the combustor ports at optimized velocities. Alternatively, the total pressure loss problem can be effectively handled through the shape optimization of the flame tube without any additional swirling or cavity effects. The present chapter is focused with these two contexts of dump diffuser flow optimization to enhance the performance of gas turbines at high altitudes.

3. METHODOLOGY

The tricky problem occurring between the compressor and the annular combustion chamber passage is an unavoidable loss in the total pressure that is caused by the substantial flow deceleration (Srinivasan et al., 1990). Further, the diffuser must be designed to limit the flow distortion as much as possible by avoiding the flow interference effects like boundary layer separation. Hence, the diffuser should be designed with minimum possible length by maintaining the area ratio and flow attachment with the combustor liner. It is a complex scenario because the reduction in the pre-diffuser length causes large scale flow separations though it is insensitive to thermal expansions. Hence, the investigation should be focused with pressure recovery and flow stability of the dump diffuser over the wide range of operating conditions. The numerical analysis is carried out with an ideal model of the dump diffuser with different gap ratios and flame tube geometries. For the latter objective, the design optimization should be carried out by optimizing the parameter static and the total pressure losses at different zones of the flame tube cowling and passage.

Though the proposed analysis is novel one in terms of different dome shapes, the localized flow recirculation in the diverging walls of the flame tube can be validated with few existing articles under some limitations. The methodology and the work plan of this chapter are illustrated below (Figure 1) with the different aspects of the numerical analysis using a block diagram.

Figure 1. Performance optimization methodology used for the dump diffuser

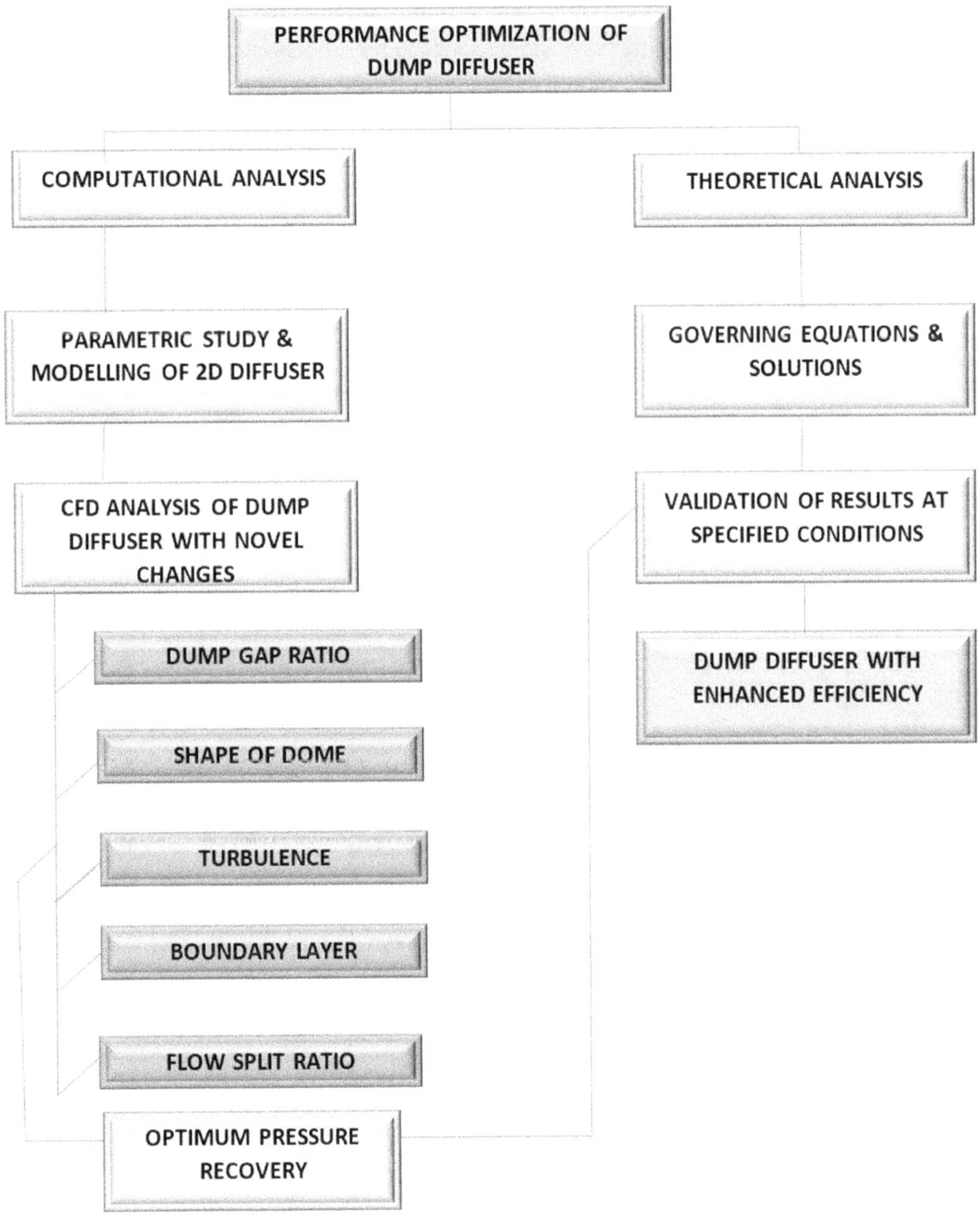

4. MODELLING AND MESH PREPARATION

The advent of modern computational techniques help to solve most complex flow problems with and without heat transfer characteristics (Aleksandrov et al., 2017). Nowadays, it is effectively used not only for linear and non-linear aerodynamic flow problems to customize the required geometric and flow parameters prior to the manufacturing process. CFD can also be used for predicting the heat transfer, mass transfer, chemical reaction, and related phenomena by solving the appropriate governing equations through the numerical process (Aleksandrov et al., 2016). As the primary intention for the usage of dump diffuser is to achieve the higher pressure recovery at high altitudes, the physical and thermodynamic flow properties across a turbojet engine is considered for the present chapter. The internal arrangements of the turbojet engine are schematically shown in Figure 2 and the air flowing through the machine shall undergo the following seven step process:

Table 1.

Step *a*-1	From far upstream, the air is received at the intake, where the velocity of the air relative to the engine is the flight velocity with significant amount of deceleration.
Step 1-2	The air velocity is decreased by the reduction in kinetic energy and it is carried to the compressor inlet through the inlet diffuser system.
Step 2-3	The air is compressed dynamically by an axial compressor and the number of stages may vary according to the required pressure ratio and type of aircraft used.
Step 3-4	The air is pressurized and "heated up" by the flow mixing process and the combustion of fuel occurs with the air mixture.
Step 4-5	The air flow is expanded through the turbine to obtain power to drive the compressor and to develop reaction forces.
Step 5-6	The air may or may not be further "heated up" by the addition and burning of more fuel in the afterburner section and allowed through the nozzle.
Step 6-7	The rapid flow expansion of exhaust gases induce more acceleration through the exhaust nozzle.

Figure 2. Schematic diagram of a turbojet engine.

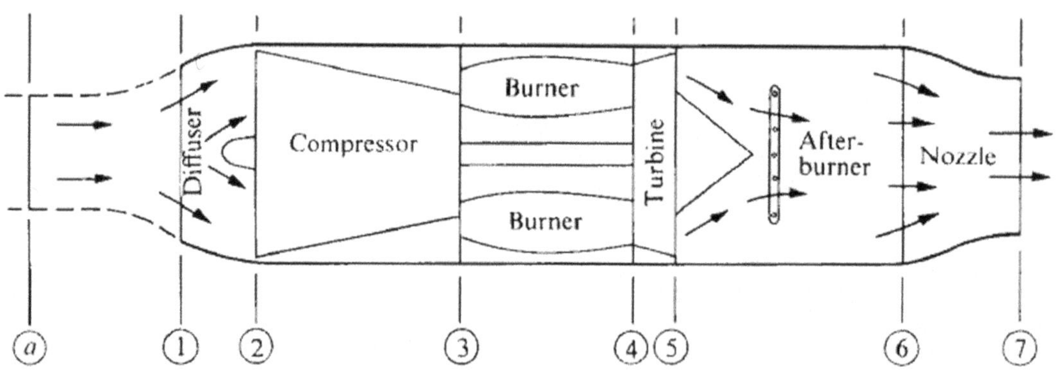

As the dump diffuser is located at the end of the axial compressor unit, it helps to decelerate the discharge coming out of the compressor and it is located at the sectional plane 3 of the turbojet engine shown in Figure 2. The specific dimensions used for the modelling process of the dump diffuser are discussed in the subsequent sections of the chapter.

4.1 Dump Diffuser Model

The parametric studies of an ideal dump diffuser model are carried out with a taper-walled flame tube geometry. The idealized physical model of the dump diffuser is illustrated in Figure 3 (a). The baseline values are selected based on the typical diffuser dimensions with fractional quantities ($h_2/h_1 = 1.95$, $l/h_1 = 3$, $H_f/h_1 = 4.8$, $d/h_1 = 2.5$, $d/h_2 = 1.1$, $2H_{sp}/h_2 = 3$, $X_s/h_1 = 2$). A 3D illustrative view of the Computer Aided Design (CAD) model of the dump diffuser with a truncated dome shape is presented in Figure 3 (b).

Figure 3. An ideal model of the dump diffuser

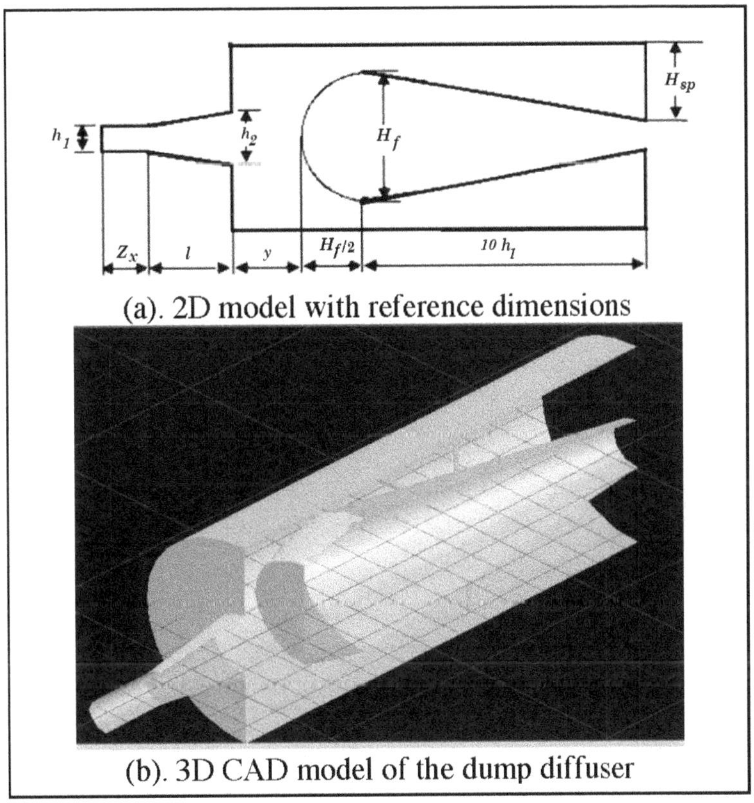

(a). 2D model with reference dimensions

(b). 3D CAD model of the dump diffuser

The reference value of h_1 is considered as 25 mm and the other values are calculated in proportion to the relation of h_1 from the base values mentioned above as follows: h_1= 25 *mm*,

h_2 = 48.75 *mm*, *l*= 75 *mm*, H_f = 120 *mm*, *d* = 53.625 *mm*, H_{sp} = 73.125 *mm and* X_s = 50 *mm.*

4.2 Theoretical Evaluation

The theoretical calculations are proceeded through the turbojet engine with the following inlet conditions: M = 0.5, $P_a = 2.27 \times 10^4$ Pa, and T_a = 216.774 K. The adiabatic efficiencies (η) are also assumed for various components according to the selected turbojet engine performance characteristics. The average specific heat ratios (γ) are also varying from 1.40 to 1.36 as the flow occurs from diffuser to nozzle respectively. Let, the fuel heating value, Q = 45,000 kJ/kg. Therefore,

$PO_{3/}PO_{2=}9; \frac{P_{04}}{P_{03}} = 0.96 ; T_{04}$=1200K

4.2.1 Compressor Inlet Conditions

With flight Mach number, $M = u/\sqrt{\gamma RT_a}$, in which (γ) is the specific heat ratio (assumed constant for the diffuser process), evaluate the stagnation temperature $T0_2$ with the following expression,

$$T_{02} = T_a\left[1+\frac{\gamma-1}{2}M^2\right] = 216.774\left[1+\frac{1.4-1}{2}(0.5)^2\right]$$

= 227.613 K

From the definition of the diffuser adiabatic efficiency ($\eta d_{)}$, the corresponding stagnation pressure can be evaluated as,

$$P_{02} = P_a\left[1+\eta_d\left(\frac{T_{02}}{T_a}\right)\right]^{\left(\frac{\gamma_d}{\gamma_d-1}\right)} = 2.27e04\left[1+0.97\left(\frac{227.613}{216.774}\right)\right]^{\left(\frac{1.4}{1.4-1}\right)}$$

=2.68 x e^{04} Pa

4.2.2 Compressor Outlet Conditions

Since the compressor pressure ratio, $P_{03}=P_{02}P_{rc}$ is specified, the outlet stagnation pressure can be determined as,

$P_{03}= P_{02}P_{rc}$ = (2.68xe^{04}) x 8 = 21.44 x e^{04} Pa

and the outlet stagnation temperature is computed as,

$$T_{03} = T_{02}\left[1+\frac{1}{\eta_c}\left\{P_{rc}^{\left(\frac{\gamma_c-1}{\gamma_c}\right)}-1\right\}\right] = 227.613\left[1+\frac{1}{0.85}\left\{8^{\left(\frac{1.37-1}{1.37}\right)}-1\right\}\right] = 429.379\text{ K}$$

In the expression of stagnation temperature, γc is the specific heat ratio related to the compression process and ηc $_i$s the compressor adiabatic efficiency.

4.2.3 Burner Fuel-Air Ratio

With specified burner outlet temperature $T_{04,}$ the fuel-air ratio can be determined with the following equation as,

$$f = \frac{\left(\frac{T_{04}}{T_{03}}\right) - 1}{\left(\frac{Q_R}{c_p T_{03}}\right) - \left(\frac{T_{04}}{T_{03}}\right)} = \frac{\left(\frac{1200}{429.379}\right) - 1}{\left(\frac{45000}{429.379}\right) - \left(\frac{1200}{429.379}\right)} = 0.0270$$

4.2.4 Turbine Inlet Pressure

As the pressure ratio across the burner P_{04}/P_{03} is the known quantity, the turbine outlet stagnation pressure can be calculated from,

$$P_{04} = P_{03}\left(\frac{P_{04}}{P_{03}}\right) = 21.44 \text{ x } e^{04}(0.96) = 20.58 \text{ x } e^{04} \text{ Pa}$$

4.2.5 Turbine Outlet Conditions

Since the turbine should supply the power required to drive the compressor (for steady adiabatic flow in both components), the temperature becomes,

$$T_{05} = T_{04} - (T_{03} - T_{02}) = 1200 - (429.379 - 227.613) = 998.234 \text{ K}$$

Then, from the definition of the adiabatic efficiency of the turbine, the stagnation pressure P_{05} may be evaluated from,

$$P_{05} = P_{04}\left[1 - \frac{1}{\eta_t}\left(1 - \frac{T_{05}}{T_{04}}\right)\right]^{\left(\frac{\gamma_t}{\gamma_t - 1}\right)} = 20.58 \text{ x } e^{04}\left[1 - \frac{1}{0.9}\left(1 - \frac{998.234}{1200}\right)\right]^{\left(\frac{1.33}{1.33-1}\right)}$$

$$= 8.943 \text{x } e^{04} \text{ Pa}$$

4.2.6 Nozzle Inlet Conditions

When no afterburner is installed, the nozzle inlet conditions become,

$T_{06}=T_{05}$ and $P_{06}=P_{05}$

Therefore, $T_{06} - 998.234$ K and $P_{06} = 8.943 \text{ x } e^{04}$ Pa

4.2.7 Nozzle Exit Conditions

By definition, the nozzle exit kinetic energy is expressed as,

$$u_e = \sqrt{2\eta_n\left(\frac{\gamma_n}{\gamma_n - 1}\right)RT_{06}\left[1-\left(\frac{P_a}{P_{06}}\right)^{\left(\frac{\gamma_n-1}{\gamma_n}\right)}\right]} = \sqrt{2\times 0.98\left(\frac{1.36}{1.36-1}\right)287\times 998.234\left[1-\left(\frac{2.27e04}{8.943e04}\right)^{\left(\frac{1.36-1}{1.36}\right)}\right]} = 803.533 \text{ m/s}$$

In which ηn is the nozzle adiabatic efficiency. This stated condition is prevailed only when the nozzle exit flow remains in the un-choked state.

4.3 Mesh Preparation and Boundary Condition

CFD can be used to calculate design mass-flow rates, pressure drops, heat transfer rates, and aerodynamic forces such as lift, drag and pitching moments. The accuracy and fidelity of modern CFD methods has significantly increased the level of design insights available to engineers throughout the design process and therefore greatly reduces companies' exposure to technical risks. By CFD techniques, designers can also validate the performance of their products in such a way that it confirms a client's specifications early in the design cycle.

The applications of CFD in design and product development generally lead to far fewer physical prototypes being essential during development, and prototype testing. Hence, it reduces the time-to-market and cost-to-market substantially with adequate safety standards. The numerical simulation of dump diffuser is an appropriate problem for the usage of CFD techniques because several pressure loss computations may not be captured even with the well-established engine testing facilities. However, the flow parameters computed through empirical equations can be utilized for assigning the initial and boundary conditions in the CFD solver settings. The straightforward quantitative comparison of theoretical values with numerical computations is impractical because of the limiting assumptions and various kind of errors involved. It becomes more reliable and safe when the heat transfer and thermodynamic characteristics of the dump diffuser are studied numerically:

1. Preprocessing: In the preprocessing phase, the element type, real constants, material properties and geometric dimensions are defined.
2. Analysis: The analysis or solution defines the type of analysis to be carried out with prescribed boundary conditions. The finite element solution will be initialized once the inlet flow parameters are defined. Here, the standard *k-ε* model is selected as it is robust and a preferred choice for initial screening of different designs, and parametric evaluations. Though it is less sensitive to adverse pressure gradients, the ease of implementation and inexpensive computation at high Reynolds numbers range make it as a suitable choice for the proposed analysis. The other turbulence models

such as k-omega (k-ω) can also be used, if the flow Reynolds number is relatively low and the flow is completely wall-bounded.

3. Post processing: The analysed cases and data will be reviewed using the graphical displays and tabular listings in the course of post-processing of results. It will be further verified against the analytical solutions.

4.4 Mesh Preparation

The mesh preparation process for the numerical simulation was carried out using ANSYS Design Modeler tool with an extensive grid independence study. The properties of geometry and mesh resolutions are tailored according to the convergence requirements with minimal uncertainties. Initially 3 million elements are generated for the 2D model of the dump diffuser with unstructured grid and zonal refinements. The advanced size functions such as proximity and curvature are also enabled in the ANSYS Fluent module with fine span angle center and smoothing. The minimum edge length is about 6.35 x 10^{-4} *m* and the growth rate is 1.2. Smooth transition inflation option is also enabled with 10 inflation layers adjacent to the dome curvature. The assigned value of y^+ is in the range of $30 < y^+ < 100$ for the customized mesh refinement and the standard *k-ε* model is more stable for the proposed problem.

4.4.1. Meshed Geometry

The unstructured mesh view of the 2D dump diffuser model is displayed in Figure 4 (a). The presented meshed view of the geometry is fixed with a dump gap ratio of 1.1. The Figure 4 (b) shows the enlarged view of the dome geometry with refined meshing and prism layers around the dome shape. It is enlarged to describe the mesh fineness near the wall and the dump region. The mesh shows that the curvature and the proximity functions are used with the advanced sizing that offers the fineness near the wall and dump surfaces.

4.5 Mesh Independence Study

The accuracy of a numerical solution depends on many things including the optimized parameters and amount of computational resources. Mesh independent study is a crucial part of numerical simulations to achieve a fairly accurate solution for the problem at the expense of least computational resources. Using an optimum mesh density, the good enough accuracy of the results would be obtained by capturing all the necessary flow features and their gradients (Xu et al., 2015). The realizable *k-ε* model also used to validate a sample case and it is observed that the stability is poor for the range of obtained y^+ values. Even though the mesh independence study is a time consuming process at the early phases of a numerical simulation, it is essential to ensure an accurate mesh refinement level for the specific problem to reach the reliable solution limits with less computing costs.

Here, in order to check the independence of the solution against the grid densities, five refinement levels are tried with the initial mesh refinement of 3 million as '*x*'. The mesh densities are changed as *x*/2 (1.5 million), *x*/3 (1 million), 1.5*x* (4.5 million), and 2*x* (6 million) respectively and the results are compared. As an indicative purpose, the velocity contours corresponding to the refinement levels of *x*/2, 1.5*x* and 2*x* are presented in Figures 5 (a), (b) & (c) respectively. The pressure coefficient at various *x* - axis positions are also captured for the refinement level of 1.5*x* through grid independence study for the dump gap ratio of 1.0 as illustrated in Figure 5(d). Interestingly, the mesh refinement level above 1.5x has not influenced the solution to the significant level and it is retained for the remainder of the numerical simulations.

Figure 4. Meshed view of the 2D dump diffuser model

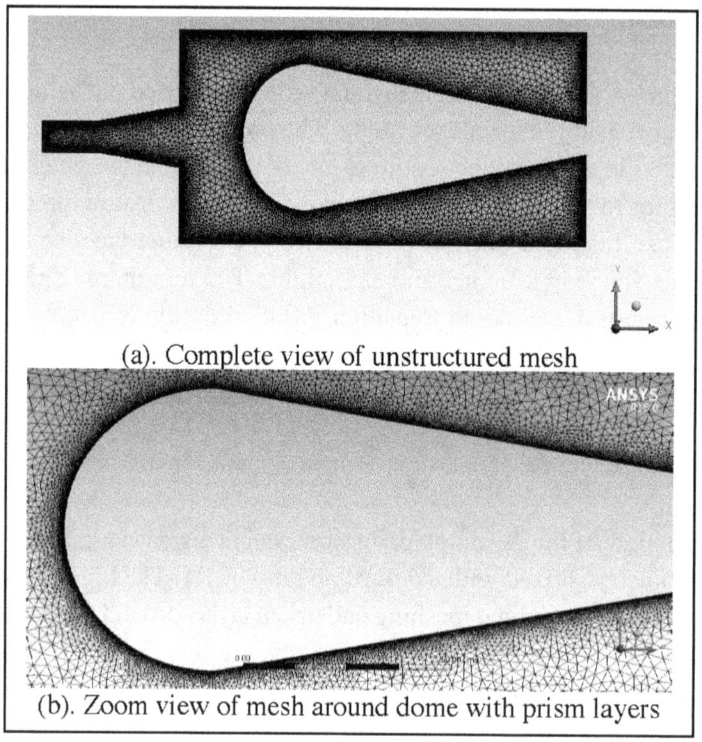

(a). Complete view of unstructured mesh

(b). Zoom view of mesh around dome with prism layers

5. RESULTS AND DISCUSSION

The results and discussion part of the chapter is divided into two major sections, namely i). The results concerning the influence of dump gap, ii). Results concerning the geometric optimization of the flame tube. The velocity inlet = 55 m/s and the pressure outlet =214.8 kPa are the initial boundary conditions assigned for the flow analysis across the dump diffuser. As compared with the prevailing operating temperatures in a turbojet engine, the temperature obtained at the exit plane of the compressor section through the theoretical calculation is about 450 K. Hence, the flow analysis is typically done at a cold flow condition as compared to the combustion temperature range of a jet engine. The dynamic pressures distributions and its contours related to the six different dump gap ratios (0.7, 0.8, 0.9, 1.0, 1.1, and 1.18) are presented below to offer the detailed insights about the dump gap optimization to minimize the total pressure losses across the dump region. The maximum dump gap ratio is restricted to 1.18 based on the various studies reported in the literature and beyond this gap ratio the total pressure losses exceed the allowable limit of 10% to 15%.

Figure 5. Velocity vectors for different grid densities

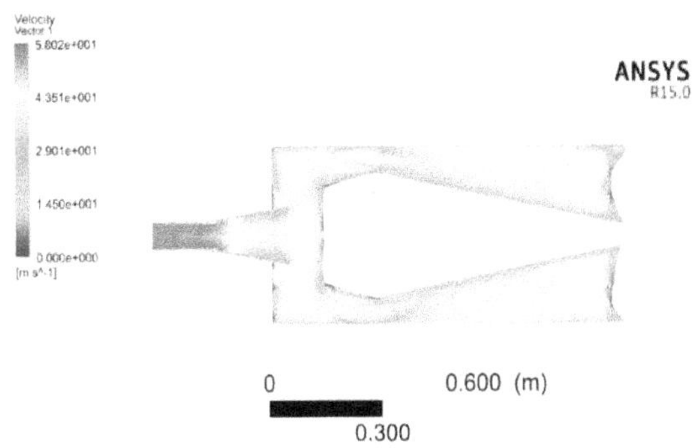

(a). Velocity distribution for 1.5 million elements

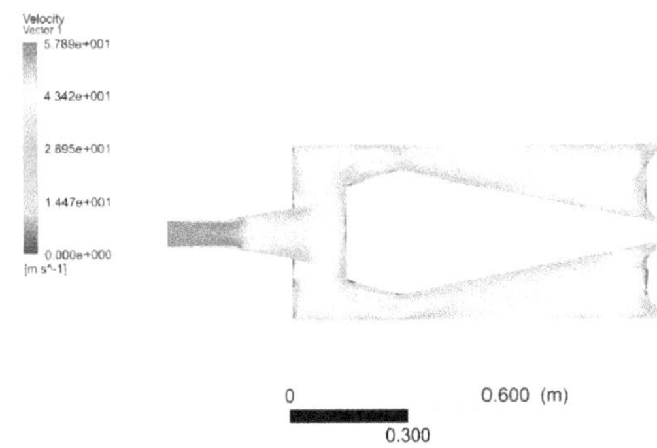

(b). Velocity distribution for 4.5 million elements

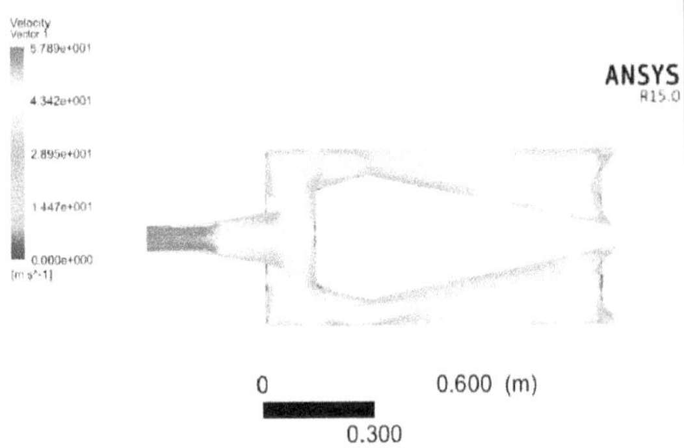

(c). Velocity distribution for 6 million elements

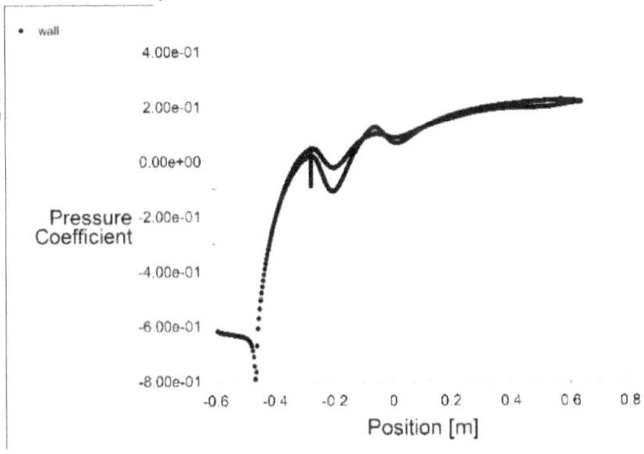

(d). Pressure coefficient Vs *x*- position plot

5.1 Influence of Dump Gap Ratio on Diffuser Performance

The dynamic pressure distributions related to the different dump gap ratios indicate the flow deceleration that is caused by the kinetic energy per unit volume of a fluid particle. Since, the fluid density and the flow speed are directly proportional to the dynamic pressure, it is increased when the density of fluid or flow velocity increases and vice versa (Carrotte et al., 1995, Carrotte et al., 1994). As the velocity of a fluid particle is the rate of change of its position with respect to a frame of reference, a reference point at the exit of the diffuser section is considered herein as a datum for the discussion. In an actual dump diffuser with hot flow conditions, the local flow characteristics such as the heat transfer and temperature also must be included for the overall diffuser optimization (Walker et al., 2009). However, the temperature at the inlet of the dump diffuser depends on the flow pressure and velocity at the exit plane of the axial compressor as computed theoretically in the earlier sections.

5.1.1 Dynamic Pressures for Dump Gap Ratios From 0.7 to 1.18:

From the dynamic pressure plots presented in Figures 6 (a) & (b) related to the dump gap ratio of 0.7 and 0.8, it is observed that the flow starts to split ahead of the dome at higher dump gap ratio. However, the further increase of dump gap (i.e., 0.9 & 1.0) tends to increase the dynamic pressure at the start of the diffuser plane as compared to the dump gap ratio 0.7 & 0.8 as illustrated in Figures 6 (c) & (d) respectively. It can be better visualized through the coloured contours presented in Figures 7 (c) & (d) respectively at the same location. Further increase of dump gap ratio results in the poor static pressure recovery and the total pressure losses exceed more than 20% as shown in Figures 6 (f) & 7 (f) respectively. Here, it is good to mention that the Reynolds number (*Re*) at the exit of the pre-diffuser region is computed as 2 x10^5 and the turbulent intensity is less than 1%.

Figure 6. Dynamic Pressure distributions for various dump gap ratios

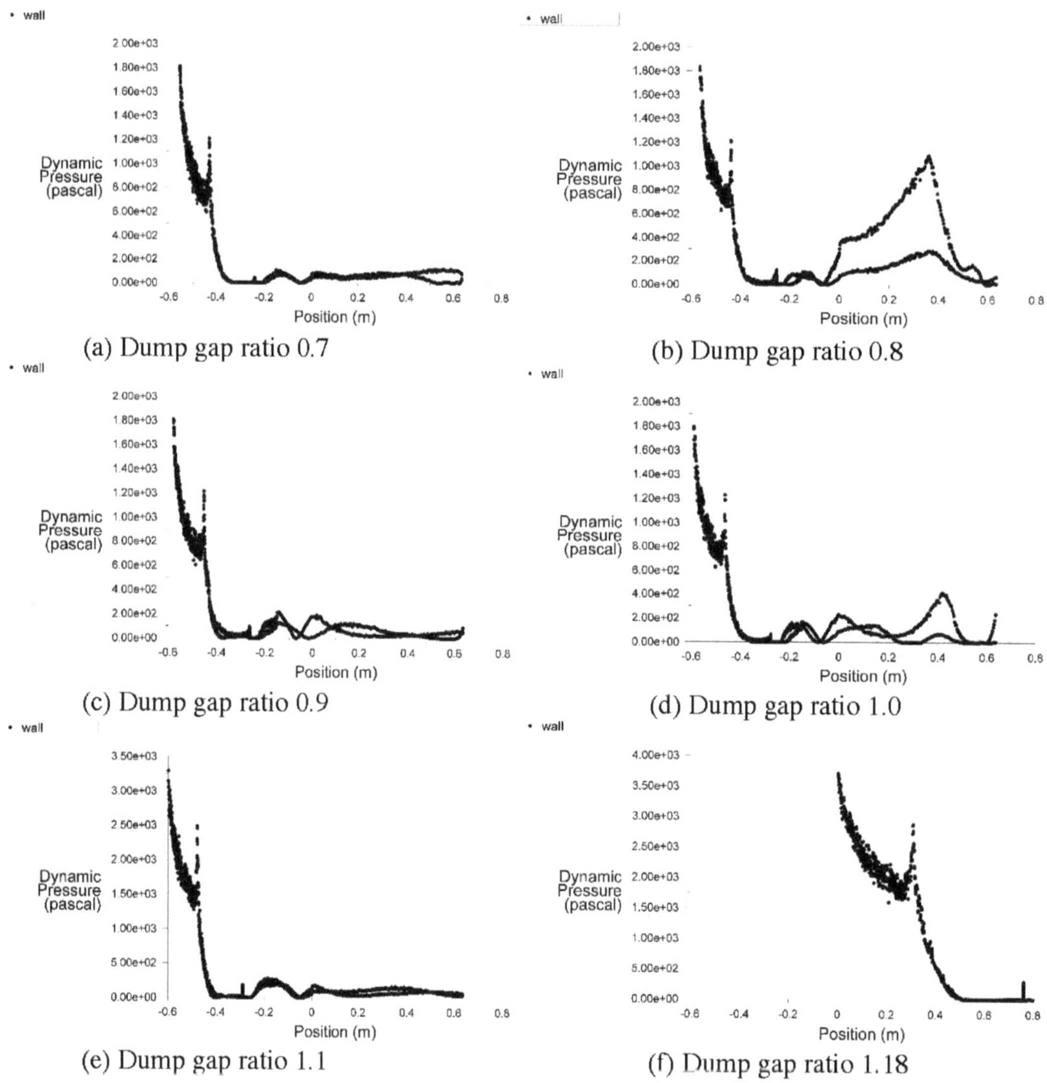

At the location of the flame tube head adequate pressure recovery must be ensured to attain the maximum combustion efficiency at high altitudes (Oates. 1989). Hence, the velocity distribution at the diffuser exit region should be maintained as low as possible which implies that the pressure is high in the downstream sections. From the dynamic pressure contours presented herein, it is observed that the dump gap ratio of 1.1 yields lower dynamic pressure at the entry of the dump diffuser with sustained pressure magnitudes along the entire section of the flame tube.

Figure 7. Dynamic Pressure contours for various dump gap ratios

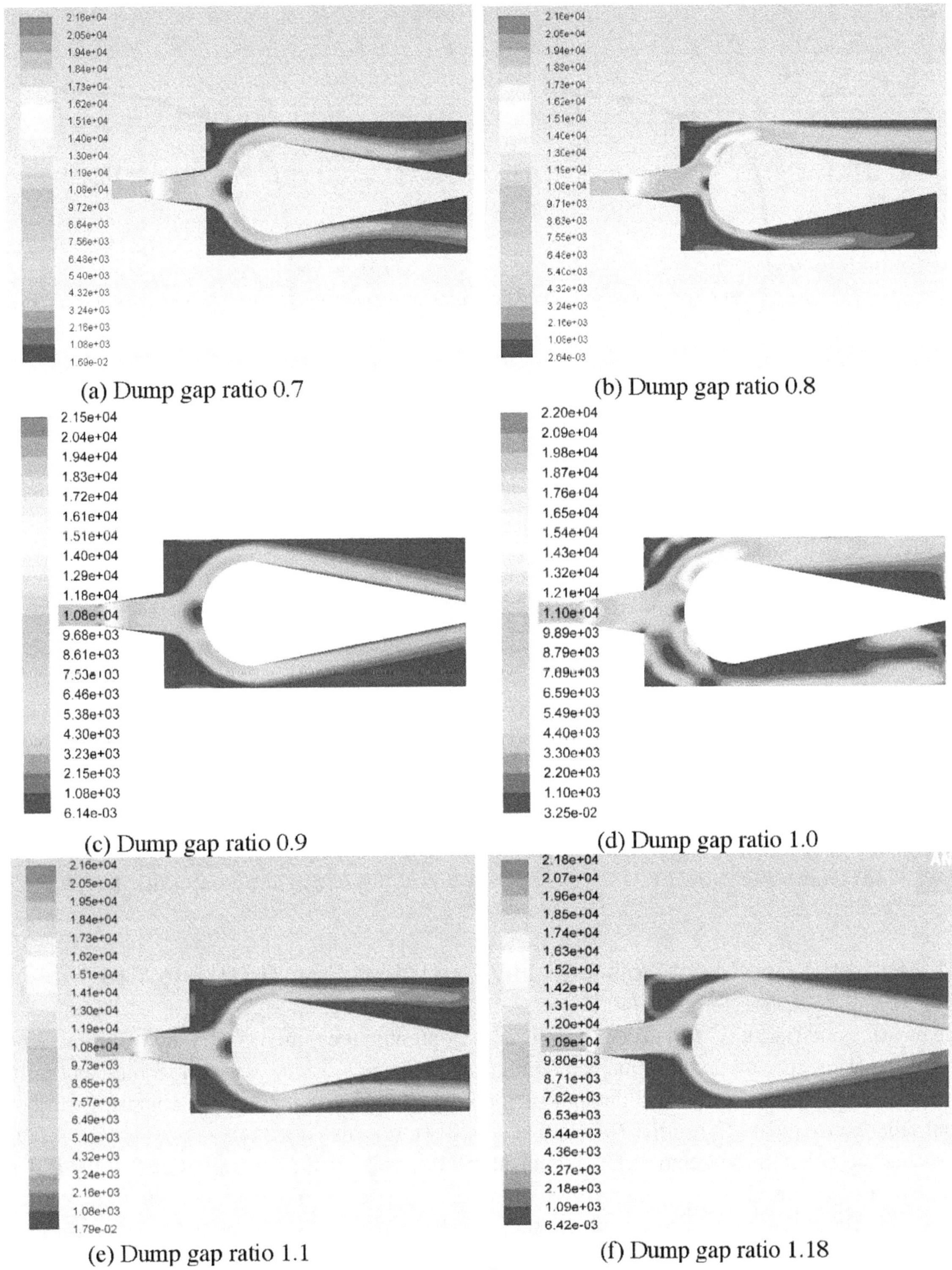

(a) Dump gap ratio 0.7

(b) Dump gap ratio 0.8

(c) Dump gap ratio 0.9

(d) Dump gap ratio 1.0

(e) Dump gap ratio 1.1

(f) Dump gap ratio 1.18

Figure 8. Pressure coefficient plots for various dump gap ratios

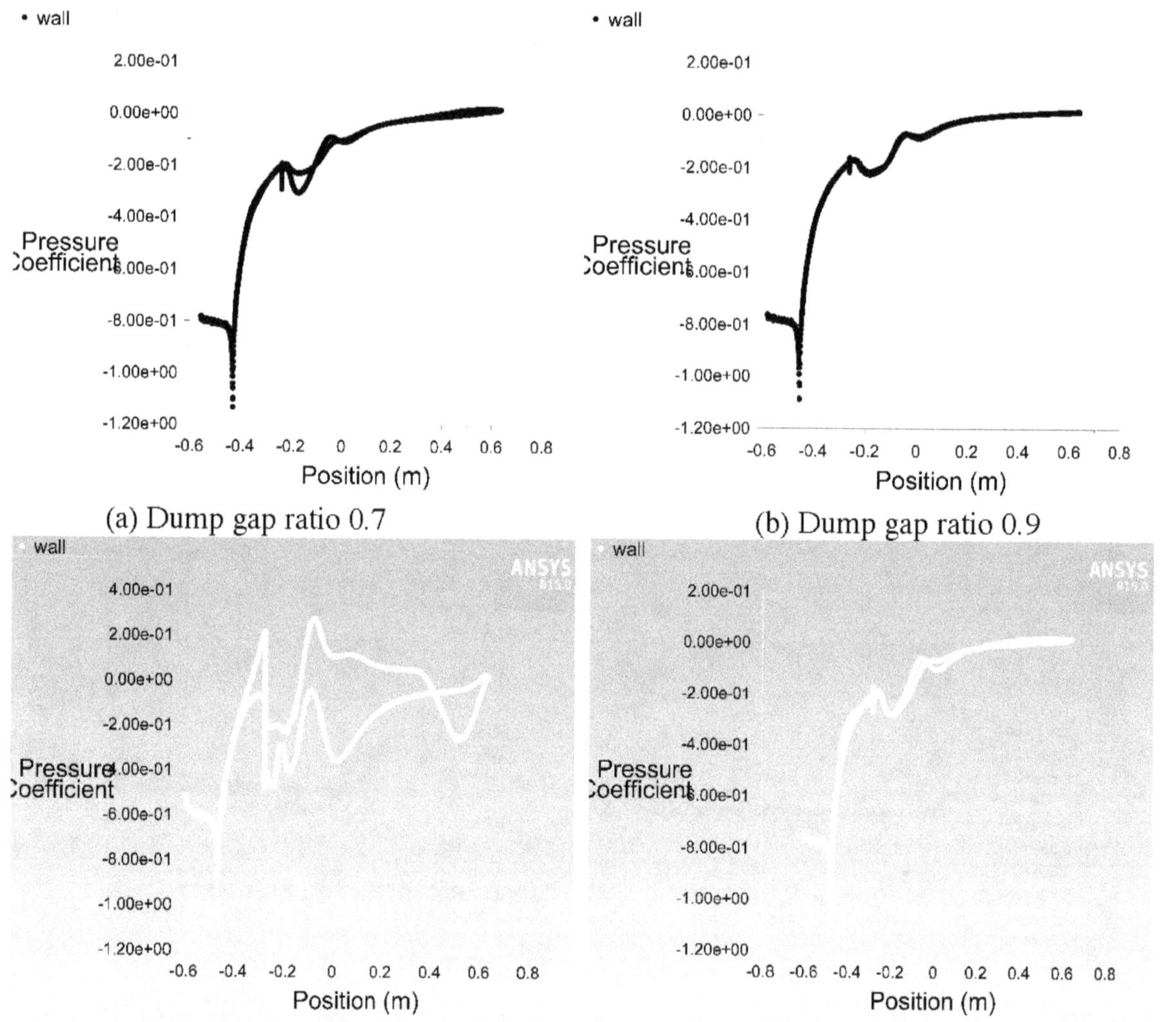

(a) Dump gap ratio 0.7

(b) Dump gap ratio 0.9

(c) Dump gap ratio 1.0

(d) Dump gap ratio 1.1

5.1.2 Pressure Coefficient Plots for Dump Gap Ratios From 0.7, 0.9 to 1.18:

The pressure coefficient (C_p) plots clearly indicate the optimum pressure recovery along the entire portion of the flame tube while the dump gap ratio is unity (Figure 8c). The essential requirement for the flame stabilization can be fulfilled through such a kind of C_p distribution that is complexly different from other conventional C_p profiles (Mariotti et al., 2014). It is also good to mention that the velocity profile at the outlet of the dump diffuser is maintained as uniform for the gap ratios 0.7, 0.9 and 1.1 because of the reduced flow separation and vortex mixing phenomena. However, the static pressure rise is significant for the dump gap ratio 1.0 that makes the choice more suitable for the performance enhancement of the dump diffuser.

5.2 Influence of Flame Tube Geometry on Diffuser Performance

The present chapter addresses few novel configurations of the flame tube that can be tailored by either horizontal or vertical truncation of an ellipse. The dump gap ratios are optimized according to the pressure recovery and flow separation characteristics with the conventional hemispherical flame tube geometry. Now, consider three different novel dome shapes for the flame tube geometry that are truncated from the base elliptical cross sections with respect to its major and minor axes as illustrated in Figures 9 (a) & (b) respectively.

Figure 9. Different dome shapes used for pressure recovery analysis

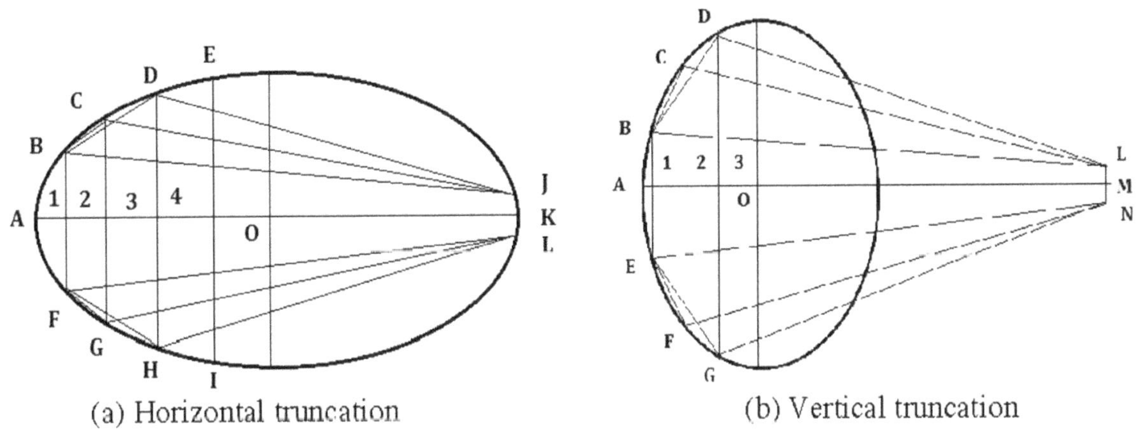

The analysis of the flame tube geometry on diffuser performance is carried out with different dome cowling geometric shapes that are tailored from an ellipse with respect to its major and minor axes as shown in Figure 9 (a). The numerical simulations are done at 9 km altitude similar to the former analysis of diffuser performance with various dump gap ratios. Though the geometric variables of the diffuser could influence several intrinsic flow properties at various inlet conditions, the static pressure recovery and amount of flow deceleration are the key parameters that are focused in this chapter. Hence, the problem description is limited within the scope of turbulence mixing and recirculation of the flow induced by various dome cowling shapes. The step-by-step numerical analysis is repeated with adequate mesh refinement by considering the prerequisite flow parameters such as dump gap ratio, inlet velocity, dome depth and profile of the pre-diffuser section. The flow properties obtained through the theoretical analysis are used to assign the boundary conditions in the course of numerical simulation for the problem investigated.

5.2.1. Total Pressure and Velocity Distributions for Various Dome Shapes (Horizontal)

The dome cowling shapes customized through Figure 9 (a) are analysed with the inlet velocity of 55 m/s. The total pressure and velocity distributions related to 3 different dome shapes are highlighted in Figure 10. Here, the total pressure losses are kept as minimum with higher pressure recovery and flow rate in the dome shape 2 as compared to the other two profiles as shown in Figures 10 (a) & (e) respectively. The dome shape 2 offers an approximate Convergent-Divergent (CD) nozzle like function that decelerates the flow better than the other horizontally tailored contours. Hence, the pressure recovery is increased about 8%-12% as observed through the contours presented in Figure 10 (c). Further, two flow recirculation zones are created adjacent to the downstream part of the flame tube head that delivers the air at optimum velocities in the intermediate region of the flame tube (Mongia et al., 2004). Alternatively, the flow velocities at the primary zone of the flame tube with shape 1 (Figure 10b) is comparatively about 20% higher than the other two configurations as displayed in Figures 10 (d) & (f) respectively.

Figure 10. Total Pressure and velocity distributions for various dome shapes (horizontal)

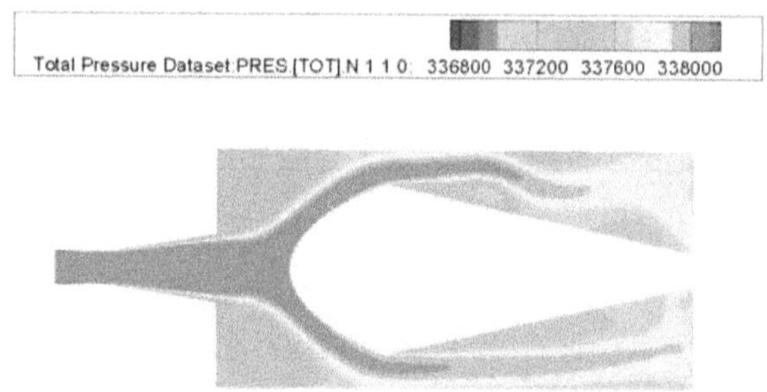

(a) Pressure for dome shape 1

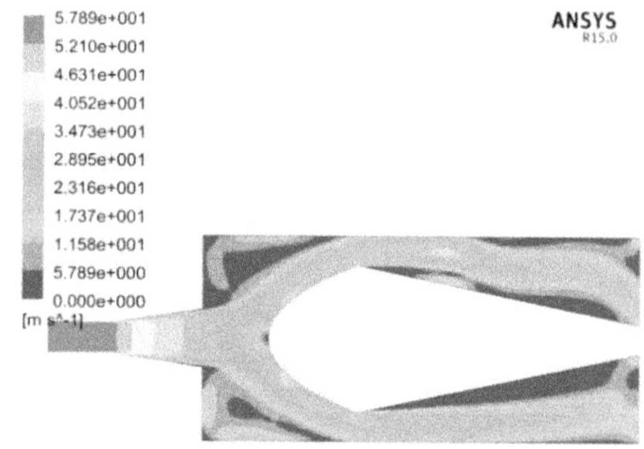

(b) Velocity for dome shape 1

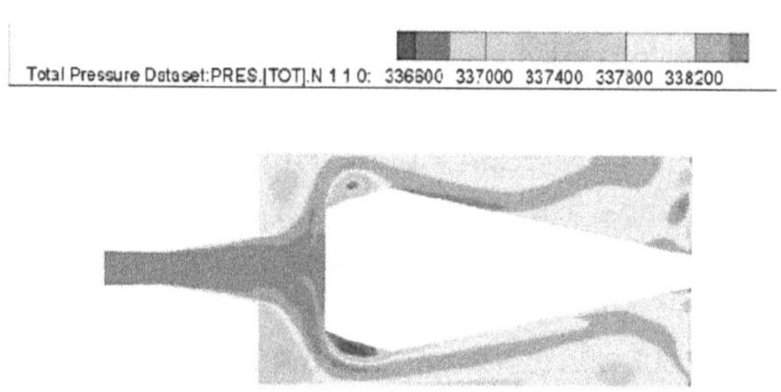

(c) Pressure for dome shape 2

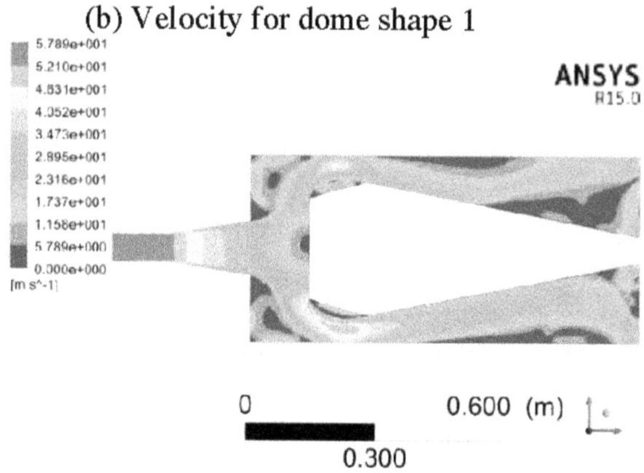

(d) Velocity for dome shape 2

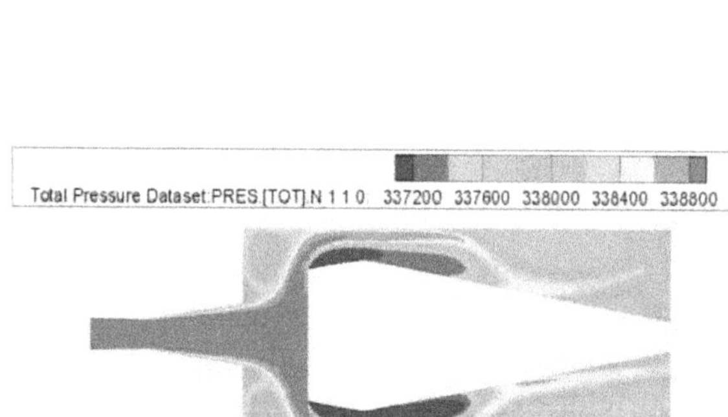

(e) Pressure for dome shape 3

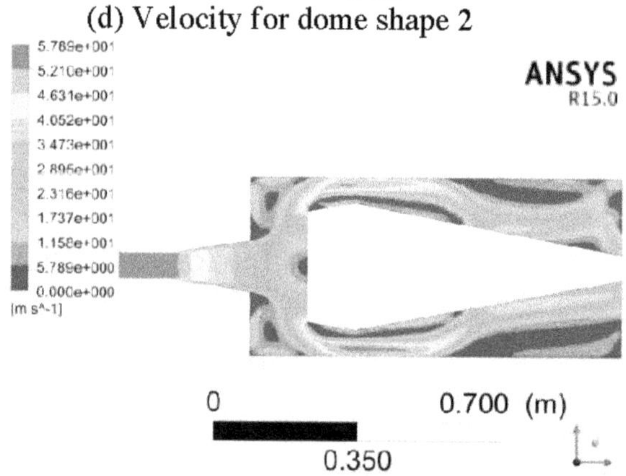

(f) Velocity for dome shape 3

5.2.2. Total Pressure and Velocity Distributions for Various Dome Shapes (Vertical)

The vertically tailored flame tube profiles have delivered significantly higher static pressure recovery than the horizontally tailored dome cowling configurations. In the traditional dump diffuser with wide angle pre-diffuser arrangement, it causes significant pressure losses while the compressor efflux is distributed through the different zones of the flame tube (Walker et al., 2004). The modern gas turbines require maximum diffusion in shorter length and hence, a hybrid arrangement such as an optimized dump gap with modified flame tube geometry is proposed in this chapter. The velocity and total pressure distributions related to the three different dome shapes are considered for the present discussion similar to the horizontally tailored dome shapes.

Figure 11. Total pressure and velocity distributions for various dome shapes (vertical)

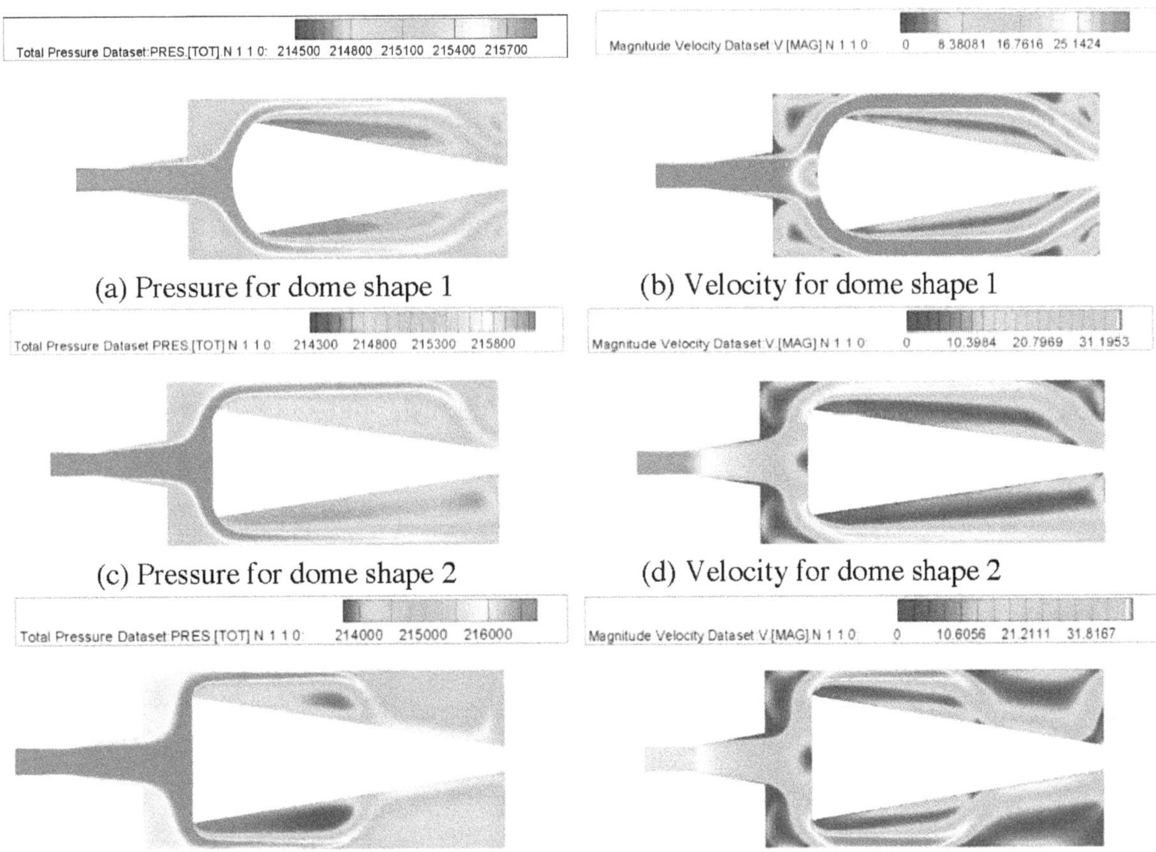

(a) Pressure for dome shape 1 (b) Velocity for dome shape 1

(c) Pressure for dome shape 2 (d) Velocity for dome shape 2

(e) Pressure for dome shape 3 (f) Velocity for dome shape 3

As the turbulent intensity at the dump gap is less than 1%, the flow separation and vortex mixing phenomena should be accomplished through the shape optimization of the dome cowling and cavities (Juhasz and Smith, 1978). Further, the present chapter is not focused with cavities and external guide vanes to achieve static pressure recovery around the deep flame tubes. Hence, the velocity and pressure fluctuations for the different dome shapes obtained through CFD simulation would serve the said purpose as shown in Figure 11. Figure 11 (a) shows minimal total pressure losses for the inlet velocity of 25 m/s and a deep recirculation zone is created around the flame tube (Figure 11b). It helps to fix the customized cavities at different zones of the flame tube to achieve satisfactory combustion performance. From the result, it is observed that the second dome shape (Figure 11c) offers a higher turbulence level (5% -20%) to increase the pressure recovery than the dome shape 3. Alternatively, the dome shape 3 offers comparatively higher inflow rate through the ports located at the dilution zone of the flame tube as shown in Figure 11 (e).

5.3 Comparison of Velocity Streamlines

The flow velocity streamlines for both horizontally and vertically tailored flame tube geometries are compared in Figure 12. It is observed from the velocity magnitude dataset that the dome cowling geometry obtained through the first horizontal and vertical cuts show two distinct flow streamline patterns as illustrated in Figure 12 (a) & (b) respectively. Specifically, two wake vortices are created at the dump gap that makes the pressure loss very minimum around the primary zone for the vertical cut hybrid dump diffuser shown in Figure 12 (b). The flow streamlines become more linear as the geometry is customized with subsequent vertical cuts with extended vortices along the entire length of the flame tube as displayed in Figure 12 (d). However, the wake vortices are created with significant turbulent intensity at isolated zones around the flame tube and dump gap of the shape 3 geometry as shown in Figure 12 (c). Hence, according to the aerodynamic design constraints of the gas turbines, the specific hybrid dump diffuser configuration can be opted with required static pressure recovery.

Figure 12. Comparison of velocity streamlines for horizontally and vertically modified domes

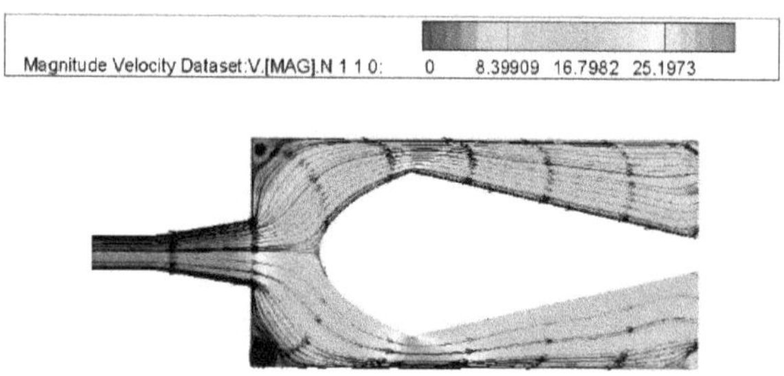

(a) Streamlines for horizontal cut shape 1

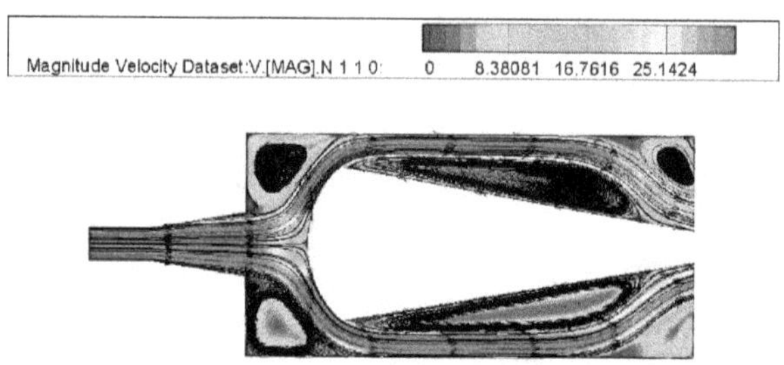

(b) Streamlines for vertical cut shape 1

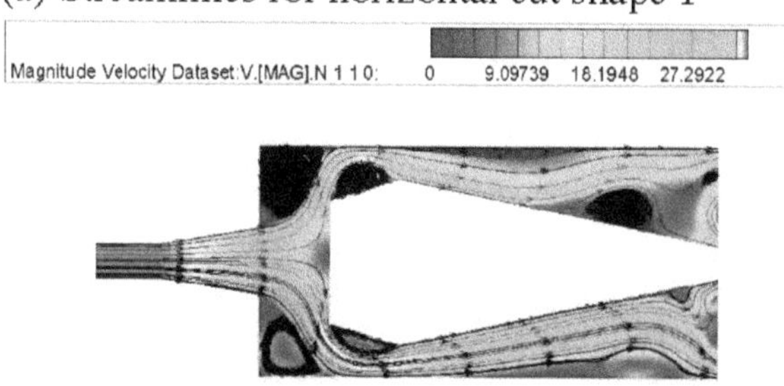

(c) Streamlines for horizontal cut shape 3

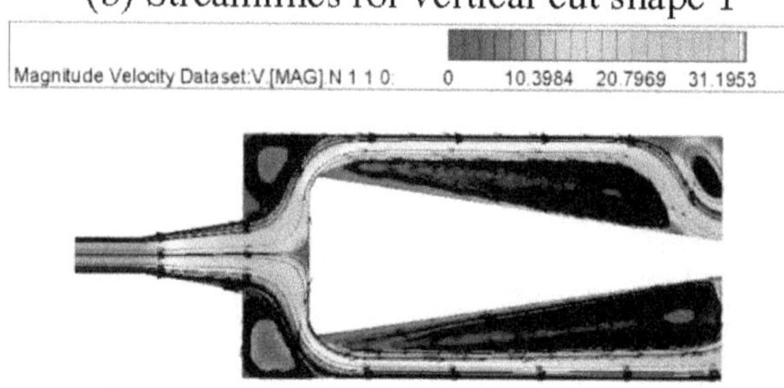

(d) Streamlines for vertical cut shape 3

Though only a few specific geometric shapes and their flow characteristics are discussed in this chapter, it enables several research opportunities to open new doors in the context of hybrid dump diffuser modelling for modern aero-gas turbines. The dynamic pressure distribution is another essential parameter that is mainly optimized through the analysis of different dump gaps. As the static pressure should be high near the entrance of the dump region, the numerically computed values are referred to a specific point in the dump gap region. From the comparative results with various dump gap ratios, it is concluded that the dump gaps 0.8 & 1.1 offer higher dynamic pressures than any other configurations at the specified point. The dump gap ratio of 1.1 offers higher vortex mixing and flow recirculation at the upstream section of the flame tube and the pressure recovery is increased about 20% at different velocities. Other flow properties also can be compared anywhere in the diffuser using the single point analogy method and the dynamic pressure computation is presented in Figure 13. The common point is chosen for all the dump gap ratios as mentioned in the inner part of Figure 13.

In addition, the vertically tailored shape 1 offers better efficiency in terms of flow mixing vortices and shape 2 obtained from the horizontal cut offers maximum pressure recovery with dump gap ratio of 1.1. The pressure distribution profiles along the length of the dump diffuser for the 3 different dome shapes (Horizontally tailored) are displayed in Figure 14. It clearly shows the optimum pressure recovery with dome shape 2 that is shown in the contour plot presented in Figure 10. Alternatively, any inner part of the diffuser or point can be assigned as a zone of interest in the Machine Learning (ML) algorithm and the flow contours are utilized to train the model. If more than 100 pressure or velocity contours available then the required data can be extracted instantly though ML techniques. Therefore, similar plots with hybrid diffuser modelling may reveal interesting outcomes from the study of pressure recovery and flow separation as discussed.

Figure 13. Comparison of dynamic pressures at the exit of diffuser for different dump gaps

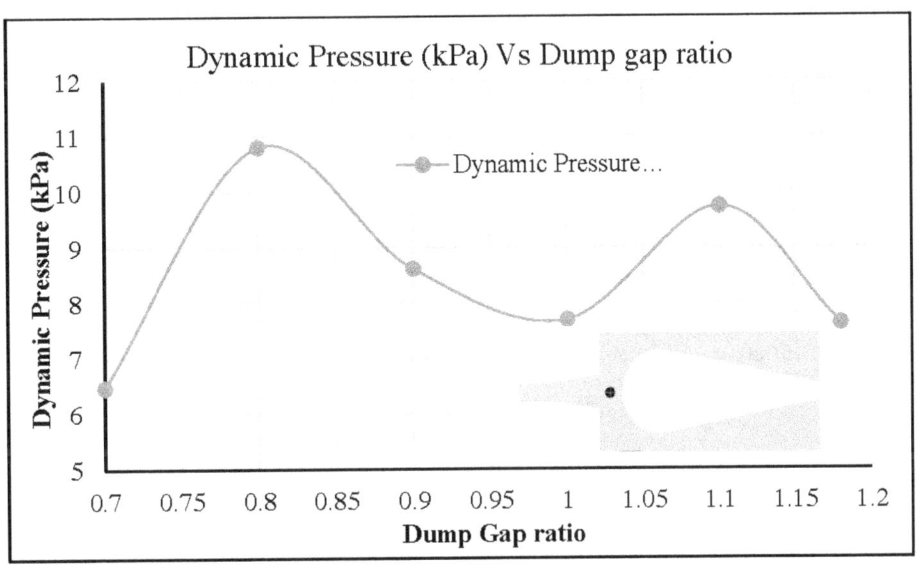

Figure 14. Comparison of pressure profiles for the 3 different dome shapes (horizontal)

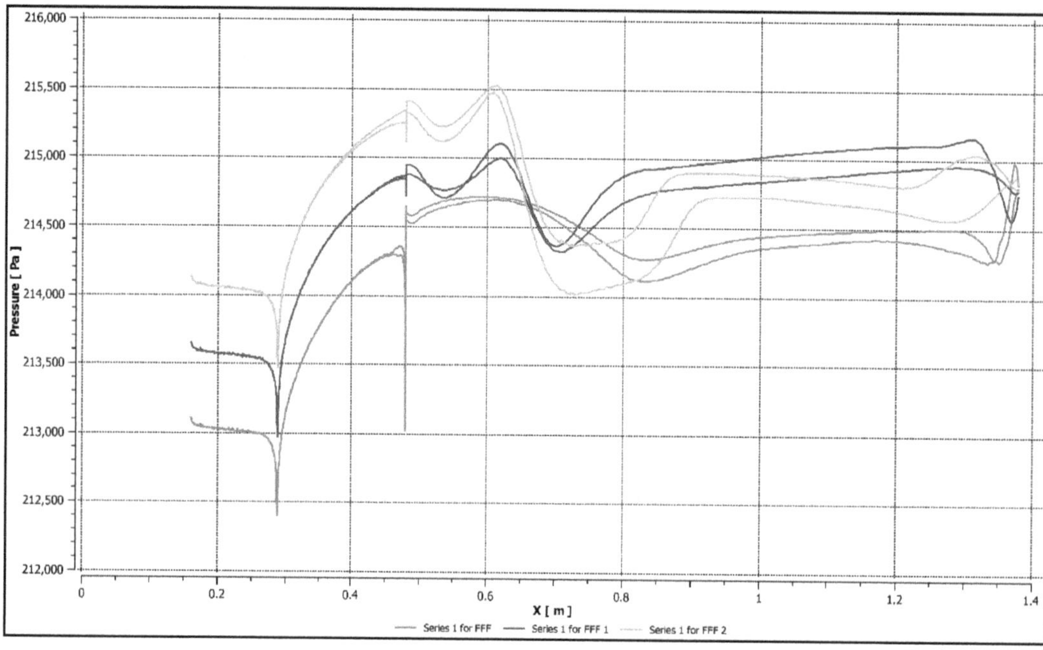

6. CONCLUSION

A detailed numerical simulation is addressed successfully with a hybrid annular dump diffuser model with different geometric modifications. Although, specifying the influences of different parameters is yet a difficult task, since they are interrelated in a complex way. Two features have been investigated in the present simulation for the recovery of pressure that are the distance between the flame tube head and the pre-diffuser exit plane (dump gap ratio) and the shape of the dome. The simulations are carried out for different dump gap ratio and for different dome shapes with velocity of 25 m/s & 55 m/s and the results are compared. The simulations are carried out for different dump gap ratios with velocity of 55 m/s and the results are compared. Similarly, the three different dome shapes are also analyzed and the optimized dome shape is presented with velocity streamlines. Finally, the result of the dump gap ratio of 1.1 offers better results as compared to other dump gap ratios and the second dome shape gives the higher pressure recovery. In future, in addition to fluid dynamic constraints and the geometry, the powerful parameters which control the recovery of pressure and the losses in a dump diffuser may also be addressed for the superior combustor performance at high altitudes.

REFERENCES

Adkins, R. C. (1975). A short diffuser with low pressure losses. *ASME J. Fluid Engng.*, *97*(3), 297–302. doi:10.1115/1.3447306

Aleksandrov, Mingazov, & Tokmovtsev. (2016). *Software for Automated Processing and Analysis of Gas Dynamics and Combustion Processes in Combustion Chambers of Gas Turbine Engines (Servis 3D)*. The Certificate on State Registration of the Computer Program no. 2016611856, KNRTU–KAI.

Aleksandrov, Sabirzyanov, & Yavkin. (2017). Influence of Simplifying a Geometrical Model of the Gas Turbine Engine Combustion Chamber on the Results of Numerical Modeling. *Russian Aeronautics, 60*(4), 575–582.

Ananda Reddy, G., & Ganesan, V. (2005). A Numerical Study of Pre-Diffuser Optimization of an Aero Gas Turbine Combustion Chamber. *Indian Journal of Engineering and Materials Sciences, 12*(August), 281–291.

Aravind, S., Sabarinathan, G., Ajith, S., Amrith, M., Sundararaj, K., & Sanal Kumar, V. R. (2019). Design of pre-diffuser cones for dump diffusers for aero gas turbine engines. AIAA Propulsion and Energy Forum and Exposition.

Carrotte, Denman, Wray, & Fry. (1994). *Detailed performance comparison of a dump and short faired combustor diffuser system*. ASME Paper, 93-GT-331.

Carrotte, J. F., Bailey, D. W., & Frodsham, C. W. (1995). Detailed Measurements on a Modern Combustor Dump Diffuser System. *Journal of Engineering for Gas Turbines and Power, 117*(4), 679–685. doi:10.1115/1.2815453

Carrotte, J. F., Denman, P. A., Wray, A. P., & Fry, P. (1994). Detailed Performance Comparison of a Dump and Short Faired Combustor Diffuser System. *Journal of Engineering for Gas Turbines and Power, 116*(3), 517–526. doi:10.1115/1.2906850

Fishenden, C. R., & Stevens, S. J. (1974). Performance of Annular Combustor-Dump diffuser. *Journal of Aircraft, 14*(1), 60–67. doi:10.2514/3.58749

Ghose, P., Datta, A., & Mukhopadhyay, A. (2013). Effect of dome shape on static pressure recovery in a dump diffuser at different inlet swirl. *International Journal of Emerging Technology and Advanced Engineering, 3*(3), 465–471.

Grewe, V., Rao, A. G., Grönstedt, T., Xisto, C., Linke, F., Melkert, J., Middel, J., Ohlenforst, B., Blakey, S., Christie, S., Matthes, S., & Dahlmann, K. (2021). Evaluating the climate impact of aviation emission scenarios towards the Paris agreement including COVID-19 effects. *Nature Communications, 12*(1), 3841. doi:10.103841467-021-24091-y PMID:34158484

Juhasz, A. J., & Smith, J. M. (1977). *Performance of High-Area-Ratio Annular Dump Diffuser Using Suction-Stabilized-Vortex Flow Control.* National Aeronautics and Space Administration.

Klein, A. (1995). Characteristics of combustor diffuser. *Progress in Aerospace Sciences, 31*(3), 171–271. doi:10.1016/0376-0421(95)00006-K

Mongia, H., Hsiao, G., Ravindra, M. V. V., & Sreedhar, P. S. V. S. (2004). Combustion Diffuser Modeling – Part V: Validation with a three passage diffuser rig data. *AIAA Paper*, 2004–4172.

Oates, G. C. (1989). *Aircraft propulsion systems. In Technology and Design*. AIAA Education Series. doi:10.2514/4.861499

Rahim, Singh, & Veeravalli. (2007). Liner dome shape effect on the annulus flow characteristics with and without swirl for a can-combustor model. *Proc. IMechE, 221*(A), 359-369.

Sanal Kumar, Sahoo, & Ragunathan. (2009). *Internal Flow Simulation of Dump Diffusers for Modern Aircraft Engines*. 45th AIAA/ASME/SAE/ASEE Joint Propulsion Conference & Exhibit, Denver, CO.

Sanal Kumar, V. R., Kim, H. D., Setoguchi, T., Matsuo, S., & Raghunathan, S. (2005). Influence of dump gap ratio on the performance characteristics of dump diffusers. *4th International Conference on Heat Transfer, Fluid Mechanics and Thermodynamics (HEFAT2005)*, 19-22.

Sanal Kumar, Sameen, Kim, Setoguchi, Mastuo, & Raghunathan. (2005). Influence Of Dump Gap On The Performance Characteristics of Dump-Diffusers. *16th International Symposium On Transport Phenomena, ISTP-16*.

Smith, J. M., & Juhasz, A. J. (1978). *Performance of a Short Annular Dump Diffuser Using Suction-Stabilized Vortices at Inlet Mach Numbers to 0.41*. NASA National Aeronautics and Space Administration Scientific and Technical Information Office.

Srinivasan, R., Freeman, G., Grahmann, J., & Coleman, E. (1990). Parametric Evaluation of the Aerodynamic Performance of an Annular Combustor Diffuser System. Paper No.AIAA-90-2163. doi:10.2514/6.1990-2163

Stevens, S. J., Wray, A. P., & Price, P. D. (1988). The Aerodynamic Performance of a Modern Vaporizing Combustor Dump Diffuser. Paper No.AIAA-88-3273. doi:10.2514/6.1988-3273

Varshney, Baig, & Hasan. (2020). Turbulent Drag Reduction on an Aircraft Wing by Wall Jets for Flow Control. *AIAA Aviation 2020 Forum*. 10.2514/6.2020-2780

Walker, A. D., Carrotte, J. F., & McGuirk, J. J. (2009). The Influence of Dump Gap on External Combustor Aerodynamics at High Fuel Injector Flow Rates. *Journal of Engineering for Gas Turbines and Power, 131*(3), 031506–031510. doi:10.1115/1.3028230

Wang, H., & Kai, H. L. (2020). Numerical Investigation of Dump Diffuser Combustor Performance at Uniform and Non-Uniform Inlet Conditions. *ASME Proceedings*. 10.1115/GT2020-15982

Xu, L., Huang, Y., Ruan, C., & Peiyong, W. F. X. (2015). Study of the Dump Diffuser Optimization for Gas Turbine Combustors. *Procedia Engineering*, 99, 828–834. doi:10.1016/j.proeng.2014.12.608

Chapter 17
Parametric Effect of Roughness Over an Airfoil

Shiva Prasad Uppu
Sandip University, India

Rathan Babu Athota
Universitat Politecnica, Spain

Sathish Kumar K.
Nehru Institute of Engineering and Technology, India

Dilip Raja N.
Vel Tech Rangarajan Dr. Sagunthala R&D Institute of Science and Technology, India

ABSTRACT

This chapter is focused on the roughness effect to evaluate the flow in order to examine the flow dynamics around airfoil for better aerodynamic efficiency. CFD analysis is done on the airfoil with circular roughness placed at two different positions, 25% and 65% chord length with two different Reynolds number. In this case, the boundary layer increased significantly due to decrease in velocity of flow resulting in increment of pressure gradient. From the computational and experimental investigation from many researchers, it is evident that adverse pressure gradient even becomes so large that the flow is forced back against the actual flow direction. In the current chapter, at 15 degrees angle of attack there was an effective increase of 65% in the aerodynamic efficiency due to roughness. There was an increase in stall angle, which refers to sudden increment of drag resulting from the aerodynamic and geometric variations over the infinite wing. With increase in Reynolds number, there is an increase in the effect of roughness at higher angles of attack.

DOI: 10.4018/978-1-6684-4230-2.ch017

INTRODUCTION

The wing sectional view is called an airfoil which can be obtained along any cross-sectional plane of the three-dimensional wing. Airfoil is designed to allow flow over the top and bottom of the wing. The design of Airfoils has allowed the broad laminar stream, like the National Advisory Committee for Aeronautics (NACA) 4, 5, 6, and 7 series, of which NACA 6 and 7 series are called explicitly as Laminar airfoils have significantly less drag at average lift coefficients than different types of airfoils (Srivastav, 2012), (Howe, 2010),(Lambert, 2006). Before the National Advisory Committee for Aeronautics (NACA) fostering these series, airfoil configuration was somewhat subjective w,ith nothing to direct the designer besides previous involvement in known shapes and trial and error with adjustments to those shapes. The Advisory Committee has proposed the design of airfoils with the same curvature on the top and bottom, which are called symmetric airfoils. The curvatures are not the same on the top and bottom of the airfoil are called Unsymmetrical airfoils.

As per the NACA, symmetrical airfoils of four series are designated as NACA 00XX, where the XX is represented as the thickness of the airfoil, and the first two digits indicate that the curve is symmetric about the midline joining the leading and trailing edge called as chord line. Be that as it may, the symmetric airfoil delivers less lift than a nonsymmetrical airfoil and generally has bothersome slow-down qualities. The helicopter's cutting edge (airfoil) should adjust to a broad scope of velocities and approaches during every insurgency of the rotor. The even airfoil conveys satisfactory execution under those exchanging conditions. Different advantages are lower cost and simplicity of development than the nonsymmetrical airfoil. Nonsymmetrical (cambered) airfoils might have a wide assortment of upper and lower surface plans. The benefits of the nonsymmetrical airfoil are expanded lift-drag proportions and more beneficial slow-down qualities. Nonsymmetrical airfoils were not utilized before in helicopters as the focal point of the pressure area moved a lot when the approach was changed. When the focus of strain moves, a bending power is applietoon the rotor edges. Rotor framework parts must be planned that would endure the bending power. Ongoing plan processes and new materials used to fabricate rotor frameworks have conquered the issues of using nonsymmetrical airfoils.

The most significant lift coefficient for respectably cambered 6-series airfoils is as high as those accomplished utilizing NACA 24-and 44-series airfoils (Graham, 1949). The NACA 230-series airfoils with thickness proportion under 20% achieve the highest, most significant lift coefficients. The most extreme lift coefficient with folds is about no different for tolerably thick 6-series airfoils for what it's worth for the NACA 23012 with flaps. Notwithstanding, the slenderer 6-series airfoils have considerably lower C_{lmax} with folds. The lift bend slant for smooth 6-series airfoils is marginally more extreme than the 24-, 44-, and 230-series airfoils. It surpasses the hypothetical worth (2π) for thin airfoils. Examination of a stream over an aerofoil has numerous applications, notwithstanding its academic interest. The significance of the impact of roughness on aerofoils in the aviation field is all around figured out. Additionally, the exhibition of rotatory hardware generally relies on the outer layer of the fixed course of airfoils. Surface roughness is a typical element of most working wing surfaces. Roughness might emerge from the assembling system, a significant administration stretch, or potentially standard collections (ice, snow, ice, salt splash)((Christopher D Griffin, 2016, Michael F., 1997, P. D. Gall, W. 2010, Zaccai, D., Bertels, F., Vos, R. 2016). Many researchers have demonstrated the impact of particle flows over the gas turbine edges, which might go from around 1.5 µm to about 10 µm (Hood 1939), (P. D. Gall, 2010), Roskam Jan, 2000), (Zaccai, D, 2016). The roughness thickness layer is around layer 1 mm. Freezing precipitation and ice formation over the wing or airfoil changes the shape of aerofoil causing general-flight mishaps every year (Zaccai, 2016). Answers for icing, like the enemy of icing and deicing, are costly and not satisfactory all the time for various reasons. Ice influences a plane's exhibition by hampering the wings' lift limit. Indeed, even in situations where the accumulation of ice isn't severe, it powers the airplane toward more fuel utilization to make up for the deficiency of lift.

SCOPE OF THE CHAPTER

An aircraft body traveling through air is exposed to a variety of forces. Of specific significance are those bodies for which the part of force inverse to the bearing of movement is little contrasted and the amount of power typical to the course of action. Bodies with this property are designated "wings" if the aspect in one course, the" thickness," is considerably more modest than the aspects in two headings ordinary to one another and the main bearing, the "chord" and the "span"; the direction of the chord is around that of the stream at the vastness, the "free stream." A segment typical to the span is known as a "wing profile" - momentarily, a "profile." We will be predominantly worried about a two-layered stream about a profile-in others words, with a stream about a wing of limitless range. In the current chapter, the authors will detail the flow parameters affecting the aerodynamics of the infinite wing by considering the factor of roughness over the wing's geometry with the help of computational tools, and roughness flow parameters are evaluated over the NACA 0012 airfoil.

Wing Design and Its Parameters

As a result of a report from the National Aeronautics and Space Administration (NACA) published in the mid-1930s, this way of thinking changed (Kundu, Kumar Ajoy. 2010, Smith, A.M.O, 1975). The authors of this landmark paper discovered significant similarities among the best-performing airfoils. The two most essential elements that impact those shapes are the tilt of the airfoil mean camber line and the thickness appropriation above and below this line. They then, at that point, introduced a progression of conditions joining these two factors that could be utilized to produce a whole group of related airfoil shapes (W.A. Timmer Duwind, 1997). As airfoil configuration became more refined, this fundamental methodology was changed to incorporate extra factors; however, these two essential mathematical qualities stayed at the core of all NACA airfoil series, as delineated beneath (Ferri, 1945), (Raymer, D, 2018).

The best way to deal with an aircraft wing starts by selecting the wing vertical area to create an aerodynamic architecture. The essential capacity of the wing is to generate lift power. An exceptional wing section area will make this called an airfoil. The wing is a three-dimensional part, while the airfoil is a cross-sectional segment of the wing. On account of the airfoil area, two different results of the airfoil, and therefore the wing, are drag and pitching moment (Chana R.1998). The wing might have a steady or a non-consistent cross-area across the wing. As per the rapid technology development and robust PCs, an airfoil plan isn't quite as hard as forty years prior. There are, as of now, several streamlined programming bundles (computational fluid dynamics) in the market that can be utilized to plan airfoil for an assortment of requirements. Not just airplane designers need to plan their airfoil, the other numerous regions that airfoil should be designed for their parts. The rundown incorporates stream motor hub blower edges, fly motor pivotal turbine edges, steam power plant hub turbine cutting edges, wind turbine propellers, radial and hub siphon impeller edges, turboprop motor propellers, diffusive and hub blower impeller edges, and enormous and little fans. (McGhee,1973, Sato, 1973, Whitcomb,1965). The efficiencies of these modern mechanical or aviation devices depend intensely on the segment of their cutting edges: "airfoil."

Figure 1. Geometrical details of the wing section/aerofoil

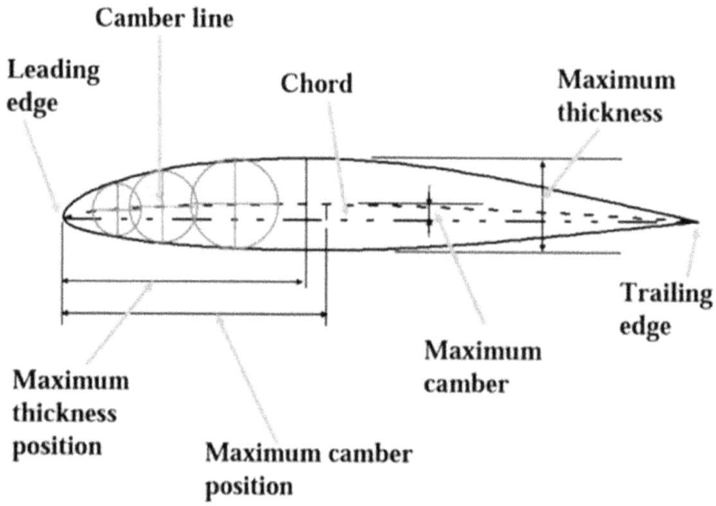

NACA Four-Digit Series: The primary collection of airfoils designed using this approach became known as the NACA Four-Digit Series. The first digit indicates the most extreme camber (m) in the level of the harmony (airfoil length), the second indicates the position of the greatest camber (p) in the tenths of the chord, and the last two numbers indicate the most significant thickness (t) of the airfoil in the level of the chord (airfoil length). A typical NACA 2415 airfoil has a maximum thickness of 15 percent and a camber of 2 percent that is found 40 percent back from the driving edge of the airfoil, for example (or 0.4c).

NACA Five-Digit Series: The NACA Five-Digit Series utilizes a similar thickness structure as the Four-Digit Series. However, the mean camber line is characterized distinctively, and the naming show is a touch more perplexing. When multiplied by 3/2, the first digit yields the plan lift coefficient (cl) in tenths. When isolated by 2, the following two numbers give the most extreme camber (p) in tenths of the chord. The last two digits again show the most significant thickness (t) in the level of connection. For instance, the NACA 23012 has the most significant thickness of 12%, a plan lift coefficient of 0.3, and a most extreme camber found 15% back from the leading edge.

NACA 1-Series or 16-Series: The 1-Series, unlike the other airfoil families we've discussed, was developed based on the airfoil hypothesis rather than mathematical correlations. When these airfoils were planned during the last part of the 1930s, many advances had been made in opposite airfoil plan strategies. The fundamental idea driving this planning approach is to indicate the ideal strain dissemination over the airfoil (this conveyance directs the lift attributes of the shape) and afterward determine the mathematical condition that delivers this tension dispersion. Thus, these airfoils were not produced utilizing a few arrangements of logical articulations like the Four-or Five-Digit Series. The 1-Series airfoils are recognized by five digits, as exemplified by the NACA 16-212. The primary digit, 1, demonstrates the series (this series was intended for airfoils with districts of scarcely supersonic stream). The 6 determines the area of least tension in tenths of harmony, for example, 60% back from the leading edge for this situation. Following a scramble, the primary digit shows the plan lift coefficient in tenths (0.2), and the last two numbers determine the most significant thickness in tenths of harmony (12%). Since the 16-XXX airfoils are the ones in particular that have at any point seen a lot of utilization, this family is frequently alluded to as the 16-Series rather than as a subset of the 1-Series (Ladson, C. L, 1975, Graham, 1949, McGhee, 1973).

Figure 2. Some typical aerofoils based on the Mach number and thickness to chord variation (Ladson, 1975; Graham, 1949; Turner, 1941).

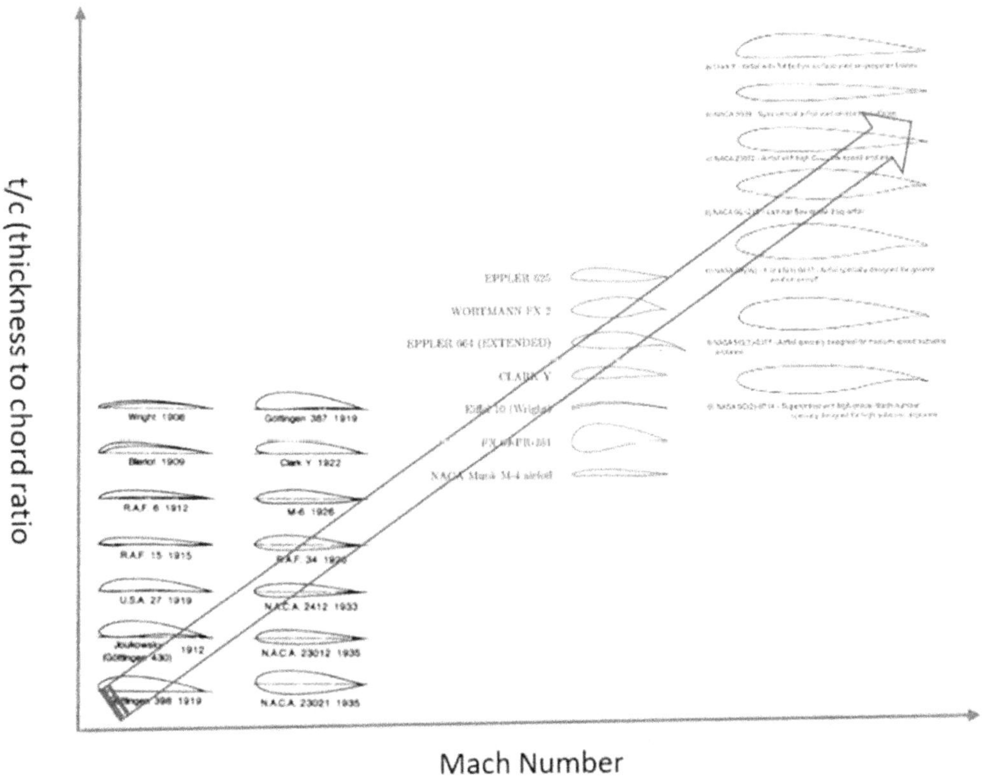

NACA 6-Series: Although NACA investigated many routes about inexact hypothetical procedures that developed the 2-Series through the 5-Series, none of these methodologies were determined to exactly offer the optimum airfoil conduct. The 6-Series was deduced using a more advanced hypothetical technique that, like the 1-Series, relied on identifying the optimum tension dispersion and advanced algebra to establish the required mathematical form. This research aimed to design airfoils that would extend the district while keeping the wind current predominantly laminar. The drag across a small range of lift coefficients can be significantly reduced in this way. The 6-Series naming convention is the most perplexing of any families discussed thus far, especially given the large variety of variants. The NACA

641-212, a=0.6, is one of the more typical models. The number 6 denotes the series in this model, indicating that this family is meant for a more notable laminar stream than the Four- or Five-Digit Series. The area of the base tension in tenths of harmony is represented by the next digit, 4. (0.4c). Addendum 1 shows that low drag may be maintained at lift coefficients 0.1 above and below the plan lift coefficient (0.2), demonstrated by the first number after the scrambling. The last two figures represent the harmony thickness, which is 12 percent. The component determined by shows the level of airfoil harmony over which the strain transmission on the airfoil is uniform, in this case, 60 percent harmony. If the quantity cannot be estimated, it is assumed that one or the conveyance is uniform throughout the airfoil.

NACA 7-Series: This was an additional attempt to increase the locations of the laminar stream across an airfoil by separating the areas of base strain on the top and lower surfaces of the airfoil. The NACA 747A315 is an example of a model. The number 7 denotes the beginning of the series, the number 4 represents the area of the base strain on the top surface in tenths of harmony (40 percent), and the number 7 denotes the area of the base tension on the lower surface in tenths of the chord (70 percent). The thickness distribution and mean line structures are demonstrated in the fourth person, a letter. Before the letters were given to families, standardized designs were developed. The fifth figure is the plan lift coefficient in tenths (0.3), while the final two digits represent the airfoil thickness in the percentage of the chord (15 percent).

NACA 8-Series: The NACA 8-Series, a last deviation from the 6- and 7-Series concept, was designed to travel at extremely high speeds. The goal was to increase laminar flow on both the top and lower surfaces of the airfoils. The NACA 835A216 is a model that uses the same naming conventions as the 7-Series. 8 indicates the series, 3 indicates the most minor strain on the upper surface in harmony tenths (0.3c), and 5 shows the slightest tension on the lower surface in harmony tenths (half); the letter A recognizes airfoils with various camber or thickness structures; 2 denotes the plan lift coefficient in harmony tenths (0.2); and 16 indicates the thickness of the airfoil, expressed in harmony levels (16 percent).

Parameter Affecting the Airfoil Performance

Airfoil geometries are broadly used in aircraft wings, compressor blades, and propellers. The choice of the airfoil is likewise vital in the plan of the aircraft wing. The reason for this chapter is to elaborate on the need of understanding the aerodynamics of wing and its parameters influencing geometrical and aerodynamic parameters, which are detailed below; this allows the user to decide the impact upsides of the boundaries of surface roughness, assault point, and velocity on streamlined execution and cooperation between the edges of an airfoil. The best boundary on the airfoil execution is surface roughness (Michael F., 1997), (Zerihan, 2001). The spotless surface airfoil execution is twice as great at the low approaches contrasted with the covered airfoil surface at both breeze speeds. The exhibition at high strategies is up to multiple times higher. The main streamlined demonstrate the boundaries to consider when choosing the right-wing profile are the C_L lift coefficient and the C_D drag coefficient. The C_L/C_D proportion is utilized to quantify the wing execution. C_L/C_D proportion changes at various assault consecrated envoys and free stream velocity.

Geometrical Parameters

1. Thickness of the Airfoil
2. Surface Roughness

Aerodynamic Parameters

1. Wind Speed
2. Angle of Attack (AOA)
3. Lift
4. Drag

Airfoils working at low Reynolds numbers have acquired prominence among the analysts and researchers because of the expanding uses of miniature air vehicles, low-speed airplanes, wind turbines, and so on through a few examinations on various airfoil boundaries, for example, nose sweep, thickness, camber and so on have been finished by analysts however the exhaustive parametric investigations of the different t/c qualities on the stream/acoustic attributes of the symmetric and asymmetric NACA airfoils are restricted in the available literary works (Ladson, 1975; Wentz, 1974). The boundaries referenced above can essentially adjust the lift and drag qualities, acoustic emanations, and so on, which eventually influence the steadiness and execution of the airplane (Chana, 1998). Subsequently, a detailed report is expected to decide the parametric conditions which give the best presentation and low commotion outflows for a scope of working conditions. Investigations of the fierce limit layer have been a significant concentration in a couple of years. Lessening fuel utilization would be fantastic if these endeavors would be economically acknowledged, where high arranged surface harshness called riblets are applied in aviation, car, and energy (Kundu, 2010,; Lambert, 2006; Turner, 1941; Prasad, 2018). There have been a lot of innovative headways in getting effective energy vehicles or frameworks, for example, the lighter composite and more eco-friendly stream motors for the flying industry, energy proficient vehicles by utilizing half and half innovation, using electric engines and hydrogen power modules for the car industry and some more. These endeavors have been helpful. However, there are more spaces for development to lessen haul by controlling surface choppiness. Skin-friction drag comes from a meager area of sluggish liquid quickly adjacent to a firm surface.

Fluid viscosity is the most critical component in determining the drag a boundary body experiences. Increasing the roughness of the wing surface can lead to an early transition of the boundary layer flow, leading to turbulent separation. The border layer will become detached at lower angles of attack if its shape changes. Because of this, the plane's total drag, or the force that opposes its forward movement, will be increased (Chana, 1998). To sustain life, an aircraft with rough wing surfaces must fly at higher angles of attack, resulting in a more significant drag force than a smooth airfoil. As a result, fuel consumption will rise because of the airplane's higher drag. However, this harmful effect is not restricted to the aircraft's wingtips. When the surface becomes more uneven, it can negatively impact health and safety. Roughness and its consequences have been studied extensively for a long time because of these and other reasons.

The aerodynamic efficiency of an airfoil determines the performance of an aircraft. The ratio of the coefficient of lift to the drag coefficient is considered aerodynamic efficiency. Flow separation is one of the main factors affecting the aerodynamic efficiency of an airfoil. Creating surface roughness over the surface of an airfoil manipulates the flow separation, altering the aerodynamic efficiency. Flow separation occurs when the boundary layer travels far enough against an adverse pressure gradient. Creating roughness at certain portions considering various parameters will impact delaying flow separation. The negative pressure gradient decreases due to surface roughness which delays the boundary layer separation. Due to this phenomenon, there is an effective decrease in pressure drag, leading to a decrement in total drag over an airfoil. As we know, pressure drag is more for laminar and less for turbulent flow; surface roughness over an airfoil makes the flow turbulent, leading to the decrease in pressure drag; hence the total drag over the airfoil is decreased.

The wind stream over an airfoil produces powers over the airfoil surface: lift, drag, and thrust force. The airspeed is lower on the "pressure" side of the airfoil and speeds up, bringing about below pressure on the "attractions" side of the airfoil. In the meantime, viscous friction between the air and the airfoil surface eases back the flow somewhat close to the surface (Howe, 2010), (Sato, 1973). The forces are characterized as follows:

Lift power (F_l): characterized oppositely to the bearing of the approaching wind stream. It results from the inconsistent tension on the upper (Suction) and lower (pressure) airfoil surfaces.

Drag power (F_d): characterized corresponding to the bearing of the approaching wind stream. This power is because of thick rubbing powers at the airfoil surface and inconsistent strain on the airfoil surfaces pointing toward and away from the approaching stream.

The current part discusses the forces acting on the body while dealing with symmetrical and unsymmetrical wing bodies; it is sufficient to evaluate the stream characteristics around a single cross-section, as the stream properties remain the same throughout the airfoil's length (Ladson, 1975). Figure. 1 shows the cross-sectional geometry: Follow-up drag-convey or lift portion number two. The pressure distribution over the airfoil surface dramatically influences the lift force and pitching moment. The drag force depends on the surface pressure and the surface shear force. Exams and evaluations consider dimensionless coefficients, which consider the varying forms and characteristics of airfoils developed for different purposes. Flows C_l eq. (1) and C_d eq. (2) were analyzed in two ways in the performance assessment of the airfoils. L (the lift force), D (the drag force), V (the wind speed), (the fluid density), S (the lift force) are some of the words used in this circumstance (the S airfoil surface area).

1. The lift is calculated by evaluating the resulting force generated by the pressure distributions in the upper surface (suction zone) and the airfoil's lower surface (pressure zone).

Figure 3. Lift coefficient curve for symmetrical and unsymmetrical aerofoil sections (Gopalarathnam Ashok, 2003), (Howe, 2010), (Sato,1973).

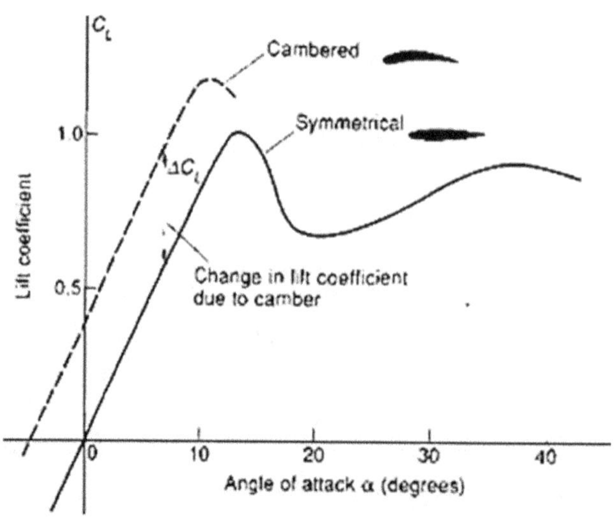

The lift coefficient C_l i defined as:

$$C_l = \frac{1}{2}\rho_\infty V_\infty^2 SL \tag{1}$$

The drag coefficient is caused by friction, pressure, and generated drag. A body submerged in a non-viscous liquid encounters no drag since it does not come into touch with the boundary layer. This is because the primary and tail edges of the profile have different levels of strain—the eddies at the tip of the sharp edge determined the generated drag.

The drag coefficient is determined as:

$$C_d = \frac{1}{2}\rho_\infty V_\infty^2 SD \tag{2}$$

Like the lift coefficient, the drag coefficient changes with the angle of attack (see Figure. 4); typically, it is substantially smaller at low angles of attack and rapidly increases at higher angles. (Howe, 2010), (Arceneaux. J. (2004).

Figure 4. Drag coefficient curve (Arceneaux. J, 2004).

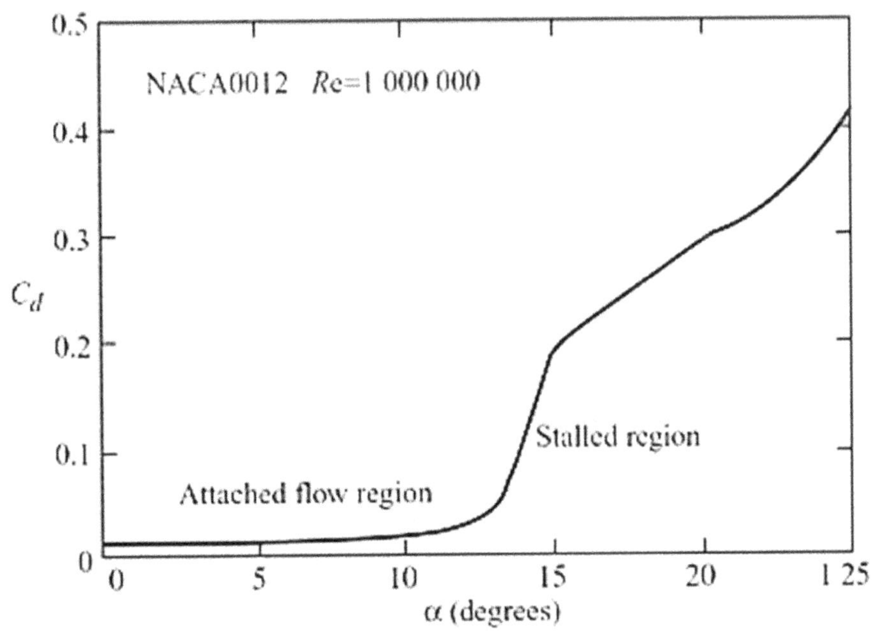

Leading-edge defilement (unpleasantness) causes huge decreases in C_{lmax} for plain and fluttered airfoils. The size of the decrease is comparative for both. The leading-edge tainting likewise decreases the lift curve slope, particularly for thicker airfoils that have the area of least pressure farther toward the back than do more slender ones (Graham, 1949; McGhee, 1979; Whitcomb, 1965; Prasad, 2018; Zerihan, 2001). These loans backing to the significance of understanding the effect of helpless surface quality and leading-edge pollution on the general attributes of the airfoils and, subsequently, the exhibition of the airplane when shaping wings.

High Lift Devices

To gain more lift on the airfoil, designers have worked on the leading and trailing edge modifications by attaching additional devices on wings or airfoils called High lift devices; these high lift devices are classified into two types: leading trailing edge devices. Slats and flaps are utilized on many commercial airplanes to produce a higher lift coefficient. Supports are streamlined constructions at the wing's leading edge, and flaps are set at the trailing edge. By controlling the distance between these constructions and the actual airfoil just as their direction, the business airplane can produce more lift at diminished velocities during take-off and landing, in this manner requiring more limited runways (Arceneaux, 2004), (Roskam, 2000). Moreover, sending these components likewise expands the drag force, which is accidental but supportive in dialing the airplane back during its arrival approach. The supports and folds are withdrawn into the wing when the aircraft is at journey elevation. In this reenactment model, we will comprehend the effect of flaps and braces on lift and drag forces. The 2D, three-component airfoil design includes a wholly sent support, actual airfoil, and fold.

Figure 5. High lift devices on the leading and trailing edges (Wentz,1974; Turner, 1941; Ferri, 1945; Riblett, 1988).

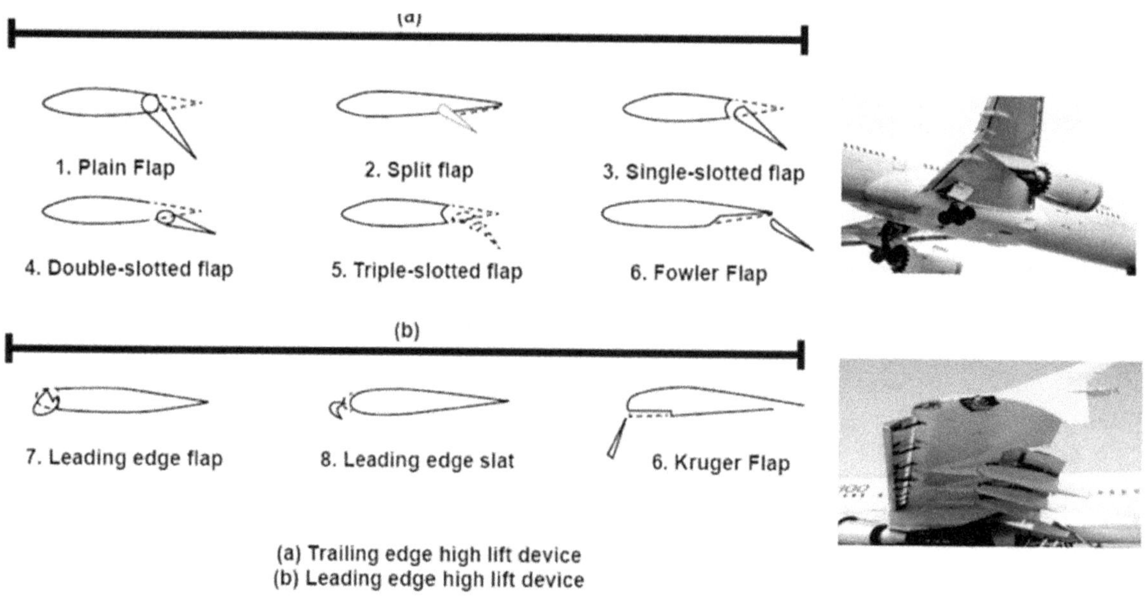

The current section introduces results for the various areas of trailing edge modifications of NACA 2412. The experimental outcomes contrast and practical tests are conveyed at MIT-MU utilizing the sub-sonic examination burrow office. In the current section, two trailing edge changes are made, as displayed in Figures 3 and 4; the length of the harmony trailing edge was chosen based on the hypothesis developed (Zerihan, 2001). To acquire the powerful outcomes, the trailing edge was changed distinctive approach in the range between 0 to 12 degrees experimentally. Applying the strategy, the results are introduced in figures which show that the primary wing and the appended following edge wing with fundamental wing or fixed-wing testing alteration, i.e., cavity and blunt.

Figure 6. Lift plots of blunt and cavity-shaped trailing edged aerofoil in slotted flap configuration (Prasad, 2018).

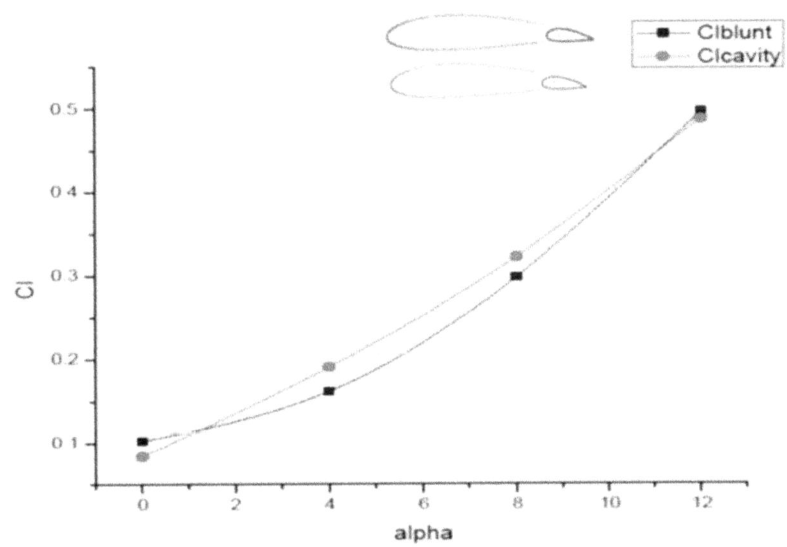

Results acquired from wind tunnel experiments show the lift and drag coefficients' improvement at various angles of attacks. The trailing edge alteration of the fundamental wing is portrayed. It shows that the unpolished has a low lift coefficient contrasted with the cavity, albeit at first, it expanded. With the increase in trailing edge approach, the lift has increased to 11 degrees with an increase in other point result has shifted by the stream actuated vortices which has impacted the cavity segment bringing about the low lift coefficient. It plainly expresses that the following rough edge has high drag at the common approach, which has reached to trivial point at the point of 4 degrees in additional augmentation the coefficient of drag has expanded. Relatively, the pain has increased with angle. These data show that a blunt trailing edge has a high drag coefficient. The trailing edge form of an aircraft wing can benefit from a change in the offset cavity and cavity. In the Slotted flap configuration, it may be argued that the equivalent is inaccurate when such changes are done.

Figure 7. Drag plots of blunt and cavity-shaped trailing edged aerofoil in slotted flap configuration (Prasad, 2018).

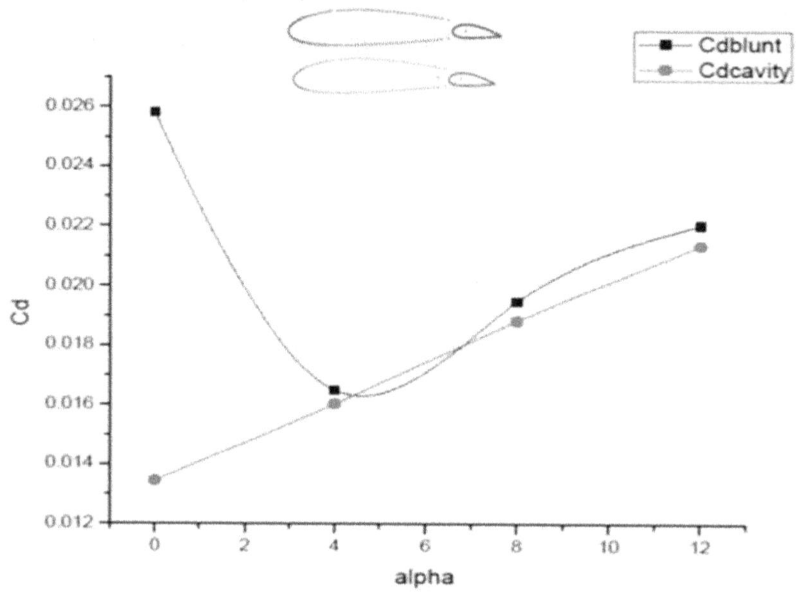

Parametric Roughness Results

This study aims to see how surface roughness affects airfoil characteristics by varying the unpleasantness type and area. Lift, drag, pressure coefficients, and stream calculations are all considered here. To further understand how surface roughness impacts airfoil streamlining properties, it's essential to look at how changes in roughness size and area affect lift and drag coefficients, hence, airfoil lifting effectiveness. Slow downtime, a critical component of the simplified features, would also be involved.

Regular occurrences like ice, ice, snow, or pitting caused by dust storms or bugs can lead to a surface level of harshness. In the real world, plane wings get more attention and are inspected more frequently than wind turbine blades, with the former being a valid requirement for public safety. Due to their proximity to the ground, wind turbine edges are more likely to suffer from surface roughness and shear due to the accumulation of debris such as ice, sludge, and salt. Adding controlled surface roughness to a model can be useful for re-creating airstream conditions, such as the transition from one layer to the next. (Dover, 1960), (P.D. Gall). Due to the severity of freezing precipitation and the form change they induce on the airfoil, general flying catastrophes occur on a massive scale every year. In 1979, ice was blamed for 49 general flying disasters, according to Luers (Dover, 1960). Deicing, the enemy of icing, and other solutions to the problem of icing might be expensive and not always sufficient for different reasons. NAACA 63-215 for wind turbines with a Reynolds number of 0.55 106 was introduced by Bloy and Roberts (McGhee, 1973). Trial and hypothetical information on Gurney-fold in wind turbine applications has also been disclosed by Kentfield and Clavelle (Roskam, 2000). They suggested that Gurney-folds be used to work on the appearance of propeller-type wind turbines. Mc Croskey(Smith,1975) has stated that NACA 0012 has the lowest drag coefficient and the highest lift coefficient at various Mach and

Reynolds numbers. The authors demonstrated there isn't a single test that can adequately characterize all of the NACA 0012 airfoil's streamlined properties or approve CFD codes.

Computational Fluid Dynamics Tool Methodology

These days, a few enterprises and designing and counseling organizations utilize the industrially accessible broadly applicable purported CFD codes for recreation of the liquid stream, hotness and mass exchange, and burning in aviation applications. The ANSYS FLUENT, ANSYS CFX, CFD++, and STAR-CCM+ are among these codes. Likewise, numerous colleges and exploration organizations apply business codes to utilize their in-house created principles. These days, open-source codes, for example, Open FOAM, are additionally uninhibitedly accessible. Likewise, there are particular codes like DPLR (information equal line unwinding) and VULCAN (thick upwind calculation for complex stream investigation) for hypersonic streams. DPLR is intended for supersonic and hypersonic streams under nonequilibrium conditions, while VULCAN is designed for interior streams in scramjet motors. Further subtleties and brief depictions of the codes can be found in (Ansys Report, 2016).

However, understanding the fundamental concepts of computational processes is necessary to apply such codes and analyze the recorded results efficiently. Likewise, significant issues are the portrayal of mind-boggling calculations and the age of reasonable networks. The business codes ordinarily have their lattice age instrument, e.g., ANSYS ICEM, however independent programming, for example, Pointwise, is likewise well known. The codes are, by and large, additionally viable with different CAD tools.

Computational Fluid Dynamics (CFD) is a mathematical method utilized to examine stream framework and related peculiarities. CFD can be perceived as a blend of physical science, math, and computational sciences, pointing to the stream field depiction (Menter, 1994), (W A Timmer). A CFD solver depends on the arrangement of coherence and energy balance conditions. Other physical and physical-synthetic peculiarities can be coupled in the reenactments, for instance, hotness and mass exchange and compound responses (homogeneous or heterogeneous frameworks). CFD turned into a virtual device for engineers generally in plan and streamlining of cycle and hardware when it presents a few benefits like expenses and time decrease of ventures (e.g., reproduction and examination of a few plan before assembling of actual models), incredible degree of stream field subtleties and expectation of stream field in severe activity conditions, where exact estimation are extravagant or even unimaginable (Ansys Report, 2016). Then again, Verification and Validation (V&V) steps are fundamental for the numerical model unwavering quality (CFD reproductions could arrive at mathematical intermingling, be that as it may, just this numerical assembly don't guarantee the sensible material science). The confirmation step comprises assessing the actual conduct of the concentrated peculiarities. In contrast, the approval step includes examining mathematical forecasts with trial information or logical arrangements when conceivable (Ansys Report, 2016), (Menter, 1994). Appropriately, CFD is a valuable asset utilized in a few areas of human knowledge, including synthetic, mechanical, aviation, and biomedical, and as of late in Microfluidics.

In this section, flow analysis was performed using ANSYS FLUENT over a NACA 0012 airfoil with circular roughness at two distinct positions and varying Reynolds numbers. NACA 0012 airfoil was designed in CATIA V5 with airfoil data. Surface roughness of circular shape was designed with a radius of 1% chord length at 25% chord length and another similar airfoil at 65% chord length. Meshing over the designed airfoil was developed pointwise, and CFD analysis was done in ANSYS FLUENT. Plots of coefficient of pressure, lift, and drag was obtained for varying attacks from 0-25 degrees with

5-degree intervals. One of the best fluid simulation tools is Ansys Fluent (Ansys Report, 2016). It has many advanced features and is known for being very accurate.

Computational Modelling

NACA 0012 airfoil was designed with a chord length of 250mm. Surface roughness of circular shape was developed over the airfoil, as shown in figure 1 and figure 2. The circular roughness is placed at 25% and 65% chord length of the airfoil. The specifications of the airfoil are mentioned below

s.no	Measuring factors	Dimension
1.	Chord Length	250 mm
2.	Span	250 mm
3.	Roughness Height	2.5 mm
4.	Roughness position	25% & 65%

Meshing

Structural meshing was done with a C – grid over the airfoil pointwise, as shown in Figures 8 and 9. Similarly, the structural mesh was done over the circular roughness, as shown in figure 10

Figure 8. C-grid structural mesh

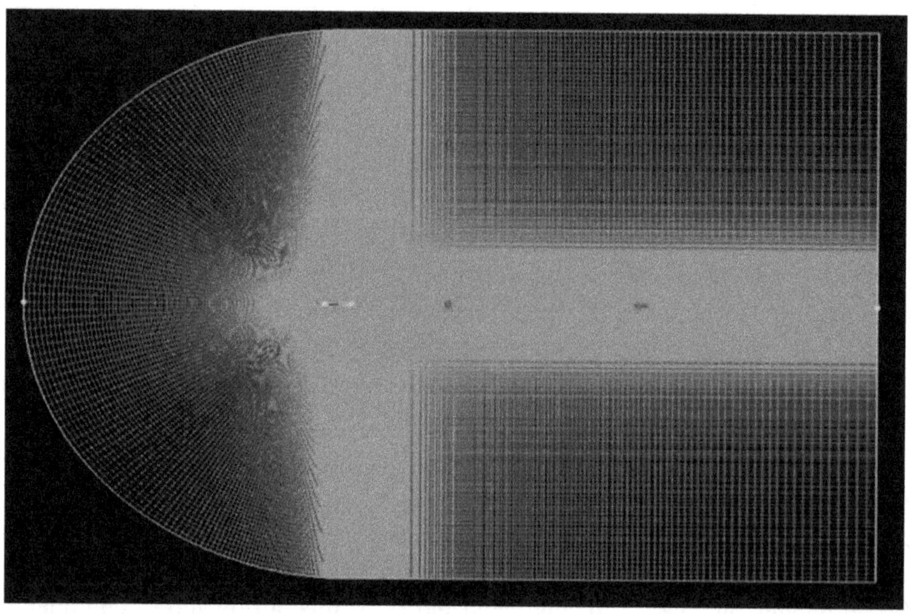

Figure 9. A mesh of the airfoil

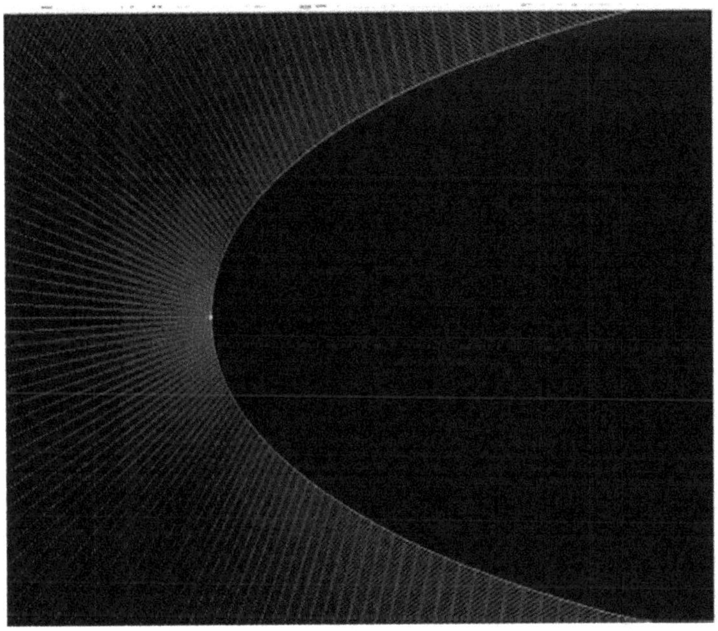

Figure 10. Mesh of circular roughness

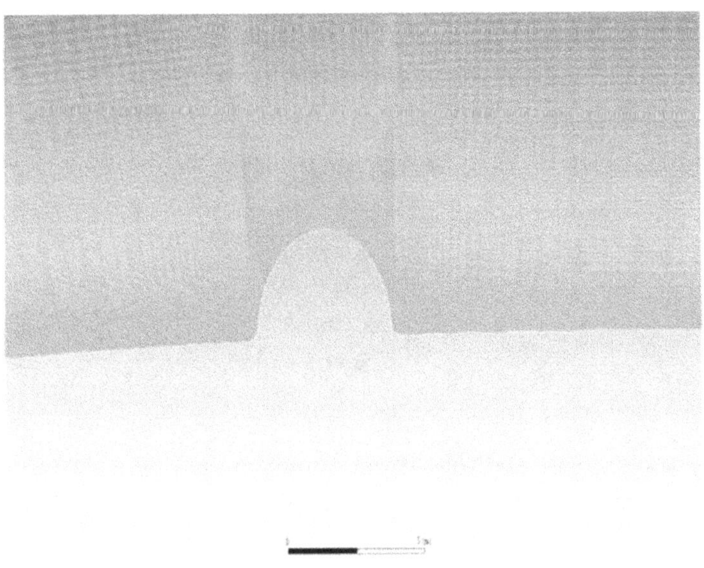

CFD Analysis Inputs

The above meshes were loaded in ANSYS FLUENT, and simulations were performed with the below boundary conditions.

The model used: k-omega – SST model

Inlet Velocity:

- 19 - Re = 3,30,000
- 23 - Re = 4,00,000

Viscosity: 1.42e-05
Method: COUPLED
Report definitions: cl & cd
Monitors: Residual monitors:1e-05
Initialization: Hybrid Initialization
Calculations are performed with 2000 iterations approx.

Analysis Results (Figures 11, 12, and 13)

Figure 11. Plain airfoil

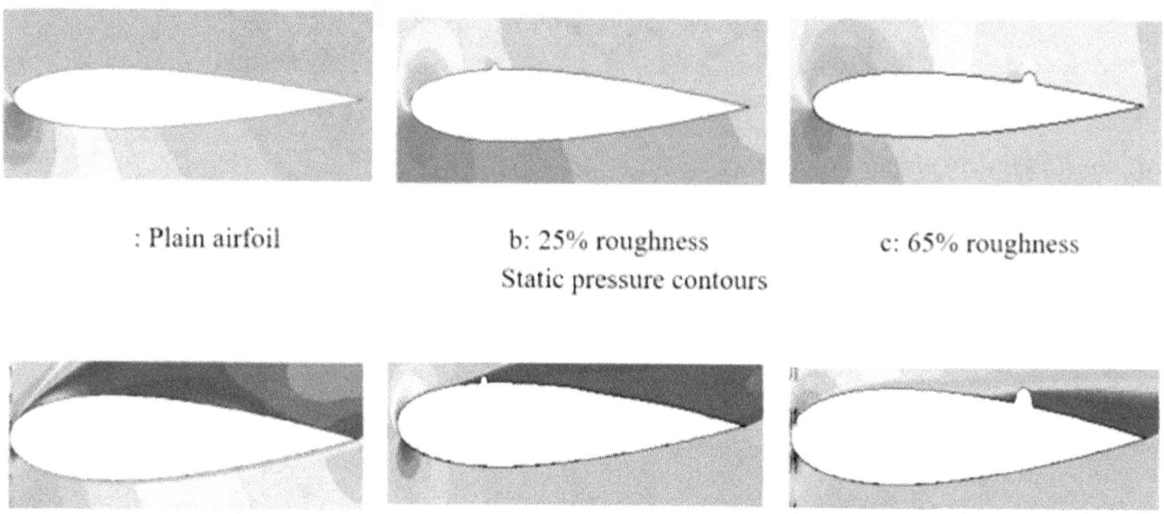

: Plain airfoil — b: 25% roughness — c: 65% roughness
Static pressure contours

d: Plain airfoil — e: 25% roughness — f: 65% roughness
Velocity contours

The contours for static pressure and velocity are shown in the figure above for plain airfoil, airfoil with roughness at 25%chord length, and airfoil with roughness at 65% chord length.

The above 2 are the comparison graphs of cl vs. alpha (figure 12) and cd vs. alpha (figure 13) for two distinct positions of roughness, one at 25% of chord length and the other at 65% chord length. The below figure 14 shows the comparison graph aerodynamic efficiency (cl/cd) vs. alpha for two distinct positions of roughness, one at 25% of chord length and the other at 65% chord length.

The above 2 are the comparison graphs of cl vs. alpha (figure 15) and cd vs. alpha (figure 16) for two different Reynolds numbers, i.e., 3,30,000 Re and 4,00,000 Re, and the circular roughness is constant at 25% chord length.

Figure 12. C_l vs Alpha (angle of attack)

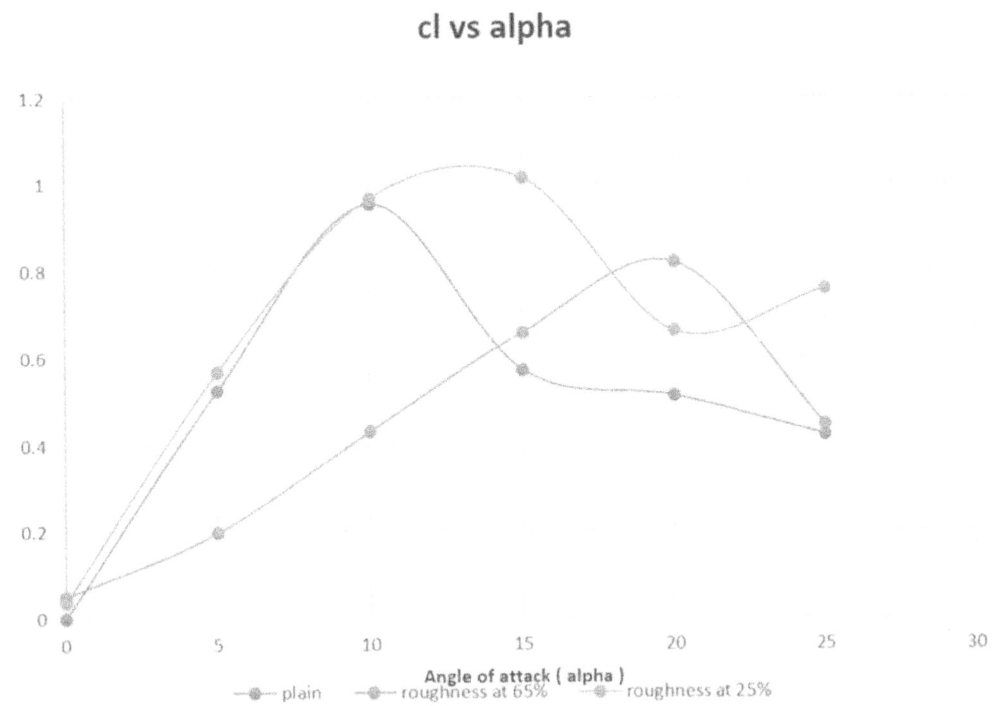

Figure 13. C_d vs Alpha (angle of attack)

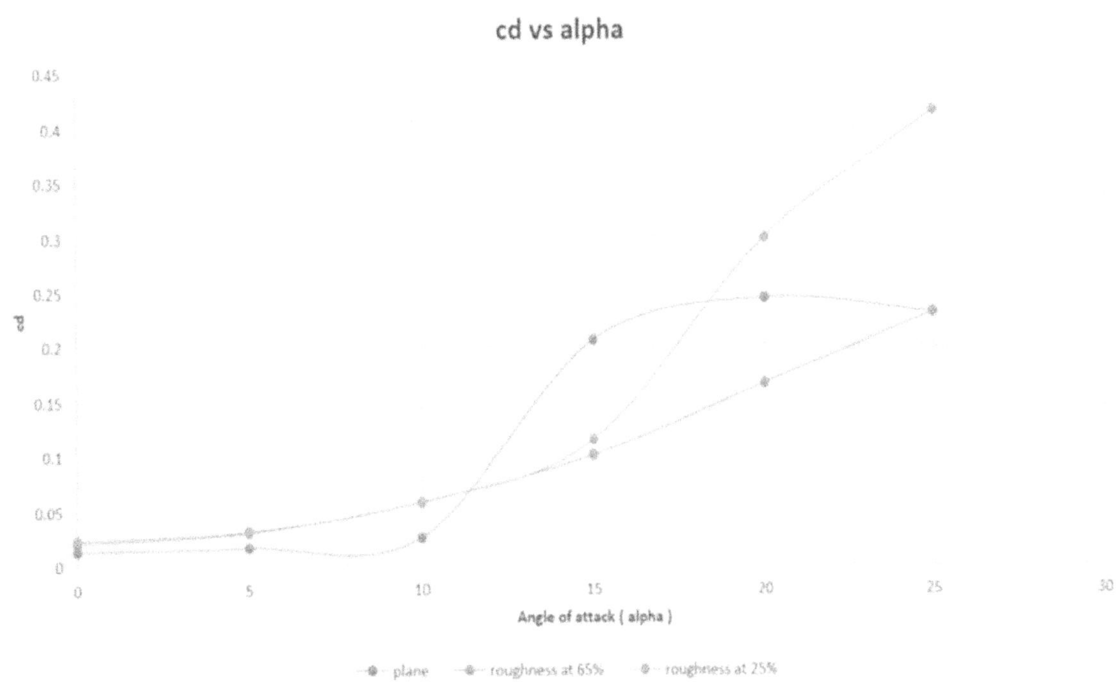

Figure 14. Aerodynamic efficiency (C_l/C_d) vs Alpha (angle of attack)

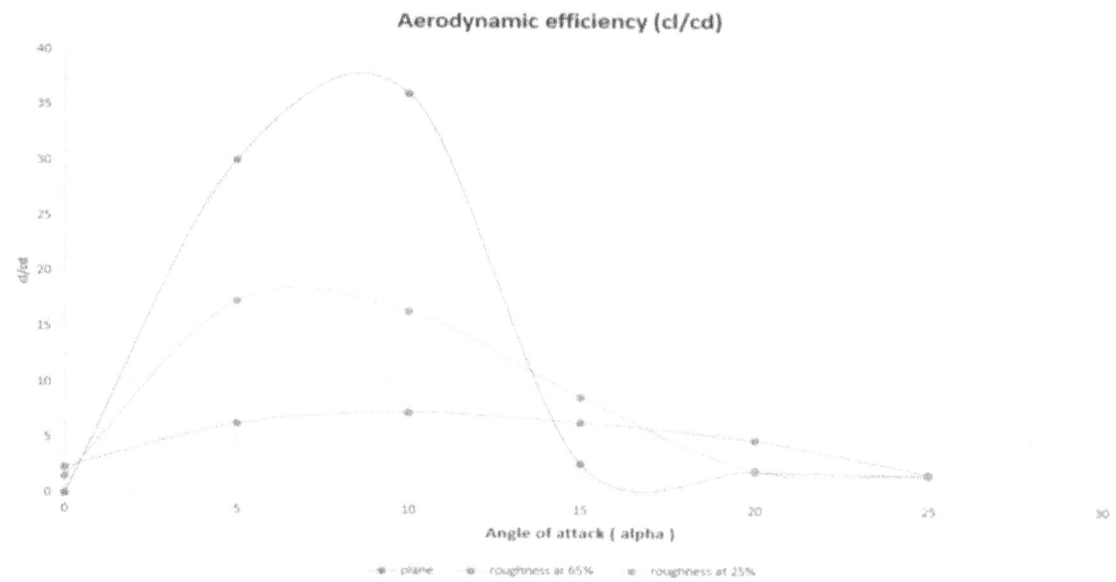

Figure 15. C_l vs. Alpha (angle of attack)

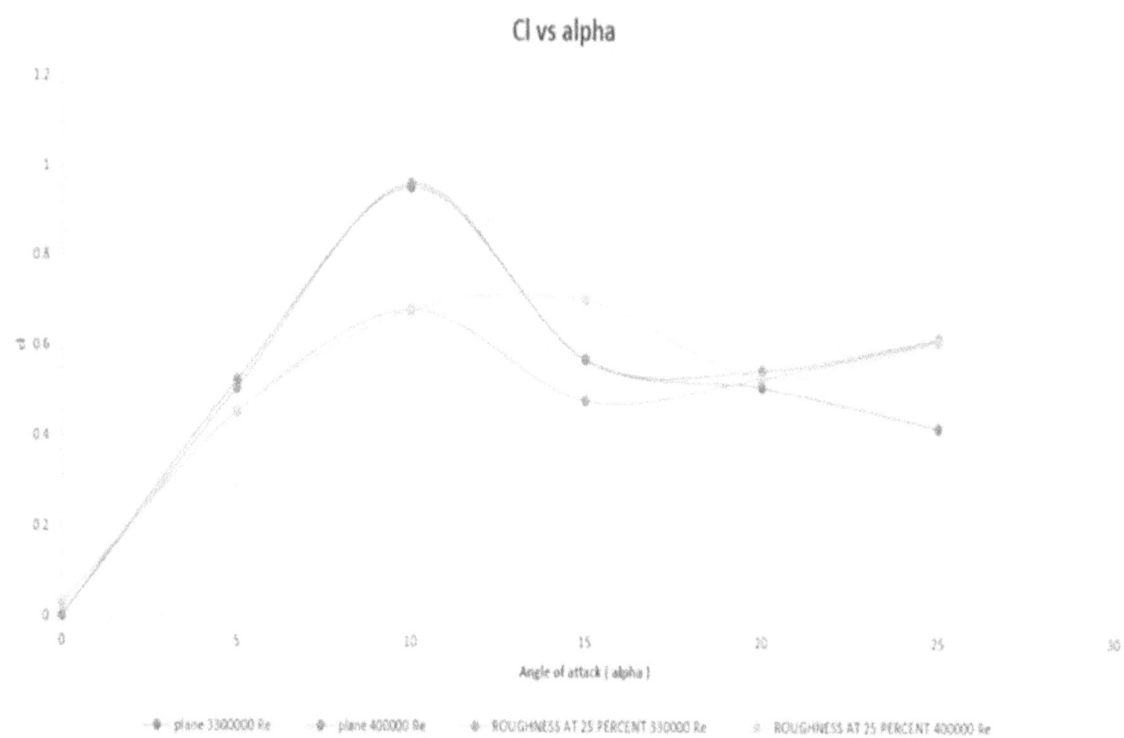

Figure 16. C_d vs Alpha (angle of attack)

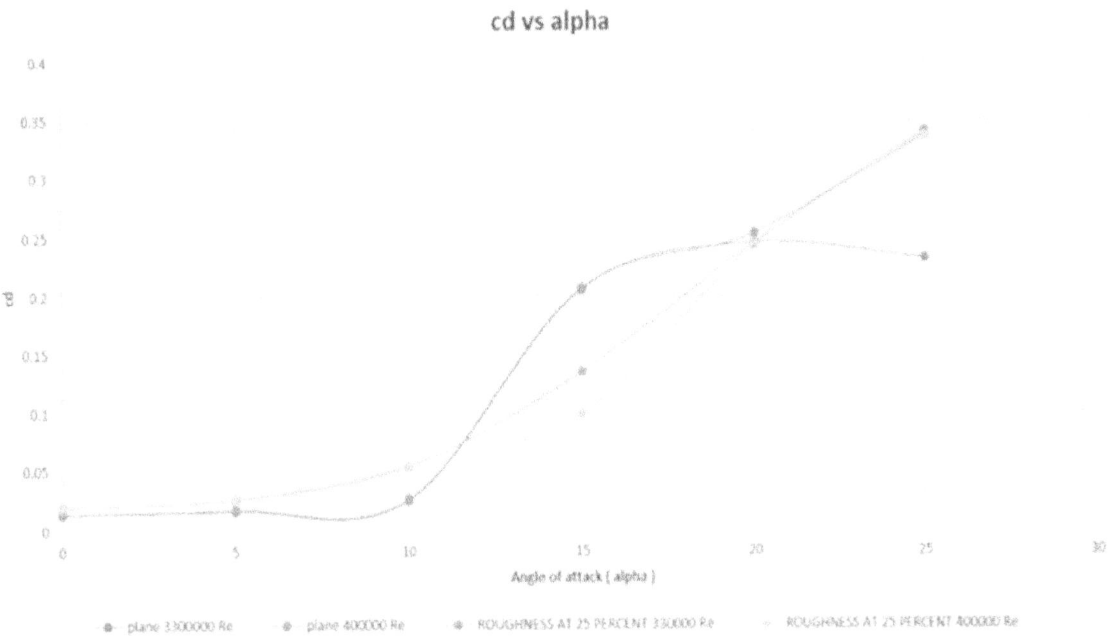

Figure 17. Aerodynamic efficiency vs alpha

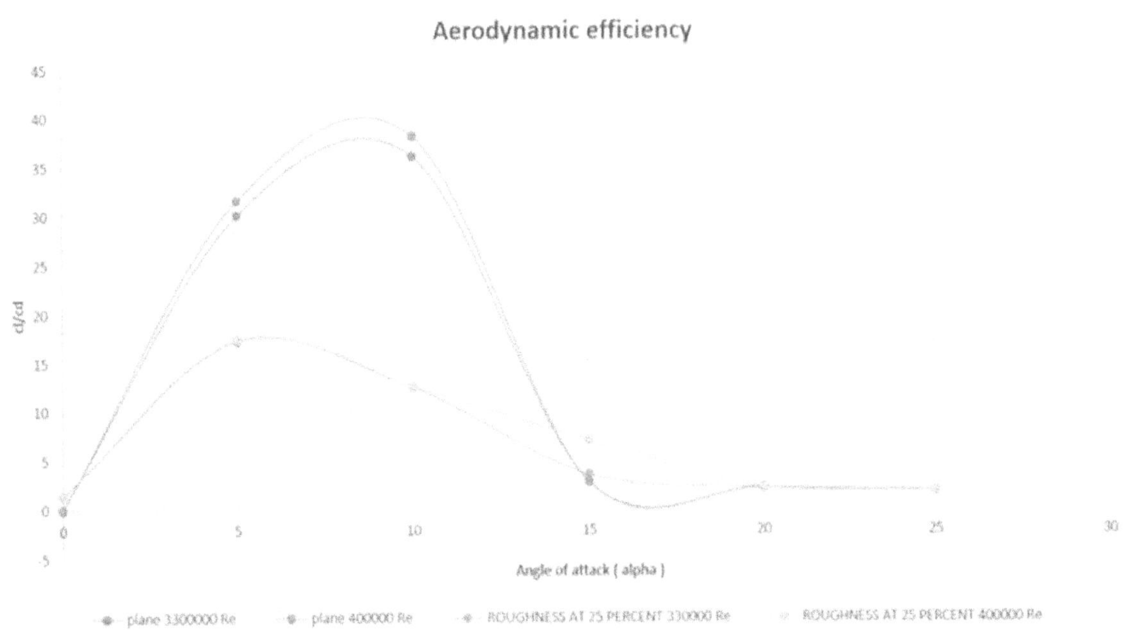

The above figure 17 shows the comparison graph aerodynamic efficiency (cl/cd) vs. alpha for two different Reynolds numbers, i.e., 3,30,000 Re and 4,00,000 Re, and the circular roughness is constant at 25% chord length.

CONCLUSION

The airfoil with roughness at varying positions was successfully designed in CATIA V5, and analysis of these airfoils was performed in ANSYS Fluent successfully. The CFD analysis for the airfoils was performed to obtain the plots of lift, drag, and aerodynamic efficiency and obtain the velocity and pressure contour distribution over the geometry. At AOA 15 degrees, there was an effective increase in the aerodynamic efficiency due to roughness. There was an increase in installing angle. Flow separation is delayed at 15 degrees AOA. Roughness at 25% chord length is more effective than roughness at 65%chord length. Aerodynamic efficiency (C_l/C_d) has been high for the high Reynolds number. At intermediate angles of attack, this effect of roughness has a detrimental impact. The flow and the airfoil's aerodynamic performance may be affected by varying surface roughness, which can be investigated further.

REFERENCES

ANSYS Fluent Theory Guide, . (2016). *Release 17.0*. ANSYS, Inc.

Arceneaux, J. (2004). Overview on drag reduction Technologies for Civil transport aircraft. *European Congress on Computational Methods in Applied Sciences and Engineering ECCOMAS*.

Ashok, G. (2003). Design of low Reynolds number airfoils with trips. *Journal of Aircraft*, *40*(4).

Chana, R., & Vishwanath, R. (1998, November-December). Base Drag Reduction Caused by Riblets on a GAW(2) Airfoil. *Journal of Aircraft*, *35*(6).

Deepanshu Srivastav, B. E. (Hons.). (2012). Flow Control over Airfoils using Different Shaped Dimples. *International Conference on Fluid Dynamics and Thermodynamics Technologies*.

Dover. (1960). *Thwaites Bryan. Incompressible Aerodynamics*. Academic Press.

Ferri, A. (1945). *NACA WR L-143 Completed Tabulation in the United States of Tests of 24 Airfoils at High Mach Numbers*. Academic Press.

Gall, P. D., Huebsch, W. W., & Roth Mayer, A. P. (2010). Dynamic roughness as a means of leading-edge separation flow control. West Virginia University.

Graham, D. J. (1949). *NACA TR-947 The Development of Cambered Airfoil Sections having Favorable Lift Characteristics at Supercritical Mach Numbers*. Academic Press.

Griffin, Huebsch, Rothmayer, & Wilhelm. (2016). *Numerical and Experimental Study on the Ability of Dynamic Roughness to Alter the Development of a Leading Edge Vortex*. Academic Press.

Harry, R. (1988). *GA Airfoils e A Catalog of Airfoils for General Aviation Use* (5th ed.). Self-Published.

Hood, M. J. (1939). *NACA TN-695 The Effects of Some Surface Irregularities on Wing Drag*. Academic Press.

Howe, D. (2010). *Aircraft conceptual design synthesis*. London Professional Engineering Publishing.

Jan & Edward. (2000). *Airplane Aerodynamics & Performance*. DAR Corporation.

Kerho & Bragg. (1997). Airfoil Boundary Layer Development and Transition with Large Leading-Edge Roughness. The University of Illinois at Urbana-Champaign.

Kundu, K. A. (2010). *Aircraft Design*. Cambridge University Press. doi:10.1017/CBO9780511844652

Ladson & Brooks Jr. (1975). *NASA TM-X-3284 Development of a Computer Program to Obtain Ordinates for NACA 4-Digit, 4-Digit Modified, 5-Digit, and 16 Series Airfoils*. Academic Press.

Lambert, S., & Morrison, J.F. (2006). *Fundamental Studies of Active Dimples*. AIAA-2006-3182.

McGhee & Beasley. (1973). *NASA TN D-7428 Low-Speed Aerodynamic Characteristics of a 17-Percent-Thick Airfoil Section Designed for General Aviation Applications*. Academic Press.

Menter, F. (1994). Two-equation eddy-viscosity turbulence models for engineering applications. *AIAA Journal*, *32*(8), 1598–1605. doi:10.2514/3.12149

Prasad, U. S., Varma, S. A., Kumar, R. S., & Vaibhav, K. M. R. S. (2018). Experimental and Numerical Analysis over the Truncated Airfoil with Slotted Flap Configuration. *Indian Journal of Science and Technology*.

Raymer, D. (2018). Aircraft design: A conceptual approach. American Institute of Aeronautics and Astronautics. https://doi.org/10.2514/4.869112.

Sato, J. (1973). *NASA TT-F-749 On Peaky Airfoil Sections*. Academic Press.

Smith, A. M. O. (1975). High-lift aerodynamics. *Journal of Aircraft*, *12*(6), 501–530.

Timmer Duwind, W. A. (1997). The effect of roughness at high Reynolds numbers on the performance of DU 97-W-300Mod. Delft University Wind Energy Institute.

Turner, H. R. Jr. (1941). *NACA WR-L-677 Lift and Drag Tests of Three Airfoil Models with Fowler Flaps*. Consolidated Aircraft Corporation.

Wentz & Seetharam. (1974). *NASA CR-2443 Development of a Fowler Flap System for a High-Performance General Aviation Airfoil*. Academic Press.

Whitcomb & Clark. (1965). *NASA TM-X-1109 An Airfoil Shape for Efficient Flight at Supercritical Mach Numbers*. Academic Press.

Zaccai, D., Bertels, F., & Vos, R. (2016). Design methodology for trailing-edge high-lift mechanisms. *CEAS Aeronaut. J.*, *7*, 521–534.

Zerihan, J. (2001). *An Investigation into the Aerodynamics of Wings in Ground Effect* [Ph.D. Thesis]. University Southampton, School of Engineering.

Chapter 18
Vehicle Aerodynamics and Different Testing Methods

Backiyaraj A.
Vel Tech Rangarajan Dr. Sagunthala R&D Institute of Science and Technology, India

Kumaran T.
Vel Tech Rangarajan Dr. Sagunthala R&D Institute of Science and Technology, India

Parthasarathy M.
Vel Tech Rangarajan Dr. Sagunthala R&D Institute of Science and Technology, India

Murugu Nachippan N.
Vel Tech Rangarajan Dr. Sagunthala R&D Institute of Science and Technology, India

Senthilkumar P. B.
Vel Tech Rangarajan Dr. Sagunthala R&D Institute of Science and Technology, India

ABSTRACT

One of the most adaptable inventions is motor-powered vehicles, especially cars. The majority of the time, safety, SFC, and operating area were considered. Vehicle makers are examining the characteristics of drag or air resistance for different body forms at various operational situations in order to increase fuel efficiency, vehicle speed, reduce wind noise, and improvise road control and vehicle steadiness when moving. The study of the rigid body traveling across the atmosphere and the interactions among its surface and the atmospheric air with varying linear acceleration and air direction is called aerodynamics. Aerodynamic drag is generally insignificant at lower vehicle speeds, but as speed increases, the air resistance magnitude also increases. This chapter deals with the external properties design considerations and testing methods of road vehicle aerodynamics like drag force. Identifying the drag force by various methods like wind tunnel tests, coast down test, and computer fluid dynamics are explained to forecast the drag coefficient.

DOI: 10.4018/978-1-6684-4230-2.ch018

1. INTRODUCTION

Nowadays fuel efficiency is one of the major factors which influences all vehicles used for transportation. The aerodynamic properties considered for vehicles like cars, Bikes also contributes to engine performance due to the drag force created on the surface. The drag force is one of the aerodynamic forces which influence vehicles motion especially speeds. Increasing aerodynamic utilities is inextricably linked to the constraints and lower fuel economy. The interest in vehicle aerodynamics, on the other hand, extends beyond external aerodynamics involve engine performance characteristics. Apart from the properties of the fluid, the vehicle power is more important to estimate performance parameters. Reynolds no is one of the important factors that have to discuss because the passing airspeed over the car is characterized by the non-dimensional no which is the ratio between inertia forces to the viscous force. The fuel economy is influenced based on the aerodynamic characteristics of drag co-efficient. The competition to increase road infrastructure and motor vehicles began. The operating speeds accelerated, and politicians saw an opportunity to impose more limitations, therefore speed limits were created. Some of the difficulties associated with velocity measurements simply do not exist in full-scale aerodynamic studies mostly on tracks. There's no need to spend money on a lower scale model because the moving wheels, rolling on the ground and airflow obstruction adjustment are all addressed. Due to the uncontrollable climate and expense constraints, this type of aerodynamic testing has improved significantly over the years. The coast down test was one of the first testing methods to determine the vehicle's drag. By the late 1990s, advances in computing and sensor technology had made it possible to measure and transfer desired forces, moments, or pressures via a wireless network at a reasonable cost. Closer inspection of the flow around a road vehicle may identify other fluid dynamic areas, such as the pillar area. When traveling in the rain, the primary pillar swirl is perfect for water precipitation, and the rearview mirror provides an oscillating wake. When travelling at high speeds, vortex noise is also caused by vortex movement near the rearview mirror. Vehicle airflow measurement and refinement is a continuous cycle and an important aspect of the automotive application that does not end when the vehicle is built. Wind tunnel testing, computational simulation and track testing are all common research and analysis instruments performed in this methodology. Experimentation was used to identify the process of drag formation and how to regulate it. The appearance of flow separation at the rear surface of automobiles was discovered to be the source of the majority of drag. Unfortunately, unlike in aeronautics, where it can be completely avoided, this aerodynamic phenomenon is an intrinsic hazard for ground vehicles and cannot be avoided. Furthermore, due to its unpredictability and sensitivity to automobile geometry, the corresponding three-dimensional flow in the wake behind a car has a complicated 3D behavior and is extremely difficult to regulate. The 3D mesh surrounding the automobile form is produced using the commercial grid generator Gambit for any specified geometry. There are around 6 million cells in all, including tetrahedrons and prisms. The mesh is highly finer in this location to correctly mimic the flow field behind the automobile; the majority of the drag coefficient is accounted for by this. The results of a simplified geometry known as the Ahmed bluff body They back up the experimental findings, claiming that the drag coefficient is very sensitive to the car's rear geometry due to the existence of separation and recirculation zones in this area, and that shape optimization may significantly lower it. The numerical tool is now ready to be utilized by automobile designers to improve the drag coefficient of future car models after this experimental validation. Moreover, the RANS algorithm is a steady-state analytical approach that might miss significant fluid physics transient properties. The computational physics method is based on molecular physics, in which Fluid motion can be controlled in the time domain by a set of "interaction rules." Several elements contribute to a vehicle's drag, and all of them must be addressed if drag is to be reduced. A fore body flow zone and a base flow region make up the outer flow over a body. By carefully designing and rising the body form in a wind tunnel, several of the separations that occur across the fore body of a complicated 3D shape may be eliminated, or at least mitigated. Small modifications to a huge number of body features can drastically reduce drag. Air deflectors, which are used to minimize truck drag, function on the same concept. The drag of the optimal two disc design was 81 percentage points lesser than those of a single disc. The body characteristic that contributes the most to total aerodynamic drag changes depending on the vehicle type. Wheel and axle drag, for example, is particularly essential on trains, and skin friction drag is much bigger than base drag, but base drag is the most critical factor in saloon vehicles. Thus, if road vehicle drag ought to be decreased to a minimal, we must first comprehend the mechanics that cause the car's base drag. The effect of a 3D bluff body's base slanting has on its drag. They experimented with the slope of a car's rear windshield and discovered that the most drag occurred when the roof was slanted at roughly 30 degrees to the horizontal. The slanted angle is an angle between the horizontal and the base. Highest drag appears to be particularly sensitive to incidence angle, with a minor difference in incidence accounting for the various slant angles for maximum drag. When turbulence is large, it is shed asymmetrically, and a recirculation type wake is created, identical with the one for the disc. However, under a range of 30 - 50 degrees, the turbulence is discharged in two powerful longitudinal vortexes that originate at the tilted base's farthest upstream point. As turbulence is pumped in through the transition zone on the upstream half of the body, the vortex develop as they trail back down the base. The lift is substantially enhanced by the existence of the vortices. In the twin vortices regime, the CD was unaffected by tiny yaw angle variations, but was highly reliant on yaw in the recirculation near wake region where CD refers to air directions. Depending on the values of Cd of the design of car structures required power is influenced, so in this study, we will discuss in detail the drag reduction in automobiles and the shape optimization. Lietz R, et al., (2000) Aerodynamic drag is another important aspect that affects vehicle stability, power consumption, and efficiency. Skin friction drag and pressure drag are the two basic components of aerodynamic drag. Pressure drag makes up the majority of total drag coefficient, which is dependent on external geometry, due to boundary layer flow separation from the back window area and the trailing area of behind vehicles. A brief review of the flow will be presented initially to comprehend some of the difficulties that had to be overcome during the construction of the new models. These fundamentals cover the dissection of several aerodynamic resistances in addition to some fluid flow around the vehicle's structure.

- Friction resistance
- Pressure resistance

1.1. Friction Resistance

The friction resistance presented in this paper is not the same as the traditional friction of the surface as defined by wall shear stress. Friction resistance on a larger macroscopic scale is discussed here, and it refers to friction caused by surface roughness, such as barriers attached to the vehicle. This form of resistance is based on the shape of the surfaces rather than the body type. The frictional resistance on the vehicle body is quite low because modern vehicles have a generally smooth body with only minor impediments. Things are different when you look at a vehicle's underbody. The underbody surface is quite rough due to many barriers such as axles, the exhaust, and the drive train. As a result, the underbody's friction resistance is substantially higher than the vehicle's body's friction resistance.

1.2 Resistance to Pressure

In comparison to shear resistance, pressure resistance is highly dependent on the vehicle's body style. There are three different body types for passenger cars: square back, fastback, and notchback are all terms for the same thing.

1.2.1. Square Back

The basic theory of flow conditions in a square back vehicle is discussed. The flow separates at the rear corners of the structure, forming a circle vortex in the vehicle's rear end it is shown in figure 1. Extremely low pressure is formed on the base plane as a result of this vortex, increasing aerodynamic drag.

Figure 1. Square back configuration

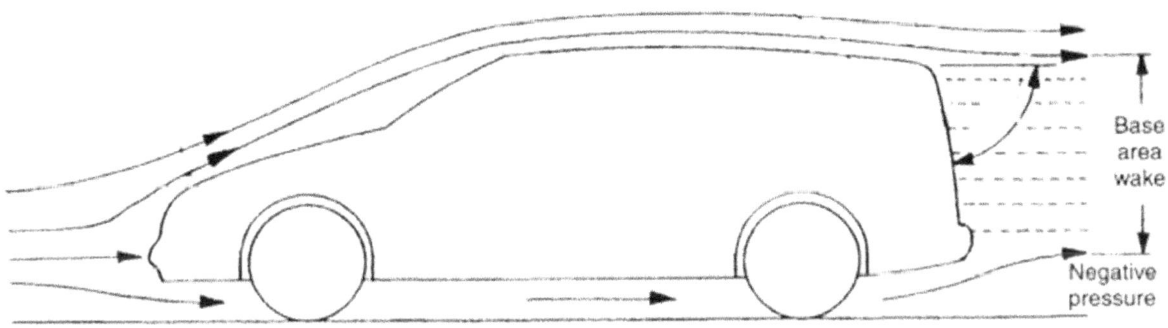

1.2.2. Fastback

The back window angle of a conventional coupe fastback is extremely flat. The vortices in longitudinal direction arise on D-pillar are most prominent flow phenomena in a fastback. in addition to these vortices, the radius is not enough between the roof and the rear window, As a result, on the back glass, a small flow separation zone is generated. Finally, in the rear end of the car, a little ring vortex emerges. All of these factors contribute to the low-pressure area on the rear window and backplane of the car. On a normal square back, there is just an increase in drag owing to the reduced condition. Mayer W et al., (2011) The low pressures in a fastback car didn't act on the vertical plane like they do in a square back, but rather on a partial one. As a result, the drag and lift are both increased. When the back axle lift is excessively large, it has a detrimental impact on stability and can become important at greater speeds.

Figure 2. Fastback configuration

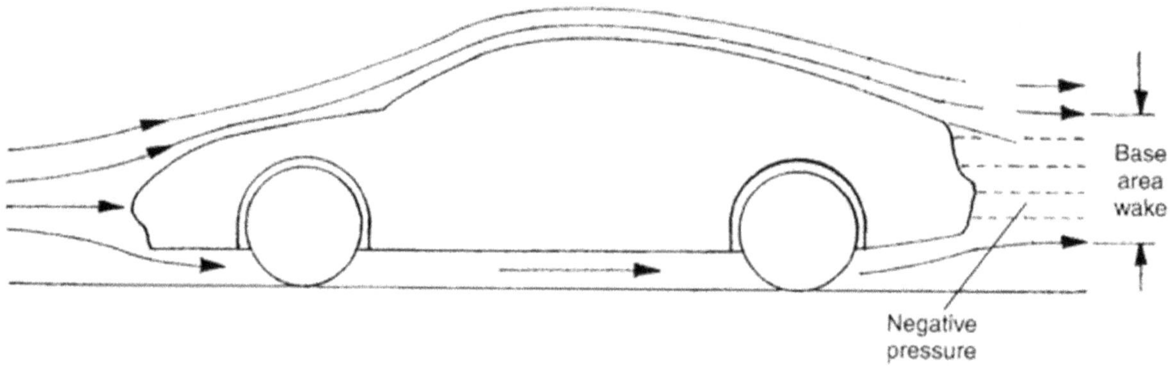

1.2.3. Notchback

A notchback vehicle's flow conditions are comparable to those of a fastback, but not identical. Two longitudinal vortices exist as well, but they are less than those found in a fastback. The vortices appear on the car deck lid after they originate on the D pillar. On the back window, A dividing zone is created by the flow along the roof that also attaches to the deck lid. Finally, in the aftermath of the car, there is a little ring vortex. These vortices, like the fastback, produce very little pressure. The detrimental aerodynamic effects are reduced because the vortices and tilted backplane are smaller. As a result, the aerodynamic resistance and lift are reduced. Reconnecting the flow on the deck lid also reduces the negative lift effect caused by low pressure. Notchbacks produce rear lift, while waggon backs produce rear down force as shown in figure 3. As a result, the two types of rear ends are assigned to different branches of the induced drag parabola. When the ground simulation is improved, the lift is often reduced, and the induced drag is altered as a result. Wickern G, et al., (2005), Wang, et al (2018) Because of the changes in the fundamental lift level, the notchback has less drag and the waggon back has more. To begin with, induced drag is only a minor component of the complex impact of improved ground simulation. When employing ground simulation, notchback vehicles do not usually display lower drag values. Square back configuration is shown in figure 3.

Figure 3. Notchback configuration

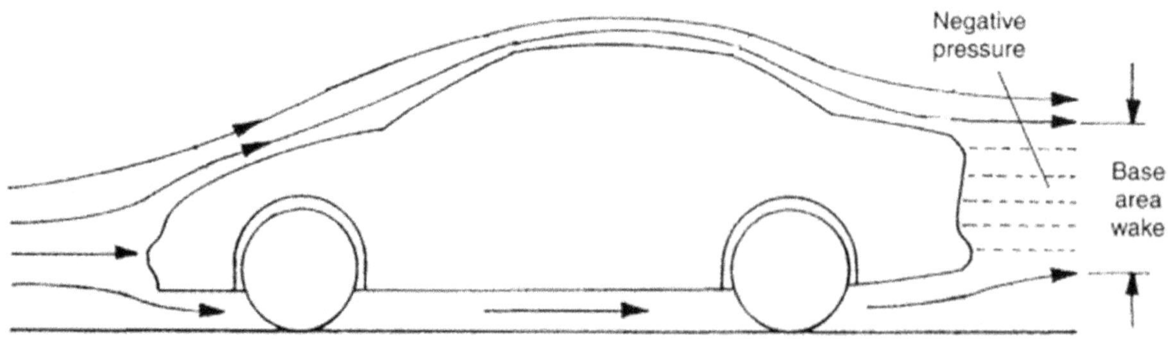

1.3. Drag

The aerodynamic influence on a racing automobile is primarily manifested through the aerodynamic and aerodynamic moment. Joseph Katz (2003) the aerodynamic force and aerodynamic moment are influenced by the vehicle's driving speed, body form, and side slip angle. National aeronautics and space administration (2021) Three vertical forces and torques that rotate around the X, Y, and Z axes, also known as six aerodynamic component forces, make up aerodynamic forces and moments these forces include drag, lift, side force, rolling moment, pitching moment, and yawing moment.

Types of Drag:

- Viscous Drag
- Induced drag
- Compressibility drag

1.3.1. Parasitic Drag

Form resistance (owing to shape), skin friction, interference, and any other non-lifting components cause parasitic drag; induced drag is created as a result of lift generation.

1.3.2. Induced Drag

When a moving body redirects the wind coming at it, it creates an aerodynamic drag force. Cars with airfoil wings that redirect air to create down force experience this drag force.

Compressibility drag:

- ◦ Due to high speed
- ◦ As the vehicle speeds up, there will be a rapid and significant rise in drag.

1.4. Drag Coefficient

Aerodynamicists utilize the drag coefficient to simulate each complicated effect of shape, tilt and flow characteristics on aircraft drag. If we rewrite the drag equation, in which the drag coefficient is calculated in terms of another variable. The Cd is calculated by multiplying the drag D by the following formula: density ρ divided by 1/2 the velocity V squared times the reference area A.

$$C_d = D / (A \text{ x } 1/2 \text{ x } \rho \text{ x } V^2)$$

The dynamic pressure q is equal to 1/2 of the density times the velocity squared. So

$$C_d = D / (q \text{ x } A)$$

The C_d is the ratio between drag force and the force generated by the q multiplied by the area. The above equation provides a method for calculating the drag coefficient. We can estimate the velocity, ρ, and A in a controlled environment (wind tunnel) and evaluate the drag created. We conclude at a value for the drag coefficient by dividing. The reference area has an impact on the actual mathematical value of the C_d that is calculated, as shown on the drag equation slide. Whenever producing drag coefficient values, it's crucial to include the reference area that was used to compute the coefficient. We can predict however much drag will be formed through using drag equations under different velocities, and a circumstances and using the drag equation. The C_d takes into account air velocity and deformability, as well as the complex relationships between object form and inclination. To make good use of the C_d, we must confirm that dynamic characteristics in our predicted and measured situations are same. Consequently, the prediction will be inaccurate. At slower speeds, viscous forces are negligible. It's crucial to match the Mach numbers in both situations at higher speeds.

1.5. The Effects of the Wind Flow Around Vehicles

Ground vehicles are subjected to external wind, and both the mean flow speed and turbulent pattern of the airflow have an impact on their aerodynamic characteristics. A vehicle will, on averages, encounter a related air flow at some modest angle of yaw over its journey life. The mean yaw angle may be computed by integrating airspeed and directional statistics with vehicle trip data. While many vehicle forebodes seem streamlined when viewed from the front, even a small amount of yaw can generate detachment from longitudinal edges and sharply. It's pointless to change a vehicle's design to reduce straight-line drag if the drag at minor degrees of yaw increases as a result. A yawed wind may also influence steering and stability, as well as produce a considerable side force on the moving vehicles in the case of lengthy vehicles like trains. The influence of air turbulence is one component of the subject of natural wind interactions with vehicles that has received less attention. P.B. Senthilkumar, et al (2021) Turbulent eddies with sizes equivalent to that of a vehicle exist near the ground, and they can interact with the body flow field, affecting boundary layer growth, separation, reattachment, and wake generation. As a result, turbulence may have a significant impact on mean flow and mean loading, and it is incorrect to think that gusts' only function is to create unstable forces. Various terrains, embankments, cuts, bridges, and other vehicles will provide turbulence to a vehicle. The comparative shear rate, defined as the R.M.S. of wind fluctuations split by average relative wind, can take any value dependent on the magnitude and wind direction. 2 / (V+ U), where U is the mean wind speed at the vehicle's top, is the relative turbulence strength faced by a vehicle going into a strong headwind. V is the vehicle speed, and 2 is the root mean square of horizontal wind speed variations. Whenever one vehicle follows as in wake of another, turbulence levels are likely to be higher, but for the sake of this discussion, 0.05 will be considered a typical number. Mean pressures are responsive to changes in turbulence intensity in the majority of situations, but the turbulence scales has minimal effect. N. Murugu Nachippan, (2019) because turbulence lowers the essential Reynolds number of a circular cylinder, several people mistakenly believes that turbulence means a rise in Reynolds number. Turbulence, on the other hand, has an influence on the flow around bodies with sharp-edged separation, with the principal consequence of speeding up the development rate of free shear layers. A.Backiyaraj, et al., (2021) the shear layers that grow faster entrap more fluid from within the divided zone, and the pressure drops even more, leading the shear film curves to climb. This can result in reattachments on downstream surfaces that would not have occurred in a good flow.

1.6. Reynolds Model Law

The ideas of numerically solving and airflow similarities are used to demonstrate that now the Reynolds number is really the only factor that affects the aerodynamics of commercial road vehicles. If viscous forces alone predominate, a model may be considered dynamically similar to the prototype if the ratio of inertial to viscous forces is the same in both the model and the prototype.

$Re = \rho VL/\mu$

Reynold's Model Makes laws that define models based on Reynold's number.

$R_{e\ prototype} = R_{e\ mode}$

$\rho_{prototype} V_{prototype} L_{prototype} / \mu_{prototype} = \rho_{model} V_{model} L_{model} / \mu_{model}$

$\rho_{prototype} V_{prototype} L_{prototype} / \rho_{model} V_{model} L_{model} \cdot 1/ \mu_{prototype} / \mu_{model} = 1$

$\rho_s V_s L_s / \mu_s = 1$

Where scale ratios are

$\rho_s = \rho_{prototype} / \rho$' $V_s = V_{prototype} / V_{mode}$l

$L_s = L_{prototype} / L_{model}$, $\mu_s = \mu_{prototype} / \mu_{model}$

Now, the Time scale ratio $T_s = L_s / V_s$
Velocity scale ratio $V_s = V_s / L_s$

Acceleration scale ratio as = V
Force scale ratio

$F_s = m_s a_s = (\rho_s A_s L_s) a_s$

$F_s = (\rho_s L_{3s})a_s$

Discharge scale ratio

$Q_s = \rho_s A_s V_s$

$Q_s = \rho_s L_{s2} V_s$

1.7. Road Vehicle Aerodynamics

The road test is conducted using the coast down method, in which the vehicle speed is increased to a predetermined level and the time taken to decelerate by 10 kilometers per hour is recorded. The vehicle is driven on a limited route with no interruptions such as braking, turning, incline, obstructions, and so on. The recorded numbers are now entered into the formulas, and the drag coefficient is calculated. The car is put through its paces in two different scenarios.

- Windows are kept opened
- Windows are kept closed

The air blower is used to analyses the flow pattern around the model, which is shot and the results are described in flow visualization. Finally, recommendations are made for reducing air drag. The variations of drag on road vehicle based on various speed is shown in figure 4. The variations aerodynamic drag rolling resistance and tire rolling resistance with respect to time shown in figure 5.

Figure 4. Variations of drag with speed

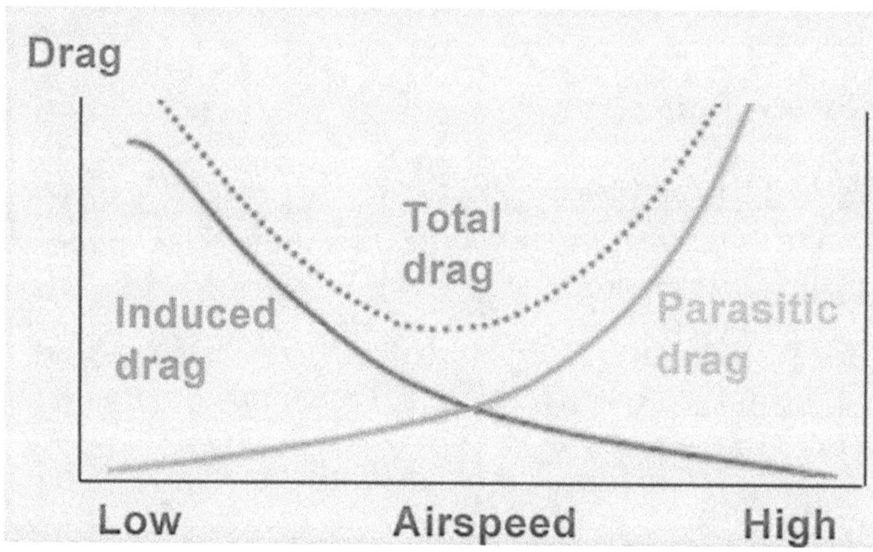

2. EVALUATION METHODS FOR ROAD VEHICLE AERODYNAMICS

The assessment and development of automobile aerodynamics is a continuous process and a vital aspect of automotive design that does not end with the vehicle's original design stage. Wind tunnel testing, computation prediction, and racetrack testing are examples of common analytical and testing methods utilized in this strategy. A wind tunnel or a quantitative model, for example, can be utilized at the original design phase even before vehicle is built, and each of these approaches may be useful over a certain requirement. Once a vehicle has now been built, it may be instrumented and tested on the track.

Figure 5. Variation of resistance with speed

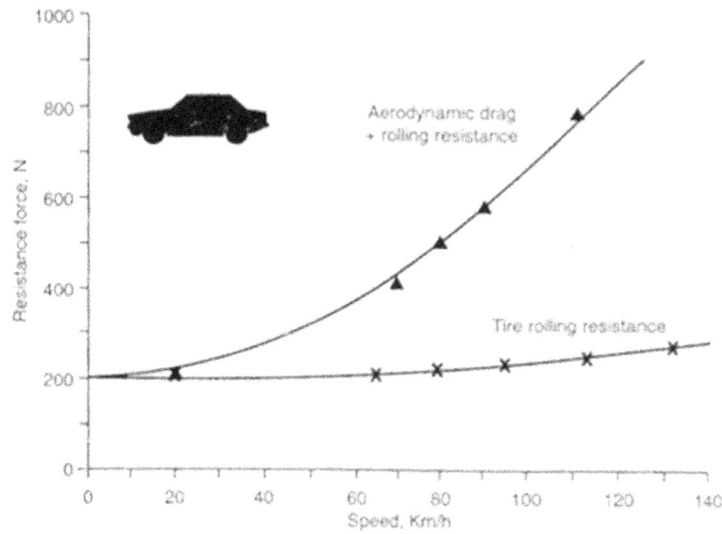

2.1. Computational Method

Due to favourable improvements in computational power and affordability, processing fluid dynamic techniques are progressively being integrated into a wide-ranging variety of manufacturing departments.

Figure 6. Computational fluid dynamics courtesy to Racecar engineering

Whenever these technologies are applied in the automotive industry, one of the benefits is the vast amount of data collected by the "solution." Unlike wind tunnel or track testing, the data from the "experiment" may be seen, investigated, and analysed several times after it is done. Furthermore, such simulated models can be developed prior to the building of an automobile and can offer data on aerodynamic stresses on different components, as well as computational fluid dynamics.

It is a common solution that depicts the surface forces around the race car body as well as the path of several streamlines. Engineers may use this data to enhance vehicle performance by lowering drag or boosting down force, as previously indicated. While computationally procedures appear to be the most appealing, they have flaws and require highly qualified aerodynamicists to carry out and understand computerized simulations. Figure 6 shows the computational fluid dynamics of formula one race car.

2.2. Wind Tunnel Testing

The convenience of testing in a highly regulated environment with a variety of equipment which does not have to be carried on the vehicle is suggested by the wind tunnel. Reduced scale models can also be tested if the vehicle has not yet been completed. Although wind tunnels were widely utilised in the building of aeroplanes, the usage of aerodynamic analysis in wind tunnels for vehicle testing raised two difficulties the first issue is how to connect the moving wheels.

Figure 7. Wind tunnel testing courtesy to Bed blog esprit design

Due to the tight fit between both the vehicle under body and the test division's immovable floor wind tunnels make it easy to test in a controlled environment using several pieces of equipment that doesn't need to be transported with the vehicle. Relatively small models can also be used to test the vehicle if it has not yet been built. Even though wind tunnels were widely used in the construction of aeroplanes, using them for car testing presented two challenges. Due to the tight fit between the vehicle body panels and the test section's fixed floor, the first issue is how to place the moving wheels.

2.3. Track Testing

Some of the challenges that come with wind tunnel testing are just not present in full-scale aerodynamics testing on the racetrack. There is no need to make a more expensive small scale model because the running wheels, moving surface, and wind tunnel obstacle corrections are all taken good care of. Naturally, you'll need a car, the climate must work together, and the expense of hiring a track and performance tuning running vehicles should not be excessive. This sort of aerodynamics testing has enhanced greatly in current years as a result of the benefits described above, as well as the uncontrolled weather and cost restrictions.

Figure 8. Track testing

2.4. Coast Down Method

The coast down test was one of the first types of testing used to assess a vehicle's drag. Despite changes in air conditions and abnormalities in the wheel rolling resistance, acceptable progressive data can be gathered. By the late 1990s, Computer and sensor technology advancements had made it feasible to measure and transfer desired forces, moments, or pressures via a wireless network at a reasonable cost. The coast down is a standard motor vehicle test that entails starting the vehicle at a fixed velocity with the vehicle non-geared and measuring the air velocity and distance reached until the vehicle comes to a complete stop. This may be done for a variety of purposes, the most frequent of which is to gather vital information regarding the vehicle's general condition and interaction with its surroundings. Ion Preda (2010) The major motive of this experiment is to discover the numerical coefficients of the resistance forces operating over the automotive at different road situations and speeds therefore that they may be replicated on proving stands to assess SFC, pollutant gasses or noise pollution, and other things. The next goal is to identify any probable irregularities in the functioning of particular vehicles subassemblies that is to verify if the technical state of the vehicle is satisfactory before subjecting it to more complex tests. The requirements for performing these types of the coast down tests are uniform.

2.4.1. The Theory Behind the Vehicle Dynamics Coasting Method

The vehicle's movement equation was created using some shortening premises and a rudimentary active model. The active model was created with the concept that a vehicle is made up of:

- mass m_t in translation motion;
- equivalent flywheel I_{nw} is the comparable flywheel for the non-driving wheels;
- equivalent flywheel I_{dw}, comparable to the driving wheels and all kinematically related rotational elements

A smooth road will be assumed for the coast-down test. The wheels are also not burdened with heavy moments because the engine is unplugged. In these circumstances, there should be no tyre-ground slide and a constant dynamic radius r_d for the vehicle wheels. As a consequence, no energy will be lost as a result of the slide. Figure 9 shows the forces acting on road vehicle. Hucho WH et al., (1990) the car co-operates with the environment while moving. As a consequence, certain forces act contrary to it. The car gravity force, weight W, Z_1 and Z_2 response forces normal to the road surface are produced, determining the rolling resistance forces F_{r1} and F_{r2} on the rear and front axles, correspondingly; the air resistance force F_a; and the driving force at the wheel F_w, which propels the vehicle. Mitscke M.et al., (1982) The principle of energy conservation will be used to get the equation for vehicle motion, It asserts that driving power Pw minus total power losses is the derivative (the simultaneous changing rate) of the car's entire kinetic motion:

$$\frac{dE_{tot}}{dt} = P_w - \sum P_{loss} \tag{1}$$

Figure 9. Forces acting on a vehicle

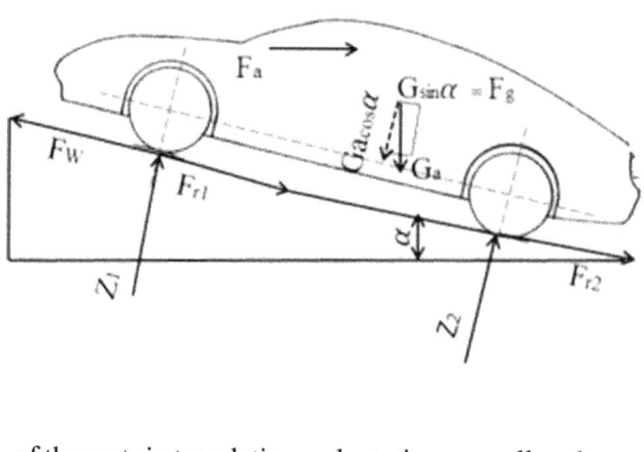

The kinetic energy E_k of the parts in translation and rotation, as well as the potential energy E_p defined by the vehicle altitude, make up the mechanical energy of the vehicle:

$$E_{tot} = E_k + E_p = E_{kt} + E_{kr} + E_p \tag{2}$$

The translational mass's kinetic energy E_k, as well as the kinetic energy of any rotating item j kinematically related to the wheels, is included in the kinetic energy E_k:

$$E_k = \frac{m_t v^2}{2} + \sum_j \frac{I_j \omega_j^2}{2} = \frac{m_i v^2}{2} + \frac{I_w \omega_w^2}{2} = \frac{m_{ap} v^2}{2} \tag{3}$$

The rotational speed of the driving wheels is given by $\omega_w = v\ r_d$. Vehicle appears to have a larger mass m_{ap} than it has.

$$m_{ap} = m_t \left(1 + \frac{I_w}{r_w^2}\right) = m_t \delta \tag{4}$$

Where $\delta = \left(1 + \frac{I_w}{m_t r_w^2}\right)$ is the coefficient is related with 1.03 to 1.04

Potential energy is considered as E_p

$$E_p = m_t \text{ g } h \tag{5}$$

Eventually g = 9.81 m/s² denoted as gravitational acceleration & also 'h' denoted as altitude

Obtaining the equation (6) by introducing the potential energy and total energy (2)&(5) respectively substitute in equ (1)

$$\frac{dE_{tot}}{dt} = m_{ap}\ v \frac{d_v}{d_t} + m_t\ \text{g} \frac{dh}{dt} = m_{ap} v\ a + m_t \text{g v} \sin\alpha = P_w - \sum P_{loss} \tag{6}$$

Where $\frac{dh}{dt} \approx \frac{ds}{dt} \sin\alpha = V \sin\alpha$ *and* a is the vehicle's acceleration.

From the equ (6)

$$m_{ap} a = \frac{P_w}{v} - \frac{\left(\sum P_{loss}\right)}{v} - m_t \text{ g} \sin\alpha = F_w - \left(R_r + R_a\right) - R_g \tag{7}$$

It is used to evaluate acceleration.

Since the engine is ungeared, the force F_w becomes negative as the vehicle coasts owing to resistance in the axle mechanisms, driveline, and transmission:

$$-F_w = R_f = M_f / rd \tag{8}$$

The corresponding friction moment given to the driving wheels is denoted by M_f. introducing the equ (8) in equ (7)

$$-a = d = (R_r + R_g + R_a + R_f)/\ map \tag{9}$$

Where the vehicle deceleration $d = -a$ during the testing

$R_{g,}$ R_r and R_a indicate grade resistance, rolling resistance and air resistance, respectively, in the equation (7)..(9), whereas R_f represents the resistance due to powertrain frictions.

The method of calculating air resistance (aerodynamic drag) takes into account not just the speed of the vehicle v, and also the wind speed is in the driving direction

$$R_a = \frac{1}{2}\rho C_d A(v+w)^2 \tag{10}$$

Where ρ is the air density and C_d is the drag coefficient. Frontal area is denoted by A, and w is the longitudinal wind velocity.

The density of the air is calculated based on the ambient conditions
which will be measured during the experiment:

$$\rho = \rho_0 \frac{P\ t_{c0} + 273}{P_0\ t_c + 273} \tag{11}$$

Where ρ_0 = 1.226 kg/m^3 is the air density under normal circumstances (t_0 = 15°C & p_0 = 760 mm Hg = 1 bar equal to 101324 Pa). As a result, the equation is used to compute R_g, which is the gravitational element normal to the ground.

$$R_g = W \sin\alpha = mt\ g \sin\alpha \tag{12}$$

The vehicle weight is denoted by W = mt g, and – is the road's longitudinal inclination angle. The hysteresis (energy loss, partly converted to heat) displayed during deflection by the combination tyre-surface of ground and the suspension, dampers and bushings is the major cause of rolling resistance. Its value may be calculated using the following formula:

$$R_r = W f \cos\alpha = mt\, g\, f \cos\alpha \tag{13}$$

Where W cosα is denoted as weight components of the vehicles vertical to the ground surface and f is the rolling resistance coefficient (dimensionless). Investigations and finite element technique simulations of tyre dynamic behavior show that as the road is uncomfortable or the increasing vehicle velocity, the f value increases. For the time being, there isn't a widely agreed equation for calculating the rolling resistance coefficients as a function of the speed of a vehicle, however for medium and small velocities, almost all sources show a linear rise with speed. The 'f' will be regarded as a linear function of vehicle speed because the driving speed is mild during cost down running.

$$f = f_0 + f_1 v \tag{14}$$

f_0 and f_1 are constants in this equation. In the event of increased inflation-pressure vehicle tyres, the technique proposes $f_1 = 0.2$ s/m, whereas f0 fluctuates depending on the surface condition of road, with an approximated minimum value of 0.0075 for good road condition. The impact of suspensions movements, which cause extra losses primarily in the suspension systems, rubber bushings, joints, and tyres, is considered as an equivalent slope to avoid, which is larger when the road surfaces are not even but remains constant with the speed of the vehicle. Because of the quick and non-linear rise in dampening forces with acceleration, this concept appears to reduce the genuine phenomena too much and should be approached with caution. Additionally, during testing, the power loss of these suspensions is likely to manifest as additional rolling resistance. The equation must be rewritten by a more sophisticated f(v) function in this scenario. The last type of resistance to examine is that caused by powertrain friction. This is the total of the frictional forces for the moving and non-driving axles:

$$R_f = \frac{M_{fd}}{r_d} + \frac{M_{fn}}{r_n} = \frac{M_f}{r_w} \tag{15}$$

The corresponding friction moment given to the driving wheels is denoted by M_f. The major causes of drivetrain frictional resistance have been discussed in greater detail in the article, as well as the effects of energy consumed for oil bubbling. If a dynamometer without rollers is available, the frictional torques on the driving and non-driving wheels, M_{fd} and M_{fn}, can be measured. Such torques are influenced (in a complicated way) by the tire rotation rates (concerning vehicle speed). The total of the transmission friction torque and rolling resistance torque may be determined if another dynamometer, this time with rollers, is also available. This total torque is proportional to the speed of the vehicle, although it has a strong linear element. Because of the severe approximations, the frictional resistance of the drivetrain will be approximated to be:

$$R_f = R_{f0} + R_{f1}v \tag{16}$$

R_{f0} is the powertrain resistance at extremely low velocities, while R_{f1} is the resistance's rising rate as velocity v increases.

Then initiating the equation (9) substitute the equations (10), (12), (13), (14) and (16), the final speed is estimated AS

$$d = \frac{\left[m_t g\left(f_0 + f_1\right)\cos\alpha + m_t g\ \sin\alpha + \frac{1}{2}\rho Cd\ A\left(v+w\right)^2 + R_{f0} + R_{f1}v \right]}{m_{ap}} = C_0 + C_1 v + C2v^2 \tag{17}$$

Where:

$$C_0 = [m_t g\left(f_0 \cos\alpha + \sin\alpha\right) + \frac{1}{2}\rho C_d\ Aw^2 + R_{f0}] / m_{ap}$$

$$C_1 = \frac{\left[m_t g\, f_1 \cos\alpha + \frac{1}{2}\rho Cd\, A w^2 + R_{f0} \right]}{m_{ap}} \tag{18}$$

$C_1 = [1 / 2\rho Cd_A]/ \text{m\_}ap$

It is observed that the unaccounted air velocity will change the rolling resistance coefficient (f) as well as the powertrain shear forces in the first two equations (18). A road that isn't exactly horizontal might have a similar unfavorable effect. If take on a flat road considering ($\alpha = 0$) and no wind ($w = 0$), simplifying the first two equations of (18)

$$C_0 = f_0 \frac{g}{\delta} + \frac{Rf0}{map}$$

$$C_1 = f_1 \frac{g}{\delta} + \frac{Rf1}{map} \tag{19}$$

3. CONCLUSION

This review of vehicle aerodynamics is related to the vehicle structure and the speed over the road and obtaining the drag reduction methods utilizing aerodynamic forces acting on the vehicle body. The various testing methods and the optimization of the parameters related to the design of the high efficient less drag vehicle have been discussed.

REFERENCES

Backiyaraj, A. (2021). Performance Improvement of Suspension Characteristics by Implementation of Active Suspension System in Two Wheelers. *International Journal of Automotive and Mechanical Engineering, 18*(3), 9113 – 9126. doi:10.15282/ijame.18.3.2021.23.0700

Hucho, W.-H. (1999). *Aerodynamik des Automobils, 5*. Springer Verlag.

Lietz, R., Pien, W., & Remondi, S. (2000). A CFD Validation Study for Automotive Aerodynamics. *SAE Technical Paper Series*. doi:10.4271/2000-01-0129

Mayer, W., & Wickern, G. (2011). The New Audi A6/A7 Family - Aerodynamic Development of Different Body Types on One Platform. *SAE International Journal of Passenger Cars. Mechanical Systems, 4*(1), 197–206. doi:10.4271/2011-01-0175

Mitscke, M. (1982). Dynamik der Kraftfahrzeuge. Springer Verlag.

Murugu Nachippan, N. (2019). Computational Analysis of Air Wings to Evaluate Downforce and Lift-Drag Ratio: Technical Note. *International Journal of Vehicle Structures and Systems*, *11*(1), 102–106. https://doi.org/10.4273/ijvss.11.1.18

National Aeronautics and Space Administration. (2021). *The drag coefficient.* Glenn Research Centre.

Preda, I., Covaciu, D., & Ciolan, G. (2010). Coast down test – Theoretical and experimental approach. Transilvania University of Brasov.

Senthilkumar, P. B., Parthasarathy, M., & Aravind, L. (n.d.). Design and analysis of a rear diffuser in a sedan car. *Materials Today: Proceedings*. doi:10.1016/j.matpr.2021.11.542

Wang, J., Li, H., Liu, Y., Liu, T., Gao, H., Tandon, P., Wisnoe, W., & bin Abdullah, M.Z. (2018). Aerodynamic research of a racing car based on wind tunnel test and computational fluid dynamics. *MATEC Web of Conferences*, *153*. 10.1051/matecconf/201815304011

Wickern, G., Wagner, A., & Zoerner, C. (2005). *Induced Drag of Ground Vehicles an It's Interaction with Ground Simulation*. SAE Technical Paper 2005-01-0872. doi:10.4271/2005-01-0872

ADDITIONAL READING

Buljac, A., Džijan, I., Korade, I., Krizmanić, S., & Kozmar, H. (2016). Automobile aerodynamics influenced by airfoil-shaped rear wing. *International Journal of Automotive Technology*, *17*(3), 377–385. doi:10.100712239-016-0039-4

Sovran, G., Morel, T., & Mason, W. T. (1978). *Aerodynamic Drag Mechanisms of Bluff Bodies and Road Vehicles The Aerodynamic Drag of Cars Current Understanding*. Unresolved Problems, and Future Prospects. doi:10.1007/978-1-4684-8434-2

APPENDIX: ACRONYMS

SFC Specific fuel consumption
C_d Drag coefficient
D Drag
V Velocity
A Area
P Density
F_r Rolling resistance forces
F_a Air resistance force
F_w Force at the wheel
E_k Kinetic Energy
E_p Potential energy

Compilation of References

Abdelghany, E., Khalil, E., Abdellatif, O., & El Harriri, G. (2016). Winglet Cant And Sweep Angles Effect On Aircraft Wing Performance. *The International Conference on Applied Mechanics and Mechanical Engineering*, 17, 1–17. 10.21608/amme.2016.35282

Abdel-Motaleb, S. A., Taylor, J. D., Hunsaker, D. F., & Coopmans, C. (2019). Comparison of induced and parasitic drag on wings with minimum induced drag. AIAA Scitech 2019 Forum. doi:10.2514/6.2019-2120

Abhijith, Krishnan, Mohan, Gokul, & Ghevarghese. (2015). Effect of Blunted Aero-Disc on Aero-Spiked Blunt Cone-Nose at High Speeds. *International Research Journal of Engineering and Technology, 2*(9).

Adkins, R. C. (1975). A short diffuser with low pressure losses. *ASME J. Fluid Engng., 97*(3), 297–302. doi:10.1115/1.3447306

AeroToolbox. (2022, May 18). *Introduction to aircraft airfoil aerodynamics.* Retrieved from https://aerotoolbox.com/intro-airfoil-aerodynamics/

Ahmad, K., Wahid, Z., & Hanouf, Z. (2019). Heat transfer analysis for Casson fluid flow over stretching sheet with Newtonian heating and viscous dissipation. *Journal of Physics: Conference Series, 1127*(1), 012028.

Ahmed, M. Y. M., & Qin, N. (2010, January–February). Drag Reduction Using Aerodisks for Hypersonic Hemispherical Bodies. *Journal of Spacecraft and Rockets, 47*(1), 62–80. doi:10.2514/1.46655

Aircraft Systems. (2022, May 27). *High speed flight – aerodynamics of Flight.* Retrieved from https://www.aircraftsystemstech.com/2017/05/high-speed-flight.html

Al Sidairi, K.A., & Rameshkumar, G.R. (2016). Design of Winglet Device for Aircraft. *International Journal of Multidisciplinary Sciences and Engineering*, 23–30.

Aleksandrov, Mingazov, & Tokmovtsev. (2016). *Software for Automated Processing and Analysis of Gas Dynamics and Combustion Processes in Combustion Chambers of Gas Turbine Engines (Servis 3D).* The Certificate on State Registration of the Computer Program no. 2016611856, KNRTU–KAI.

Aleksandrov, Sabirzyanov, & Yavkin. (2017). Influence of Simplifying a Geometrical Model of the Gas Turbine Engine Combustion Chamber on the Results of Numerical Modeling. *Russian Aeronautics, 60*(4), 575–582.

Amendola, G., Dimino, I., Concilio, A., Andreutti, G., Pecora, R., & Cascio, M. L. (2018). Preliminary design process for an adaptive winglet. *International Journal of Mechanical Engineering and Robotics Research*, *7*(1), 83–92.

Analysis, W., & Deakin, D. (2011). Simulate To Innovate. *Aerospace (Basel, Switzerland)*, (July), 32–33.

Ananda Reddy, G., & Ganesan, V. (2005). A Numerical Study of Pre-Diffuser Optimization of an Aero Gas Turbine Combustion Chamber. *Indian Journal of Engineering and Materials Sciences*, *12*(August), 281–291.

Anand, V., & Christov, I. C. (2021). Revisiting steady viscous flow of a generalized Newtonian fluid through a slender elastic tube using shell theory. *ZAMM-Journal of Applied Mathematics and Mechanics. Zeitschrift für Angewandte Mathematik und Mechanik, 101*(2), e201900309.

Ananthasayanam, M. R., & Narasimha, R. (1987). A proposed international tropical reference atmosphere up to 1000 km. *Advances in Space Research, 7*(10), 117–131. doi:10.1016/0273-1177(87)90084-6

Anderson, J. D. (1989). Hypersonic and High Temperature Gas Dynamics. The McGraw-Hill Company.

Anderson, J. D., & Bowden, M. L. (2005). Introduction to Flight (6th ed.). McGraw-Hill Higher Education.

Anderson, J. D., Jr. (1989). Hypersonic and High Temperature Gas Dynamics. The McGraw-Hill Company.

Anderson, D. M., & Davis, S. H. (1993). Two-fluid viscous flow in a corner. *Journal of Fluid Mechanics, 257*, 1–31.

Anderson, J. D., & Bowden, M. L. (2005). *Introduction to flight*. McGraw Hill Book Company.

ANSYS Fluent Theory Guide, . (2016). *Release 17.0*. ANSYS, Inc.

Aravind, S., Sabarinathan, G., Ajith, S., Amrith, M., Sundararaj, K., & Sanal Kumar, V. R. (2019). Design of pre-diffuser cones for dump diffusers for aero gas turbine engines. AIAA Propulsion and Energy Forum and Exposition.

Arceneaux, J. (2004). Overview on drag reduction Technologies for Civil transport aircraft. *European Congress on Computational Methods in Applied Sciences and Engineering ECCOMAS*.

Ashok, G. (2003). Design of low Reynolds number airfoils with trips. *Journal of Aircraft, 40*(4).

Avelar, A. C., Fisch, G., & Marciotto, E. (2012). Atmospheric Boundary Layer Simulation in a Wind Tunnel for Analysis of the Wind Flow at the Centro de LanÃ § amento de AlcÃĒntara. In *4th AIAA Atmospheric and Space Environments Conference* (p. 2930). Academic Press.

Azlin, M. A., Taib, C. M., Kasolang, S., & Muhammad, F. H. (2011, July). CFD analysis of winglets at low subsonic flow. In *Proceedings of the World Congress on Engineering* (*Vol. 1*, pp. 6-8). Academic Press.

Backer, J. A., Lowe, C. P., Hoefsloot, H. C. J., & Iedema, P. D. (2005). Poiseuille flow to measure the viscosity of particle model fluids. *The Journal of Chemical Physics, 122*(15), 154503. doi:10.1063/1.1883163 PMID:15945641

Backiyaraj, A. (2021). Performance Improvement of Suspension Characteristics by Implementation of Active Suspension System in Two Wheelers. *International Journal of Automotive and Mechanical Engineering, 18*(3), 9113 – 9126. doi:10.15282/ijame.18.3.2021.23.0700

Bahumanyam, P. K. (2014). Evaluation of Novel Wing design for UAV. *Proceeding of Comsol Conference*.

Bai, S. (1959). *Introduction to the theory of compressible flow*. van Nostrand.

Barzegar Gerdroodbary, M. (2014, September). Numerical analysis on cooling performance of counter flowing jet over aerodisked blunt body. *Shock Waves, 24*(5), 537–543. doi:10.100700193-014-0517-4

Beckwith, T. G., Buck, N. L., & Marangoni, R. D. (1969). *Mechanical measurements* (Vol. 5). Addison-Wesley.

Beechook, A., & Wang, J. (2013, September). Aerodynamic analysis of variable cant angle winglets for improved aircraft performance. In *2013 19th International Conference on Automation and Computing* (pp. 1-6). IEEE.

Bejan, Charles, & Lorente. (2014). The evolution of airplanes. *Journal of Applied Physics, 116*(4).

Benard, N., & Moreau, E. (2010). Capabilities of the dielectric barrier discharge plasma actuator for multi-frequency excitations. *Journal of Physics*, 43(14), 1–14.

Berry, S. A., Horvath, T. J., Greene, F. A., Kinder, G. R., & Wang, K. C. (2006). Overview of boundary layer transition research in support of orbiter return to flight. *AIAA Paper*, 2006–2918.

Bharathi, Balamanikandan, Paul, Sundar, & Vikash. (2015). Effect of Chevron on Transonic Nozzle. *International Journal of Engineering Research & Technology*, 4(4), 1413–1417.

Blackmore, D., Ting, L., & Knio, O. (2007). Studies of perturbed three vortex dynamics. *Journal of Mathematical Physics*, *48*(6), 065402.

Bourdin, P., Gatto, A., & Friswell, M. I. (2006). The application of variable cant angle winglets for morphing aircraft control. In *Collection of Technical Papers - AIAA Applied Aerodynamics Conference*. American Institute of Aeronautics and Astronautics Inc. 10.2514/6.2006-3660

Bourdin, P., Gatto, A., & Friswell, M. I. (2008). Aircraft control via variable cant-angle winglets. Journal of Aircraft, 414–423. doi:10.2514/1.27720

Bourdin, P. (2002). Planform effects on lift-induced drag. In *20th AIAA Applied Aerodynamics Conference*. American Institute of Aeronautics and Astronautics Inc. 10.2514/6.2002-3151

Braun, E. M., Lu, F. K., & Wilson, D. R. (2009). Experimental research in aerodynamic control with electric and electromagnetic fields. *Progress in Aerospace Sciences*, *45*(1-3), 30–49. doi:10.1016/j.paerosci.2008.10.003

Bravo-Mosquera, Cerón Muñoz, & Catalano. (2019). Design and computational analysis of a closed non-planar wing aircraft coupled to a boundary layer ingestion propulsion system. In AIAA Propulsion and Energy Forum and Exposition. American Institute of Aeronautics and Astronautics Inc.

Brazeal, J., Wuetcher, A., & Wang, X. (2020). Revisit of Hypersonic Small-Disturbance Theory for Perfect Gases. *AIAA 2020-0102. AIAA Scitech 2020 Forum.*

Britan, A., Karpov, A. V., Vasilev, E. I., Igra, O., Ben-Dor, G., & Shapiro, E. (2004). Experimental and Numerical Study of Shock Wave Interaction with Perforated Plates. *Journal of Fluids Engineering*, *126*(3), 399–409. doi:10.1115/1.1758264

Buckholz, R. H. (1986). *The functional role of wing corrugations in living systems*. Academic Press.

Bushnell, D. M. (2003). Aircraft drag reduction - A review. In *Proceedings of the Institution of Mechanical Engineers, Part G: Journal of Aerospace Engineering*. Professional Engineering Publishing Ltd. 10.1243/095441003763031789

Bushnell, D. M., & Tuttle, M. H. (1979). Survey and Bibliography on Attainment of Laminar Flow Control. In *Air Using Pressure Gradient And Suction*. NASA Reference Publication.

Butler, D. S. (1960). The numerical solution of hyperbolic systems of partial differential equations in three independent variables. *Proceedings of the Royal Society of London. Series A, Mathematical and Physical Sciences*, *255*(1281), 232–252. doi:10.1098/rspa.1960.0065

Calvert, J. G. (1990). Glossary of atmospheric chemistry terms (Recommendations 1990). *Pure and Applied Chemistry*, *62*(11), 2167–2219. doi:10.1351/pac199062112167

Carrotte, Denman, Wray, & Fry. (1994). *Detailed performance comparison of a dump and short faired combustor diffuser system*. ASME Paper, 93-GT-331.

Carrotte, J. F., Bailey, D. W., & Frodsham, C. W. (1995). Detailed Measurements on a Modern Combustor Dump Diffuser System. *Journal of Engineering for Gas Turbines and Power*, *117*(4), 679–685. doi:10.1115/1.2815453

Carrotte, J. F., Denman, P. A., Wray, A. P., & Fry, P. (1994). Detailed Performance Comparison of a Dump and Short Faired Combustor Diffuser System. *Journal of Engineering for Gas Turbines and Power*, *116*(3), 517–526. doi:10.1115/1.2906850

Cavcar, M. (2000). The international standard atmosphere (ISA). Anadolu University.

Cayzac, R., Carette, E., Denis, P., & Guillen, P. (2011). Magnus effect: Physical origins and numerical prediction. *Journal of Applied Mechanics*, *78*(5), 051005. doi:10.1115/1.4004330

Cerón-Muñoz. Catalano, & Coimbra. (2008). Passive, active, and adaptative systems for wing vortex drag reduction. *ICAS Secretariat - 26th Congress of International Council of the Aeronautical Sciences 2008*, *4*, 1537–1548.

Chana, R., & Vishwanath, R. (1998, November-December). Base Drag Reduction Caused by Riblets on a GAW(2) Airfoil. *Journal of Aircraft*, *35*(6).

Chandler, D. (2019). *MIT and NASA engineers demonstrate a new kind of airplane wing*. Academic Press.

Chang, P. K. (2014). *Separation of flow*. Elsevier.

Chanthavane, S. L. (2021). *Space Radiation Interactions in Earth's Atmosphere, Low Earth Orbit, and Deep Space* (Doctoral dissertation). Rice University.

Chattot, J. J. (2004). Analysis and design of wings and wing/winglet combinations at low speeds. *AIAA Paper*, *13*(October), 2256–2265. doi:10.2514/6.2004-220

Chattot, J. J. (2006). Low-speed design and analysis of wing/winglet combinations including viscous effects. *Journal of Aircraft*, *43*(2), 386–389. doi:10.2514/1.15349

Cheng, H. K. (1963). Recent Advances In Hypersonic Flow Research. *AIAA Journal*, *1*(2), 295–310.

Cho, Y. C., & Shyy, W. (2011). Adaptive flow control of low-Reynolds number aerodynamics using dielectric barrier discharge actuator. *Progress in Aerospace Sciences*, *47*(7), 495–521. doi:10.1016/j.paerosci.2011.06.005

Chushkin, P. I. (1968). Numerical method of characteristics for three-dimensional supersonic flows. *Progress in Aerospace Sciences*, *9*, 41–122. doi:10.1016/0376-0421(68)90004-3

CoS. (2010). *Guide: Guide to Reference and Standard Atmosphere Models* (AIAA G-003C-2010 (2016)).

Cosin, R., Catalano, F. M., Correa, L. G. N., & Entz, R. M. U. (2010). Aerodynamic analysis of multi-winglets for low speed aircraft. *27th Congress of the International Council of the Aeronautical Sciences 2010*, 1622–1631.

Crittenden, T. M., Shlyubsky, D., & Glezer, A. (2004). Combustion-driven jet actuators in reversed flow configurations. *Proceedings of the second AIAA flow control conference*, 1-12. doi:10.2514/4.479892

d'Humières, G., & Stollery, J. L. (2010, February). Drag reduction on a spiked body at hypersonic speed. *Aeronautical Journal*, *114*(1152), 113–119. doi:10.1017/S0001924000003584

Daidzic, N. E. (2015). Efficient general computational method for estimation of standard atmosphere parameters. *International Journal of Aviation, Aeronautics, and Aerospace*, *2*(1), 3.

Daidzic, N. E. (2019). On atmospheric lapse rates. *International Journal of Aviation, Aeronautics, and Aerospace*, *6*(4), 2.

Dakka, S., & Johnson, O. (2019). Aerodynamic design and exploration of a blended wing body aircraft at subsonic speed. *International Journal of Aviation, Aeronautics, and Aerospace*, *6*(5). doi:10.15394/ijaaa.2019.1411

Daniel, C., & Petre, C. (2013). Acoustic characteristics of the flow over different shapes of nozzle chevrons. *ICAS Bulletin Cojan*, *5*(3), 53–59. doi:10.13111/2066-8201.2013.5.3.6

de Sá, C. V. S., & von Borries Lopes, A. (2021). An instructive derivation of d'Alembert's paradox. *European Journal of Physics*, *43*(1), 015004. doi:10.1088/1361-6404/ac329a

Deepanshu Srivastav, B. E. (Hons.). (2012). Flow Control over Airfoils using Different Shaped Dimples. *International Conference on Fluid Dynamics and Thermodynamics Technologies*.

Dhawan, S. J., & Narasimha, R. (1958). Some properties of boundary layer flow during the transition from laminar to turbulent motion. *Journal of Fluid Mechanics*, *3*(4), 418–436.

Dinesh, Venkatesan, DV, SK, & Kumar. (2014). Diagnostic Investigation of Aircraft Performance at Different Winglet Cant Angles. *International Journal of Mechanical, Aerospace, Industrial and Mechatronics Engineering*, *8*(12), 1987–1996.

Dippold, V. (2008). CFD Analyses and Jet-Noise Predictions of Chevron Nozzles with Vortex Stabilization. Glenn Research Center.

Douglas, E. (1964). Wall A Study Of Hypersonic Aircraft. *Fourth National Airport Conference*.

Dourado, E., & Kotrous, M. (2016, July 20). *Airplane Speeds Have Stagnated for 40 Years*. Mercatus Center, George Mason University. Retrieved from mercatus.org/publications/technology-and-innovation/airplane-speeds-have-stagnated-40-years

Dover. (1960). *Thwaites Bryan. Incompressible Aerodynamics*. Academic Press.

Dunbar, B. (1996, September). *High-speed research (HSR) – smoothing the flow over Supersonic Wings*. NASA. Retrieved from https://www.nasa.gov/centers/langley/news/factsheets/HSR-Wings.html

Eghlima, Z., & Mansour, K. (2017). Drag reduction for the combination of spike and counterflow jet on blunt body at high Mach number flow. *Acta Astronautica*, *133*, 103–110.

Elert, G. (n.d.). *Shock waves*. The Physics Hypertextbook. Retrieved from https://physics.info/shock/

Eller, D., & Heinze, S. (2005). Approach to Induced Drag Reduction with Experimental Evaluation. *Journal of Aircraft*, *42*(6), 1478–1485. doi:10.2514/1.11713

Fan, B., & Huang, J. (2021). Coupled Fluid-Thermal Investigation on Drag and Heat Reduction of a Hypersonic Spiked Blunt Body with an Aerodisk. *Aerospace 2022*, *9*, 19.

Fan, Z. (2010). Measurement of Aerodynamic Forces and Moments in Wind Tunnels. In *Encyclopedia of Aerospace Engineering*. John Wiley & Sons, Ltd. doi:10.1002/9780470686652.eae078

Fazle Rabbi, M., Nandi, R., & Mashud, M. (2015). Induce Drag Reduction of an Airplane Wing. *American Journal of Engineering Research*, *4*(6), 219–223.

Federal Aviation Administration & United States Federal Aviation Administration. (2009). *Pilot's handbook of aeronautical knowledge*. Skyhorse Publishing Inc.

Ferri, A. (1945). *NACA WR L-143 Completed Tabulation in the United States of Tests of 24 Airfoils at High Mach Numbers*. Academic Press.

Field, J. E. (1991). The physics of liquid impact, shock wave interactions with cavities, and the implications to shock wave lithotripsy. *Physics in Medicine and Biology*, *36*(11), 1475–1484. doi:10.1088/0031-9155/36/11/007 PMID:1754618

Fishenden, C. R., & Stevens, S. J. (1974). Performance of Annular Combustor-Dump diffuser. *Journal of Aircraft*, *14*(1), 60–67. doi:10.2514/3.58749

Fletcher, D. G. (2014) *Fundamentals Of Hypersonic Flow-Aerothermodynamics* RtoAvt Lecture Series On "Critical Technologies For Hypersonic Vehicle Development", Held At The Von Kármán Institute, Rhode-St-Genèse, Belgium.

Gad-el-Hak, M. (2000). Flow control: Passive, active, and reactive flow management. Cambridge University Press.

Gall, P. D., Huebsch, W. W., & Roth Mayer, A. P. (2010). Dynamic roughness as a means of leading-edge separation flow control. West Virginia University.

Gaurav, S., & Jain, K. K. (2014). Numerical Investigation of Fluid Flow and Aerodynamic Performance of a Dragonfly Wing Section for Micro Air Vehicles (MAVs) Applications. *Int J Innovat Scient Res*, *92*, 285–292.

Ghose, P., Datta, A., & Mukhopadhyay, A. (2013). Effect of dome shape on static pressure recovery in a dump diffuser at different inlet swirl. *International Journal of Emerging Technology and Advanced Engineering, 3*(3), 465–471.

Glezer, A., & Amitay, M. (2002). Synthetic jets. *Annual Review of Fluid Mechanics*, *34*(1), 503–529. doi:10.1146/annurev.fluid.34.090501.094913

Goh, E. G., & Wan Nik, W. B. (2018). A generalized model for viscosity as a function of shear rate. *Journal of Engineering and Applied Sciences (Asian Research Publishing Network)*, *13*(9), 3219–3223.

Goldfield, E. C. (2018). *Bioinspired Devices: Emulating Nature's Assembly and Repair Process*. Harvard University Press.

Goldstein, S. (1948). On laminar boundary-layer flow near a position of separation. *The Quarterly Journal of Mechanics and Applied Mathematics*, *1*(1), 43–69.

Graham, D. J. (1949). *NACA TR-947 The Development of Cambered Airfoil Sections having Favorable Lift Characteristics at Supercritical Mach Numbers*. Academic Press.

Greenblatt, D., Goksel, B., Rechenberg, I., Schule, C. Y., Romann, D., & Paschereit, C. O. (2008). Dielectric barrier discharge flow control at very low flight Reynolds numbers. *AIAA Journal*, *46*(6), 1528–1541. doi:10.2514/1.33388

Grewe, V., Rao, A. G., Grönstedt, T., Xisto, C., Linke, F., Melkert, J., Middel, J., Ohlenforst, B., Blakey, S., Christie, S., Matthes, S., & Dahlmann, K. (2021). Evaluating the climate impact of aviation emission scenarios towards the Paris agreement including COVID-19 effects. *Nature Communications*, *12*(1), 3841. doi:10.103841467-021-24091-y PMID:34158484

Griffin, Huebsch, Rothmayer, & Wilhelm. (2016). *Numerical and Experimental Study on the Ability of Dynamic Roughness to Alter the Development of a Leading Edge Vortex*. Academic Press.

Grigore, C., Virgil, S., & Daniel-Eugeniu, C. (2014). Acoustic control study of turbofan nozzles with triangular chevrons. *INCAS Bulletin*, *6*(1), 13–20. doi:10.13111/2066-8201.2014.6.1.2

Guerrero, J., Sanguineti, M., & Wittkowski, K. (2018). CFD Study of the Impact of Variable Cant Angle Winglets on Total Drag Reduction. *Aerospace (Basel, Switzerland)*, *5*(4), 126. doi:10.3390/aerospace5040126

Gülhan, A., Willems, S., & Neeb, D. (2021, November 12). Shock Interaction Induced Heat Flux Augmentation in Hypersonic Flows. *Experiments in Fluids*, *62*(12), 242. doi:10.100700348-021-03336-y

Han, H. G., & Jiang, Z. (2018). Hypersonic Flow Field Reconfiguration and Drag Reduction of Blunt Body with Spikes and Sideward Jets. *International Journal of Aerospace Engineering*, *2018*, 7432961. doi:10.1155/2018/7432961

Harfash, A. J., & Meften, G. A. (2020). Poiseuille Flow with Couple Stresses Effect and No-slip Boundary Conditions. *Journal of Applied and Computational Mechanics*, *6*(Special Issue), 1069–1083.

Harish, V., & Kumar, C. (2016). International Journal of Current Engineering and Technology Drag Reduction using Suction Slit to the Main Wing. *International Journal of Current Engineering and Technology, 6*(4).

Harry, R. (1988). *GA Airfoils e A Catalog of Airfoils for General Aviation Use* (5th ed.). Self-Published.

Henderson, W. P., & Holmes, B. J. (1989). Induced drag - Historical perspective. SAE Technical Papers. doi:10.4271/892341

Hoffman, J., & Johnson, C. (2010). Resolution of d'Alembert's paradox. *Journal of Mathematical Fluid Mechanics, 12*(3), 321–334. doi:10.100700021-008-0290-1

Holt, M. (2012). *Numerical methods in fluid dynamics*. Springer Science & Business Media.

Hood, M. J. (1939). *NACA TN-695 The Effects of Some Surface Irregularities on Wing Drag*. Academic Press.

Hossain, A., Rahman, A., Hossen, J., Iqbal, P., Shaari, N., & Sivaraj, G. K. (2011a). Drag reduction in a wing model using a bird feather like winglet. *Jordan Journal of Mechanical and Industrial Engineering, 5*(3), 267–272.

Houghton, E. L., & Carpenter, P. W. (2003). *Aerodynamics for engineering students*. Elsevier.

Ho, W. H., & New, T. H. (2013, December). CFD Analysis of Bio-Inspired Corrugated Aerofoils. *Proceedings of Eleventh International conference of Fluid Dynamics*.

Ho, W. H., & New, T. H. (2017). Unsteady numerical investigation of two different corrugated airfoils. *Journal of Aerospace Engineering, 231*(13), 2423–2437. doi:10.1177/0954410016682539

Howe, D. (2010). *Aircraft conceptual design synthesis*. London Professional Engineering Publishing.

Huang, Liu, & Xia. (n.d.). Drag reduction mechanism induced by a combinational opposing jet and spike concept in supersonic flows. *Acta Astronautica*.

Hucho, W.-H. (1999). *Aerodynamik des Automobils, 5*. Springer Verlag.

Hunn, D. (n.d.). *The Road to Hypersonics -Key Challenges, Advantages and Disadvantages*. Lockheed Martin Missiles and Fire Control.

International Civil Aviation Organization (ICAO). (1993). *Manual of the ICAO Standard Atmosphere: extended to 80 kilometres (262 500 feet)* (Vol. 7488). International Civil Aviation Organization. Atmospheric and Space Environments.

Jaganraj,, R., Boopathy,, G., & Varun,, V., & Vigneshwar. (2015). Experimental investigation of passive flow control on bluff bodies. *International Journal of Applied Engineering Research, 108*(2), 19793–19798.

Jaganraj, R., Boopathy, G., & Varun, V. (2015). Experimental Investigation of Passive Flow Control on Bluff Bodies. *International Journal of Applied Engineering Research: IJAER, 10*(8), 19793–19798.

Jahanmiri. (2012). *Active Flow Control, review report*. Academic Press.

Jan & Edward. (2000). *Airplane Aerodynamics & Performance*. DAR Corporation.

Jenifer, N., & Preethi, S. (2016). Numerical Analysis of Chevron Nozzle with Various Configurations for Noise Reduction. *International Journal of Innovative Research in Science, Engineering and Technology, 5*(5).

Jiang, Z., Liu, Y., Han, G., & Zhao, W. (2009). Experimental demonstration of a new concept of drag reduction and thermal protection for hypersonic vehicles. *Lixue Xuebao, 25*(3), 417–419. doi:10.100710409-009-0252-8

Jie, Yaoa, & Shan. (2018). Numerical investigation on drag and heat reduction mechanism of combined spike and rear opposing jet configuration. *Acta Astronautica*.

John, B., Bhargava, D., Punia, S., & Rastogi, P. (2021). Drag and Heat Flux Reduction using Counterflow Jet and Spike - Analysis of their Equivalence for a Blunt Cone Geometry at Mach 8. *Journal of Applied Fluid Mechanics*, *14*(2), 375–388.

John, D. A. (2017). *Modern Compressible Flow*. Mc Graw Hill.

Johnson, W. (1986). The Magnus Effect—Early investigations and a question of priority. *International Journal of Mechanical Sciences*, *28*(12), 859–872. doi:10.1016/0020-7403(86)90032-9

Joseph, D. D. (2003). Viscous potential flow. *Journal of Fluid Mechanics*, *479*, 191–197.

Joslin, R. D., & Miller, D. N. (2009). Fundamentals and applications of modern flow control. *Progress in Astronautics and Aeronautics*.

Juhasz, A. J., & Smith, J. M. (1977). *Performance of High-Area-Ratio Annular Dump Diffuser Using Suction-Stabilized-Vortex Flow Control*. National Aeronautics and Space Administration.

Kaleeswaran, P., & Shanmughasundaram, P. (2016). Experimental and Statistical analysis on the noise reduction using chevron nozzle in supersonic free jet, U.P.B. *Scientific Bulletin, Series D*, 78(3), 21–30.

Kalimuthu & Mehta. (2008). Experimental investigation on spiked body in hypersonic flow. *The Aeronautical Journal*.

Kalimuthu, R., Mehta, R. C., & Rathakrishnan, E. (2010, January–February). Drag Reduction for Spike Attached to Blunt-Nosed Body at Mach 6. *Journal of Spacecraft and Rockets*, *47*(1), 219–222. doi:10.2514/1.46023

Katyal, A., & Srivastava, S. (2013). Hypersonic Aerodynamics of Aerospace Vehicle Design: Basic Approach and Study. *Advances in Aerospace Science and Applications, 3*(3), 209-214.

Kaushik, V., Mahore, M., & Patil, S. (2018). Analysis of Dimpled Wing of an Aircraft. *International Journal of Engineering Development and Research, 6*.

Kaushik, M. (2019). The Standard Atmosphere. In *Theoretical and Experimental Aerodynamics* (pp. 3–25). Springer. doi:10.1007/978-981-13-1678-4_1

Kenyon, K. E. (2016). On the Magnus Effect. *Nature and Science*, *8*(02), 49–52. doi:10.4236/ns.2016.82006

Kerho & Bragg. (1997). Airfoil Boundary Layer Development and Transition with Large Leading-Edge Roughness. The University of Illinois at Urbana-Champaign.

Kim & Kim. (2020). Drag Reduction Effect by Counter-flow Jet on Conventional Rocket Configuration in Supersonic/ Hypersonic Flow. *Journal of Aerospace System Engineering, 14*(4), 18-24.

Klein, A. (1995). Characteristics of combustor diffuser. *Progress in Aerospace Sciences*, *31*(3), 171–271. doi:10.1016/0376-0421(95)00006-K

Knight, D., & Kianvashrad, N. (2021). Simulation of Hypersonic Shock Wave Laminar Boundary Layer Interactions. *11th Ankara International Aerospace Conference*.

Kroo, I. (2005b). Nonplanar wing concepts for increased aircraft efficiency. Advanced Concepts for Future Civil Aircraft, 1–29.

Kroo, I. (2001). Drag due to Lift: Concepts for Prediction and Reduction. *Annual Review of Fluid Mechanics*, *33*(1), 587–617. doi:10.1146/annurev.fluid.33.1.587

Kundu, K. A. (2010). *Aircraft Design*. Cambridge University Press. doi:10.1017/CBO9780511844652

Kurzke, J., & Halliwell, I. (2018). Gas Properties and Standard Atmosphere. In *Propulsion and Power* (pp. 613–618). Springer. doi:10.1007/978-3-319-75979-1_14

Labitzke, K. (1982). On the interannual variability of the middle stratosphere during the northern winters. *Kisho Shushi. Dai2shu*, *60*(1), 124–139.

Ladson & Brooks Jr. (1975). *NASA TM-X-3284 Development of a Computer Program to Obtain Ordinates for NACA 4-Digit, 4-Digit Modified, 5-Digit, and 16 Series Airfoils*. Academic Press.

Lambert, S., & Morrison, J.F. (2006). *Fundamental Studies of Active Dimples*. AIAA-2006-3182.

Lanchester, F. W. (1907) Aerial flight, Vol. 1: Aerodynamics. Archibald Constable & Co.

Levy, D. E., & Seifert, A. (2009). Simplified dragonfly airfoil aerodynamics at Reynolds numbers below 8000. *Physics of Fluids*, *21*(7), 071901. doi:10.1063/1.3166867

Lietz, R., Pien, W., & Remondi, S. (2000). A CFD Validation Study for Automotive Aerodynamics. *SAE Technical Paper Series*. doi:10.4271/2000-01-0129

Lissaman, P. B. S. (1983). Low-Reynolds-number airfoils. *Annual Review of Fluid Mechanics*, *15*(1), 223–239. doi:10.1146/annurev.fl.15.010183.001255

Lin, Y., & Jiang, Z. (2013, March). Conceptual Study on Non-ablative TPS for Hypersonic Vehicles. *AIAA Journal*, *51*(3).

Mason, W. H. (2018). *Hypersonic Aerodynamic*. Configuration Aerodynamics.

Mayer, W., & Wickern, G. (2011). The New Audi A6/A7 Family - Aerodynamic Development of Different Body Types on One Platform. *SAE International Journal of Passenger Cars. Mechanical Systems*, *4*(1), 197–206. doi:10.4271/2011-01-0175

McGhee & Beasley. (1973). *NASA TN D-7428 Low-Speed Aerodynamic Characteristics of a 17-Percent-Thick Airfoil Section Designed for General Aviation Applications*. Academic Press.

Mehta, R.C. (2000). Numerical heat transfer study over spiked-blunt body at Mach 6.8. *J Spacecraft & Rockets, 37*(5), 700 – 701.

Mehta, R. C. (2010, April). Numerical simulation of the flow field over conical, disc and flat spiked body at Mach 6. *Aeronautical Journal*, *114*(1154), 225–236. doi:10.1017/S0001924000003675

Menezes, Saravanan, Jagadeesh, & Reddy. (2002). Aerodynamic Drag Reduction Using Aerospikes For Large Angle Blunt Cone Flying At Hypersonic Mach Number. *22nd AIAA Aerodynamic Measurement Technology and Ground Testing Conference*.

Menezes, . (2003, October). Experimental Investigations of Hypersonic Flow over Highly Blunted Cones with Aerospikes. *AIAA Journal*, *41*(10).

Menter, F. (1994). Two-equation eddy-viscosity turbulence models for engineering applications. *AIAA Journal*, *32*(8), 1598–1605. doi:10.2514/3.12149

Meriwether, J. W., & Gerrard, A. J. (2004). Mesosphere inversion layers and stratosphere temperature enhancements. *Reviews of Geophysics*, *42*(3).

Mitscke, M. (1982). Dynamik der Kraftfahrzeuge. Springer Verlag.

Mohebbi, F., & Sellier, M. (2014). On the Kutta condition in potential flow over airfoil. *Journal of Aerodynamics*.

Mongia, H., Hsiao, G., Ravindra, M. V. V., & Sreedhar, P. S. V. S. (2004). Combustion Diffuser Modeling – Part V: Validation with a three passage diffuser rig data. *AIAA Paper*, 2004–4172.

Moreau, E. (2007). Airflow control by non-thermal plasma actuators. *Journal of Physics. D, Applied Physics*, *40*(3), 605–636. doi:10.1088/0022-3727/40/3/S01

Morimoto, N., Aso, S., & Tani, Y. (2014). Reduction Of Aerodynamic Heating And Drag With Opposing Jet Through Extended Nozzle In High Enthalpy Flow. *29th Congress of the International Council of the Aeronautical Society*.

Mureithi, E. W., Mwaonanji, J. J., & Makinde, O. D. (2013). *On the boundary layer flow over a moving surface in a fluid with temperature-dependent viscosity*. Academic Press.

Murphree, Z. R., Yüceil, K. B., Clemens, N. T., & Dolling, D. S. (2007). Experimental studies of transitional boundary layer shock wave interactions. *Collection of Technical Papers - 45th AIAA Aerospace Sciences Meeting, 20*, 13782–13792. 10.2514/6.2007-1139

Murugu Nachippan, N. (2019). Computational Analysis of Air Wings to Evaluate Downforce and Lift-Drag Ratio: Technical Note. *International Journal of Vehicle Structures and Systems*, *11*(1), 102–106. https://doi.org/10.4273/ijvss.11.1.18

National Aeronautics and Space Administration. (2021). *The drag coefficient.* Glenn Research Centre.

National Research Council (US) Transportation Research Board, Committee for a Study of Public-Sector Requirements for a Small Aircraft Transportation System, & National Research Council. (2002). *Future flight: A review of the small aircraft transportation system concept*. Author.

New, T. H., Chan, Y. X., Koh, G. C., Hoang, M. C., & Shi, S. (2014). Effects of corrugated aerofoil surface features on flow-separation control. *AIAA Journal*, *52*(1), 206–211. doi:10.2514/1.J052398

Ning, A., & Kroo, I. (2008, August). Tip extensions, winglets, and c-wings: conceptual design and optimization. In *26th AIAA Applied Aerodynamics Conference* (p. 7052). 10.2514/6.2008-7052

NPTEL-Aerospace. (n.d.). *Module 1: Hypersonic Atmosphere Lecture 1: Characteristics of Hypersonic Atmosphere*. Author.

Oates, G. C. (1989). *Aircraft propulsion systems. In Technology and Design*. AIAA Education Series. doi:10.2514/4.861499

Okamoto, M., Yasuda, K., & Azuma, A. (1996). Aerodynamic characteristics of the wings and body of a dragonfly. *The Journal of Experimental Biology*, *199*(2), 281–294. doi:10.1242/jeb.199.2.281 PMID:9317808

Ou, M., Yan, L., Huang, W., Li, S., & Li, L. (2018). Detailed parametric investigations on drag and heat flux reduction induced by a combinational spike and opposing jet concept in hypersonic flows. *International Journal of Heat and Mass Transfer*, *126*, 10–31. Advance online publication. doi:10.1016/j.ijheatmasstransfer.2018.05.013

Owczarek, J. A. (1964). *Fundamentals of Gas Dynamics* (1st ed.). International Textbook Company.

Peng. (2015). Research on aerodynamic characteristic of the hypersonic projectile with drag-reduction spike. *Kexue Jishu Yu Gongcheng*, *15*(9), 142–147.

Peterson, B. (2020, December 14). *These New Planes Could Change the Way You Travel.* Conde Nast Traveler. Retrieved from https://www.cntraveler.com/story/new-planes-that-will-change-the-way-you-travel

Prager, S. (1961). Viscous flow through porous media. *Physics of Fluids*, *4*(12), 1477–1482.

Prandtl, L. (1918). *Tragftii geltheorie, 1. Mit-teilung*. Nachr Ges Wiss Gottingen.

Prandtl, L. (1921). Uber die Eindringungsfestigkeit (Harte) plastischer Baustoffeund die Festigkeit von Schneiden. *Zeitschrift für Angewandte Mathematik und Mechanik*, *1*, 15–21. doi:10.1002/zamm.19210010102

Prasad, U. S., Ajay, V. S., Rajat, R. H., & Samanyu, S. (2017, May). Aerodynamic Analysis Over Double Wedge Airfoil. *IOP Conference Series. Materials Science and Engineering, 197*(1), 012076. doi:10.1088/1757-899X/197/1/012076

Prasad, U. S., Varma, S. A., Kumar, R. S., & Vaibhav, K. M. R. S. (2018). Experimental and Numerical Analysis over the Truncated Airfoil with Slotted Flap Configuration. *Indian Journal of Science and Technology.*

Preda, I., Covaciu, D., & Ciolan, G. (2010). Coast down test – Theoretical and experimental approach. Transilvania University of Brasov.

Rahim, Singh, & Veeravalli. (2007). Liner dome shape effect on the annulus flow characteristics with and without swirl for a can-combustor model. *Proc. IMechE, 221*(A), 359-369.

Rajendrasing, R., Girka, Y., Zaharenko, V., Mallikarjuna, V., & Boopathiraja, A. (2015). *Aerodynamics and Thermal Features of Reentry Spacecraft with Blunt Body Capsule.* Aerodynamics, Dynamics, Ballistics and Flight Control of Aircraft. Aviation and Space Equipment and Technology No. 2 (119) UDC 532.58.

Rakich, J. V. (1969). *A method of characteristics for steady three-dimensional supersonic flow with application to inclined bodies of revolution.* National Aeronautics and Space Administration.

Rakich, J. V., & Cleary, J. W. (1970). Theoretical and experimental study of supersonic steady flow around inclined bodies of revolution. *AIAA Journal, 8*(3), 511–518. doi:10.2514/3.5698

Raymer, D. (2018). Aircraft design: A conceptual approach. American Institute of Aeronautics and Astronautics. https://doi.org/10.2514/4.869112.

Roache, P. J. (1976). *Computational fluid dynamics.* Computational Fluid Dynamics.

Roe, P. (2005). Thin Shock Layer Theory Revisited. *4th AIAA Theoretical Fluid Mechanics Meeting.* 10.2514/6.2005-5194

Rolf. (1977). Fluctuating Pressure Environment of a Drag Reduction Spike. *J. Spacecraft, 14*(12).

Ros, M. (2018, February 16). *What will passenger planes look like in 2068?* CNN Travel. Retrieved from https://edition.cnn.com/travel/article/passenger-planes-future-look/index.html

Roth, J. R., Sherman, D. M., & Wilkinson, S. P. (1998). Boundary layer flow control with a one atmosphere uniform glow discharge surface plasma. *AIAA Journal*, 1–28.

Russo, G. P. (2011). *Aerodynamic measurements: from physical principles to turnkey instrumentation.* Academic Press.

Sadraey, M. H. (2012). Wing Design. In M. H. Sadraey (Ed.), *Aircraft Design: A Systems Engineering Approach* (pp. 161–264). John Wiley & Sons. doi:10.1002/9781118352700.ch5

Sanal Kumar, Sahoo, & Ragunathan. (2009). *Internal Flow Simulation of Dump Diffusers for Modern Aircraft Engines.* 45th AIAA/ASME/SAE/ASEE Joint Propulsion Conference & Exhibit, Denver, CO.

Sanal Kumar, Sameen, Kim, Setoguchi, Mastuo, & Raghunathan. (2005). Influence Of Dump Gap On The Performance Characteristics of Dump-Diffusers. *16th International Symposium On Transport Phenomena, ISTP-16.*

Sanal Kumar, V. R., Kim, H. D., Setoguchi, T., Matsuo, S., & Raghunathan, S. (2005). Influence of dump gap ratio on the performance characteristics of dump diffusers. *4th International Conference on Heat Transfer, Fluid Mechanics and Thermodynamics (HEFAT2005)*, 19-22.

Sanderson, S. R., & Sturtevant, B. (1995). Shock Wave Interactions in Hypervelocity Flow. *Shock Waves @ Marseille I*, 69–74. doi:10.1007/978-3-642-78829-1_10

Sandham, N. D. (2016). Effects of Compressibility and Shock-Wave Interactions on Turbulent Shear Flows. *Flow, Turbulence and Combustion, 97*(1), 1–25. doi:10.100710494-016-9733-6

Sarow, S. A. (2020, June). Flows of Viscous Fluids in Food Processing Industries: A review. *IOP Conference Series. Materials Science and Engineering, 870*(1), 012032.

Sasi Kumar, M., Abirami, K., Sandhiya, K., Vijay, G., & Vishnu Varthan, S. (2018). Noise reduction and thrust enhancement in various modified chevron nozzle. *International Journal of Development Research, 8*(1), 18540-18544.

Sato, J. (1973). *NASA TT-F-749 On Peaky Airfoil Sections*. Academic Press.

Sauerwein, H. (1967). Numerical calculations of multidimensional and unsteady flows by the method of characteristics. *Journal of Computational Physics, 1*(3), 406–432. doi:10.1016/0021-9991(67)90048-4

Senthilkumar, Mudholkar, & Sanjay. (n.d.). Comparative Study on Aerodynamic Drag Reduction of a Blunt Nose Body using Aerospike and Aerodisk – Numerical Approach. *IOP Conference Series: Materials Science and Engineering.*

Senthilkumar, P. B., Parthasarathy, M., & Aravind, L. (n.d.). Design and analysis of a rear diffuser in a sedan car. *Materials Today: Proceedings*. doi:10.1016/j.matpr.2021.11.542

Shang, J. J., & Yan, H. (2020). High-Enthalpy Hypersonic Flows. *Advances in Aerodynamics*. doi:10.1186/s42774-020-00041-y

Sharma, A.S. (2020). Steady Generalized Plane Couette Flow of Viscous Incompressible Fluid between Two Porous Parallel Plates through Porous Medium with Magnetic Field. *International Journal of Creative Research Thoughts, 8*(12).

Shyy, W., Lian, Y., Tang, J., Viieru, D., & Liu, H. (2008). Aerodynamics of low Reynolds number flyers. Cambridge University Press.

Sikroria, T., Soria, J., & Ooi, A. (2018, December). A novel method for determination of convective velocity of coherent structures in high speed flows. In *Proc. 21st Australasian Fluid Mechanics Conf* (pp. 1-4). Academic Press.

Sivasankar, G. A., & Balamurugan, S. (2017). Virtual Analysis of Chevron Nozzle Lobe Design Modification. *Asian Journal of Research in Social Sciences and Humanities, 7* (1), 79–86. doi:10.5958/2249-7315.2017.00007.7

Smith, S. C. (1996). *A Computational and Experimental Study of Nonlinear Aspects of Induced Drag*. NASA Report No 3598.

Smith, A. M. O. (1975). High-lift aerodynamics. *Journal of Aircraft, 12*(6), 501–530.

Smith, J. M., & Juhasz, A. J. (1978). *Performance of a Short Annular Dump Diffuser Using Suction-Stabilized Vortices at Inlet Mach Numbers to 0.41.* NASA National Aeronautics and Space Administration Scientific and Technical Information Office.

Smith, M. J., Komerath, N., Ames, R., Wong, O., & Pearson, J. (2001). Performance analysis of a wing with multiple winglets. *19th AIAA Applied Aerodynamics Conference*. 10.2514/6.2001-2407

Srinath & Reddy. (2010). Experimental Investigation of the Effect of Aerospike Geometry on Drag Reduction and Heat Transfer Rates at Hypersonic. *International Journal of Hypersonics, 2.*

Srinivasan, R., Freeman, G., Grahmann, J., & Coleman, E. (1990). Parametric Evaluation of the Aerodynamic Performance of an Annular Combustor Diffuser System. Paper No.AIAA-90-2163. doi:10.2514/6.1990-2163

Stevens, S. J., Wray, A. P., & Price, P. D. (1988). The Aerodynamic Performance of a Modern Vaporizing Combustor Dump Diffuser. Paper No.AIAA-88-3273. doi:10.2514/6.1988-3273

Stull, R. B. (2015). *Practical meteorology: An algebra-based survey of atmospheric science*. Academic Press.

Sun, Huang, Ou, Zhang, & Li. (2019). A survey on numerical simulations of drag and heat reduction mechanism in supersonic/hypersonic flows. *Chinese Journal of Aeronautics*.

Suvagiya, M., & Sharma, S. D. (2018). The effect of geometric parameters of chevron nozzle on generation of streamwise vortices in high subsonic jets. *20th Annual CFD Symposium*, 9-10.

Talay, T. (1975). *Introduction to the aerodynamics of flight* [NASA SP-367]. Langley Research Center. Retrieved from https://practicalaero.com/wp-content/uploads/2010/04/NASA-SP-367.pdf

Talay, T. A. (1975). Introduction to the Aerodynamics of Flight (No. NASA-SP-367). Academic Press.

Tamai, M., Wang, Z., Rajagopalan, G., Hu, H., & He, G. (2007, January). Aerodynamic performance of a corrugated dragonfly airfoil compared with smooth airfoils at low Reynolds numbers. In *45th AIAA aerospace sciences meeting and exhibit* (p. 483). 10.2514/6.2007-483

Taylor, T. D. (1974). *Numerical Methods for Predicting Subsonic, Transonic and Supersonic Flow*. Advisory Group for Aerospace Research and Development Paris (France).

Taylor, G. I., & Maccoll, J. W. (1933). The air pressure on a cone moving at high speeds.—II. *Proceedings of the Royal Society of London. Series A, Containing Papers of a Mathematical and Physical Character*, *139*(838), 298–311. doi:10.1098/rspa.1933.0018

Thayer, J. P., & Livingston, J. M. (2008). Observations of wintertime arctic mesosphere cooling associated with stratosphere baroclinic zones. *Geophysical Research Letters*, *35*(18).

Tillmark, N., & Alfredsson, P. H. (1992). Experiments on transition in plane Couette flow. *Journal of Fluid Mechanics*, *235*, 89–102.

Timmer Duwind, W. A. (1997). The effect of roughness at high Reynolds numbers on the performance of DU 97-W-300Mod. Delft University Wind Energy Institute.

Tomasi, C., Vitake, V., & De Santis, L. V. (1998). Relative optical mass functions for air, water vapour, ozone and nitrogen dioxide in atmospheric models presenting different latitudinal and seasonal conditions. *Meteorology and Atmospheric Physics*, *65*(1), 11–30.

Tong, F., Duan, J., & Li, X. (2021). Characteristics of wall-shear stress fluctuations in shock wave and turbulent boundary layer interaction. *Journal of Turbulence*, *22*(12), 761–783. doi:10.1080/14685248.2021.1974466

Tsuchiya, T., Kanamori, M., & Takahashi, T. (2019). Numerical analysis of linear wave propagation in the atmosphere with temperature gradient for Mach cutoff reproduction. *Japanese Journal of Applied Physics, 58*(SG), SGGB01.

Turner, H. R. Jr. (1941). *NACA WR-L-677 Lift and Drag Tests of Three Airfoil Models with Fowler Flaps*. Consolidated Aircraft Corporation.

Tu, W., & Jin, H. D. (2014). Numerical simulation study on missile antidrag spike drag reduction. *Computational Simulation*, *31*(4), 87–91.

Urzay, J. (2020). The Physical Characteristics Of Hypersonic Flows Center For Turbulence Research. Stanford University.

Varshney, Baig, & Hasan. (2020). Turbulent Drag Reduction on an Aircraft Wing by Wall Jets for Flow Control. *AIAA Aviation 2020 Forum*. 10.2514/6.2020-2780

Virk, D., & Virk K. (2018). A Review on Magnus Effect and its Applications. *International Journal of Information and Computing Science, 5*(8).

Viviani, A., Pezzella, G., & Golia, G. (2012). Aerothermodynamic Field Past a Reentry Capsule for Sample Return Missions. *28th International Congress of the Aeronautical Sciences.*

Von Helmholtz, H. (1868). Über discontinuierliche Flüssigkeits-Bewegungen. Monatsberichte derKöniglichen Preussische Akademieder Wissenschaften zu Berlin, 23, 215–228.

Von Karman, T. (2003). Compressibility effects in aerodynamics. *Journal of Spacecraft and Rockets, 40*(6), 992–1011. doi:10.2514/2.7046

Von Zahn, U., Fiedler, J., Naujokat, B., Langematz, U., & Krüger, K. (1998). A note on record-high temperatures at the northern polar stratopause in winter 1997/98. *Geophysical Research Letters, 25*(22), 4169–4172.

Walker, A. D., Carrotte, J. F., & McGuirk, J. J. (2009). The Influence of Dump Gap on External Combustor Aerodynamics at High Fuel Injector Flow Rates. *Journal of Engineering for Gas Turbines and Power, 131*(3), 031506–031510. doi:10.1115/1.3028230

Wang, H., & Kai, H. L. (2020). Numerical Investigation of Dump Diffuser Combustor Performance at Uniform and Non-Uniform Inlet Conditions. *ASME Proceedings.* 10.1115/GT2020-15982

Wang, J., Li, H., Liu, Y., Liu, T., Gao, H., Tandon, P., Wisnoe, W., & bin Abdullah, M.Z. (2018). Aerodynamic research of a racing car based on wind tunnel test and computational fluid dynamics. *MATEC Web of Conferences, 153.* 10.1051/matecconf/201815304011

Wentz & Seetharam. (1974). *NASA CR-2443 Development of a Fowler Flap System for a High-Performance General Aviation Airfoil.* Academic Press.

Whitcomb & Clark. (1965). *NASA TM-X-1109 An Airfoil Shape for Efficient Flight at Supercritical Mach Numbers.* Academic Press.

Wickern, G., Wagner, A., & Zoerner, C. (2005). *Induced Drag of Ground Vehicles an It's Interaction with Ground Simulation.* SAE Technical Paper 2005-01-0872. doi:10.4271/2005-01-0872

Wirz, H. J. (1975). *Progress in Numerical Fluid Dynamics.* Springer-Verlag. doi:10.1007/3-540-07408-2

Wlezien. (2000). Micro Adaptive Flow Control. *ICON 2000 Congress*, 2101-1 - 2101-9.

Xu, L., Huang, Y., Ruan, C., & Peiyong, W. F. X. (2015). Study of the Dump Diffuser Optimization for Gas Turbine Combustors. *Procedia Engineering*, 99, 828–834. doi:10.1016/j.proeng.2014.12.608

Yadavn, R., Velidi, G., & Guven, U. (2014). Aerothermodynamics of generic re-entry vehicle with a series of aerospikes at nose. *Acta Astronautica, 96*, 1–10. doi:10.1016/j.actaastro.2013.11.015

Yao, Y. F., & Sandham, N. D. (2002). DNS of turbulent flow over a bump with shock/boundary-layer interactions. In W. Rodi & N. Fueyo (Eds.), *Engineering Turbulence Modelling and Experiments 5* (pp. 677–687). Elsevier. doi:10.1016/B978-008044114-6/50065-X

Yuan, C., Song, J., & Liu, M. (2019). Comparison of compressible and incompressible numerical methods in simulation of a cavitating jet through a poppet valve. *Engineering Applications of Computational Fluid Mechanics, 13*(1), 67–90. doi:10.1080/19942060.2018.1552202

Zaccai, D., Bertels, F., & Vos, R. (2016). Design methodology for trailing-edge high-lift mechanisms. *CEAS Aeronaut. J., 7*, 521–534.

Zerihan, J. (2001). *An Investigation into the Aerodynamics of Wings in Ground Effect* [Ph.D. Thesis]. University Southampton, School of Engineering.

Zhang, M. H. (2015). *Atmospheric Circulation and Wind Systems*. Wiley. doi:10.1002/9781118900116.ch9

Zheng, Y. Y. (2014). Supersonic flow field investigation of the effect of a mechanical spike with and without counterflow jet *J. Heat and Mass Transfer*, *10*(1), 75–83.

Zucker, R. D., & Biblarz, O. (2019). *Fundamentals of gas dynamics*. John Wiley & Sons.

Zucrow, M. J., & Hoffman, J. D. (1976). *Gas Dynamics* (Vols. 1-2). Academic Press.

About the Contributors

Naren Shankar Radhakrishnan received BE degree in Mechanical Engineering from Madha Engineering College, Chennai, India, in 2007 and ME and PhD degrees in Aeronautical Engineering from the Madras Institute of Technology (MIT), Anna University, Chennai, India, in 2009 and 2017, respectively. From 2009, he served as an Assistant Professor in various private engineering colleges. Since April 2018, he has been working as Associate Professor in the Department of Aeronautical Engineering, Vel Tech Rangarajan Dr Sagunthala R&D Institute of Science and Technology. He has completed a funded project on Autonomous Underwater Vehicle, funded by Naval Science and Technology Limited, DRDO under the Centre for Acquisition of Research Services (CARS) Scheme. Currently, he is working on a project funded by DST SERB under Teachers Associateship for Research Excellence (TARE) under the title Optimising Supersonic Co-Axial Nozzle design for Mixing Enhancement. Additionally he is working on an internally funded project on Hypersonic Shock Tunnel for NanoTechnological Applications. He has published 16 SCI/SCOPUS Indexed journals and conferences and filed three patents. He has produced one Ph.D. and five scholars are currently pursuing Ph.D. His current research interests include high-speed jets, aerodynamics, propulsion and unconventional energy engineering. Dr Naren is an Associate Member of the Aeronautical Society of India. Naren Shankar Radha Krishnan is the corresponding author and can be contacted at: rnarenshankar@gmail.com

* * *

Rathan Babu Athota is currently pursuing his doctoral study at Department of Physics, Universitat Politecnica, Spain.

Joseph Manuel D. is an Associate Professor in the Department of Mechanical Engineering at Velammal Institute of Technology affiliated to Anna University, Chennai, where he has been a faculty member since 2017. He is in teaching profession for more than 12 years. His research interests lie in the area of Composites, 3D Printing, Additive Manufacturing, Research Methodologies etc. He published 10 research articles in various national and international journals and Published Patent also.

Mala D. received her Ph.D degree in Mechanical Engineering from ANNA UNIVERSITY INDIA. Her doctoral research focused on Nano-heat transfer. She had been working as an Assistant Professor in the Department of Mechanical, University College of Engendering Panruti. Her research interest lie in the fields of Thermal Engineering and Heat Transfer. She has published over 20 papers in peer reviewed journals and conferences.

Madhankumar G. completed his B.E Aeronautical Engineering in 2010 and M.E Aeronautical Engineering at 2013. Currently pursuing Ph.D Mechanical Engineering in Anna University, Chennai. His areas of interest are Aerodynamics, Acoustics and Computational Fluid Dynamics. He had 8 years of teaching experience and published 6 research papers.

Habib Gürbüz graduated from the Department of Mechanical Education at Gazi University in 2002. He received his M.Sc. degree from Department of Mechanical Education of Dumlupınar University in 2005. He obtained his PhD degree from the Department of Mechanical Engineering Suleyman Demirel University in 2010. Currently, he is an Assoc. Prof Dr. at Suleyman Demirel University. His main research interests are internal combustion engine, combustion, alternative fuels, hydrogen energy, waste heat recovery, thermoelectric, intelligent traffic control systems, new vehicle technologies.

Muralidharan Gurunathan currently working as Assistant Professor in the Department of Aeronautical Engineering at Surya Group of Institutions, Vikiravandi, Villupuram, Tamilnadu, India. He Has Published a book titled "Aero Engineering Thermodynamics" and his areas of interest are, Aero engineering Thermodynamics, Aircraft Propulsion, Aerodynamics and Computational Fluid Dynamics.

Bibal J. V. is presently associated with Indian Institute of Information Technology, Kottayam, India as an Assistant Professor. She obtained her B.E, M.E and Ph.D degrees in the domain of cloud computing from Anna University, India. Her research interests are Cloud computing, Big Data Analytics and Machine Learning. Presently, she is involved in the development of various cloud-based societal applications. She is a life member of ISTE, India.

Kumar K. M. completed his B.E mechanical engineering-2000, university of Madras, ME-CAD from Anna University-2007,Ph. D Mechanical engineering- Anna university- 2017. His research areas of interest are Materials, Manufacturing, Machining and composite material. He had 21 years of teaching experience and published more than 20 research publications.

Parthasarathy M. has completed M.E and PhD degrees in Automobile Engineering from Madras Institute of Technology Campus, Anna University, Chennai and Bachelor of Engineering from Ranipettai Engineering College, Vellore. Dr M. Parthasarathy is working as an Associate Professor in the Department of Automobile Engineering, Vel Tech University at Avadi, Chennai. He has more than 7 years of teaching and research experience. Dr M. Parthasarathy research areas are Internal Combustion Engine, Sustainable Fuel, Pollution Control, Homogenous Charge Compression Ignition Engine and Dual fuel Combustion. He has published more than 40 articles in highly reputed International Journals. He has also presented 20 papers at International and national conferences. He has more than 850 citations and a good h-Indexed (14) by Scopus. He is the reviewer of Environmental Science and Pollution Research and Journal of Research in Environmental Science and Toxicology. He has been supervising more than 10 research scholars and master students. He is a member of the professional bodies of the Society of Automotive Engineers (SAE) and the International Association of Engineers (IAENG).

Ramzania M. is a research scholar at Department of Aeronautical Engineering, Anna University Regional Campus Tirunelveli.

Rajkamal M. D. is an Assistant Professor in the Department of Mechanical Engineering at Velammal Institute of Technology affiliated to Anna University, Chennai, where he has been a faculty member since 2013. He is in teaching profession for more than 12 years. His research interests lie in the area of Computational Fluid Dynamics (CFD), Thermal Sciences, Internal Combustion Engine (IC Engine),3D Printing, Additive Manufacturing, Research Methodologies etc. He published almost 25 research articles in various national and international journals and conferences. He holds 2 international Australian Patent Grants and 3 published Indian Patents which signifies his academic credentials. He also co-authored 4 books which is being used by the first year to final year students, He has published Some videos in the YouTube Channel for the Subjects Like Engineering Graphics and Dynamics of Machines. He is a member of International Association of Engineers (IAENG) and Institute of Research Engineers and Doctors.

Dilip N. is Associate Professor at Vel Tech Rangarajan Dr. Sagunthala R&D Institute of Science & Technology, Chennai.

N. Murugu Nachippan has completed M.E degree in Production engineering from Thiagarajar college of engineering, Madurai and B.E degree in Automobile engineering from Sri Venkateswara College of Engineering, Chennai. N. Murugu Nachippan is working as Assistant Professor in the Department of Automobile Engineering, Vel Tech Rangarajan Dr. Sagunthala R&D Institute of Science and Technology at Avadi, Chennai. He has more than 7 years of teaching and research experience. His research interests are Internal Combustion Engine, Sustainable Fuel, Pollution Control and composite materials He has published more than 20 articles in Scopus indexed journals. He has also presented 10 papers in the International and national conferences. He is member in professional bodies of the International Association of Engineers (IAENG).

Vishal Naranje is an associate professor in the Department of Mechanical Engineering at Amity University Dubai. He had 20 years of teaching and research experience in Indian and foreign universities. His research interest concerns mainly in Design of Expert Systems, Digital Manufacturing, Decision making tools, Micro-machining. He has published more than 50 research articles, books chapters in SCI, Scopus indexed journals and conferences etc. He has received research grants from various funding agencies in India. He has completed few industrial consultancy projects in India and UAE. He is reviewer for the journals published by Springer, Elsevier, Taylor and Francis etc.

Murugan Ponnusamy is an educationalist as well as an Institution builder with teaching along with administrative experience. He is working as Assistant Professor in the Department of Mechanical Engineering, Jimma Institute of Technology, Jimma University, Ethiopia. He graduated in B.E (Mechanical Engineering) from University of madras in 1998. He completed M.E. from Sathyabama University in 2005. He received Doctor of Philosophy (Ph.D) in Mechanical Engineering from College of Engineering Guindy, Anna University in 2020. He is having more than two decades of experience in academic, administration, research in various Engineering Colleges. His expert areas are Thermal Engineering, Engineering Thermodynamics, Heat and Mass transfer, Refrigeration and Air-conditioning, Fluid Mechanics and Fluid Power Systems . He has articles in Scopus/SCI indexed journals to his credit. He has international patent grants publications. He has delivered guest lectures and keynote speeches on vario. us occasions. He received awards and rewards for his accomplishments in administration

Irish Angelin S. is currently a PhD research scholar in the Department of Aeronautical Engineering at Vel Tech Rangarajan Dr. Sagunthala R&D Institute of Science and Technology, India. Her interest is in analyzing the flow characteristics involved in hypersonic Aerodynamics. She has worked in high-speed flow. Her current research is concerned with High-Speed Jets. She also has 10 months of research experience from Vikram Sarabhai Space Centre (VSSC), Thiruvananthapuram during which she was working on the payload fairing of a rocket forebody at transonic speeds. She has worked as an Assistant Professor in the Aeronautical Department for 7 years. She has received the young scientist award and the best teacher award.

Kaliappan S. is an Associate Professor in the Department of Mechanical Engineering at Velammal Institute of Technology affiliated to Anna University, Chennai, where he has been a faculty member since 2011. He is the Head of the Mechanical and Mechatronics Engineering Department. He completed his Ph.D. at Anna University, Chennai, his postgraduate studies in Internal Combustion Engineering at College of Engineering Gundy, Anna University, Chennai and his undergraduate studies at R.V.S College of Engineering & Technology, Dindigul. He is in teaching profession for more than 23 years.His research interests lie in the area of Computational Fluid Dynamics(CFD), Thermal Sciences, Internal Combustion Engine(IC Engine), Automotive system, Research Methodologies etc.He has served as technical committee head for roughly fifteen conferences and workshop and worked as the reviewers and Editorial Board Member for several National and International Journals. He published almost 35 research articles in various national and international journals and conferences, organized nearly 20 STTPs, FDPs, Conferences and other technical events. He holds 15 international Australian Patent Grants and 3 published Indian Patents which signifies his academic credentials. He also co-authored 7 books which is being used by the first year to final year students. He is Life member in Indian Society for Technical Education (LMISTE), International Association of Engineers (IAENG) and also a fellow member in Universal Association of Mechanical and Aeronautical Engineers (UAMAE), Institute Of Research Engineers And Doctors (theIRED). He got Young Researcher Award in InSc awards-2021 for his research journal of title "Numerical Investigation of Sinusoidal and Trapezoidal Piston Profiles for an IC Engine", which is published in the Journal of Applied Fluid Mechanics.

Socrates S. is an Assistant Professor in the Department of Mechanical Engineering at Velammal Institute of Technology affiliated to Anna University, Chennai, where he has been a faculty member since 2013. He is in teaching profession for more than 8 years. His research interests lie in the 3D Printing, Additive Manufacturing, Research Methodologies etc. He published almost 8 research articles in various national and international journals and conferences. He holds 5 international Australian Patent Grants and 3 published Indian Patents which signifies his academic credentials.

Mothilal T. completed master degree and doctoral from Anna University, published more than 30 research articles in journals of high repute, specialized in thermal, alternate fuels, CFD, received several domain awards from NPTEL, four patents granted, 21 years of teaching experience.

Senthilkumar T. S. is working as Assistant Professor/Mechanical at SreeSowdambika College of Engineering, Chettikurichi. His research area is tribology behavior of composites. For past 3 years he was working in this area.

Shiva Prasad Uppu is presently working in the field of fluid flows internal & external flows Research include Computational Aero Acoustics, CFD, Experimental Aerodynamics, Fluid Mechanics, Hypersonic Aerodynamics, Propulsion and Combustion. Contributed fourteen Journal Publications, and received Excellent Young Researcher Award of the year 2018.

Vignesh Kumar V. is working as Assistant Professor/Mechanical at St. Joseph College of Engineering, Chennai. His research area Tribology behavior of Composites. For past 4 years he was working in this area.

Balaji Venkatesan is an educationalist as well as an Institution builder with teaching along with administrative experience. He is working as Professor and Head of the Department of Mechanical Engineering, Loyola Institute of Technology (Affiiated to Anna University) He graduated in B.E (Mechanical Engineering) from Madurai Kamaraj University in 1998. He completed M.E. from College of Engineering Guindy, Anna University in 2006. He received Doctor of Philosophy (Ph.D) in Mechanical Engineering from College of Engineering Guindy, Anna University in 2017. He is having more than two decades of experience in academic, administration, research in various Engineering Colleges. His expert areas are Alternate Fuel, Thermal Engineering, Engineering Thermodynamics, Heat and Mass transfer and Optimization. He has articles in Scopus/SCI indexed journals to his credit. He has international patent grants and national patent publications. He has delivered guest lectures and keynote speeches on various occasions. He received awards and rewards for his accomplishments in administration.

Index

D

E

F

G

H

I

J

K

L

M

N

P

Q

R

S

T

V

W

www.ingramcontent.com/pod-product-compliance
Lightning Source LLC
LaVergne TN
LVHW060357300926
849162LV00006B/1292